Advances in Intelligent Systems and Computing

Volume 1350

AF806305

The series "Advances in Intelligent Systems and Computing" contains publications on theory, applications, and design methods of Intelligent Systems and Intelligent Computing. Virtually all disciplines such as engineering, natural sciences, computer and information science, ICT, economics, business, e-commerce, environment, healthcare, life science are covered. The list of topics spans all the areas of modern intelligent systems and computing such as: computational intelligence, soft computing including neural networks, fuzzy systems, evolutionary computing and the fusion of these paradigms, social intelligence, ambient intelligence, computational neuroscience, artificial life, virtual worlds and society, cognitive science and systems, Perception and Vision, DNA and immune based systems, self-organizing and adaptive systems, e-Learning and teaching, human-centered and human-centric computing, recommender systems, intelligent control, robotics and mechatronics including human-machine teaming, knowledge-based paradigms, learning paradigms, machine ethics, intelligent data analysis, knowledge management, intelligent agents, intelligent decision making and support, intelligent network security, trust management, interactive entertainment, Web intelligence and multimedia.

The publications within "Advances in Intelligent Systems and Computing" are primarily proceedings of important conferences, symposia and congresses. They cover significant recent developments in the field, both of a foundational and applicable character. An important characteristic feature of the series is the short publication time and world-wide distribution. This permits a rapid and broad dissemination of research results.

Indexed by DBLP, EI Compendex, INSPEC, WTI Frankfurt eG, zbMATH, Japanese Science and Technology Agency (JST), SCImago.

All books published in the series are submitted for consideration in Web of Science.

More information about this series at http://www.springer.com/series/11156

Jessnor Arif Mat Jizat · Ismail Mohd Khairuddin ·
Mohd Azraai Mohd Razman ·
Ahmad Fakhri Ab. Nasir ·
Mohamad Shaiful Abdul Karim ·
Abdul Aziz Jaafar · Lim Wei Hong ·
Anwar P. P. Abdul Majeed ·
Pengcheng Liu · Hyun Myung ·
Han-Lim Choi · Gian-Antonio Susto
Editors

Advances in Robotics, Automation and Data Analytics

Selected Papers from iCITES 2020

Springer

Editors
Jessnor Arif Mat Jizat
Faculty of Manufacturing and Mechatronic Engineering Technology
Universiti Malaysia Pahang
Pekan, Malaysia

Mohd Azraai Mohd Razman
Faculty of Manufacturing and Mechatronic Engineering Technology
Universiti Malaysia Pahang
Pekan, Malaysia

Mohamad Shaiful Abdul Karim
College of Engineering
Universiti Malaysia Pahang
Gambang, Malaysia

Lim Wei Hong
UCSI University
Kuala Lumpur, Malaysia

Pengcheng Liu
University of York
York, UK

Han-Lim Choi
Department of Aerospace Engineering
Korea Advanced Institute of Science and Technology
Daejeon, Korea (Republic of)

Ismail Mohd Khairuddin
Faculty of Manufacturing and Mechatronic Engineering Technology
Universiti Malaysia Pahang
Pekan, Malaysia

Ahmad Fakhri Ab. Nasir
Faculty of Manufacturing and Mechatronic Engineering Technology
Universiti Malaysia Pahang
Pekan, Malaysia

Abdul Aziz Jaafar
Faculty of Manufacturing and Mechatronic Engineering Technology
Universiti Malaysia Pahang
Pekan, Malaysia

Anwar P. P. Abdul Majeed
Faculty of Manufacturing and Mechatronic Engineering Technology
Universiti Malaysia Pahang
Pekan, Malaysia

Hyun Myung
School of Electrical Engineering
Korea Advanced Institute of Science and Technology
Daejeon, Korea (Republic of)

Gian-Antonio Susto
Department of Information Engineering
University of Padua
Padova, Italy

ISSN 2194-5357 ISSN 2194-5365 (electronic)
Advances in Intelligent Systems and Computing
ISBN 978-3-030-70916-7 ISBN 978-3-030-70917-4 (eBook)
https://doi.org/10.1007/978-3-030-70917-4

This Springer imprint is published by the registered company Springer Nature Switzerland AG
The registered company address is: Gewerbestrasse 11, 6330 Cham, Switzerland

Preface

The International Conference on Innovative Technology, Engineering and Sciences 2020 (iCITES 2020), is the second edition of the conference series organized by Universiti Malaysia Pahang through its Alumni Society in an effort to promote key innovation in the following overarching themes and individual symposia, i.e. green and frontier materials, innovative robotics and automation, renewable and green energy, sustainable manufacturing as well as data analytics. The conference is aimed at building a platform that allows relevant stakeholders to share and discuss their latest researches, ideas and survey reports from theoretical to practical standpoint of the aforementioned fields.

ICITES2020 received more than 170 submissions. All submissions were reviewed in a single-blind manner, and the best 40 papers recommended by the reviewers are published in this volume. The editors would like to thank all the authors who submitted their papers as the papers are of good quality and represented good progress in industrial and robotic vision, motion control, autonomous mobile robots, intelligent sensors and actuators, multi-sensor fusion, deep learning and approaches and data processing.

The editors also would like to thank Assoc. Prof. Han-Lim Choi, Jamie Steel, Dr. Rabiu Muazu Musa, Dr. Miles Stopher, Assoc. Prof. Dr. Kazem Reza Kashyzadeh, Jee Kwan Ng for delivering their keynote speeches at the conference. They had to bring a new perspective on cutting-edge issues especially in the fields of robotics, automation and data analytics.

The editors hope that readers find this volume informative. We thank Springer for undertaking the publication of this volume. We also would like to thank the conference organization staff and the members of the international program committees for their hard work.

Contents

Editor Biographies

Dipl. Ing. (FH) Jessnor Arif Mat Jizat is a researcher at Innovative Manufacturing, Mechatronics & Sports Laboratory, Faculty of Manufacturing and Mechatronic Engineering Technology in Universiti Malaysia Pahang (UMP). He completed his master's degree at UMP and diploma (FH) at Hochschule Karlsruhe, Germany. His research interest includes machine learning, robotics, robotic vision and sports engineering.

Ismail Mohd Khairuddin is a lecturer at Universiti Malaysia Pahang. He received his bachelor's degree in Mechatronics Engineering from Universiti Teknikal Malaysia Melaka (UTeM) in 2010 and was awarded a master's degree in Mechatronics and Automatic Control from Universiti Teknologi Malaysia in 2012. His research interests include rehabilitation robotics, mechanical and mechatronics design, mechanisms, control and automation, bio-signal processing as well as machine learning.

Dr. Mohd Azraai Mohd Razman is a senior lecturer at Universiti Malaysia Pahang. He graduated from the University of Sheffield, UK, before he obtained his M.Eng. and Ph.D. from Universiti Malaysia Pahang (UMP) in Mechatronics Engineering. His research interest includes optimization techniques, control systems, signal processing, instrumentation in aquaculture, sports engineering as well as machine learning.

Ahmad Fakhri bin Ab. Nasir is a senior lecturer at Universiti Malaysia Pahang (UMP). He received his bachelor's degree in Information Technology from Universiti Malaya and a master's degree in Manufacturing Engineering from Universiti Malaysia Pahang. He pursued his Ph.D. at the Universiti Sultan Zainal Abidin. He has published several articles and actively doing research related to computer vision, pattern recognition, image processing, machine learning as well as parallel computing.

Mohamad Shaiful Abdul Karim received his B.Eng. (Electrical and Electronics), M. Eng. (Advanced Science and Engineering) degree and D. Eng. (Advanced Electrical, Electronic and Computer Systems) from Ritsumeikan University, Japan, in 2011, 2013 and 2016, respectively. In 2016, he joined Universiti Malaysia Pahang as a senior lecturer at College of Engineering. He is currently engaged in the research of microwave engineering, communication and biomedical medical devices.

Abdul Aziz bin Jaafar is an associate professor and researcher at the Faculty of Manufacturing and Mechatronic Engineering Technology, Universiti Malaysia Pahang. He received B.Eng (Mechanical/System) from Universiti Putra Malaysia and Ph.D. (Mechanical Engineering) from University of Bath, UK. His research interest is mainly on fluid flow and heat transfer, applications of light thin material on emergency and recreational shelters subjected to dynamic loading.

Dr. Wei Hong Lim is currently an assistant professor and a researcher at UCSI University. He obtained his B.Eng. (Hons) Mechatronic Engineering and Ph.D. in Computational Intelligence from Universiti Sains Malaysia, Penang, Malaysia. He was attached to the Intelligent Control Laboratory in National Taipei University of Technology, Taiwan, as the postdoctoral researcher from 2015 to 2017 and as the visiting researcher in 2019.

Anwar P. P. Abdul Majeed graduated with a first-class honours bachelor's degree from Universiti Teknologi MARA (UiTM), Malaysia. He obtained his master's degree Imperial College London, UK, before receiving his Ph.D. from Universiti Malaysia Pahang (UMP). He is a chartered engineer (C.Eng.) at the Institution of Mechanical Engineering (IMechE), UK.

Pengcheng Liu received the B.Eng. degree in measurement and control, the M.Sc. degree in control theory and control engineering from the Zhongyuan University of Technology, China, and the Ph.D. degree in robotics and control from Bournemouth University, UK. He is currently a lecturer (Tenured Assistant Professor) at the Department of Computer Science, University of York, UK. . He is an associate editor of IEEE Access, and he received the Global Peer Review Awards from Web of Science in 2019 and the Outstanding Contribution Awards from Elsevier in 2017.

Prof. Hyun Myung received the B.S., M.S. and PhD degrees in electrical engineering from the Korea Advanced Institute of Science and Technology (KAIST), Daejeon, Korea. Since 2008, he has been a professor at the Department of Civil and Environmental Engineering, KAIST, and he is the head of the KAIST Robotics Program. From 2019, he is a professor at the School of Electrical Engineering. He led the development of the world-first robots such as jellyfish removal robot (JEROS) and CAROS (wall-climbing drones).

Han-Lim Choi received his M.S. in Aerospace Engineering from the Korea Advanced Institute of Science and Technology (KAIST), Daejeon, Korea, in 2020. He then pursued his PhD in Aeronautics and Astronautics, Massachusetts Institute of Technology (MIT), USA. He is currently serving as an associate professor at the Department of Aerospace Engineering, KAIST, Daejeon. His research interest includes information-theoretic decision-making for cyber-physical systems, planning and control for multi-agent systems, air and space vehicle guidance and control as well as environmental sensing systems.

Dr. Gian Antonio Susto received the M.S. degree in control systems engineering and the Ph.D. degree in information engineering from the University of Padova, Italy, in 2009 and 2013, respectively. He is currently an assistant professor at the University of Padova and a chief data scientist and founder at Statwolf Limited, Dublin, Ireland. His research interests include manufacturing data analytics, machine learning, gesture recognition and partial differential equations control.

Multilanguage Speech-Based Gender Classification Using Time-Frequency Features and SVM Classifier

Taiba Majid Wani[1], Teddy Surya Gunawan[1,2(✉)], Hasmah Mansor[1], Syed Asif Ahmad Qadri[1], Ali Sophian[3], Eliathamby Ambikairajah[2], and Eko Ihsanto[4]

[1] Electrical and Computer Engineering Department, International Islamic University Malaysia, Gombak, Malaysia
tsgunawan@iium.edu.my
[2] School of Electrical Engineering and Telecommunications, UNSW, Sydney, Australia
[3] Mechatronics Engineering Department, International Islamic University Malaysia, Gombak, Malaysia
[4] Electrical Engineering Department, Universitas Mercu Buana, Jakarta, Indonesia

Abstract. Speech is the most significant communication mode among human beings and a potential method for human-computer interaction (HCI). Being unparallel in complexity, the perception of human speech is very hard. The most crucial characteristic of speech is gender, and for the classification of gender often pitch is utilized. However, it is not a reliable method for gender classification as in numerous cases, the pitch of female and male is nearly similar. In this paper, we propose a time-frequency method for the classification of gender-based on the speech signal. Various techniques like framing, Fast Fourier Transform (FFT), auto-correlation, filtering, power calculations, speech frequency analysis, and feature extraction and formation are applied on speech samples. The classification is done based on features derived from the frequency and time domain processing using the Support Vector Machines (SVM) algorithm. SVM is trained on two speech databases Berlin Emo-DB and IITKGP-SEHSC, in which a total of 400 speech samples are evaluated. An accuracy of 83% and 81% for IITKGP-SEHSC and Berlin Emo-DB have been observed, respectively.

Keywords: Gender classification · Pre-processing · Fast-Fourier Transform (FFT) · Support Vector Machine (SVM)

1 Introduction

The study of speech signals and the different techniques utilized for their process is speech processing. Speech processing is implemented in various applications such as speech recognition, speech compression, speech synthesis, speech coding, and speaker recognition technologies [1], amidst speech recognition is an imperative one. A large amount of information can be gathered from a speech signal like gender, words, dialect, emotion, and age that could be utilized for various applications, and gender classification

J. A. Mat Jizat et al. (Eds.): iCITES 2020, AISC 1350, pp. 1–10, 2021.
https://doi.org/10.1007/978-3-030-70917-4_1

is one of them. Gender classification is a crucial task in Human-Computer Interactions (HCI) like personal identification [2]. Any machine that has the proficiency of classification of gender can be utilized in many areas, e.g., speaker indexing [3], speaker diarization [4], HCI [5], annotation, and multimedia database [6]. In recent years, gender classification is given priority in the research world as gender inhibits great and unique details involving females and males' social activities.

Human voices are uncommon among all the creatures generating sounds since every single wave exhibits a distinct frequency. For sound and voice analysts, recognizing human gender based on the voice has been a challenging task for enhancing HCI systems, particularly developing customized dialogue systems that depend on gender speech [7]. In human speech, physiological dissimilarities like vocal tract length or thickness, glottis, and variations in speaking styles can be identified for recognizing gender [8]. Most of the studies on gender classification are based on acoustic features that depend on the detailed evaluation of the fundamental frequency. Commonly, female speakers have higher fundamental frequency and formant frequency, and the formant frequency differences are lesser than the fundamental frequency differences between females and males [9]. Nevertheless, fallacious evaluation of fundamental frequency might lead to a considerable reduction in the validation of detecting gender.

Besides, in different traditional speech features like Mel-frequency cepstral coefficients (MFCC) [10], relative spectral PLP coefficients (RASTA-PLP) [11], linear predictive cepstral coefficients (LPCC) [12] and linear predictive coefficients (LPC) [13], are used for the classification of gender. Recently in [7, 14–16], various deep learning algorithms have been utilized for the recognition of gender-based voice. This paper has proposed a preprocessing method involving a few techniques like framing, FFT, auto-correlation, filtering, speech frequency analysis, and power calculation to classify male and female voices. The classification is done based on features derived from the frequency and time domain processing using the Support Vector Machines (SVM) algorithm. We have considered only three features for each domain to analyze the accuracy when a small number of features are taken. The audio samples are taken from two datasets: Berlin Emotional database (Emo-DB) [17] and Indian Institute of Technology Kharagpur Simulated Emotion Hindi Speech Corpus (IITKGP-SEHSC) [18].

2 Related Works

Due to the development and advancement in the gender classification techniques, significant improvement has been reported in the machine's interactional and perception abilities and has led to several inherent utilizations in immense application scopes [19]. In [20], the authors proposed a feature set based on Pitch-Range (PR) to classify gender and age. Three age groups of males, i.e., young, middle-aged, and senior groups, were examined and applied for female voices. For the evaluation process, two different classification techniques were employed, Support Vector Machines (SVM) and *k*-Nearest Neighbor (kNN). The presented method achieved the highest accuracy rates as compared to the previous state-of-art models. Only the young male speaker's group achieved a lower accuracy rate in terms of classification.

For the extraction of voiced frames from speech signals, Zero Crossing Rate (ZCR) and Short-Term Energy (STE) were utilized in [21]. A total of 400 speech signals were examined taken from the IIIT-H database, 200 samples of male voices, and 200 females. Two components of feature vector MFCC's and pitch were considered, and the classification was carried by SVM. Around 99.5% of the accuracy rate was achieved for the identification of gender.

In [22], the authors proposed a framework for the classification of gender by extracting basic features from a speech like energy, MFCC, and pitch. TIMIT database was used for the experiment, in which 140 male speech samples and 140 female speech samples were evaluated. Around 20% of data was used for testing and 80% for training. For the classification processing, SVM was used, which yielded an accuracy of 96.45%. An evaluation of higher-order statistics and speech analysis was carried out in [23] to compare and classify males and females based on their voices. Higher-order statistics are the varied parameters of spectral descriptors by which the spectral analysis is done. The calculation was done on higher-order statistics like spectral entropy, spectral slope, spectral flatness, spectral centroid, and spectral slope. The classification was the basis of voiced and unvoiced speech, lower formants, and peakiness of speech.

Recently in [24], the accuracy of 100% was achieved using MFCC, machine learning algorithm (J 48), and Vector Quantization (VQ). The proposed classification system tested and trained on the database consisting of 2270 voice samples of celebrities, out of which 1138 were of males and 1132 of females. Nevertheless, this paper aims to develop a fundamental and practical voice-based gender classification system based on the preprocessing technique. The classification is carried out using the SVM algorithm.

3 Proposed Methodology

The proposed framework consists of preprocessing techniques including framing, Fast Fourier Transform (FFT), auto-correlation, filtering, power calculations, speech frequency analysis, feature extraction, formation, and classification.

3.1 Pre-processing, Framing, Fast Fourier Transform, Filtering, Feature Selection

Pre-processing involves the acquisition of the input signal, silence removal, noise removal, and normalization. In framing, the audio signals are to be framed or segmented. The signal is framed according to the window size N, which can be varied as 512, 1024, or 2048 samples, as shown in Eq. (1).

$$n_f = \left\lfloor \frac{F_S}{N} \right\rfloor \tag{1}$$

where n_f is the number of frames, F_s is the sampling frequency, and N is the window size. FFT transforms N frame samples from the time domain to the frequency domain, as shown in Eq. (2).

$$y_i(n) = \sum_{k=1}^{N} S_i(k)h(k)e^{2\pi/N} \tag{2}$$

where, k is the length of the FFT, $S_i(k)$ is the frequency domain signal, $h(k)$ is the window size with N samples long.

Filtering performs the convolution of the FFT output, and the coefficients are generated by a Gaussian window, as shown in Eq. (3).

$$z[n] = \sum_{i=0}^{N} a_i \cdot x[n-i] \tag{3}$$

where $x[n]$ is the input signal, $z[n]$ is the output signal, N is the filter order and a_i is the Gaussian filter coefficients. Moreover, power spectrum calculation could be performed on the speech signal as shown in Eq. (4).

$$F_p = \left| y^2\left(1 : \frac{n_f}{2} + 1\right) \right| \tag{4}$$

where F_p is the total power of the input signal, y^2 is the smoothening filter output and n_f is the number of frames. Besides, for a particular coefficient, the input signal is autocorrelated with itself, as shown in Eq. (5).

$$\phi = E[x(n)x(n-m)] \tag{5}$$

where ϕ is the estimated autocorrelated output, $x(n)$ is the input signal, and m is the signal delay. Next, the power spectrum of the speech signal is analyzed as shown in Eq. (6).

$$F_x = \frac{F_s}{fr + t_y - 1} \tag{6}$$

where $f_1 = \frac{F_s}{500}$ and $f_2 = \frac{F_s}{50}$ are the maximum amplitude at 500 and 50 Hz, respectively. The autocorrelation in the range of f_1 and f_2 is then can be calculated as $\phi = \phi(f_2 + 1 : 2 \times f_2 + 1)$. Then, $[\phi_{max}, t_y] = max(\phi(f_1 : f))$, where t_y is the sampling rate and F_x in Eq. (6) yields the output value that is used to classify the gender. Finally, the crucial phase in the classification of gender is feature selection. Speech consists of numerous emotions and features, and one cannot state with certainty, which set of features must be modeled and thus making a requirement for the utilization of feature selection.

3.2 Classifier

The classifier used is SVM, which is a supervised learning technique. SVM is best suited for the binary classification as it separates the two different classes using a hyperplane. The design of the inner product, i.e., the kernel, is essential in utilizing the SVM's successfully. After completing the training process, when the input sample is subjected to testing, the output identifies the input speech sample either of female or male.

For the classification of gender-based on voice, the first step is the preprocessing involving the acquisition of the input audio signal, normalizing, silence internal removal & noise removal. The signal is then framed, i.e., the number of samples needs are framed to perform FFT. FFT is done to convert the time domain signal into the frequency domain and for better analysis. The convolutional is performed between the FFT output, and the coefficients generated are by the Gaussian window technique. Finally, the absolute power of the signal is obtained. Thus, final features are extracted based on the power present at different frequencies. Also, parallel time-domain processing is done to obtain the separate set of features, performing autocorrelation on the input time-domain signal, i.e., the input speech signal is autocorrelated with itself for a coefficient. The typical features for male will have a fundamental frequency in-between 85–180 Hz, and for a female, it would be 165 to 255. Thus, the entire feature extraction technique is based on extracting key features present around these frequencies. The features are combined from both the time and frequency domain and used for training and classification. The SVM model is used for training and classification, as shown in Fig. 1.

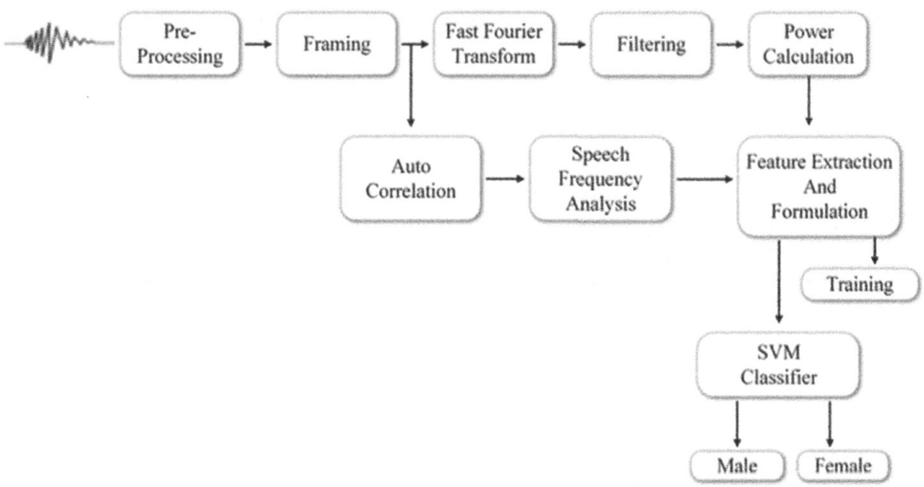

Fig. 1. The proposed framework for gender classification using speech signals

4 Results and Discussion

4.1 Multilanguage Databases

Two databases have been used for the experiment, including the Berlin Emotional Database (Emo-DB) and Indian Institute of Technology Kharagpur Simulated Emotion Hindi Speech Corpus (IITKGP-SEHSC). A total of 400 speech samples have been evaluated for the experiment.

Berlin Emo-DB is an acted German speech dataset and is publicly available. The database consists of audio files recorded by five males and five females and contains 535 speech samples. For this experiment, 200 samples were considered from this database, 93 of male speech, and 107 of female speech. On the other hand, IITKGP-SEHSC consists of 10 professional artists (five male and five female). The total number of utterances in the database is 12000 samples. Around 200 speech samples were taken from this database for the evaluation process, including 103 male speeches and 97 female speeches.

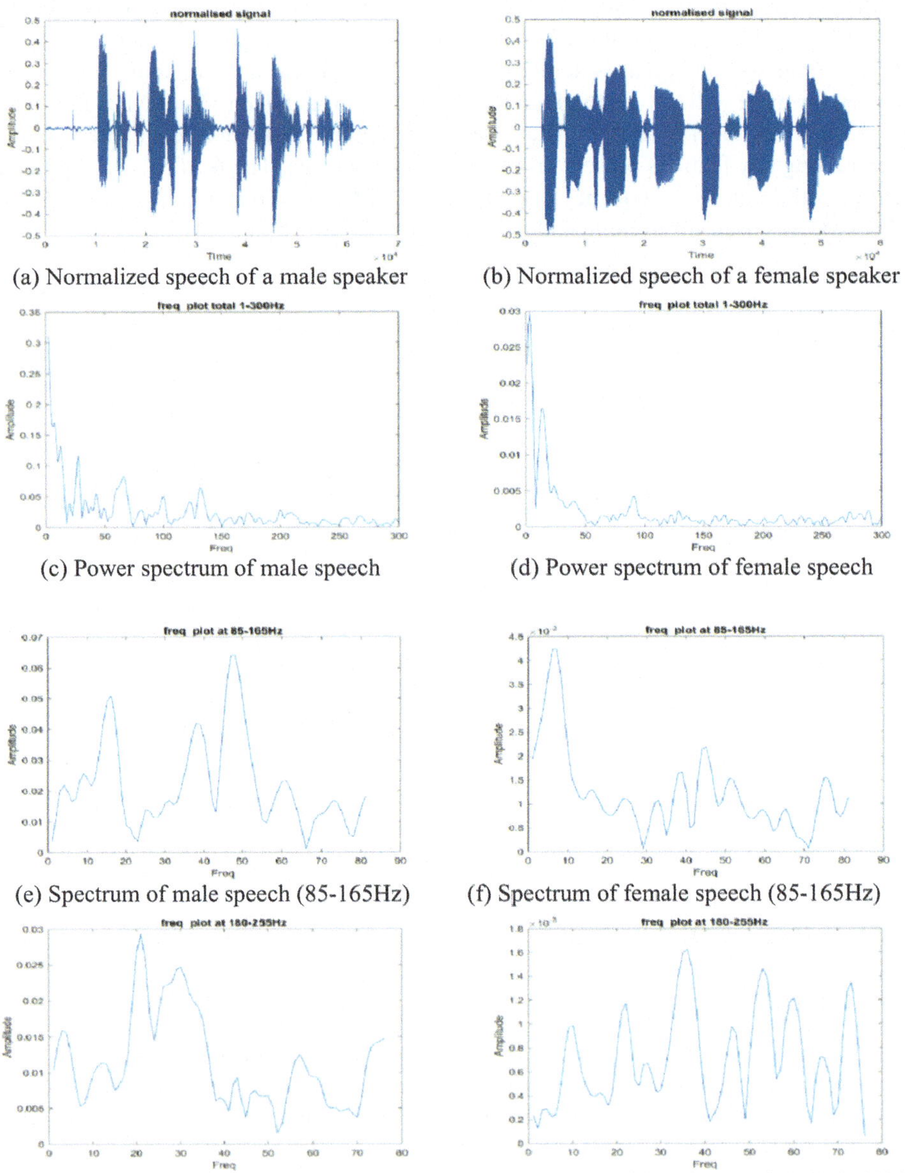

(a) Normalized speech of a male speaker
(b) Normalized speech of a female speaker
(c) Power spectrum of male speech
(d) Power spectrum of female speech
(e) Spectrum of male speech (85-165Hz)
(f) Spectrum of female speech (85-165Hz)
(g) Spectrum of male speech (180-255Hz)
(h) Spectrum of female speech (180-255Hz)

Fig. 2. Speech signals and their power spectrums in different frequency ranges

4.2 Experimental Setup

The software used was Matlab version 2020a for the implementation of the proposed architecture. Two datasets have been considered, including Berlin Emo-DB and IITKGP-SEHSC, for the evaluation process. From Emo-DB and IITKGP-SEHSC databases, 107 and 93 speech samples of females and males were taken from each database. A total of 400 samples were taken, and the data was divided into 70% for training and 30% for testing. The training was performed on a laptop with i5–8250 CPU with 8 GB onboard memory. Two experiments were performed. The first experiment involved the training on Emo-DB and the second on the IITKGP-SEHSC database.

4.3 Experiment on Speech Analysis of Male and Female Speech Signals

Figure 2 shows the same sentences uttered by male and female speakers and its power spectrum plotted in the range of pitch frequency (up to 300 Hz). Figure 3 shows the signal power in the time domain autocorrelation plots.

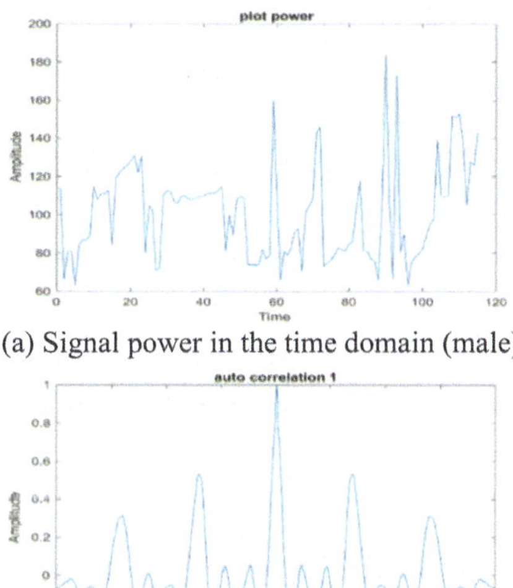

(a) Signal power in the time domain (male)

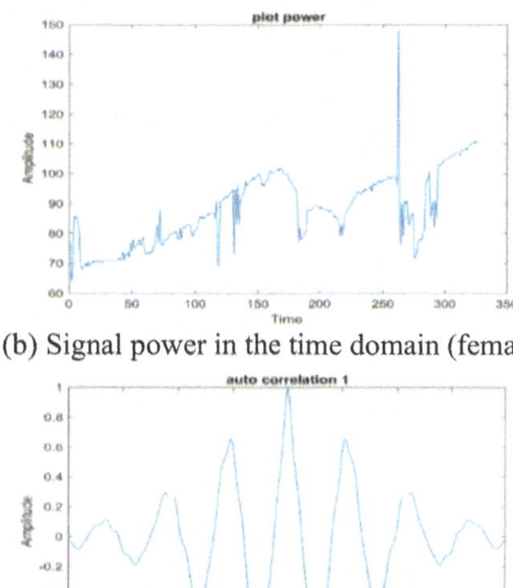

(b) Signal power in the time domain (female)

(c) Autocorrelation for male speech

(d) Autocorrelation for female speech

Fig. 3. Speech signals power in the time domain, and its autocorrelation signals

The difference between male and female speech signals can be observed in the time and frequency domain from these plots. Therefore, the unique characteristic can be used as the features for gender classification. In the next section, SVM will be trained and used as the classifier.

4.4 Experiment on Gender Classification Using SVM

Table 1 and Table 2 show both time domain and frequency domain feature extracted at male and female fundamental frequencies components for Emo-DB and IITKGP-SEHSC, respectively. The features are derived from frequency and time domain processing. F1, F2, and F3 correspond to power components at the fundamental frequency for male, female, and total sum in the frequency domain, respectively. Time-domain features T1 and T2 correspond to power for male and female speech at their fundamental frequencies, and T3 is the max amplitude of the autocorrelation output, respectively.

Table 1. Time-domain and frequency-domain features for Berlin EmoDB

Sample	Frequency domain features			Time domain features			Actual	Predicted
	F1	F2	F3	T1	T2	T3		
1	0.78	0.56	8.34	10.82	107.23	163.00	Male	Male
2	12.55	11.80	143.00	0.93	152.30	209.00	Male	Female
3	31.46	16.67	125.44	1.67	156.33	231.20	Male	Female
4	14.70	7.00	57.20	16.80	44.60	173.00	Male	Male
5	37.40	35.80	364.90	8.66	255.40	209.00	Male	Male
6	0.14	0.64	1.74	0.00	28.25	234.00	Male	Female
7	4.22	16.40	142	12.70	64.00	186.00	Male	Male
8	6.10	18.47	201.00	7.20	8.80	182.00	Female	Female
9	0.17	0.51	2.70	3.25	1.85	216.00	Female	Female
10	6.74	6.79	82.50	7.65	194.50	232.00	Female	Female
11	0.24	0.25	4.25	4.50	124.60	206.00	Female	Female

Multiple feature extraction is applied to the speech samples taken from the two databases and can be observed. From the Table 1 and Table 2, the time-domain features (T1 and T2) and frequency-domain feature (F1 and F2) are the power component for male and female varies and depends on a certain threshold. The threshold is derived based on the signal strength at fundamental frequencies corresponding to the male and female range. A simple classifier code is built based on these threshold values. In Eq. (6), gender is classified based on the output value F_x, which is the autocorrelated output. If its value falls in the range of 80–175 Hz and 175–255 Hz, it is classified as male and female, respectively. Nevertheless, this value depends on the recording environment and other parameters and thus, cannot be considered for all speech samples of the dataset.

The dataset is split into training and testing datasets with a ratio of 70% to 30%. The combined time and frequency domain features are trained and classified using SVM. As shown in Table 1 and 2, the classification results obtained were 83% for the IITKGP-SEHSC database and 81% for Berlin Emo-DB.

Table 2. Time-domain and frequency-domain features for IITKGP-SEHSC

Sample	Frequency domain features			Time domain features			Actual	Predicted
	F1	F2	F2	T1	T2	T3		
12	0.16	0.21	1.80	35.60	53.20	164.00	Female	Female
13	0.02	0.02	0.48	20.00	41.10	187.00	Male	Male
14	0.02	0.02	0.51	11.69	9.50	152.00	Male	Male
15	0.03	0.03	0.75	3.30	26.60	218.00	Female	Female
16	0.02	0.02	0.71	0.68	53.80	219.00	Female	Female
17	0.03	0.35	0.08	11.40	30.00	114.00	Male	Male
18	0.04	0.03	0.79	17.40	10.30	181.00	Male	Male
19	0.03	0.03	0.73	16.40	9.30	161.00	Male	Male
20	0.02	0.02	0.52	6.90	38.80	227.90	Female	Female
21	0.01	0.01	0.43	5.80	34.21	222.40	Female	Female

5 Conclusions and Future Works

In this paper, we have put forward a preprocessing technique for gender classification. The features are derived from frequency and time domain processing, and several parameters are calculated. Two databases Berlin Emo-DB and IITKGP-SEHSC, are used, and 200 speech samples are taken from each of the databases, i.e., 400 samples are taken consisting of male and female speeches. For the classification process, the Support Vector Machine (SVM) was used 83% of the recognition rate was obtained for IITKGP-SEHSC and 81% for Berlin EMO-DB. In future work, we will use spectrograms and Convolutional Neural Networks (CNNs) for the classification process in order to improve the recognition rate and hence performance for the gender classification system.

Acknowledgments. The authors would like to express their gratitude to the Malaysian Ministry of Education (MOE), which has provided research funding through the Fundamental Research Grant, FRGS19–076-0684 (FRGS/1/2018/ICT02/UIAM/02/4). The authors would like to acknowledge support from International Islamic University, University of New South Wales, and Universitas Mercu Buana.

References

1. Dabrowski, A., Marciniak, T.: Audio signal processing. In: The Computer Engineering Handbook (2001)
2. Breslin, S., Wadhwa, B.: Gender and human-computer interaction. In: The Wiley Handbook of Human Computer Interaction Set (2017)
3. Sedaaghi, M.H.: A comparative study of gender and age classification in speech signals. Iran. J. Electr. Electron. Eng. **5**(1), 1–12 (2009)

4. Doukhan, D., Carrive, J., Vallet, F., Larcher, A., Meignier, S.: An Open-Source Speaker Gender Detection Framework for Monitoring Gender Equality (2018). https://doi.org/10.1109/ICASSP.2018.8461471
5. Zhang, W., Smith, M.L., Smith, L.N., Farooq, A.: Gender and gaze gesture recognition for human-computer interaction. Comput. Vis. Image Underst. **149**, 32–50 (2016). https://doi.org/10.1016/j.cviu.2016.03.014
6. Harb, H., Chen, L.: Voice-based gender identification in multimedia applications. J. Intell. Inf. Syst. **24**, 179–198 (2005). https://doi.org/10.1007/s10844-005-0322-8
7. Alkhawaldeh, R.S.: DGR: gender recognition of human speech using one-dimensional conventional neural network. Sci. Program. (2019). https://doi.org/10.1155/2019/7213717
8. Simpson, A.P.: Phonetic differences between male and female speech. Linguist. Lang. Compass **3**, 621–640 (2009). https://doi.org/10.1111/j.1749-818X.2009.00125.x
9. Vorperian, H.K., Kent, R.D., Lee, Y., Bolt, D.M.: Corner vowels in males and females ages 4 to 20 years: fundamental and F1–F4 formant frequencies. J. Acoust. Soc. Am. **146**, 3255–3274 (2019). https://doi.org/10.1121/1.5131271
10. Archana, G.S., Malleswari, M.: Gender identification and performance analysis of speech signals (2015). https://doi.org/10.1109/GCCT.2015.7342709
11. Zeng, Y.M., Wu, Z.Y., Falk, T., Chan, W.Y.: Robust GMM based gender classification using pitch and RASTA-PLP parameters of speech (2006). https://doi.org/10.1109/ICMLC.2006.258497
12. Yucesoy, E., Nabiyev, V.V.: Comparison of MFCC, LPCC and PLP features for the determination of a speaker's gender (2014). https://doi.org/10.1109/siu.2014.6830230
13. Yusnita, M.A., Hafiz, A.M., Fadzilah, M.N., Zulhanip, A.Z., Idris, M.: Automatic gender recognition using linear prediction coefficients and artificial neural network on speech signal (2018). https://doi.org/10.1109/ICCSCE.2017.8284437
14. Buyukyilmaz, M., Cibikdiken, A.O.: Voice gender recognition using deep learning (2016). https://doi.org/10.2991/msota-16.2016.90
15. Raahul, A., Sapthagiri, R., Pankaj, K., Vijayarajan, V.: Voice based gender classification using machine learning (2017). https://doi.org/10.1088/1757-899X/263/4/042083
16. Pondhu, L.N., Kummari, G.: Performance analysis of machine learning algorithms for gender classification (2018). https://doi.org/10.1109/ICICCT.2018.8473192
17. Burkhardt, F., Paeschke, A., Rolfes, M., Sendlmeier, W., Weiss, B.: A database of German emotional speech (2005)
18. Koolagudi, S.G., Reddy, R., Yadav, J., Rao, K.S.: IITKGP-SEHSC: Hindi speech corpus for emotion analysis (2011). https://doi.org/10.1109/ICDECOM.2011.5738540
19. Xu, W., Zhuang, Y., Long, X., Wu, Y., Lin, F.: Human gender classification: a review. Int. J. Biom. **8**, 275–300 (2016). https://doi.org/10.1504/ijbm.2016.10003589
20. Barkana, B.D., Zhou, J.: A new pitch-range based feature set for a speaker's age and gender classification. Appl. Acoust. **98**, 52–61 (2015). https://doi.org/10.1016/j.apacoust.2015.04.013
21. Gupta, M., Bharti, S.S., Agarwal, S.: Support vector machine based gender identification using voiced speech frames (2016). https://doi.org/10.1109/PDGC.2016.7913219.
22. Chaudhary, S., Sharma, D.K.: Gender identification based on voice signal characteristics (2018). https://doi.org/10.1109/ICACCCN.2018.8748676
23. Qadri, S.A.A., Gunawan, T.S., Wani, T., Alghifari, M.F., Mansor, H., Kartiwi, M.: Comparative analysis of gender identification using speech analysis and higher order statistics (2019). https://doi.org/10.1109/ICSIMA47653.2019.9057296
24. Shareef, M.S., Abd, T., Mezaal, Y.S.: Gender voice classification with huge accuracy rate. Telkomnika (Telecommun. Comput. Electron. Control) **18**, 2612–2617 (2020). https://doi.org/10.12928/TELKOMNIKA.v18i5.13717

Affective Computing for Visual Emotion Recognition Using Convolutional Neural Networks

Arselan Ashraf[1], Teddy Surya Gunawan[1,2](✉), Farah Diyana Abdul Rahman[1], Ali Sophian[3], Eliathamby Ambikairajah[2], Eko Ihsanto[4], and Mira Kartiwi[5]

[1] Electrical and Computer Engineering Department, International Islamic University Malaysia, Gombak, Malaysia
tsgunawan@iium.edu.my
[2] School of Electrical Engineering and Telecommunications, UNSW, Sydney, Australia
[3] Mechatronics Engineering Department, International Islamic University Malaysia, Gombak, Malaysia
[4] Electrical Engineering Department, Universitas Mercu Buana, Jakarta, Indonesia
[5] Information Systems Department, International Islamic University Malaysia, Gombak, Malaysia

Abstract. Affective computing is a developing interdisciplinary examination field uniting specialists and experts from different fields, from artificial intelligence, natural language processing to intellectual and sociologies. The thought behind affective computing is to give computers the aptitude of insight that will, in general, comprehend human feelings. Notwithstanding these victories, the field needs hypothetical firm establishments and efficient rules in numerous regions, especially in feeling demonstrating and developing computational models of feeling. This exploration manages affective computing to improve the exhibition of Human-Machine Interaction. This work's focal point is to distinguish the emotional state of a human utilizing deep learning procedure, i.e., Convolutional Neural Networks (CNN) containing parameters like three convolution layers, pooling layers, learning rates, two fully connected layers, batch normalizations, and dropout ratios. The Warsaw Set of Emotional Facial Expression Pictures dataset has been utilized to build up a feeling acknowledgment model, which will have the option to perceive five facial feelings, including happy, sad, anger, surprise, and neutral. The database was selected based on its validation study of facial display photographs. Dataset was split based on a 65:35 ratio for training and testing/validation, respectively. The proposed framework design and the strategy has been discussed in this paper alongside the experimental findings. Our model's recognition accuracy came out to be 80% and 83.33% for validation set 1 and validation set 2, respectively. The performance parameters have also been evaluated in terms of the confusion matrix, recall, and precision.

Keywords: Affective computing · Artificial intelligence · Emotion recognition · Convolutional Neural Networks

J. A. Mat Jizat et al. (Eds.): iCITES 2020, AISC 1350, pp. 11–20, 2021.
https://doi.org/10.1007/978-3-030-70917-4_2

1 Introduction

Affective computing is one of the most dynamic exploration subjects, moreover, having progressively full consideration. The rise in interest in this field is due to promising applications in many fields, for example, computer-generated reality, shrewd reconnaissance, and emotion recognition [1]. In past years, there has been an increase in research works on emotion recognition via artificial intelligence. An emotion recognition model was build using a multi-modal approach [2]. The work was carried on different pictures and video information bases, including the ADFES-BIV information base. The ADFES-BIV, KNN+LBP, and LBP+SVM strategies were placed in various cell sizes, i.e., 8, 16, 32, and 64. The best exactness for LBP+KNN came out to be 87.44% by utilizing cell size 32. Moreover, video supports portraying three levels of forces of enthusiastic explanations were proposed in [3] with 69% validation accuracy.

Facial activities pass on feelings like this pass on an individual's character, state of mind, and expectations. Feelings ordinarily rely on the facial highlights of a person alongside the voice. Many machine learning techniques are present to create an emotion recognition system. However, this examination will focus on image-based feeling acknowledgment utilizing deep learning. Image-based feeling acknowledgment is multidisciplinary and incorporates fields like brain science, emotional registering, and human-computer interaction. The message's primary bit is the outward appearance, which builds up 55% of the overall impression [4]. There must be legitimate element edges of the outward appearance inside the extension to make a well-fitted model for image-based feeling acknowledgment [5]. Rather than utilizing traditional methods, deep learning gives an assortment of exactness, learning rate, and expectation. Convolutional Neural Networks (CNN) is a deep learning strategy that has offered help and a stage for breaking down visual symbolism.

Convolution is the primary utilization of a channel to an input that results in the activation. Repeated utilization of a comparative channel to an input achieves a guide of establishments called a component map, indicating the regions and nature of a perceived component in contribution, for instance, an image [6]. Deep learning has made incredible progress in perceiving feelings, and CNN is the notable profound learning strategy that has accomplished noteworthy execution in picture preparation [7]. Inspired by deep learning, this research aims to formulize the image-based emotion recognition model.

2 Proposed Visual Emotion Recognition System

Emotion recognition can be performed using facial features extracted from the images [8]. The proposed model performs initial pre-processing of the input images, including cropping, RGB to grey color conversion, histogram equalization, as shown in Fig. 1. The next step is to feed those images to the image input layer of CNN, which extracts different features from input images and predicts the five emotions. In this work, Images were processed, and final output images were saved in the dataset as.mat files in Matlab, which could be quickly loaded, trained, and tested instead of performing the repetitive steps like pre-processing and face detection every time. The CNN model is trained on an image dataset, namely the Warsaw Set of Emotional Facial Expression Pictures.

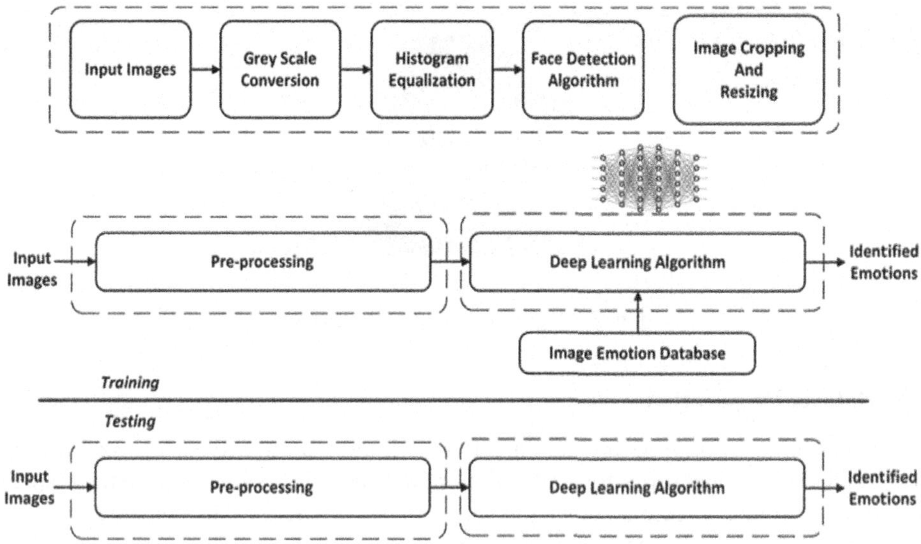

Fig. 1. Proposed visual emotion recognition algorithm

3 Implementation

While implementation, the first stage focuses on image pre-processing steps. The proposed algorithm is implemented in Matlab.

3.1 Image Processing

Image acquisition is defined as acquiring a picture from some source, usually any folder or other source for preparation. Since pictures are not of uniform sizes, its resized to a worth $M \times N$, where M and N can be varied dependent on the outcomes required [9]. For post handling, RGB pictures should be changed over into a greyscale design. A grayscale image makes handling a lot less complicated [10], as shown in Fig. 2. Meanwhile, a picture histogram showcases the graphical portrayal of the difference appropriation in a computerized picture [11].

For this research, the face detection is using the Viola-Jones algorithm [12] has been incorporated in this work. A detected face is then cropped to acquire a more extensive and distinct look of the facial picture [13].

3.2 Emotion Image Database

For this research, an image-based dataset has been considered to develop the proposed emotion recognition system. Warsaw Set of Emotional Facial Expression Pictures (WSE-FEP) [14] contains 210 top-notch pictures of 30 people. They show six essential feelings (happy, fear, disgust, anger, sadness, surprise), and neutral presentation.

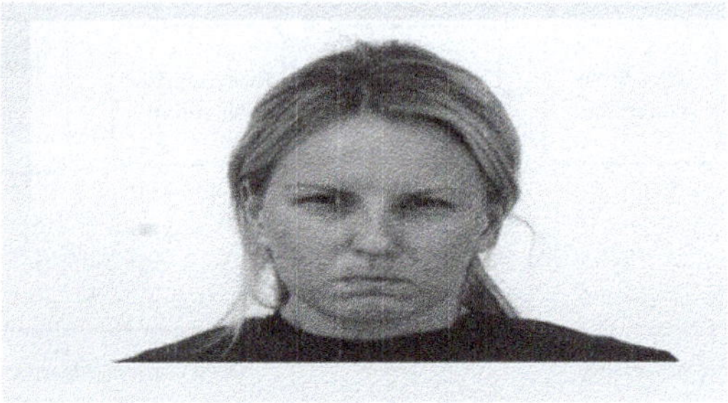

Fig. 2. A grayscale converted image.

3.3 Dataset Preparation

The first step is to arrange images in different folders as it will be instrumental in developing a command script for each emotion, which can read images from the respective emotion folder. Figure 3 shows the snapshots of separate folders created for each emotion. This process is done for all other emotions, i.e., sad, neutral, and surprised.

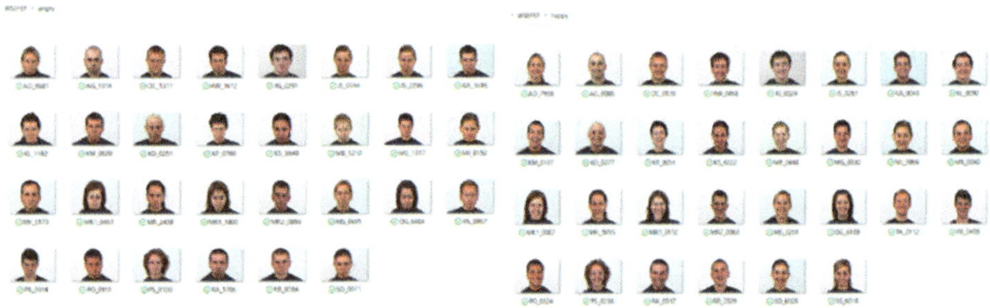

Fig. 3. Sample of an arranged set of images for angry and happy emotion

3.4 Convolutional Neural Networks (CNN)

CNN is a various leveled neural organization that comprises of an assortment of layers in a grouping. A typical model ordinarily comprises a few convolutional layers where the visual substance (for example, facial highlights) are spoken to as a bunch of highlight maps got in the wake of convolving the contribution with an assortment of channels are found out during the preparation stage. Pooling layers might be acquainted after convolutional layers with amass greatest initiation highlights from convolutional include maps. Because of pooling, the spatial goal of these guides is diminished. It also has fully connected layers where every neuron of the info layer is associated with each neuron in the layer. At last, a SoftMax layer plays out the last grouping task dependent on this portrayal.

The proposed CNN architecture is shown in Fig. 4. The input of the network is a 200×200 face image generated from the input dataset. Next, there are four convolutional

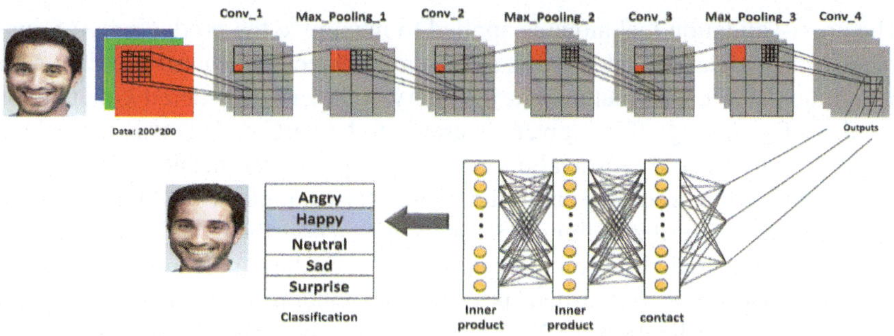

Fig. 4. Configuration of convolutional neural network

layers with polling applied to each layer. The convolutional layer consists of various components like filter size, padding, and stride. The first convolutional layer consists of filter sizes 3 × 16, the second 3 × 32, and the third 3 × 64. Every layer is connected by rectified linear units (ReLU) and a max-pooling layer of 3 × 3 with stride 2 followed by two fully connected layers. A dropout layer follows two fully connected layers with a dropout ratio of 50% to avoid overfitting.

4 Results and Discussion

This experiment was carried on the Warsaw Set of Emotional Facial Expression Pictures (WSEFEP) dataset. The integrated development environment (IDE) used was Matlab 2020a with Deep Learning and Signal Processing Toolboxes. The maximum epochs for training were set to 100, while the maximum iterations were set to 300. The hardware used was a laptop with Intel Core i7 8th generation, NVidia GTX 1050 GPU, with 8 GB RAM.

Dataset emotion labeling has been shown in Table 1. The model was trained for 300 iterations with 100 epochs. The sample size of the files was limited. Therefore, cross-validation was performed to enhance the results in terms of accuracy, precision, and recall. The dataset is loaded and split into training and validation datasets. Of which, about 65% accounts for training and 35% for testing/validation.

Table 1. Emotion labelling

Emotion	Files	Labels
Angry	30	1
Happy	30	2
Neutral	30	3
Sad	30	4
Surprise	30	5

The cross-validation technique is applied to maximize the productivity of our data samples. Cross-validation is defined as the resampling technique used to assess AI models on a finite dataset. The method has a solitary boundary considered k that alludes to the number of gatherings that a given dataset is to be part of. This way, the technique is frequently called k-fold cross-validation. When a particular incentive for k is picked, it could be utilized instead of k about the model, for example, $k = 2$ turning out to be 2-cross-validation. To utilize a finite example to gauge how the model will act in general when used to make predictions on information not used during the model's preparation.

For our model $k = 2$, the data is resampled randomly into two sets, including validation set 1 and validation set 2. The proportion of the data set for training is 65%, and the data set for validation/testing is 35% for both sets. The data splitting for two validation sets can be illustrated, as shown in Fig. 5.

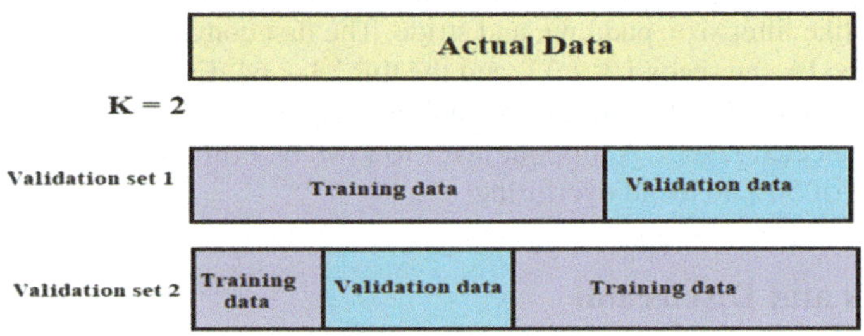

Fig. 5. CNN configuration for training and validation

Validation Set 1: The results in terms of accuracy, precision, recall, and confusion matrix for the the validation set 1 are shown in Fig. 6 and 7.

The accuracy for set 1 came out to be 83.33% on 300 iterations with 100 epochs. The elapsed time was 30 s. The validation accuracies for emotions: anger, happy, and surprise were highly encouraging with 100% attainment. However, emotions: neutral and sad showed low performance on validation data due to limited sample data for these emotions in the opted database for training. The performance of these emotions can be further boosted in upcoming works by implementing sufficient data resources. For clear visualization of precision and recall, Table 2 demonstrates them for all five emotions on the validation set 1.

Validation Set 2: The results in terms of accuracy, precision, recall, and confusion matrix for the validation set 2 are shown in Fig. 8 and 9.

The accuracy for set 2 came out to be 80% on 300 iterations with 100 epochs. The elapsed time was 30 s. The validation accuracy for emotions: anger and happy were highly encouraging with 100% attainment and 96.7% for surprise emotion. Nevertheless, emotions: neutral and sad showed low performance on validation data due to limited sample data for these emotions in the opted database for training. The performance of these emotions can be further boosted in upcoming works by implementing sufficient

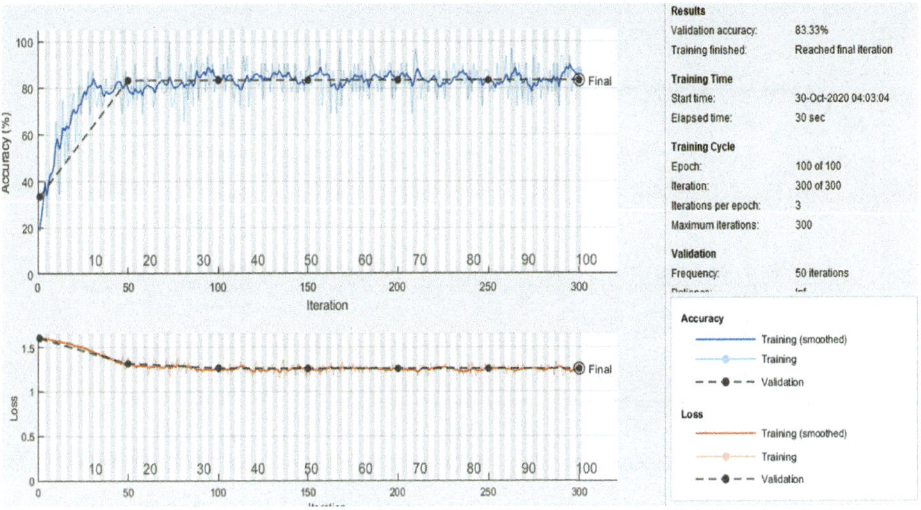

Fig. 6. Accuracy for validation set 1.

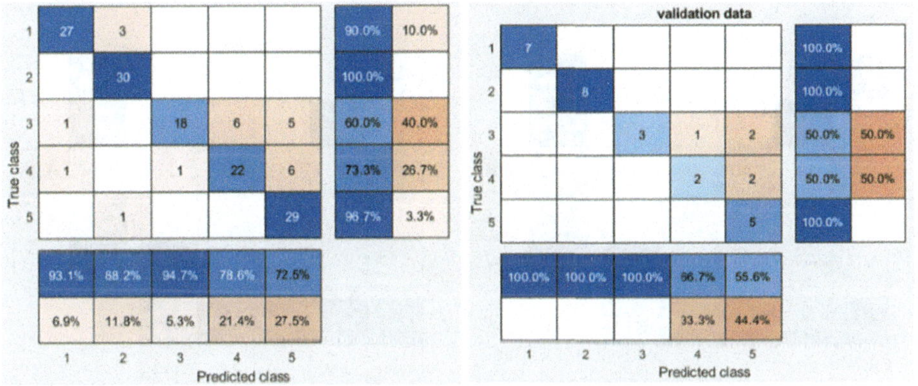

Fig. 7. Confusion matrix, recall, and precision for total and validation data for set 1.

Table 2. Recall and precision for validation data on set 1

Emotion	Recall	Precision
Angry	1	1
Happy	1	1
Neutral	0.5	1
Sad	0.5	0.667
Surprise	1	0.556

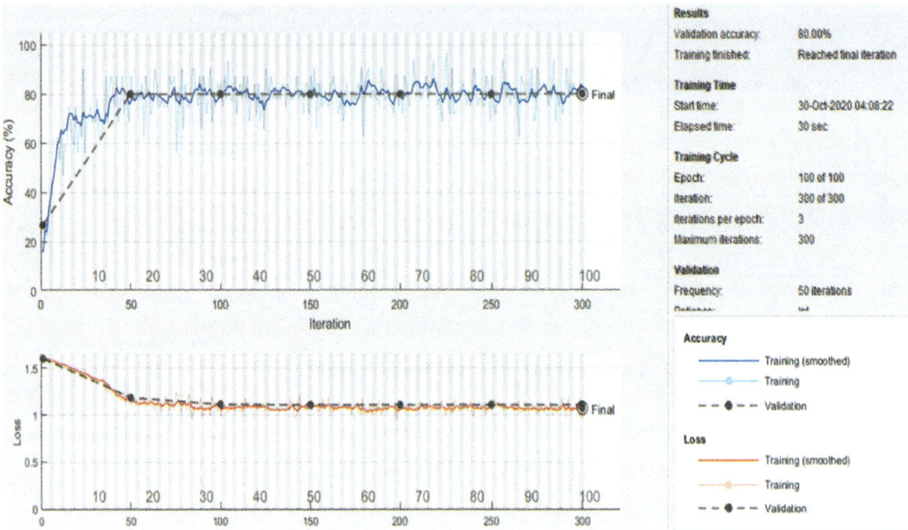

Fig. 8. Accuracy for validation set 2.

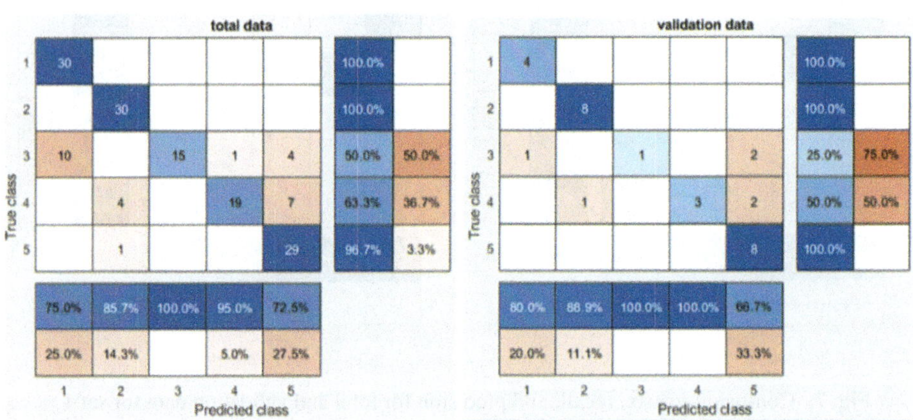

Fig. 9. Confusion matrix, recall, and precision for total and validation data for set 2.

data resources. For clear visualization of precision and recall, Table 3 demonstrates them for all five emotions on the validation set 2.

Our work results are an example of successful research accomplishment in the field of visual emotion recognition. This research was carried out for five emotions: happy, sad, neutral, angry, and surprised. With more emotions, the performance parameters usually start to decline, but with the introduction of cross-validation on our data samples, we achieved promising results. By involving cross-validation, our research got the upper hand over others to resample our data and obtain a new set of results. We got the liberty to select an outperformed set for the demonstration of results. The results from our research

Table 3. Recall and precision for validation data on set 2

Emotion	Recall	Precision
Angry	1	0.80
Happy	1	0.889
Neutral	0.25	1
Sad	0.5	1
Surprise	1	0.667

are benchmarked with some recent related works shown in Table 4. The accuracy of our proposed algorithm is outperformed other traditional algorithms.

Table 4. Benchmarking with recent related work

Authors-Year	Accuracy (%)
Olszanowski, M., Pochwatko, et al. (2015)	82.0
Goma Mohamed Salem Najah, (2017) accuracy for LBP+KNN	78.0
Goma Mohamed Salem Najah, (2017) accuracy for LBP+SVM	80.0
Our Work	**83.3**

5 Conclusions and Future Works

Affective computing is one of the most dynamic exploration subjects, moreover, having progressively full consideration. This paper presents the details about developing an image-based emotion recognition system using the affective computing technique. The proposed system, along with the methodology required in building the architecture, has been discussed. This research's image dataset is the Warsaw Set of Emotional Facial Expression Pictures (WSEFEP). Further, the experimental findings have been shown, of which accuracy came out to be 83.33% for the validation set 1 and 80% for the validation set 2. Other performance paraments like confusion matrix, precision, and recall have been shown with figures' help. The extension for future upgrades is engaging in this field. Distinctive multi-model deep learning procedures can be utilized alongside various structures to improve the presentation boundaries. Aside from perceiving the feelings only, there can be further expansion of the emotional intensity scale. This may assist with foreseeing the intensity of the perceived feeling. Additionally, multi-sources of data can be utilized in upcoming works. For instance, speech, visual and textual inputs can be utilized to plan a model alongside the utilization of multiple datasets.

Acknowledgments. The authors would like to express their gratitude to the Malaysian Ministry of Education (MOE), which has provided research funding through the Fundamental Research Grant,

FRGS19-076-0684 (FRGS/1/2018/ICT02/UIAM/02/4). The authors would like to acknowledge support from International Islamic University, University of New South Wales, and Universitas Mercu Buana.

References

1. Tao, J., Tan, T.: Affective Computing: A Review, pp. 981–995. Springer (2005)
2. Najah, G.M.S.: Emotion estimation from facial images (Masters dissertation), ResearchGate (2017)
3. Wingenbach, T.S., Ashwin, C., Brosnan, M.: Correction: validation of the amsterdam dynamic facial expression set – bath intensity variations (ADFES-BIV): a set of videos expressing low, intermediate, and high intensity emotions. PLOS ONE **11**(12), e0168891 (2016). https://doi.org/10.1371/journal.pone.0168891
4. Wu, C.-H., Lin, J.-C., Wei, W.-L.: Survey on audiovisual emotion recognition: databases, features, and data fusion strategies. APSIPA Trans. Signal Inf. Process. **3**, e12 (2014)
5. Gunawan, T.S., Ashraf, A., Riza, B.S., Haryanto, E.V., Rosnelly, R., Kartiwi, M., Janin, Z.: Development of video-based emotion recognition using deep learning with Google Colab. TELKOMNIKA (Telecommun. Comput. Electron. Control) **18**(5), 2463 (2020). https://doi.org/10.12928/telkomnika.v18i5.16717
6. Ashraf, A., Gunawan, T.S., Riza, B.S., Haryanto, E.V., Janin, Z.: On the review of image and video-based depression detection using machine learning. Indonesian J. Electr. Eng. Comput. Sci. **19**(3), 1677 (2020). https://doi.org/10.11591/ijeecs.v19.i3.pp1677-1684
7. Cai, Y., Zheng, W., Zhang, T., Li, Q., Cui, Z., Ye, J.: Video based emotion recognition using CNN and BRNN. In: Communications in Computer and Information Science Pattern Recognition, pp. 679–691 (2016)
8. Fan, Y., Lam, J.C.K., Li, V.O.K.: Video-based emotion recognition using deeply-supervised neural networks. In: Proceedings of 2018 on International Conference on Multimodal Interaction, ICMI 2018 (2018)
9. Trussell, H., Vrhel, M.: Introduction. Fundamental of Digital Imaging, pp. 1–6 (2008)
10. Shamir, A., Avidan, S.: Seam carving for content-aware image resizing. ACM Trans. Graph (2007)
11. Gonzalez, R., Woods, R.: Digital Image Processing. Addison-Wesley Publishing Company (1992)
12. Viola, P., Jones, M.: Robust real-time object detection. In: IJCV, pp. 1–3 (2001)
13. Boubenna, H., Lee, D.: Image-based emotion recognition using evolutionary algorithms. Biol. Inspired Cogn. Archit. **24**, 70–76 (2018)
14. Olszanowski, M., Pochwatko, G., Kuklinski, K., Scibor-Rylski, M., Lewinski, P., Ohme, R.K.: Warsaw set of emotional facial expression pictures: a validation study of facial display photographs. Front. Psychol. **5** (2015)

Speech Emotion Recognition Using Deep Neural Networks on Multilingual Databases

Syed Asif Ahmad Qadri[1], Teddy Surya Gunawan[1,2](✉), Taiba Majid Wani[1], Eliathamby Ambikairajah[2], Mira Kartiwi[3], and Eko Ihsanto[4]

[1] Electrical and Computer Engineering Department, International Islamic University Malaysia, Gombak, Malaysia
tsgunawan@iium.edu.my
[2] School of Electrical Engineering and Telecommunications, UNSW, Sydney, Australia
[3] Information Systems Department, International Islamic University Malaysia, Gombak, Malaysia
[4] Electrical Engineering Department, Universitas Mercu Buana, Jakarta, Indonesia

Abstract. The research community's ever-increasing interest in studying human-computer interactions (HCI), systems deducing, and identifying a speech signal's emotional aspects has emerged as a hot research topic. Speech Emotion Recognition (SER) has brought the development of automated and intelligent analysis of human utterances to reality. Typically, an SER system focuses on extracting the features from speech signals such as pitch frequency, formant features, energy-related and spectral features, tailing it with a classification quest to understand the underlying emotion. The key issues pivotal for a successful SER system are driven by the proper selection of proper emotional feature extraction techniques. In this paper, Mel-frequency Cepstral Coefficient (MFCC) and Teager Energy Operator (TEO) along with a new proposed Feature Fusion of MFCC and TEO referred to as Teager-MFCC (TMFCC) is examined over a multilingual database consisting of English, German and Hindi languages. Deep Neural Networks have been used to classify the different emotions considered, happy, sad, angry, and neutral. Evaluation results show that the proposed fusion TMFCC with a recognition rate of 92.7% outperforms TEO and MFCC. With TEO and MFCC configurations, the recognition rate has been found as 88.5% and 90.0%, respectively.

Keywords: Speech Emotion Recognition · Mel-frequency Cepstral Coefficient (MFCC) · Teager Energy Operator (TEO) · Deep Neural Networks (DNN)

1 Introduction

Automatic speech recognition, or simply speech recognition, is the technology that manages methodologies and procedures to perceive speech signals [1]. It was discovered that voice could be the next mechanism for speaking with machines, mainly computer-based systems. A requirement for gathering emotions from spoken utterances increment exponentially. Emotion recognition manages the investigation of inducing emotions, techniques utilized for gleaning [2]. Facial expressions and speech signals can perceive

J. A. Mat Jizat et al. (Eds.): iCITES 2020, AISC 1350, pp. 21–30, 2021.
https://doi.org/10.1007/978-3-030-70917-4_3

emotion. Different techniques have been evolved to detect emotions, such as neural networks, signal processing, computer vision, and machine learning. Emotion Recognition is acquiring its prevalence in research [3].

Recently, much attention has been given to emotion recognition using speech, as it aims to analyze the emotional states through speech signals. This area is referred to as Speech Emotion Recognition (SER) and is explicated as the classification of processing methodologies, and simultaneously the speech signals are classified for the detection of emotions embedded in them [4]. An SER system requires a speech-corpora (dataset), identifying different speech features and an appropriate choice of a classification model. Local features of SER frameworks are examined in the four classifications that are spectral features, voice quality features, Prosodic Features, and Teager Energy Operator (TEO) [5]. In recent times, SER systems' efficacy has been enhanced by using a fusion of different features. The feature fusion of 2 stages was implemented in [6]. Initially, a proper selection of features was done and are fused. Two techniques, sequential floating forward selection (SFFS) and sequential forward selection (SFS) were implemented to eliminate dimensionality arising due to the fusion of features. For the classification, four methods were applied, k-Nearest Neighbor (KNN), Regularized Discriminant Analysis (RDA), Linear Discriminant Analysis (LDA), and Support Vector Machines (SVM). These algorithms were trained on two speech databases, Spanish and Berlin EMO-DB. The evaluation process revealed that SFFS performed better than SFS. It was found that 70% accuracy was obtained for LDA, 80% for KNN, and 90% for RDA and SVM. A stressed speech emotion recognition (SSER) using a feature fusion of TEO and Linear prediction coefficient (LPC) known as T-LPC was proposed in [7]. Gaussian Mixture Model (GMM) was utilized for the classification of emotions of Berlin emotional database. Four experiments were performed, one using LPC, the second with pitch feature extraction, the third with LPC + pitch feature extraction, and the last with T-LPC's proposed method. The proposed method outperformed the other extraction methods with an accuracy of 88% and 82.7% for females and males.

Classification of speech emotion recognition can be carried out in two ways: (a) traditional classifiers and (b) Deep Learning classifiers. Traditional classifiers include GMM, Hidden Markov Model (HMM), Artificial Neural Network (ANN), and SVM. Deep Learning classifiers include Deep Neural Networks (DNN), Convolutional Neural Networks (CNN), Recurrent Neural Networks (RNN), and Long Short-Term Memory (LSTM). Many works have recently been put-forward, where deep learning techniques have proved to be best suited for emotion recognition over traditional techniques [8]. Different tasks like feature extraction, feature selection, and classification of such features to recognize emotions should be performed by an effective system [5]. DNN is utilized to recognize emotions in [9], an end-to-end deep learning-based real-time speech emotion recognition system is developed. DNN was trained on spectrograms generated from two databases, SAVEE, and eNTERFACE. An accuracy of 60.53% for eNTERFACE and 59.7% for SAVEE was obtained. DNN was trained on MSP-Podcast corpus in [10]. Valance, dominance, and arousal scores were predicted using the regression model, and different training parameters were investigated alongside batch normalization, residual network, and activation functions. The experimental results obtained were that the model's performance was enhanced when the size of the training set was increased, and with an increase in the number of layers, batch normalization is needed.

Now, this paper uses DNN for the classification of emotions. A multilingual database consisting of English, German and Hindi languages is formulated, and Mel-frequency Cepstral Coefficient (MFCC) and Teager Energy Operator (TEO) features are extracted from the speech signals along with a new-fangled fusion of MFCC and TEO referred to as Teager-MFCC (TMFCC). The three databases used are, the English corpus is Toronto emotional speech set (TESS) [11], the German corpus is Berlin Emo-DB [12], and the Hindi corpus is Indian Institute of Technology Kharagpur Simulated Emotion Hindi Speech Corpus (IITKGP-SEHSC) [13]. For this paper, only four emotions have been considered, happy, sad, neutral, and angry.

2 Proposed Speech Emotion Recognition Algorithm

In this paper, we propose a new feature fusion extraction technique of Mel-frequency cepstral coefficient (MFCC) and Teager energy operator (TEO) known as TMFCC for the recognition of emotions. For the classification of emotions, DNN has been used. The proposed algorithm block diagram is shown in Fig. 1.

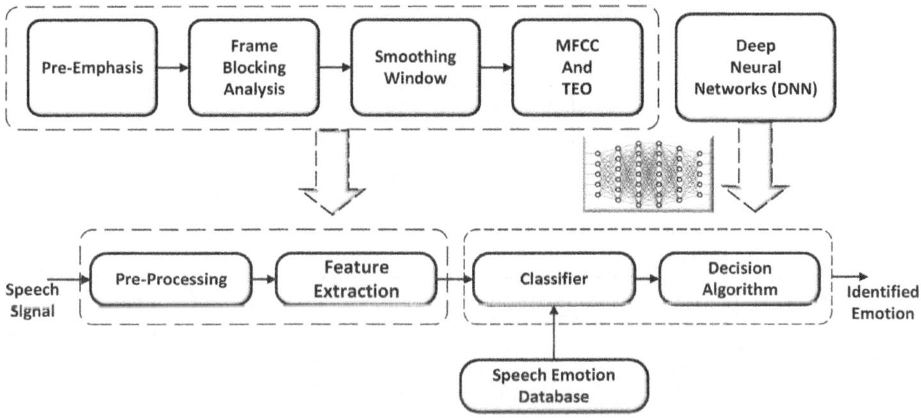

Fig. 1. The proposed algorithm for SER using feature fusion of TMFCC and DNN

Initially, a multilingual database consisting of Toronto emotional speech set (TESS) and English database, Berlin Emo-DB a German database, and Indian Institute of Technology Kharagpur Simulated Emotion Hindi Speech Corpus (IITKGP-SEHSC), a Hindi database is formulated. The speech signals of this multilingual database under a pre-processing technique for noise removal and silence removal. The speech signals are framed before windowing is applied to extract two main features, MFCC and TEO. These two features are fused and form TMFCC for the better classification of emotions. After the feature extraction, these features are subjected to a deep learning classifier, DNN, and finally, the emotions are identified.

2.1 The Feature Fusion TMFCC

Combined MFCC and TEO are used for feature extraction. A combination of features from MFCC and TEO is used to train the neural network. MFCC features set extracted from the speech signal. Then the speech signal is first divided into non-uniform sub-bands in Mel-scale using a multi-rate filter bank. Next, the Teager energies of the sub-signals are estimated. Finally, the feature vector is constructed by the combination of both. TMFCC feature extraction is utilized to detect four different emotions, happy, sad, angry, and neutral, from the speech signal (Fig. 2).

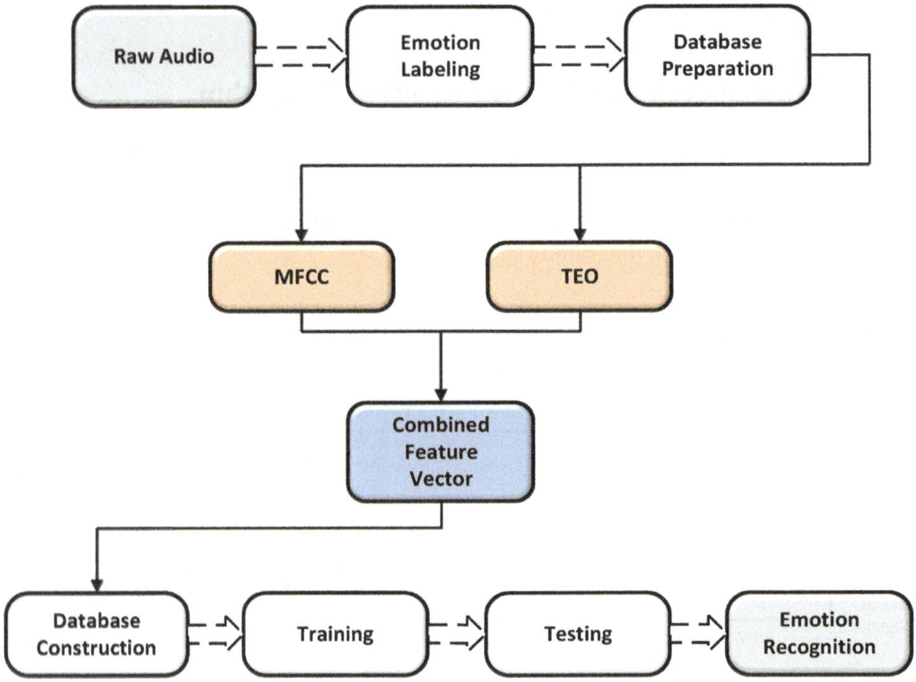

Fig. 2. Implementation flow for the fusion of MFCC and TEO

2.2 Mel-Frequency Cepstral Coefficient (MFCC)

The MFCC is the most efficient spectral representation of speech signals for the extraction of emotional state. MFCC is the inverse-FFT of the log of the frequency-warped spectrum. Frequencies of the spectrum are wrapped because the ear hairs are not evenly spaced in the frequency domain. The Mel-frequency scale approximates that uneven spacing and is given by Eq. (1).

$$mel(f) = 2595 * log10\left(1 + \frac{f}{100}\right) \tag{1}$$

where $mel(f)$ is Mel frequency for a given frequency, f in Hz.

The ear consists of the cochlea that primarily contains more filters at low frequency and very few filters at a higher frequency, mimicking Mel-filters. Therefore, MFCC aims to convert time-domain signals to the frequency domain signal by mimicking the cochlea function using Mel-filters. Mel-filters are the filter banks consisting of uniform overlapping filters with a constant bandwidth of 100 and various center frequencies.

2.3 Teager Energy Operator (TEO)

The Teager energy operator is a nonlinear operator that was introduced to calculate instantaneous energy with an improved signal to noise ratio (SNR) [14]. There is always some noise associated with the recording and processing equipment. With TEO, the noisy part present in the signal is suppressed while calculating its corresponding energy. Contrary to this, the typical squared energy operator takes the input signal and the noise present and finds the energy.

TEO provides a measure of the energy of a speech signal. It is a simple algorithm that is derived for enabling on-the-fly calculation of the energy required. TEO is also used to calculate the energy of a signal. The TEO for a continuous-time signal is defined as:

$$\psi(x(t)) = \frac{d}{dt}x^2(t) - x(t)\left(\frac{d^2}{dt^2}x(t)\right) \tag{2}$$

whereas the TEO for a discrete signal is defined as

$$\psi(x(n)) = x^2(n) - x(n+1)x(n-1) \tag{3}$$

where $x(n)$ is the sampled speech signal, ψ is the Teager energy operator.

2.4 Deep Neural Networks

DNN is supervised learning based on ANN computing methods. Instead of having a single hidden layer, a deep neural network has multiple hidden layers that help the network understand and emulate more complex and abstract behaviors. DNN has been used with great success in recognizing emotions [21]. Deep networks have recently illustrated representative and discriminative learning capabilities over many applications in several years. DNN creates emotion state probability distributions after taking the standard acoustic features as input. DNN is trained to predict the probabilities of each emotional state. The DNN can be treated as a segment-level emotion recognizer. An exciting characteristic of DNNs is that they can learn high-level invariant features from raw data.

3 Results and Discussion

3.1 Databases

Three databases have been used for the evaluation process of the proposed SER system. Three corpora of emotions are in German, English, and Hindi. The English corpus is

Toronto's emotional speech set (TESS), the German corpus is the well-known Berlin Emo-DB, and The Hindi corpus is the Indian Institute of Technology Kharagpur Simulated Emotion Hindi Speech Corpus (IITKGP-SEHSC). Four emotions have been considered for this paper, which are anger, happiness, neutral, and sad. The total utterances used in this paper are shown in Table 1.

Table 1. Total utterances for multilingual databases

Emotion	German	English	Hindi	Total
Anger	100	100	150	350
Happiness	100	100	150	350
Neutral	63	100	150	313
Sad	79	100	150	329

3.2 Experimental Setup

A simple performance system was used for processing, i.e., a multicore system with Intel Core i5 5200U 2.20 GHz, 8 GBytes RAM, and 1TBytes hard disk, installed with Windows 10 operating system and Matlab 2018b with signal processing and deep learning toolbox. Other running applications were minimized during the simulation as much as possible. Three experiments were carried out using only the MFCC feature, the second one using only TEO feature, and the third one feature fusion, TMFCC. DNN was trained on a given multilingual dataset. The data was divided into the ratio of 70:15:15, where 70% was used for training, 15% for testing, and 15% for validation. We have selected 26 MFCC features and used the final matrix with 1343 feature vectors.

3.3 Experimental Result on MFCC

Table 2 represents the results for MFCC Feature Extraction Technique. The loop was run for various neural network configurations. As it can be seen from Table 2, the best performance obtained for MFCC for the [4750 856 154 28] configuration with four hidden layers is 90% with training time 168.3 s. The worst performance is observed for the [10] configuration with a recognition rate of 65% with one hidden layer in 0.60 s of training time. The training time of [7751 2278 670 197 58 17] configuration is highest among all the configuration taken, i.e., 771.21 s, but has the second-best performance in terms of recognition rate, which is 85.9% with six hidden layers. The least training time is observed for [30] configuration with a recognition rate of 81.3% with one hidden layer in 0.45 s.

Table 2. Performance results for MFCC, TEO, and Feature Fusion TMFCC

Neural network configuration	Hidden layer(s)	Recognition rate (%) and Training time (s)					
		MFCC (%)	Training time	TEO (%)	Training time	TMFCC (%)	Training time
[10]	1	65	0.60	61.7	1.77	67	1.09
[20]	1	79.4	0.49	59.6	0.67	71.2	0.384
[30]	1	81.3	0.45	69.8	0.87	73	0.455
[20 10]	2	74.2	0.59	55	0.79	71.8	0.809
[10 20 15]	3	75.5	1.15	66.7	1.03	59.9	1.21
[10 20 30 10]	4	76.7	0.626	51.4	1.02	60	1.5
[40 35 20 15 10]	5	77.5	1.88	64.3	3.58	75.1	3.25
[200 100 50 10]	4	83.8	9.91	76	6.7	70.3	11.57
[500 200 100 10]	4	84	20	64.9	12.95	76.8	16.5
[1000 300 100 50 10]	5	85	29.6	72.2	38.01	82.1	38.6
[363]	1	82.5	3.62	57.3	5.16	73	6
[1515 87]	2	82.3	16.24	75.9	20.78	66.7	16.57
[3094 363 43]	3	84	63.91	68.4	145.89	75.6	76.2
[4750 856 154 28]	4	**90**	**168.3**	71.6	267.07	**92.7**	**151.8**
[6320 1515 363 87 21]	5	85.8	322.145	**88.5**	**533.33**	76	382
[7751 2278 670 197 58 17]	6	85.9	771.21	70.5	645.02	77.7	734.7

3.4 Experimental Result on TEO

Table 3 shows the results of the TEO feature extraction technique. The loop was run for various neural network configurations. In Table 2, for TEO, the best performance is obtained for the [6320 1515 363 87 21] configuration with five hidden layers is 88.5% with training time 533.33 s. The worst performance is observed for the [10 20 30 10] configuration with a recognition rate of 51.4% in 1.02 s of training time with four hidden layers. The second-best performance in terms of recognition rate is observed for the [200 100 50 10] configuration, 76% in 6.7 s with four hidden layers. The training time of [7751 2278 670 197 58 17] configuration is the highest among all configurations simulated, i.e., 645.02 s, with a recognition rate of 70.5% with six hidden layers. The least training time observed for the [20] configuration with a recognition rate of 59.6% with one hidden layer is 0.67 s.

3.5 Experimental Result on Fusion Features (TMFCC)

Table 3 shows the results for fusion (TMFCC) feature extraction techniques. The loop was run for various neural network configurations. In Table 2, for Fusion (MFCC+TEO), the best performance is obtained for the [4750 856 154 28] configuration with four hidden layers is 92.7% with training time 151.8 s. The worst performance is observed for the [10 20 15] configuration with a recognition rate of 59.9% in 1.21 s of training time with three hidden layers. The training time of [7751 2278 670 197 58 17] configuration is highest among all the configuration taken, i.e., 734.7 s, but has the second-best performance in terms of recognition rate, which is 77.7% with six hidden layers. The least training time is observed for the [20] configuration with a recognition rate of 71.2% with one hidden layer in 0.384 s.

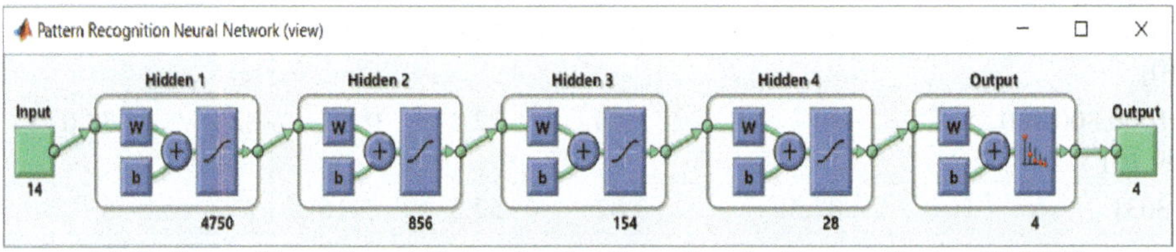

(a) Optimum Deep Neural Networks Configuration for Feature Fusion TMFCC

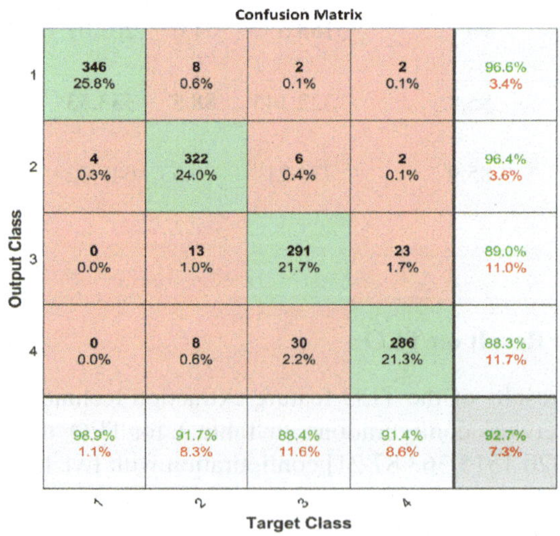

(b) Confusion Matrix for TMFCC

Fig. 3. Optimum configuration for TMFCC and its performance

3.6 Performance Comparison and Optimum Configuration

Table 3 shows the comparison of performance between MFCC, TEO, and TMFCC. It is evident by the classification accuracy. Emotions are detected best by proposed Feature Fusion TMFCC, followed by TMFCC and TEO, as shown in Table 3. The network configuration has little effect on the recognition rate for MFCC and Fusion, except for TEO.

It is clear from Table 3 and detailed in Fig. 3 that TMFCC can be utilized to classify emotions yielding high recognition accuracy. The main reason that TMFCC provides the best results is that the fusion consists of both MFCC features and TEO features. MFCC is highly accurate in representing the short-time power spectrum envelope, and TEO is best suited for recognizing energy-emotions. Thus, the fusion of both these features resulted in better recognition than the particular feature extraction technique of MFCC and TEO.

Table 3. Comparison of MFCC, TEO and Feature Fusion (TMFCC)

Feature	DNN configuration	Training time (s)	Recognition rate (%)
MFCC	[4750 856 154 28]	168.3	90
TEO	[6320 1515 363 87 21]	533.33	88.5
FUSION	[4750 856 154 28]	151.8	**92.7**

4 Conclusions and Future Works

In this paper, DNN is trained on a multilingual database consisting of Toronto's emotional speech set (TESS), Berlin Emo-DB, and Indian Institute of Technology Kharagpur Simulated Emotion Hindi Speech Corpus (IITKGP-SEHSC). Three feature extraction have been used, MFCC, TEO, and proposed Feature Fusion TMFCC. The comparison determined the proposed Fusion TMFCC outshined both TEO and MFCC with a 92.7% recognition rate, while for TEO and MFCC, it was 88.5% and 90.0%, respectively. This paper's outcome would assist information on a pragmatic emotional speech recognition implementation driven by a wiser selection of underlying feature extraction techniques. In future works, the recognition accuracy for feature fusion can be improved by using different structural, neural, and statistical pattern classifiers, or only a classifier fusion method can be evaluated for SER.

Acknowledgments. The authors would like to express their gratitude to the Malaysian Ministry of Education (MOE), which has provided research funding through the Fundamental Research Grant, FRGS19-076-0684 (FRGS/1/2018/ICT02/UIAM/02/4). The authors would like to acknowledge support from International Islamic University, University of New South Wales, and Universitas Mercu Buana.

References

1. Lu, X., Li, S., Fujimoto, M.: Automatic speech recognition. In: SpringerBriefs in Computer Science (2020)
2. Dzedzickis, A., Kaklauskas, A., Bucinskas, V.: Human emotion recognition: Review of sensors and methods. Sensors **20**(3), 592 (2020). https://doi.org/10.3390/s20030592
3. Furui, S., Kikuchi, T., Shinnaka, Y., Hori, C.: Speech-to-text and speech-to-speech summarization of spontaneous speech. IEEE Trans. Speech Audio Process. **12**(4), 401–408 (2004). https://doi.org/10.1109/TSA.2004.828699
4. Swain, M., Routray, A., Kabisatpathy, P.: Databases, features and classifiers for speech emotion recognition: a review . Int. J. Speech Technol. **21**, 93–120 (2018). https://doi.org/10.1007/s10772-018-9491-z
5. El Ayadi, M., Kamel, M., Karray, F.: Survey on speech emotion recognition: features, classification schemes, and databases. Pattern Recogn. **44**(3), 572–587 (2011). https://doi.org/10.1016/j.patcog.2010.09.020
6. Kuchibhotla, S., Vankayalapati, H., Anne, K.: An optimal two stage feature selection for speech emotion recognition using acoustic features. Int. J. Speech Technol. **19**(4), 657–667 (2016). https://doi.org/10.1007/s10772-016-9358-0
7. Bandela, S.R., Kishore Kumar, T.: Emotion recognition of stressed speech using teager energy and linear prediction features (2018). https://doi.org/10.1109/ICALT.2018.00107
8. Shrestha, A., Mahmood, A.: Review of deep learning algorithms and architectures. IEEE Access (2019). https://doi.org/10.1109/ACCESS.2019.2912200
9. Fayek, H.M., Lech, M., Cavedon, L.: Towards real-time speech emotion recognition using deep neural networks (2015). https://doi.org/10.1109/ICSPCS.2015.7391796
10. Abdelwahab, M., Busso, C.: Study of dense network approaches for speech emotion recognition (2018). https://doi.org/10.1109/ICASSP.2018.8461866
11. Tillman, T.W., Carhart, R.: An expanded test for speech discimination utilizing CNC monosyllabic words (1966)
12. Burkhardt, F., Paeschke, A., Rolfes, M., Sendlmeier, W., Weiss, B.: A database of German emotional speech (2005)
13. Koolagudi, S.G., Reddy, R., Yadav, J., Rao, K.S.: IITKGP-SEHSC: Hindi speech corpus for emotion analysis (2011). https://doi.org/10.1109/ICDECOM.2011.5738540
14. Zhou, G., Hansen, J.H.L., Kaiser, J.F.: Nonlinear feature based classification of speech under stress. IEEE Trans. Speech Audio Process. **9**, 201–216 (2001). https://doi.org/10.1109/89.905995

Prototype Development of Graphical Pattern Security System on Raspberry Pi

Teddy Surya Gunawan[1,2](✉), Fatin Nabilah Nasir[1], Mira Kartiwi[3], and Nanang Ismail[4]

[1] Electrical and Computer Engineering Department, International Islamic University Malaysia, Gombak, Malaysia
tsgunawan@iium.edu.my
[2] School of Electrical Engineering and Telecommunications, UNSW, Sydney, Australia
[3] Information Systems Department, International Islamic University Malaysia, Gombak, Malaysia
[4] Electrical Engineering Department, UIN Sunan Gunung Djati, Bandung, Indonesia

Abstract. Password is a primary security mechanism that consists of a secret phrase created using alphabetic, numeric, alphanumeric, and symbolic characters, or a combination. It is used to restrict access to a system, application, or service to only those users who have the authorization to use the system or device. Because of increased computing power, modern passwords must be very long and complicated, which will make them hard to remember. Some research studies show that it is easier for people to remember visual passwords instead of textual ones. Therefore, this research aims to enhance the authentication process's security using the graphical pattern unlock approach on Raspberry Pi. The grid size of 5×5 was selected in the implementation, as it provides the tradeoff between security strength and an easy pattern to remember. The proposed system could be connected to the smart home system, which can enhance their security. Prototype validation revealed that the prototype is working as intended, and the authentication process took around 1.2 s to complete.

Keywords: Smart home · Security enhancement · Graphical pattern · Raspberry Pi

1 Introduction

Nowadays, passwords hold a crucial role in our society as they are implemented in many different confidential situations. The passwords have been derived into various approaches like Personal Identification Number (PIN) codes, fingerprints, pattern locks, and passwords with the common goal to protect users' private information. Due to the numerous amounts of sensitive data stored on the devices, the need for security increases, which leads to the authentication process as an essential topic for research. Screen pattern locks are used as a protection mechanism to avoid sensitive information leakage from any device [1].

J. A. Mat Jizat et al. (Eds.): iCITES 2020, AISC 1350, pp. 31–38, 2021.
https://doi.org/10.1007/978-3-030-70917-4_4

The graphic pattern is one of the examples of a graphical password mechanism used for an authentication process. A graphic pattern is a pattern that needs to be connected at least four contact points without lifting and avoiding any intermediate points [2]. The pattern itself is in various grid size patterns as 3×3 size with nine nodes interface is commonly used in the different authentication process, i.e., the smartphone's graphical pattern. In [3], if the number of connected nodes must be at least four points, and each node can be touched only once, then the possible pattern is around 38911.

As stated by [4], the length and complexity of one text-based password determine how secure the code is, yet humans have difficulties recalling them. Users tend to select simpler passwords in terms of length, which are in the weak password category, without considering their security. It is also easier for users to remember a graphical approach than a text-based approach like PIN passwords, as described in [5].

One method to improve the security strength of a graphical pattern is by expanding the grid size or increasing the number of contact points [6], using embedded pattern strength [7], changing pattern layout arrangement [5], multi-layered drawing [8], rotation [9], or use a strength meter [10]. Many kinds of research have been conducted on improving the strength of a graphical pattern on a smartphone. However, not much of the research focused on improving a smart home system's security, like [11, 12]. Therefore, this paper aims to develop a fast and accurate graphics pattern recognition system implemented on Raspberry Pi. The target is to be able to validate and authenticate a user within 1 s processing time. Our proposed system's primary input device is an LCD touchscreen that functions as both the input and output at the same time. Moreover, the primary output device is the magnetic lock, which is controlled by Raspberry Pi.

2 Design of Graphical Pattern Security System

2.1 Hardware Design

Raspberry Pi is a single-board computer that can run various operating systems, either Linux or non-Linux based [13]. Examples of Linux-based operating systems are Raspbian, which is the most recommended operating system for Raspberry Pi, Android Things, an embedded version of the Android operating system, and Arch Linux ARM. Meanwhile, an example of non-Linux based operating systems is Windows 10 IoT Core, HelenOS, RISC OS Pi, and Haiku.

Figure 1 shows the proposed block diagram of the graphical pattern security system using Raspberry Pi. The Raspberry Pi is connected to the LCD touchscreen, which is shown in Fig. 2. Raspberry Pi 3 has a quad-core processor, 1 GB RAM, dan built-in graphics [13]. The relay circuit is connected to Raspberry Pi and provides an interface to the high voltage magnetic lock. The status could be shown using LED indicators or further send messages using an internet connection or as part of a smart home system. Finally, the overall power required, including magnetic lock, is less than 3 A, so that a compact power supply, i.e., a smartphone charger, could be used as the power supply. The LCD touch screen has the function as output as well as input.

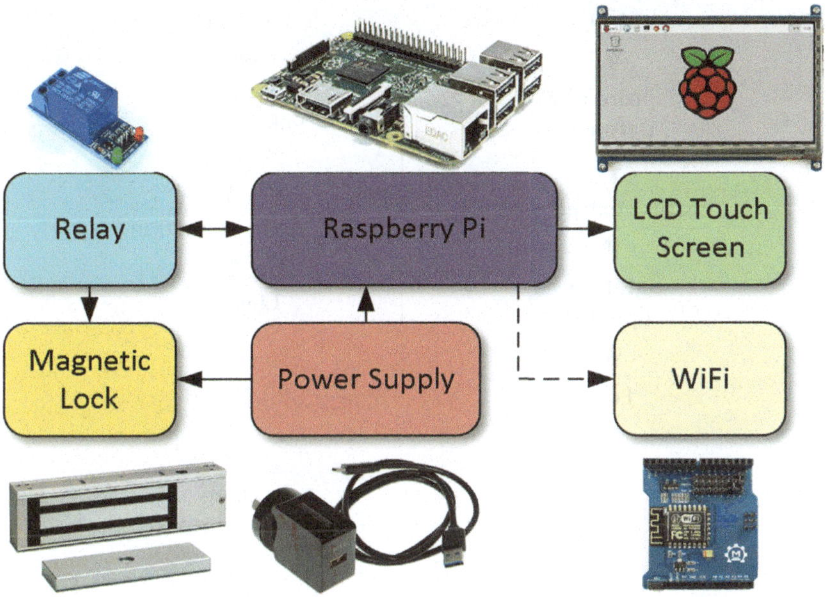

Fig. 1. Block diagram of the proposed hardware

2.2 Software Design

Microsoft Visual Studio is an integrated development environment (IDE) used to develop programs and applications such as web applications and mobile applications. It uses Microsoft software development platforms, i.e., Windows API and Windows Forms. It supports different kinds of programming languages, and it allows users to edit and debug the code. Few examples of built-in language available include C, C#, Visual Basic,.Net, and JavaScript. After developing the program in Visual Studio, it needs to be copied into and run on Raspberry Pi 3. First, the mono repository package needs to be added to the system. As Raspbian Stretch is chosen to be installed for this research; thus, the package repository for Raspbian is added.

There are three central graphical security systems, including feature extraction, classifier, and training database. During the feature extraction phase, the system will extract the input pattern's behavior, i.e., the starting and ending node chosen by the user, the length of the pattern, and the distance between each selected node. Then, the classifier categorizes the complexity of the pattern based on the pattern behavior. The classifier is also used to compare the input pattern with the enrolled pattern in the database. Based on the flow process between the classifier and the database, if the input pattern matches the database's enrolled pattern, the system will send feedback to the decision rule.

Figure 2 shows the flowchart of user registration, database creation, and login verification. In this research, the graphical approach used to develop the pattern layout is the expanded grid size approach proposed by [6]. The pattern layout is created in a 5 $\times$ 5 grid size.

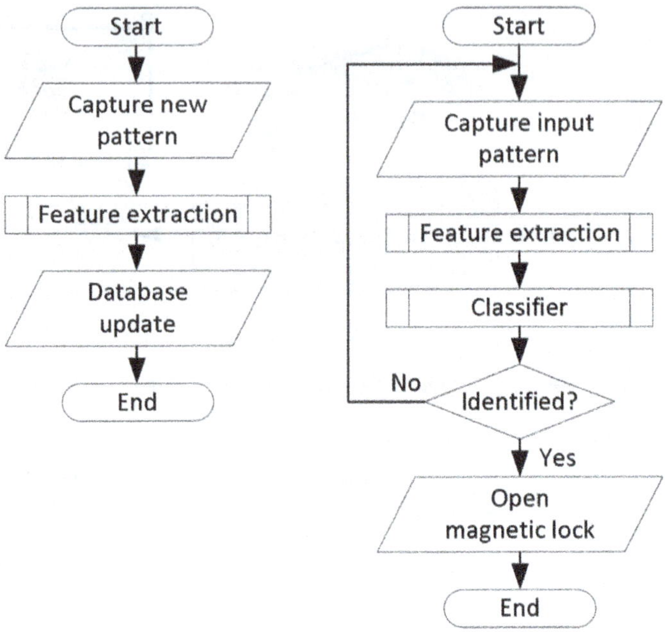

(a) Registration (b) Authentication

Fig. 2. System flowchart for registration and authentication

3 Implementation

After the codes have been compiled in the Visual Studio, the executable file must be copied into the Raspberry Pi 3. Then, the Mono repository needs to be added to the system. The mono-devel package also needs to be installed in the code compilation system, as shown in Table 1.

Table 1. Installation of Mono repository on Raspberry Pi 3

```
sudo apt install apt-transport-https dirmngr
sudo apt-key adv --keyserver hkp://keyserver.ubuntu.com:80 --
recv- keys 3FA7E0328081BFF6A14DA29AA6A19B38D3D831EF
echo "deb https://download.mono-project.com/repo/debian sta-
ble-raspbianstretch main" | sudo tee
/etc/apt/sources.list.d/mono-official-stable.list
sudo apt update
sudo apt install mono-devel
```

A prototype is needed to test the coding and evaluate the design. The process involved is including software development, which is done by using Microsoft Visual Studio, and hardware development, which uses Raspberry Pi 3 and its components. C# language is used to code all the instructions, programs, and graphical user interface (GUI).

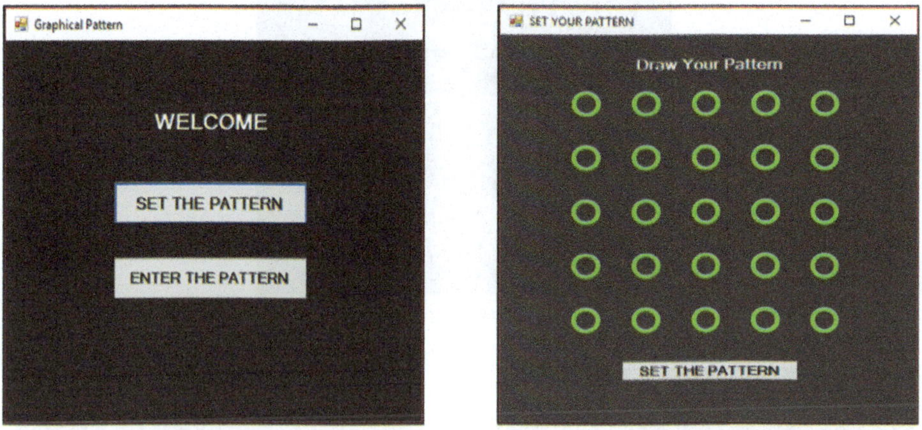

(a) Start Page (b) Pattern Set

Fig. 3. GUI design for start page and pattern set functions

Figure 3 shows the user interface of the Start Page and Pattern Set functions. The layout pattern is created in the bigger size, i.e., 5 × 5 grid size. The pattern layout is created in a bigger size as it is one of the security enhancement approaches. As the pattern grid size increases, the number of possible patterns also increases. Hence, there would be a high possibility of the complexity of user-chosen patterns to increase. Some functions have been set on the lock screen pattern to help the user draw a better pattern. For instance, the user cannot create the closed-loop pattern, i.e., the node cannot be contacted more than once.

Moreover, the user is allowed to reverse the gesture while drawing the pattern. This function helps the user to create a better pattern with a better complexity. The coordinates of each node are saved as they will be used during the authentication process. The coordinates of the saved pattern are compared with the coordinates of the inputted pattern.

Figure 4(a) shows the user interface of the Pattern Input function. During the authentication process, the user authenticates the input pattern with the pattern that has been created earlier. The coordinates of the inputted pattern are recorded and compared with the saved pattern. Next, the Valid Pattern message will be displayed if the coordinates of both patterns match, as shown in Fig. 4(b). Otherwise, the Invalid Pattern message will be displayed, as shown in Fig. 4(c).

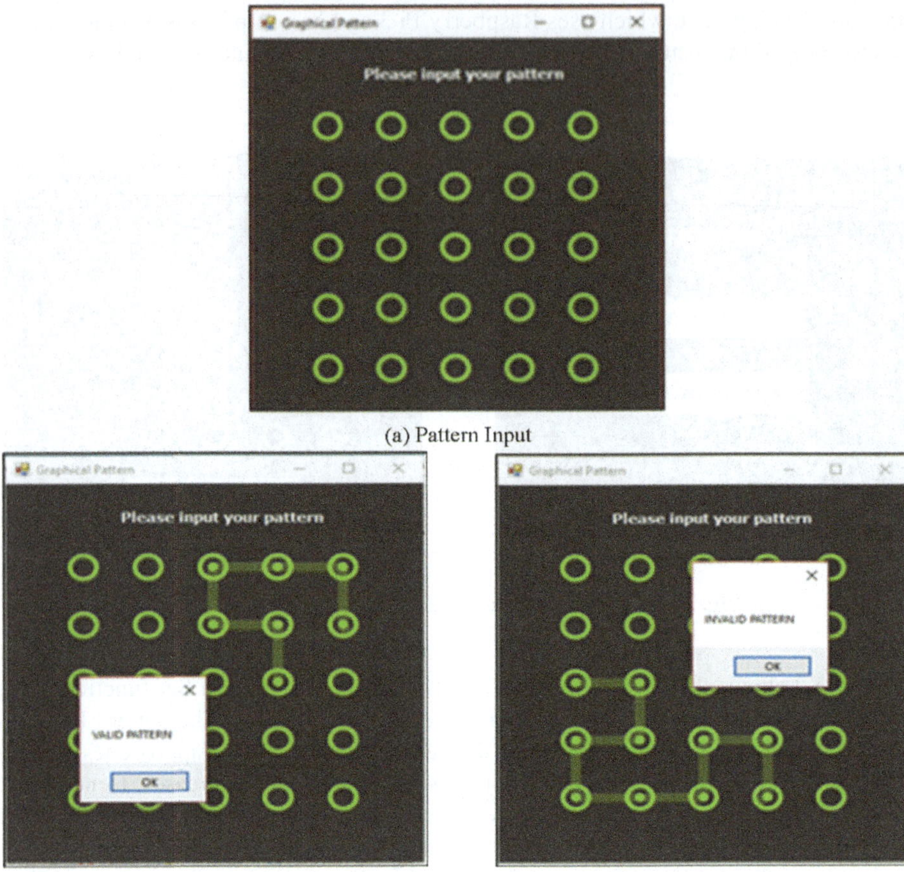

(a) Pattern Input

(b) Valid Pattern Input

(c) Invalid Pattern Input

Fig. 4. GUI design for pattern input, valid, and invalid pattern input

4 Prototype Validation

Figure 5 (a) and (b) shows the actual Start Page and user interface of the Pattern Input function. It is the actual implementation of Fig. 3, previously. Pattern Input will capture and validate the user's graphical pattern password. Several evaluations have been conducted in this phase, in which we found that there is no lagging experienced by the user. The authentication time to authenticate registered users is 1.2 s on average when tested in 10 experiments.

Figure 5 (c) and (d) shows the actual valid and invalid pattern message. It is the actual implementation of Fig. 4, previously. The system will validate the entered pattern to be checked with the database's registered user's pattern. If it is the correct pattern, it will open the magnetic lock. While if it is incorrect, it will provide the user with three more attempts before it triggers the alarm.

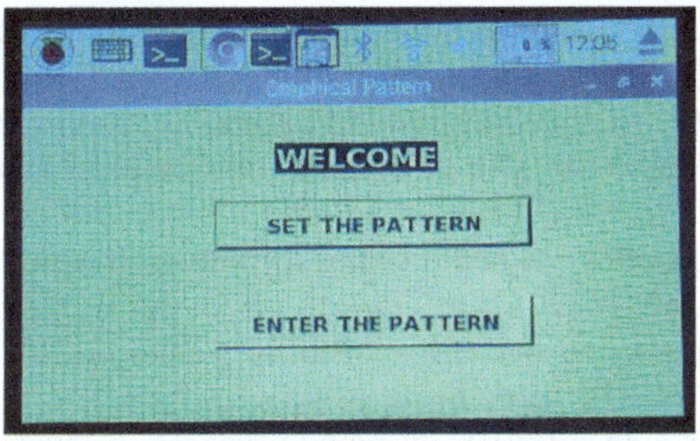

(a) Actual Start Page GUI

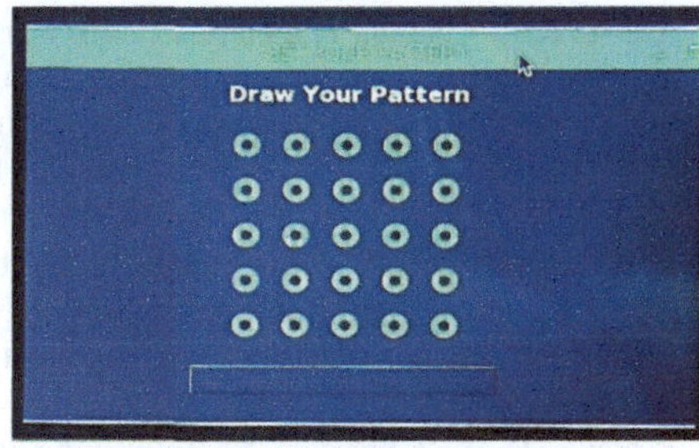

(b) Actual Pattern Input GUI

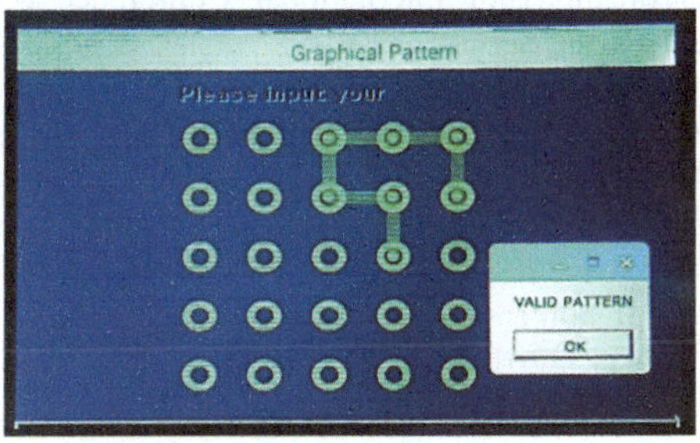

(c) Actual Valid Pattern Message GUI

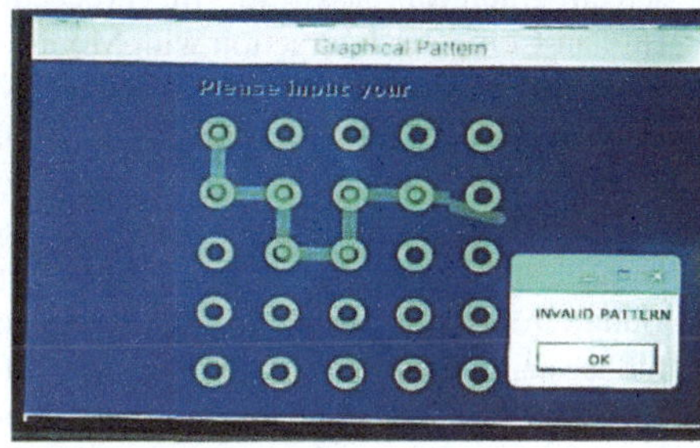

(d) Actual Invalid Pattern Message GUI

Fig. 5. Actual prototype GUI

5 Conclusions and Future Works

This paper has presented the development of a graphical pattern security system on Raspberry Pi. First, the hardware and software requirements have been designed so that it can fulfill the specifications. Next, a 5 × 5 pattern grid was implemented in Raspberry Pi using C#. Results showed that our proposed prototype works as intended with an authentication time around 1.2 s on average. Further research can be conducted, including implementation with various pattern grid size, implementation of pattern strength meter, and integration with the smart home system.

Acknowledgments. The authors would like to express their gratitude to the Malaysian Ministry of Education (MOE), which has provided research funding through the Fundamental Research Grant, FRGS19–076-0684 (FRGS/1/2018/ICT02/UIAM/02/4). The authors would like to acknowledge support from International Islamic University, University of New South Wales, and UIN Sunan Gunung Djati.

References

1. Li, Q., Dong, P., Zheng, J.: Enhancing the security of pattern unlock with surface EMG-based biometrics. Appl. Sci. **10**(2), 541 (2020)
2. Aviv, A.J., et al.: Smudge attacks on smartphone touch screens. Woot **10**, 1–7 (2010)
3. Løge, M.D.: Tell me who you are and I will tell you your unlock pattern. NTNU (2015)

4. De Angeli, A., et al.: Is a picture really worth a thousand words? Exploring the feasibility of graphical authentication systems. Int. J. Hum. Comput. Stud. **63**(1–2), 128–152 (2005)
5. Uellenbeck, S., et al.: Quantifying the security of graphical passwords: the case of android unlock patterns. In: Proceedings of the 2013 ACM SIGSAC Conference on Computer and Communications Security (2013)
6. Aviv, A.J., Kuber, R., Budzitowski, D.: Is bigger better when it comes to android graphical pattern unlock? IEEE Internet Comput. **21**(6), 46–51 (2017)
7. Siadati, H., et al.: Fortifying android patterns using persuasive security framework. In: UBICOMM 2015, p. 81 (2015)
8. Chiang, H.-Y., Chiasson, S.: Improving user authentication on mobile devices: a touch-screen graphical password. In: Proceedings of the 15th International Conference on Human-Computer Interaction with Mobile Devices and Services (2013)
9. Chakrabarti, S., Landon, G.V., Singhal, M.: Graphical passwords: drawing a secret with rotation as a new degree of freedom. In: Proceedings of the Fourth IASTED Asian Conference on Communication Systems and Networks. Citeseer (2007)
10. Sun, C., Wang, Y., Zheng, J.: Dissecting pattern unlock: the effect of pattern strength meter on pattern selection. J. Inf. Secur. Appl. **19**(4–5), 308–320 (2014)
11. Gunawan, T.S., et al.: Prototype design of smart home system using internet of things. Indonesian J. Electr. Eng. Comput. Sci. **7**(1), 107–115 (2017)
12. Gunawan, T.S., et al.: Performance evaluation of smart home system using internet of things. Int. J. Electr. Comput. Eng. **8**(1), 400 (2018)
13. Gunawan, T.S., et al.: Development of face recognition on raspberry pi for security enhancement of smart home system. Indonesian J. Electr. Eng. Inform (IJEEI) **5**(4), 317–325 (2017)

Development of Automatic Obscene Images Filtering Using Deep Learning

Abdelrahman Mohamed Awad[1], Teddy Surya Gunawan[1,2(✉)], Mohamed Hadi Habaebi[1], and Nanang Ismail[3]

[1] Electrical and Computer Engineering Department, International Islamic University Malaysia, Gombak, Malaysia
tsgunawan@iium.edu.my

[2] School of Electrical Engineering and Telecommunications, UNSW, Sydney, Australia

[3] Electrical Engineering Department, UIN Sunan Gunung Djati, Bandung, Indonesia

Abstract. Because of Internet availability in most societies, access to pornography has become a severe issue. On the other side, the pornography industry has grown steadily, and its websites are becoming increasingly popular by offering potential users free passes. Filtering obscene images and video frames is essential in the big data era, where all kinds of information are available for everyone. This paper proposes a fully automated method to filter any storage device from obscene videos and images using deep learning algorithms. The whole recognition process can be divided into two stages, including fine detection and focus detection. The fine detection includes skin color detection with YCbCr and HSV color spaces and accurate face detection using the Adaboost algorithm with Haar-like features. Moreover, focus detection uses AlexNet transfer learning to identify the obscene images which passed stage one. Results showed the effectiveness of our proposed algorithm in filtering obscene images or videos. The testing accuracy achieved is 95.26% when tested with 3969 testing images.

Keywords: Convolutional neural networks · Adaboost algorithm · Haar-like features · Skin detection · Pornographic · Image filtering · AlexNet

1 Introduction

In the past decade, there were many filtering techniques used for the detection of pornography. In 1996, content-based filtering was proposed in [1]. Along with that, there are many blocking and filtering obscene content through media content (videos and images), Web indexing, and browsing. Keyword-based, blacklist-based, and content-based techniques are the categories of pornographic image filtering techniques [2]. Classification of pornographic images and videos is an unavoidable issue, but it is vaguely defined to some extent. Some define the pornographic frame or image as the images containing a full naked body, some define them as the images with massive skin exposed, and others categorize an image as a pornographic image if it has shown a human's private part.

J. A. Mat Jizat et al. (Eds.): iCITES 2020, AISC 1350, pp. 39–49, 2021.
https://doi.org/10.1007/978-3-030-70917-4_5

Presently, pornography regulations exist in different countries, and accessing pornography websites is prohibited. Youth's entry into pornography is simple, leading to moral degradation and social problems [2]. In most countries, the lack of control and regulation on available information on the Internet has also become a significant concern for society. More than 4 million sexual content-based websites on the Internet are around 10% of all websites [3]. Moreover, Netflix, Twitter, and Amazon have fewer unique visitors than adult websites [4]. For that reason, anti-pornography filtering is essential to prevent the harmful impact of pornography content from reaching and affect our societies. Therefore, this paper aims to develop automatic obscene images and video filtering using deep learning algorithms. Two stages were proposed, including fine detection (skin and face detection) and focus detection. The aim is to develop an efficient obscene image or video filtering, which requires minimal computation but with high detection accuracy.

2 Obscene Image Filtering Techniques

Blocking and filtering obscene content through media content (videos and images) and Web indexing and browsing are crucial. Several techniques are currently used for filtering pornographic images and videos, including keyword-based, blacklist-based, and content-based techniques [2].

2.1 Keyword-Based Technique

It is one of the static filtering techniques which is effectively used in Web page analysis and filtering. Keyword-based systems attempt to recognize pictures by analyzing the text names or the text surrounding them on the website. Keyword analysis can be used so that a web page can be classified into pages with pornographic content or not. The static nature language model is used for web page modeling [2]. If the applied keyword appears on a Web page, the Web page would be blocked. This method is quite challenging to implement on a large scale because of two significant problems, the sensitivity to keywords and the difficulty of adding more keywords in all languages [5].

2.2 Blacklist-Based Technique

The URLs of pornography websites are collected and added to a blacklist. Once one of these URLs is accessed, an inspection software blocks this access. However, the number of pornography sites is increased unprecedented every day. Hence, manually includes in a static blacklist of all adult websites is impractical. The blacklist must be updated daily to overcome this problem, so this kind of software would be more efficient [6].

2.3 Content-Based Technique

Content-based procedures assess pictures by coordinating picture content examination. The pioneering work was conducted by [1] when they used both texture properties and color to initiate a classifier for human skin location. After locating an image's human skin, these locations were passed to a geometric direct in light of body plans. Adult and non-adult pictures are used to set up a neural framework classifier of skin and non-skin classes [2].

By examining an image or a video (by extracting its key-frames) content-based technique can identify whether it contains obscene content or not. Several methods use this technique to identify nudity, which uses an approach based on the human structure model, but it is inefficient. They need to have completely naked people in the picture or the frame and simple background with simple colors. Nowadays, pornographic images have a variety in the background, scale, scenario, and human action [6]. The feature-based approach used a classifier based on the majority voting of a frame sample [7]. The convolutional neural networks (CNN) are used to classify the frames classified. The approach emphasizes the extraction of the feature from the picture [8]. Most of the feature-based approach uses bag-of-feature and CNN [9]. However, the bag-of-features approach has a high false-positive rate [10].

3 Proposed Obscene Image Filtering Algorithm

Skin tone is a crucial feature of pornographic image and video detection. A skin pixel can be classified based on the texture or color of the pixel. Assuming that the skin color belongs to a specific range of colors called (i.e., thresholds), this assumption allowed many techniques to use color-based techniques to classify pornographic images. In this method, the color thresholds for different skin colors are first learned from a series of marked training pictures. The pixel in the image is then compared to the criteria to figure out if the pixel is a skin pixel or not. Classifying whether the video connotations obscene frames or not start by extracting the keyframe, then the individual frame or an image has to be changed YCrCb and HSV color as a second step, YCrCb model is used because it is luma-independent [11].

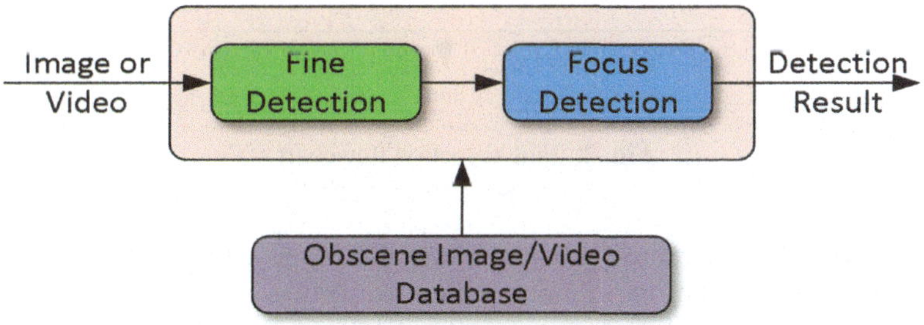

Fig. 1. Block diagram of the proposed system

In this research, we will use two levels of filtering, as shown in Fig. 1. The first one is *fine detection*, which uses a color-based technique to identify either the image or a video containing obscene videos. The second stage is *focus detection*, which passes the images with more skin regions into a trained neural network based on AlexNet transfer learning, identifying the images safely and efficiently.

3.1 Fine Detection Stage

This stage's primary focus is to identify the skin region and indicate if this skin is only on the face area or body area. The input for this stage is the images or videos (keyframe extracted as an image). The YCrCb mask will mask images, and the HSV mask then compiles the two masks to get the skin region. The segmented images with skin region will be passed to the face detection algorithm to check if there is a face or not. If there is no face, the skin region is a body region so that the image will be passed to the next stage, i.e., focus detection. If the image has a face, the face area will be removed from the calculation. If there is still a skin region representing more than 5% of the total area, the image will be passed to focus detection. If not, the image will be classified as a safe image, as shown in Fig. 2.

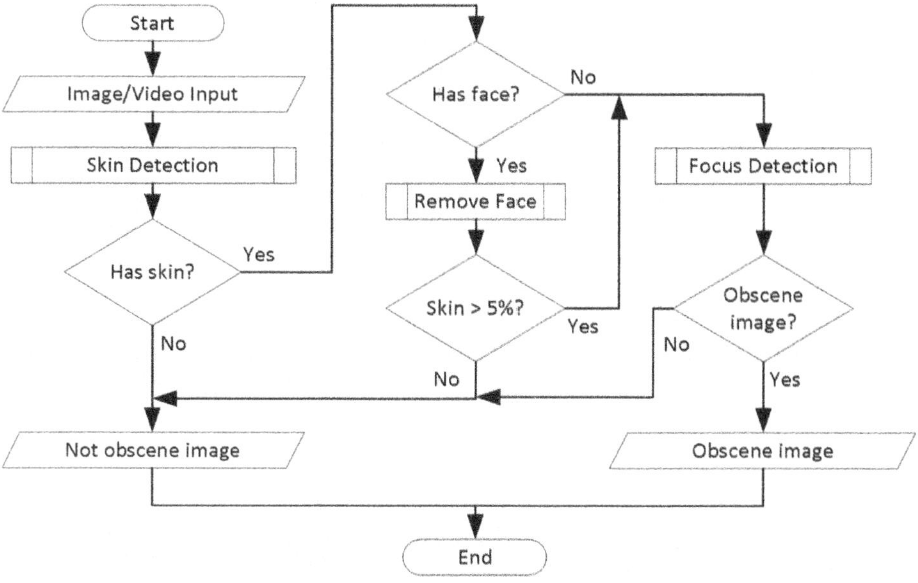

Fig. 2. Fine detection flowchart

In this paper, the keyframe extraction algorithm proposed in [12] will be modified and used. For implementation, FFmpeg-python was utilized to extract the keyframes. The second step of video-based pornographic detection and the first step of image-based pornographic detection is converting the frame or the image to the YCrCb color space. In this color space, Y represents luminance information, Cr and Cb are color-difference components that are storing the color information of an image [13]. YCbCr color model results in a better performance than RGB in skin detection because it is luma-independent. YCbCr values can be obtained from RGB color space using Eq. (1), as stated [11].

$$Y = 0.299R + 0.287G + 0.11B;\ Cr = R - Y;\ Cb = B. \qquad (1)$$

There are large areas of naked skin in pornographic images. For this reason, skin color is the most stable attribute in pornographic images. Which makes the exact reason that skin color detection is indispensable for pornographic image recognition. This paper uses the skin-color model proposed in [14] to detect the skin-color regions of images, which will use both HSV and normalized RGB (YCbCr) models.

Detecting facial landmarks is a subset of the shape prediction problem. Taking the detection speed into account, the face detection algorithm that combines the Adaboost algorithm with Haar-like features is chosen, not affected by skin-color detection results. For the facial images, the skin-color areas are fundamentally composed of the faces. As for the pornographic images, skin-color areas are usually larger than the facial areas. Using this prior knowledge, we can determine if the image is a facial image or not. The facial areas are denoted as *Facial Region* (*FR*), and skin-color areas are denoted as *Skin Region* (*SR*). If $SR - FR < 5\%$, the image will be judged as the facial image, as shown in Fig. 3.

3.2 Focus Detection Stage

After applying *fine detection,* the images will pass through a pre-trained convolutional neural network to identify if the image is safe or not, as shown in Fig. 3. This research is using transfer learning using the AlexNet neural network. AlexNet is a convolutional neural network with eight layers trained on more than a million images. In transfer learning, the original layers are used with a slight change in the number of output classes and then restrained with the new dataset.

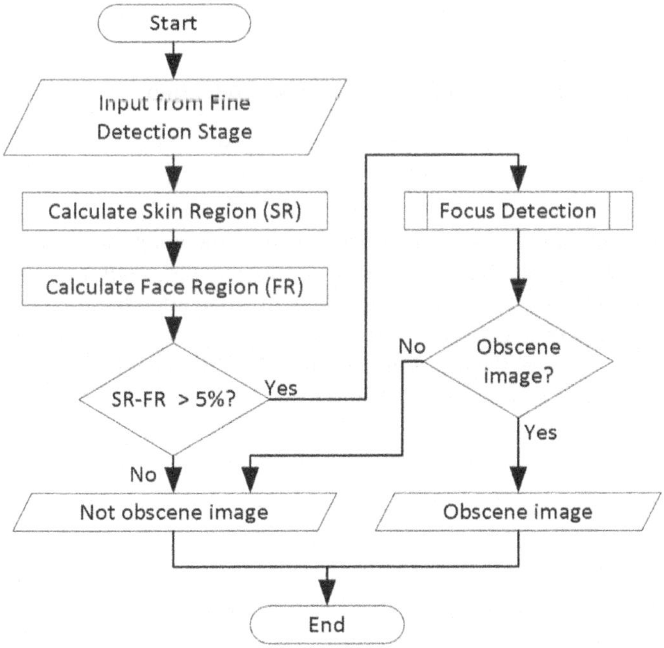

Fig. 3. Focus detection flowchart

4 Results and Discussion

4.1 Dataset

The primary dataset used was collected by [15], which contains nearly 80 h of 400 pornographic and 400 non-pornographic videos. The keyframe of the videos was extracted to create 6500 porn frames and 7000 non-porn frames. For porn images, 3500 images were added to the dataset, and 3000 images were added to the safe images to create 10000 images for each category. The dataset was divided into 70% training data, 10% validation data, and 20% testing data.

4.2 Experiment on Fine Detection Stage: Skin Detection

In the skin detection stage, two masks were applied to each image. Then the two masks were compiled to get results with more accuracy. This stage results are under the effect of two main factors: image lighting and skin-like pixels (Table 1).

By looking at image numbers 1 and 2, we can see that these two images have complex backgrounds with skin-like pixels. The skin percentage is higher than what it is supposed to be. We can conclude that skin detection using both YCbCr and HSV masks give good detection with the pictures that do not have skin-like pixels. Nevertheless, it gives a high skin percentage in the pictures that have complex backgrounds or skin-like pixels. The image lighting also affects the number of skin pixels in the image. If the light is high, the skin color will be changed to another lighter color to be considered a normal pixel. All images will be passed to face detection except image number 3 because all images have some skin regions percentage.

4.3 Experiment on Fine Detection Stage: Face Detection

A Python library based on the Adaboost algorithm with Haar-like features is used to detect the pictures' face area in the face detection step. After detecting the face area in the image, the pixel number is reduced from the previous step's skin region. If the skin region represents more than 5% of the total skin area, the image will proceed to the next stage, i.e., focus detection. Otherwise, it will be considered a safe face image.

Table 2 shows the results of using face detection and showing the skin area without a face. Images 1, 2, and 3 will be passed to the next stage, i.e., focus detection, because they met the criterion, which has a skin region >5% of the total image size. On the other hand, image 5 considered safe images because skin regions <5%.

Table 1. Example of skin detection

#	Original image	YCbCr mask	HSV mask	Compiled mask	% Skin
1					22.7 %
2					44.1 %
3		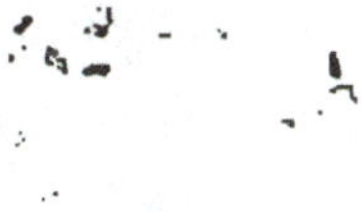			0.0%
4	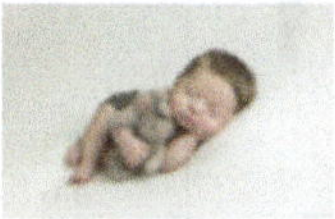				17.5%
5					20.2 %

4.4 Testing on the Fine Detection Stage

For research purposes, the fine detection stage was tested with 1000 images categorized as Table 3. A safe image is an image that does not contain any pornographic features. The safe images were divided into two categories: hard and easy. Hard represents the images with skin-alike pixels, bad lighting, or large skin areas, but they are not obscene images, such as babies' images. The easy image did not have any skin pixels. The unsafe image is the image that showed exposed body skin. If the body's skin is in a large area, it is considered an easy unsafe image. Otherwise, it is a hard unsafe image. The hard safe image has a large number of false detection because of skin-alike pixels or lighting problems. For the hard unsafe images, five images are considered a safe picture because the exposed body skin exists in a small area.

Table 2. Example of face detection

#	Original image	Skin Region	% Skin	Face area	% Skin without Face
1			22.7 %		10.4%
2			44.1 %		41.4%
3			17.5%		12.5%
5			20.2 %		4.6 %

Table 3. Testing data distribution and its detection results

	Safe images		Unsafe images	
	Hard	Easy	Hard	Easy
Number of images	250	250	250	250
Detection result				
Correct identification	160 (64%)	240 (96%)	207 (82.8%)	235 (94%)
Incorrect identification	90 (36%)	10 (4%)	43 (17.2%)	25 (10%)

4.5 Experiment on the Focus Detection Stage

After applying fine detection, the images will be classified by a convolutional neural network. This CNN is trained using the transfer learning method using the Alexnet that has eight layers. In transfer learning, the original layers are used with a change in the number of output classes and then retrained with the new dataset, as shown in Fig. 4.

Only two classes are used as output: safe and unsafe images. Safe represents an image that is not pornographic. Unsafe represent an image that is pornographic.

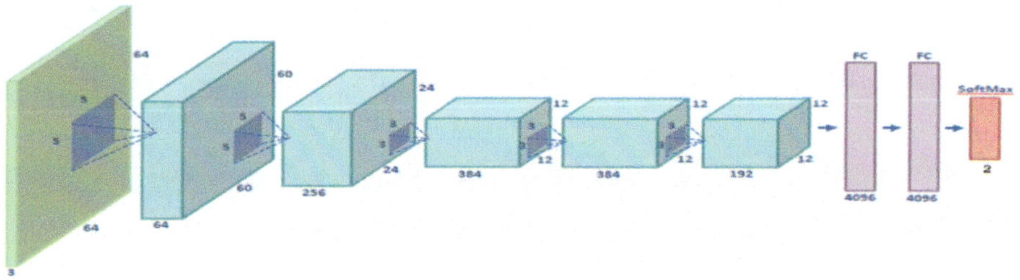

Fig. 4. Modified AlexNet structure for transfer learning of the focus detection stage

The dataset used for training contains 20000 images. Half of them are safe, and the other half contains unsafe images. For obscene image detection, the safe dataset selected from over 10000 images varied from human images, non-human images, babies' images, animals' images, images with complex backgrounds, and images with typical backgrounds. The dataset was divided into 70% training, 20% testing, and 10% validation.

Table 4. Confusion matrix

Actual	Predicted	
	Safe	*Unsafe*
Safe	TP: 1901 (94.6%)	FP: 109 (5.4%)
Unsafe	FN: 79 (4.0%)	TN: 96.0%

The training is done using Matlab, which took 3 h using a local machine with 540 iterations over the data. This training's resulting accuracy was 98.5%, with a loss of less than 0.01, and validation accuracy of 98.19%, as depicted in Fig. 5. The testing data represent 20% of the total data, which equals 3969 images. The results of this testing are shown in the confusion matrix in Table 4. The overall accuracy of the testing was 95.26%.

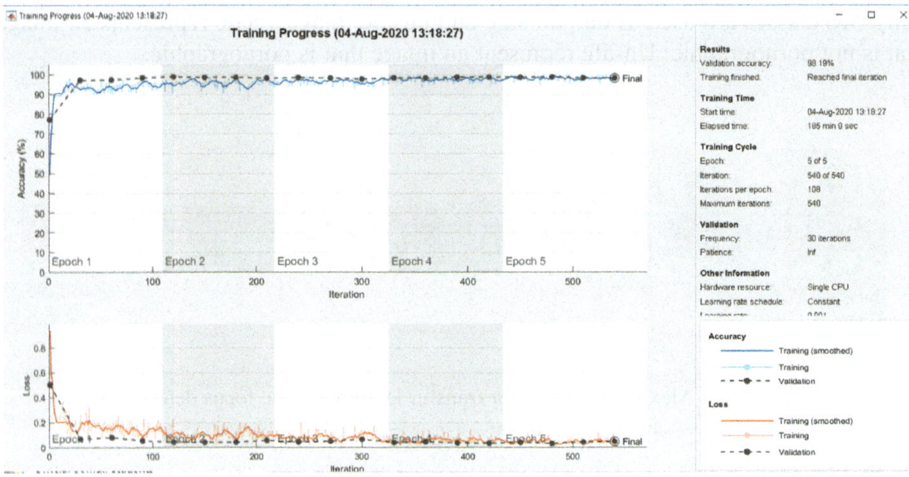

Fig. 5. Training progress report

5 Conclusions and Future Works

We have proposed an obscene image filtering system that can be fast and accurate. There are two stages in this algorithm, including fine and focus detection. Fine detection consists of two steps: skin detection followed by face detection. This stage's results were fast but not entirely accurate due to some lighting and skin-like pixels related problems. A transfer learning using AlexNet was done as the focus detection stage. The final accuracy achieved is around 95.25% when testing with 3969 images. Moreover, the false positive and false negative were 5.4% and 4.0%, respectively. Future works include improving the skin detection algorithm for skin-like pixels and bad lighting images, improving the face detection algorithm, and use other image datasets for training and testing.

Acknowledgments. The authors would like to express their gratitude to the Malaysian Ministry of Education (MOE), which has provided research funding through the Fundamental Research Grant, FRGS19-076-0684 (FRGS/1/2018/ICT02/UIAM/02/4). The authors would like to acknowledge the International Islamic University and the University of New South Wales's support.

References

1. Fleck, M.M., Forsyth, D.A., Bregler, C.: Finding naked people. In: European Conference on Computer Vision. Springer (1996)
2. Zaidan, A., et al.: An automated anti-pornography system using a skin detector based on artificial intelligence: a review. Int. J. Pattern Recognit. Artif. Intell. **27**(04), 1350012 (2013)
3. Ahmed, F., Shafiq, M.Z., Liu, A.X.: The Internet is for porn: measurement and analysis of online adult traffic. In: 2016 IEEE 36th International Conference on Distributed Computing Systems (ICDCS). IEEE (2016)
4. Kleinman, A.: Each Month Porn Sites Get More Visitors Than Netflix, Amazon and Twitter Combined (2013). https://www.huffpost.com/entry/internet-porn-stats_n_3187682

5. Nikam, A.V.: Locating and detecting nipple for pornographic image identification. Int. J. Comput. Sci. Eng. **6**(7), 1343–1347 (2018)
6. Nian, F., et al.: Pornographic image detection utilizing deep convolutional neural networks. Neurocomputing **210**, 283–293 (2016)
7. Moustafa, M.: Applying deep learning to classify pornographic images and videos. arXiv preprint arXiv:1511.08899 (2015)
8. Wang, Y., Jin, X., Tan, X.: Pornographic image recognition by strongly-supervised deep multiple instance learning. In: 2016 IEEE International Conference on Image Processing (ICIP) (2016)
9. Jin, X., Wang, Y., Tan, X.: Pornographic image recognition via weighted multiple instance learning. IEEE Trans. Cybern. **49**(12), 4412–4420 (2018)
10. Lv, L., et al.: Pornographic images detection using high-level semantic features. In: 2011 Seventh International Conference on Natural Computation. IEEE (2011)
11. Kolkur, S., et al.: Human skin detection using RGB, HSV and YCbCr color models. arXiv preprint arXiv:1708.02694 (2017)
12. Yan, J., Gao, X.: Pornographic video detection with MapReduce. Int. J. Mach. Learn. Cybern. **9**(12), 2105–2115 (2018)
13. Chelali, F.Z., Cherabit, N., Djeradi, A.: Face recognition system using skin detection in RGB and YCbCr color space. In: 2015 2nd World Symposium on Web Applications and Networking (WSWAN). IEEE (2015)
14. Hajraoui, A., Sabri, M.: Face detection algorithm based on skin detection, watershed method and gabor filters. Int. J. Comput. Appl. **94**(6), 33–39 (2014)
15. Avila, S., et al.: Pooling in image representation: the visual codeword point of view. Comput. Vis. Image Underst. **117**(5), 453–465 (2013)

Development of Colorization of Grayscale Images Using CNN-SVM

Abdallah Abualola[1], Teddy Surya Gunawan[1,2](✉), Mira Kartiwi[3], Eliathamby Ambikairajah[2], and Mohamed Hadi Habaebi[1]

[1] Electrical and Computer Engineering Department, International Islamic University Malaysia, Gombak, Malaysia
tsgunawan@iium.edu.my
[2] School of Electrical Engineering and Telecommunications, UNSW, Sydney, Australia
[3] Information Systems Department, International Islamic University Malaysia, Gombak, Malaysia

Abstract. Nowadays, there is a growing interest in colorizing many grayscales or black and white images dating back to before the colored camera for historical and aesthetic reasons. Image and video colorization can be applied to historical images, natural images, astronomical photography. This paper proposes a fully automated image colorization using a deep learning algorithm. First, the image dataset was selected for training and testing purposes. A convolutional neural network (CNN) was designed with several layers of convolutional and max pooling. Support Vector Machine (SVM) regression was used at the final stage. The proposed algorithm was implemented using Python with Keras and Tensorflow libraries in Google Colab. Results showed that the proposed system could predict the colored image from the training process's learning knowledge. A survey was then conducted to validate our findings.

Keywords: Grayscale image · Color image · Convolutional neural networks · SVM regression

1 Introduction

Nowadays, we have many grayscale or black and white images that date back to before the colored cameras. Converting grayscale images into colored images is a complex, multilayered process that requires high computational requirements. The goal is to convert one channel of image data, i.e., grayscale image, to a Red-Green-Blue (RGB) image, a three-channel image data. The colorization problem has two extents that have to be modeled, and they are the input space, which represents the feature of the grayscale images.

Many types of research have been conducted in converting grayscale images to colored images. A semi-manual colorization method was proposed in [1], which still requires users to assign the right color to a specific area or patch. Neural networks and several image processing techniques were proposed in [2]. Three neural networks output

J. A. Mat Jizat et al. (Eds.): iCITES 2020, AISC 1350, pp. 50–58, 2021.
https://doi.org/10.1007/978-3-030-70917-4_6

were combined into one output, which transfers colors from the source image to the target image. Similar reference images were used in [3], but it was limited to superpixel representation where it extracted features from reference and target images. It supported spatial coherence but had inaccuracy when it comes to thin objects' boundaries.

Image colorization based on modeling was proposed in [4], but the model can not distinguish global objects. In [5], the best colors were selected using energy minimization, variational formulation, and a patch-based method, in which the color data was extracted from a reference image. In [6], 22 convolutional layers were used with the ImageNet dataset [7] to predict each pixel histogram to deal with the multi-model technique. A Generative Adversarial Network (GAN) [8] has been used to colorize the SUN dataset's natural grayscale images [9]. Even though many researchers and developers created various models and achieved pleasing results, however there quality and computational time could be further improved. There are many proposed solutions, but the most immense challenge is automating the process where no user interference is required, especially for a considerable number of images.

2 Proposed Colorization Using CNN-SVM

The goal that needs to be achieved is to use a grayscale image as an input, a one-dimension image, and convert it to a three-dimensional RGB image. Figure 1 shows the proposed algorithm. First, the input images are down-sampled to 256 × 256 resolution to make the training more straightforward and reduce the memory requirement. Regardless of the small resolution, the network should learn smooth colorization, as observed in [6]. To colorized output, a conventional spline interpolation algorithm is used. After having the input, the network converts the image to the CIELab color space. This color space is easy to deal with, and it is designed to match the human eye perception of images. Lab color space consists of luminance (L), the grayscale part of an RGB image. At the same time, a and b are the color channels. The a channel controls colors between green and red, and b channel controls colors between blue and yellow.

First, the RGB color space can be transformed to CIEXYZ color [10] as defined in CIE standards and is shown in Eq. (1). Then, the CIEXYZ color can be transformed into CIELab color space, as shown in Eq. (2).

$$\begin{bmatrix} X \\ Y \\ Z \end{bmatrix} = \frac{1}{0.17697} \begin{bmatrix} 0.49000 & 0.31000 & 0.20000 \\ 0.17697 & 0.81240 & 0.01063 \\ 0.00000 & 0.01000 & 0.99000 \end{bmatrix} \begin{bmatrix} R \\ G \\ B \end{bmatrix} \tag{1}$$

$$L^* = 116f\left(\tfrac{Y}{Y_n}\right) - 16 \qquad a^* = 500\left(f\left(\tfrac{X}{X_n}\right) - f\left(\tfrac{Y}{Y_n}\right)\right)$$
$$b^* = 200\left(f\left(\tfrac{Y}{Y_n}\right) - f\left(\tfrac{Z}{Z_n}\right)\right) f(t) = \begin{cases} \sqrt[3]{t} & \text{if } t > \delta^3 \\ \frac{t}{3\delta^2} + \frac{4}{29} & \text{otherwise} \end{cases} \tag{2}$$

where $t = \frac{Y}{Y_n}$ and $\delta = \frac{6}{29}$. L channel values range from 0 to 100, and a and b range from −128 to 127. Because not every combination of these ranges becomes visible RGB color, we consider only the range −110 to 110. However, some combinations do not map back to RGB. In these cases, the range changes to RGB space (0, 255) [11].

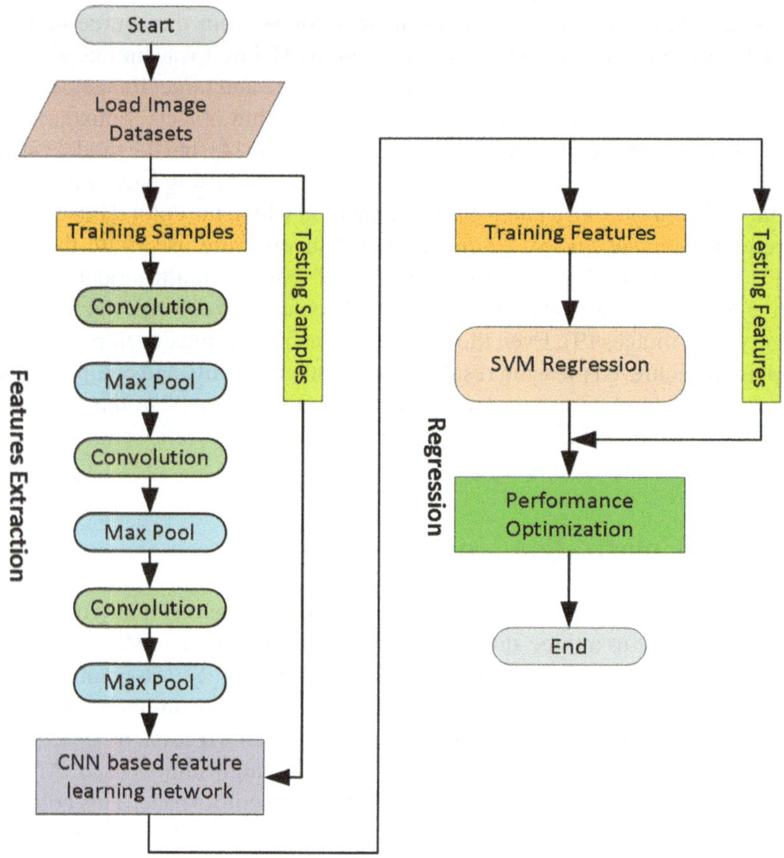

Fig. 1. Proposed colorization algorithmusing CNN-SVM

The approach to estimating *a* and *b* channels is to deal with the problem as a classification task. This approach is about estimating every pixel color histogram for the target image. The color histogram is equal to the probability distribution of the network prediction for every pixel. The SoftMax activation function will do the normalization of the output to give an appropriate distribution.

After estimating *ab* channels, we upsampled the channels using spline interpolation to get the original dimensions. An extra convolutional layer will be inserted after the prediction layer is activated by the SoftMax function for colorization in Lab space. The prediction calculation will be in the form of:

$$ab = \sum\nolimits_i p_i \cdot q_i \tag{3}$$

where p is the predicted probabilities, and q is the canonical 2D *ab* points selected on the grid.

A loss function in every neural network calculates prediction errors and obtains useful gradients to be used in the update function to update the network's weights. Since the process is predicting probability distribution, the Kullback-Leibler divergence function is applied. This function act as a distance function of two probability distributions. Moreover, the loss function can be enhanced by using a prior boosting reweighing factor called class rebalancing. The loss function becomes:

$$L\left(\widehat{P}, P\right) = -\sum_{h,w} v\left(P_{h,w}\right) \sum_{q}\left(\log\left(\widehat{P}_{h,w,q}\right) - \log\left(P_{h,w,q}\right)\right) \tag{4}$$

where $\widehat{P}$ is the color histogram of network prediction for pixels, P is the color histogram of the ground truth image, h,w is the indices of the pixels, and v is the weighting term which is defined by [12]:

$$\begin{gathered} v\left(P_{h,w}\right) = w_{q^*}^{*}, q^* = \arg\max P_{h,w,q} \\ w \approx \frac{1}{(1-\lambda)r + \frac{\lambda}{313}} \end{gathered} \tag{5}$$

where r is the empirical prior distribution of ab pairs calculated over the training set, and λ is a term used to create a mixture of the training set prior probability and uniform distribution.

3 Results and Discussion

3.1 Experimental Setup

The dataset used in training is ImageNet, which has around one million images [7]. These images were split into three groups, training, validation, testing. Around 10000 images were used for validation, 10000 images were used for testing, and the rest was used for training. This separation is to ensure unbiased results. Each image had to go through some preparations. First, convert the image at full resolution to CIELab color space and pull out the L channel. Second, resize the original image to the network input size, which is 224×224. Then pull back the resized L channel into the network.

In this research, the Python programming language will be used. Python will be applied to an open-source library for neural networks, including Keras and Tensorflow, using Google Colab. For training purposes, the most common dataset ImageNet will be used. ImageNet dataset contains RGB and grayscale images with over 10 thousand categories and over 3 million images. It is considered the most suitable dataset that can be used for training and testing.

3.2 Colorization Experiments

Figure 2 shows the example of successful image coloring cases alongside the grayscale input and the ground truth image. It can be observed that some colors are not similar to the original images, but that is still acceptable where the output images are still considered realistic. Figure 3 presents an impressive result that has been achieved on images from outside the dataset (the testing process). However, the proposed system still can not correctly colorize some types of images. The possible reason is the lack of variety of colors in the training images, which leads to a slight bias towards some colors.

Fig. 2. Example of succesful colorization in the testing process

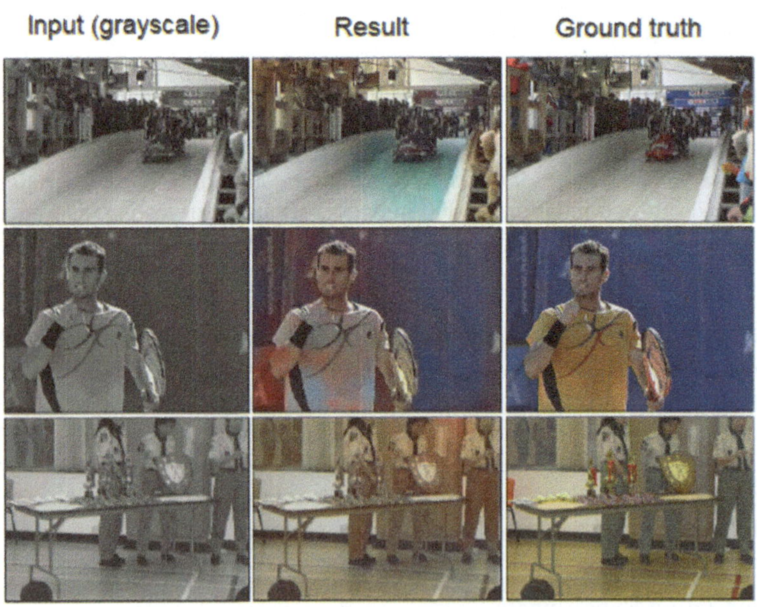

Fig. 3. Example of unsuccessful colorization process

3.3 Historical and 4K Natural Image Colorization Experiment

Some old grayscale historical images have been used to view the results and decide if the output is realistic to test the real-life images. Figure 4 presents some colored legacy images that have been colored using the proposed system. Moreover, 4K high-resolution grayscale images were colorized using our proposed system, and the result is shown in Fig. 5.

Fig. 4. Results of colorization of legacy images

Fig. 5. Example of 4k natural images colorization

3.4 Performance Comparison

Figure 6 shows the difference between our proposed system results and the algorithm proposed in [13]. Besides, Table 1 shows a detailed comparison between our proposed system and the deep colorization module.

Fig. 6. Performance comparison

Table 1. Performance comparison

	Deep colorization [13]	Proposed system
Algorithm	• Extract feature sets • Fully convolutional network • Three layers of neural network regressor • Joint bilateral filter	• CNN and SVM regression
Learning	• Extract features • Train FCN • Train regressor	• Train CNN and SVM from pixel to color distribution • Tune single parameters on the validation
Dataset	• Sun dataset • 2688 images for training • 1334 images for testing • Limited variety of scenes	• ImageNet dataset • 1 million images for Training • 10000 images for testing • Diverse categories and scenes
Run-time	• 4.9 s per image • MATLAB implementation	• 21.1 ms perimage • Caffe (Python) on K40 GPU

A survey has been done on 7 engineering students. The survey included 15 colorization images as the output from our proposed system. Alongside these images, the ground truth images were presented as well. The first part of the survey asks participants which image is the original image using the first eight out of 15 images. The second part of the survey asks participants to determine whether the presented image (the last seven out of 15 images) is real or synthetic (colorized by our algorithm).

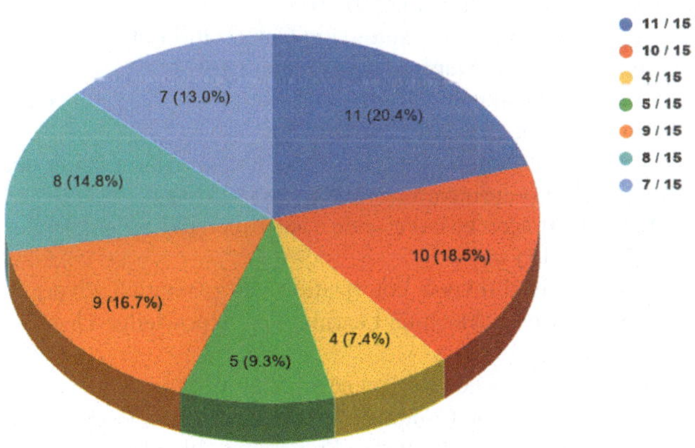

Fig. 7. Qualitative analysis of the colorized images

Figure 7 shows the summary of the people's opinions, where each question represents one point for guessing right. The highest score was 11 out of 15, while the lowest score is 4 out of 15. We are interested in the lower score, which means our proposed algorithm produces a realistic colorized image. If we take a score of 8 as the median (correctly identified original versus colorized images), we found that 3 participants favored our algorithm, 1 participant cannot decide, and 2 participants were not convinced with our proposed algorithm.

4 Conclusions and Future Works

This research has presented a method to colorize images using a deep convolutional neural network with well-chosen appropriate configuration parameters and functions. Python with Keras and Tensorflow libraries were used in the implementation. Google Colab with GPU was used for training and testing the proposed algorithms. Quantitative and qualitative analysis of the results showed that our algorithm could produce realistic colored images from the original grayscale images. Future works include training with other image datasets, different deep learning configurations, and algorithm optimization.

Acknowledgments. The authors would like to express their gratitude to the Malaysian Ministry of Education (MOE), which has provided research funding through the Fundamental Research Grant, FRGS19-076-0684 (FRGS/1/2018/ICT02/UIAM/02/4). The authors would like to acknowledge the International Islamic University and the University of New South Wales's support.

References

1. Bugeau, A., Ta, V.-T.: Patch-based image colorization. In: Proceedings of the 21st International Conference on Pattern Recognition (ICPR2012). IEEE (2012)
2. Karlik, B., Sariöz, M.: Coloring grayscale image using artificial neural networks. In: 2009 2nd International Conference on Adaptive Science & Technology (ICAST). IEEE (2009)
3. Gupta, R.K., et al.: Image colorization using similar images. In: Proceedings of the 20th ACM International Conference on Multimedia (2012)
4. Sousa, A., Kabirzadeh, R., Blaes, P.: Automatic Colorization of Grayscale Images. Department of Electrical Engineering, Stanford University (2013)
5. Bugeau, A., Ta, V.-T., Papadakis, N.: Variational exemplar-based image colorization. IEEE Trans. Image Process. **23**(1), 298–307 (2013)
6. Yu, F., et al.: LSUN: construction of a large-scale image dataset using deep learning with humans in the loop. arXiv preprint arXiv:1506.03365 (2015)
7. Russakovsky, O., et al.: Imagenet large scale visual recognition challenge. Int. J. Comput. Vis. **115**(3), 211–252 (2015)
8. Cao, Y., et al.: Unsupervised diverse colorization via generative adversarial networks. In: Joint European Conference on Machine Learning and Knowledge Discovery in Databases. Springer (2017)
9. Xiao, J., et al.: Sun database: large-scale scene recognition from abbey to zoo. In: 2010 IEEE Computer Society Conference on Computer Vision and Pattern Recognition. IEEE (2010)
10. Gonzalez, R.C., Woods, R.E.: Digital Image Processing, 4th edn. Pearson (2018)
11. Ruderman, D.L., Cronin, T.W., Chiao, C.-C.: Statistics of cone responses to natural images: implications for visual coding. JOSA A **15**(8), 2036–2045 (1998)
12. Zhang, R., Isola, P., Efros, A.A.: Colorful image colorization. In: European Conference on Computer Vision. Springer (2016)
13. Deshpande, A., Rock, J., Forsyth, D.: Learning large-scale automatic image colorization. In: Proceedings of the IEEE International Conference on Computer Vision (2015)

Numerical Assessment of the Effects of Rooftop PVs on Ambient Air Temperature

Asmaa Zaz[1(✉)], Mohammed Ouassaid[1], Mohammed Bakkali[2], and Mounir Ghogho[2]

[1] Engineering for Smart and Sustainable Systems Research Centre, Mohammadia School of Engineers (EMI), Mohammed V University (UM5), Rabat, Morocco
asmaazaz2017@gmail.com
[2] ICT Research Laboratory (TICLab), International University of Rabat (UIR), Rabat, Morocco

Abstract. With the increasing deployment of photovoltaic panels (PVs) in buildings' rooftops within urban environments, architects and engineers beyond considering electricity production benefits, they also look at patterns of pedestrian thermal comfort and urban climates. We aim in this paper to assess the impact of PVs on the micro climate. We adopted a micro scale study and we chose a 44 × 44 m^2 area in the heart of the city of Casablanca, the Moroccan economic capital. Through numerical predictions, two simulations are run, the first one is the reference simulation corresponding to the actual state of our study area (without PVs), and the second is the one fully covered by PVs. A comparison between the two simulations shows that PVs cause a warming effect during the evening and night-time.

Keywords: Rooftop PV · Urban microclimate · Numerical modelling · PV impact

1 Introduction

From various renewable energy sources, solar energy has gathered significant attention in urban environments. Among all the emerging solar technologies, photovoltaic panels (PVs) is one of the most affordable and promising approaches that can meet future energy needs and alleviate environmental problems [1]. A large spread of innovative materials is proposed to be integrated in the building envelope [2–4], or to replace conventional materials [5–7]. Most commonly, PVs are installed on building rooftops [8].

As well-known, a major portion of the transmitted irradiation is not converted to electric current, because of the manufacturing defects and physical limitations of the PV cells [9]. The best laboratories reported currently that solar energy to electrical power conversion efficiency for mono-Si PV cells was about 26.7% [10]. It means that the remaining unutilized absorbed energy does not get dissipated to the surroundings fast enough, it just builds up inside the PV panel and consequently increases the temperature of the entire panel [11].

J. A. Mat Jizat et al. (Eds.): iCITES 2020, AISC 1350, pp. 59–68, 2021.
https://doi.org/10.1007/978-3-030-70917-4_7

Based on that, we see clearly that coupling between the building and PV system breaks the original energy balance of the building envelope [12]. This balance includes the incoming and outgoing short-wave and long-wave radiation, anthropogenic heat, heat storage in buildings and surfaces, evapotranspiration, and heat advection by wind [13]. All these parameters alter the urban micro climate, which strongly influences the wellbeing of cities inhabitants [14]. Therefore, urban planners need to address the urban micro climate in their land use strategies and modelling [15].

Several works have pointed out the need to study in detail the effect of albedo, emittance and sub-roof resistance in defining the roof's heat gain or loss in PV installations [16]. Some studies suggest that PVs induce day and night time cooling effects. For example, a Weather Research and Forecasting (WRF) simulation conducted in Osaka City in Japan [17], showed that during daytime, cooling effect on roof surfaces is estimated to 2.1 °C. During night time, cooling effect on roofs is approximated to 1.4 °C. Other studies prove the opposite by claiming that photovoltaic panels increase temperatures within their surroundings.

The aim of the present paper is to investigate and analyze, through numerical modelling, the thermal impact of rooftop PVs on local temperatures. More specifically, the model ENVI-met 4.4.5 was applied in one particular area of Casablanca city. For comparison-wise, we conducted the same analysis with and without PVs.

2 Methodology

The 3D model ENVI-met 4.4.5 was chosen for this study because of its ability to model and predict the outdoor thermal climate, taking into account the energy exchange processes at the outer envelope of a building. It can estimate the detailed weather conditions for each second of the day for each wall and façade segment of the building or at any point within the model's domain.

As shown in the Fig. 1, this model uses a building geometry extracted from a GIS file as input, as well as properties of building materials. Further inputs require appropriate weather data for every hour of the day, describing temperature, relative humidity, air pressure, solar irradiation and wind direction and speed.

Two simulations are run: the first one considering buildings with conventional roofs and facades, and the second one considering the same building materials as the previous one but adding PVs to rooftops. The model evaluated air temperature in each case. Then we calculated the difference (T2 – T1) in order to conclude the effect of PVs on the outdoor thermal comfort.

3 Scale Study

Computational simulations can be employed to study urban microclimates at different spatial scales. They differ particularly in their purpose, data collection and processing methods. The American Meteorological Society (AMS) provides the following definitions:

- Macroscale or synoptic scale: the scale of atmospheric motions with a typical range of many hundreds of kilometers.

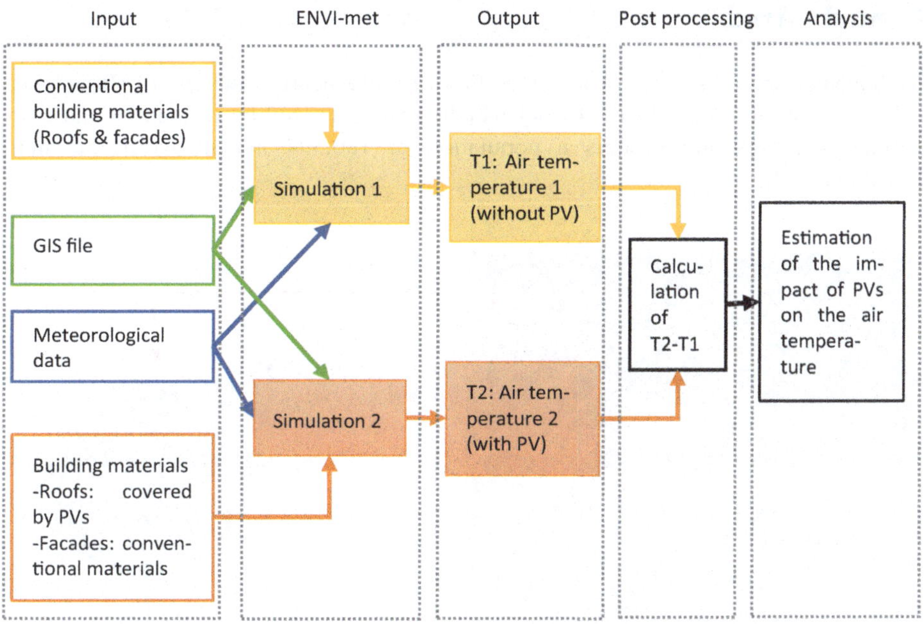

Fig. 1. Diagram of the methodology adopted in this paper

- Mesoscale: the scale of atmospheric phenomena having horizontal scales ranging from a few to several hundred kilometers.
- Microscale: the scale of atmospheric motions in spatial scales of 2 km or less.

Table 1 highlights some relevant building physics' information including the building scale, building layouts and different components ranging from construction materials to human behavioral patterns [18]. While this level of detailed information is generally not always involved in meteorological studies, they are of particular importance to urban physics and mainly how different urban surfaces provoke changes in local environmental settings.

Table 1. Spatial scales in climate modeling, with typical horizontal dimensions.

Spatial scale	Mesoscale	Microscale	Building	Component	Material/Human
Distance	<200 km	<2 km	<100 m	<10 m	<1 m

We have favored in our study the building scale, with an area of 44 * 44 m^2. We had a particular interest on this as it allows us to quantify properly the physical behavior of rooftops PVs and their detailed interactions with the buildings and the atmosphere nearby. This guides us to response to some of the big challenges we might encounter namely climate, energy and health issues.

4 Study Area

We have chosen a 44 × 44 m^2 area (see Fig. 2) in the heart of the city of Casablanca, and more precisely the Maârif district which is characterized by residential and commercial buildings and shelters a population of 180,394 inhabitants in the main downtown [19].

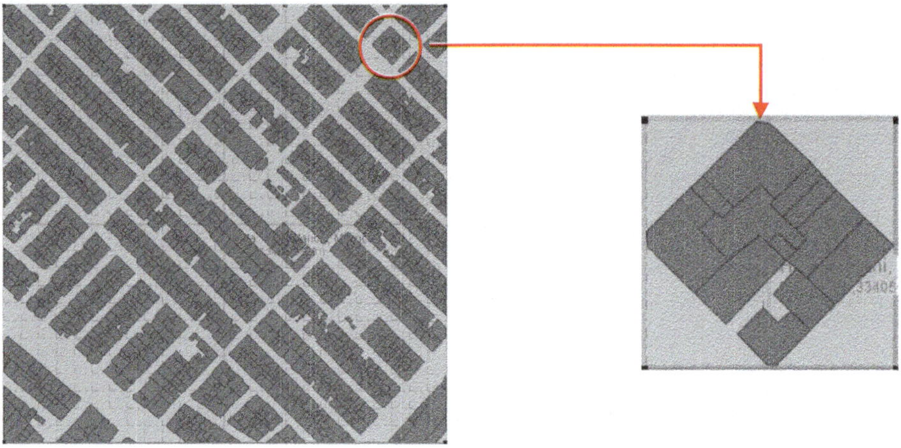

Fig. 2. Buildings of 500 * 500 m^2 of Maârif district, the red circle presents the study area. Buildings of our study area (44 × 44 m^2)

From a Geographic Information System (GIS) layer, we digitized the study area with a resolution of 2 m * 2 m * 2 m as shown in Fig. 3. The vertical extension should be at least three times as high as the tallest building, to provide open space above buildings in order to avoid turbulence at the top of the model volume. In fact, space at the top is needed for wind flow and convection triggered by the buildings to be stabilized again. Otherwise, it is pressed downwards causing the whole simulation to give instable results. Then, we convert our 2.5 D area to a 3D one by applying the appropriate height data to each building.

5 Meteorological Data

From an annual meteorological data extending from June 2018 to June 2019, we chose May 31st, 2019 for numerical modelling, as the value of recorded solar irradiation reached its peak around 1139 Wh/m^2 at 13:00. The total simulation time was 24 h, starting from 6:00 a.m. These weather conditions were taken by solar radiation data (SoDa). So, we considered the wind speed (at 10 m) equal to 3,46 m/s as the average speed of the day. Then, we put a time series of temperature (at 2 m) and relative humidity (at 2 m) with a time step of 1 h.

Limited information is available relating the panels' temperature with humidity. Studies reported that with rising humidity levels, panels' power output decrease, and the panel temperature got reduced. This is happening due to the generation of water droplets around the panels' surface at high humidity levels, which acts as a cooling agent and as a result the panel surface temperature reduces [20].

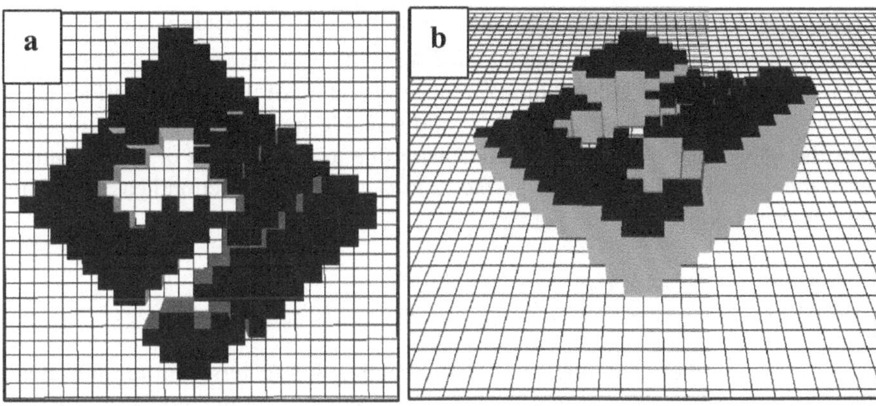

Fig. 3. Digitization of the study area (44 × 44 m^2) after the implementation of PVs. The blue color refers to the PV panels. The grey color represents the conventional wall materials. a- View from above. b- 3D view

6 Materials Properties

Walls and roofs in ENVI-met are represented as a sandwich of materials. Then, they are defined by the physical properties of their components, such as thickness, emissivity, specific heat, thermal conductivity and density of materials. The further values required are those of absorption, transmission and reflection, which must add up to a value of one within each material as shown in the Eq. (1):

$$\tau + \alpha + \varphi = 1 \tag{1}$$

where τ is the transmission, α is the absorption and φ is the reflection.

In the same way, PV panels are defined as an outer layer of the roofs, separated by an air gap. So, we considered panels with around 3.8 mm. From the lower side towards the upper side exposed to the sun, we have a back sheet of 0.1 mm thickness, a layer of an encapsulant material typically Epoxy-Vinyl-Acetate, EVA of 0.2 mm thickness, solar cells of 0.3 mm thickness separated in their plane by few millimeters of encapsulant, another layer of EVA like the previous one, and finally a glass cover of 3 mm thickness [21].

7 Outdoor Energy Balance

The outside temperatures are estimated using the energy balance as shown in the Eq. (2):

$$EB = Q^{abs}_{sw,net} + \epsilon\left(Q_{lw} - \sigma T^4\right) - H_w - LE_w - G_w \cong 0 \tag{2}$$

with $Q^{abs}_{sw,net}$ as the net shortwave radiation absorbed by the façade's surface, ϵ as the emissivity of the facade, σ as the Stefan-Boltzmann Constant, $\epsilon(Q_{lw} - \sigma T^4)$ as the longwave radiation balance depending on the temperature of the outside surface and the incoming longwave radiation Q_{lw}, H_w as the sensible heat flux into the atmosphere, LE_w as the latent heat flux into the atmosphere due to evaporation or condensation of water at the outside of the façade's surface and G_w as the conduction heat flux from or to the adjacent node inside the wall/roof [21].

8 Results

First, we compute the air temperature at the height of 3 meters, before and after the implementation of PVs. The results shown in Figs. 4 and 5 are the model state at 6 p.m. on the 31st May, and Fig. 6 illustrates the air temperature differences. The arrows in Fig. 4 show the direction of wind at this time. In fact, knowing where the wind is coming from gives an important insight into which air would be affected by photovoltaic panels. So, we see clearly in Fig. 4 that only the winds who crossed the PV panels have changed.

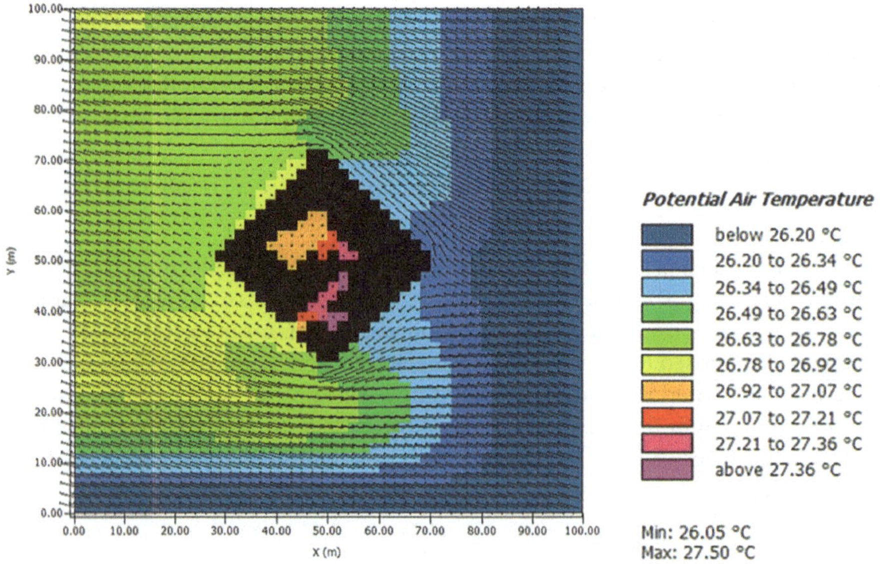

Fig. 4. Air temperature (z = 3 m) after the implementation of PVs at 6 p.m.

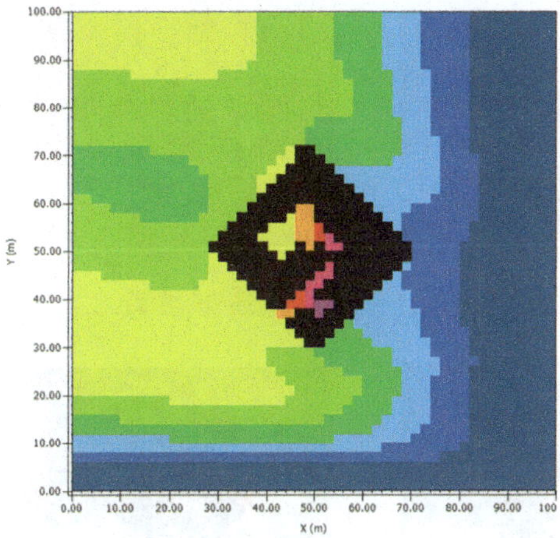

Fig. 5. Air temperature (z = 3 m) before the implementation of PVs at 6 p.m

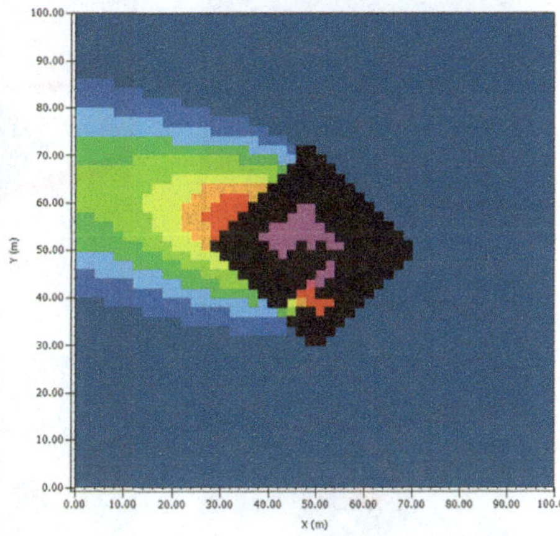

Fig. 6. Air temperature difference at 6 p.m.

To compare the results during the day, we present the air temperature differences, between the two scenarios with and without PVs over several hours.

In the morning, the air temperature was similar for both cases as shown in the Fig. 7, illustrating the 0 °C difference at 6 a.m. This result was roughly the same at 9 a.m. and at noon.

In the afternoon, starting from 1 p.m. the air temperature surrounding the buildings with integrated PVs was warmer than that encircling the conventional ones. This increase was estimated to up to 0.02 °C at 1p.m., 0.06 °C at 3 p.m. and up to 0.15 °C at 6 p.m.

At night, this warming effect reaches its maximum value of 0.20 °C at 9 p.m., then it starts to drop, but is still significant and attains 0.19 °C at 11 p.m., 0.18 °C at midnight and 0.15 °C at 3 a.m. of the next day.

9 Analysis and Future Studies

As demonstrated, adding PVs to the building roofs impacts the air temperature by altering the way that incoming energy is reflected back to the atmosphere or absorbed, stored, and reradiated [22]. As PV panel temperatures are consistently warmer than most exposed roof surfaces with the exception of very dark membrane roofs and have twice the surface area at the top and the bottom for convective heat transfer, it is reasonable to expect that PVs implemented on rooftops may warm up the urban environment [23]. For more detailed results, we should not only consider the average wind speed as we initially did, but a logarithmic wind speed profile. Sometimes indeed, the speed can be great enough to make a natural ventilation and cancel any air related heating effects. Over other time periods, the speed was much lower which makes the warming effect greater. Another parameter that should be considered is the building

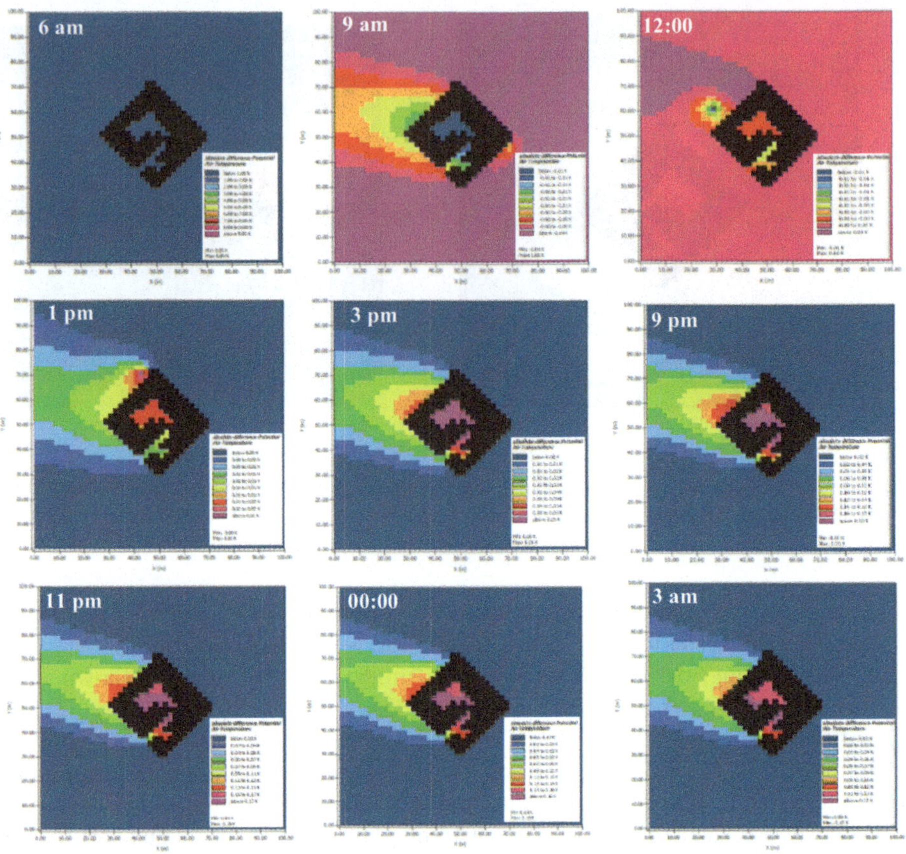

Fig. 7. Air temperature difference during the 31st May from 6 a.m. to 3 a.m. next day.

design which have a direct impact on the wind flow, level of exposure to the sky, thus having implications in daylight availability, solar radiation heat gain and perceived openness [24]. Also, we should consider the surrounding buildings of our study area, because this may change different parameters like reducing wind speed, causing shading effect or supplement heat, especially if they are also covered by PVs.

10 Conclusion

We quantified the difference in air temperature before and after the implementation of PVs on the rooftops. The results showed that morning's temperatures are roughly identical in the two scenarios. Contrariwise, a specific warming was occurred during the evening and night-time emphasizing a little Heat Island effect, going up to 0.20 °C. This value is expected to be more important if we implement PVs on a larger area like a whole borough or a city. As the overall effects of various parameters on the urban micro climate, it is extremely complicated to define the impact of rooftop PVs on the thermal environment, because it may exhibit different patterns. Thus, follow-up studies on the urban forms in various cities and regions with different climatic characteristics are required to discover further information on the impact of PVs.

11 Limitation

An exact implementation of solar panels is not available yet in ENVI-met. So, the modeling of rooftop PVs was approximated to an outer layer of the building roofs, and PVs were defined by their material properties like having a very low albedo. Thereby, these simplifications don't take into account the energy production.

References

1. Jung, S.K.: Optimal control methods for PV-integrated shading devices. Doctoral dissertation. University of Michigan (2014)
2. Wanphen, S., Nagano, K.: Experimental study of the performance of porous materials to moderate the roof surface temperature by its evaporative cooling effect. Build. Environ. **44**, 338–351 (2009). https://doi.org/10.1016/j.buildenv.2008.03.012
3. Ökte, A.N., Karamanis, D.: A novel photo responsive ZnO-fly ash nanocomposite for environmental and energy applications. Appl. Catal. B: Environ. **142–143**, 538–552 (2013)
4. Karamanis, D., Kyritsi, E., Krimpalis, S., Vardoulakis, E., Gorgolis, G., Kapsalis, V., Mihalakakou, G., Ökte, N.: Cooling roofs through low temperature solar-heat transformations in hydrophilic porous materials. Adv. Build. Energy Res. **7**(2), 235–243 (2013)
5. Synnefa, A., Santamouris, M., Akbari, H.: Estimating the effect of using cool coatings on energy loads and thermal comfort in residential buildings in various climatic conditions. Energy Build. **39**, 1167–1174 (2007). https://doi.org/10.1016/j.enbuild.2007.01.004
6. Kolokotsa, D., Santamouris, M., Zerefos, S.C.: Green and cool roofs' urban heat island mitigation potential in European climates for office buildings under free floating conditions. Sol. Energy **95**, 118–130 (2013). https://doi.org/10.1016/j.solener.2013.06.001
7. Santamouris, M.: Cooling the cities—a review of reflective and green roof mitigation technologies to fight heat island and improve comfort in urban environments. Sol. Energy **103**, 682–703 (2014). https://doi.org/10.1016/j.solener.2012.07.003
8. Oliver, M., Jackson, T.: Energy and economic evaluation of building-integrated photovoltaics. Energy **26**(4), 431–439 (2001)
9. Alharbi, F.H., Kais, S.: Theoretical limits of photovoltaics efficiency and possible improvements by intuitive approaches learned from photosynthesis and quantum coherence. Renew. Sustain. Energy Rev. **43**, 1073–1089 (2015)
10. NREL. NREL 2018. http://www.nrel.gov/. Accessed 15 Mar 2018
11. Aly, S.P., Ahzi, S., Barth, N., Abdallah, A.: Using energy balance method to study the thermal behavior of PV panels under time-varying field conditions. Energy Convers. Manage. **175**, 246–262 (2018)
12. Wang, D., Qi, T., Liu, Y., Wang, Y., Fan, J., Yue Wang, H.D.: A method for evaluating both shading and power generation effects of rooftop solar PV panels for different climate zones of China. Sol. Energy **205**, 432–445 (2020)
13. Oke, T.R.: Street design and urban canopy layer climate. Energy Build. **11**(1–3), 103–113 (1988)

14. Kuttler, W.: Stadtklima. Teil 1: Grundzüge und Ursachen. Beitragsserie: Klimaänderung und Klimaschutz, pp. 1–13 (2004)
15. Heldens, W., Heiden, U., Esch, T., Dech, S.: Potential of hyperspectral data for urban micro climate analysis (2010)
16. Kapsalis, V., Karamanis, D.: On the effect of roof added photovoltaics on building's energy demand. Energy Build. **108**, 195–204 (2015)
17. Ma, S., Goldstein, M., Pitman, A.J., Haghdadi, N., MacGill, I.: Pricing the urban cooling benefits of solar panel deployment in Sydney, Australia (2017)
18. Blocken, B.: Computational Fluid Dynamics for urban physics: importance, scales, possibilities, limitations and ten tips and tricks towards accurate and reliable simulations. Build. Environ. **91**, 219–245 (2015)
19. Office of the High Commissioner for Planning: General Population and Housing Census (2004)
20. Tripathi, A.K., Ray, S., Aruna, M., Prasad, S.: Evaluation of solar PV panel performance under humid atmosphere. In: Materials Today: Proceedings (2020)
21. Simon, H.: Modeling urban microclimate: development, implementation and evaluation of new and improved calculation methods for the urban microclimate model ENVI-met (2016)
22. Barron-Gafford, G.A., Minor, R.L., Allen, N.A., Cronin, A.D., Brooks, A.E., Pavao-Zuckerman, M.A.: The photovoltaic heat island effect: larger solar power plants increase local temperatures (2016)
23. Pham, J.V., Baniassadi, A., Brown, K.E., Heusinger, J., Sailor, D.J.: Comparing photovoltaic and reflective shade surfaces in the urban environment: effects on surface sensible heat flux and pedestrian thermal comfort. Urban Climate **29**, 100500 (2019)
24. Zhang, J., Le, X., Shabunko, V., Tay, S.E.R., Sun, H., Lau, S.S.Y., Reindl, T.: Impact of urban block typology on building solar potential and energy use efficiency in tropical high-density city. Appl. Energy **240**, 513–533 (2019)

Sliding Mode Control of Onboard Energy Storage System for Railway Braking Energy Recovery

Sadiq Eziani(✉) and Mohammed Ouassaid

Mohammadia School of Engineers (EMI), Engineering for Smart and Sustainable Systems Research Center, Mohammed V University, Rabat, Morocco
sadiq.eziani@gmail.com, ouassaid@emi.ac.ma

Abstract. In this study, a sliding mode controller (SMC) is constructed in order to control an onboard energy storage system ESS ensuring regenerative braking energy (RBE) saving in a railway traction system. The complete mathematical model of a supercapacitor (SC) Based energy storage system (ESS) is developed. Then, the SMC scheme is designed as to enhance both of precision and rapidity of the system. To illustrate the efficiency of the proposed controller, numerical simulations are carried out in MATLAB/SIMULINK environment. The simulation results validate the efficiency of the proposed controller, it shows that the SMC ensures a good reference tracking performance, and reduces the integral time absolute error (ITAE) by 14.39% and 21.81% compared to the classical Proportional Integral (PI) and fuzzy self-tuning PI (FSTPI) controllers, respectively. In addition, the overshoot is reduced by 18.94% and 40.28% compared to FSTPI and PI; respectively. Furthermore, a faster transient response is recorded.

Keywords: Sliding mode control · Energy efficiency · Energy storage system · Supercapacitor · Braking energy recovery

1 Introduction

Due to environmental problems related to climate change facing the world, energy efficiency has gained a widespread attention from the railway transportation sector actors, with the aim of reducing greenhouse gas emissions, while supporting the continuous development of railway transportation sector. The improvement in energy efficiency of railway transport sector not only has environmental benefits, but also economic rewards. In fact, it increases power system reliability and stability through decentralization and reduces the necessary contracted power [1]. In this scope, different technologies and strategies are put under study, including timetable optimization [2], driver advisory system implementation [3]. Nevertheless, recuperation and improvement of RBE still the greatest choice [4].

J. A. Mat Jizat et al. (Eds.): iCITES 2020, AISC 1350, pp. 69–80, 2021.
https://doi.org/10.1007/978-3-030-70917-4_8

Generally, the most of trains have the possibility of using regenerative braking sooner than mechanical braking [4]. Therefore, when the train is decelerating the electric motor converts the available kinetic energy into electric energy, this process is called RBE. This energy is handled as needed. Therefore, it could redirect to catenary, stored using ESSs, or it can be sent back to the power grid by creating a reverse path using reversible substation [5].

The outstanding development of energy storage device technologies and power electronic converters enabled the ESS to be the greatest solution to increase railway energy efficiency when it is combined with regenerative braking [6]. Consequently, SC energy storage device is widely used when compared to chemical batteries and flywheels. Indeed, SC has a long life cycle, high power density and a short charging and discharging time which make them able to absorb the amount of energy available during train braking.

Onboard ESSs enable the train to save its RBE and reuse it in the next acceleration, this storage technology allows peak power reduction, line voltage stabilization and line loses reduction. Real-time implementations of onboard ESSs record from 18% to 24% of energy saving [4]. As regards to wayside energy storage, the ESS is installed along the railway line, and it absorbs braking energy from any vehicle on the line and deliver it when it is needed for vehicles acceleration. In addition, wayside ESS affords also reduction of voltage sag problem, which is a voltage reduction under certain limits. Therefore, wayside ESS helps to regulate line voltage and protect the equipment connected to the feeding line. Real-time implementations of wayside ESS record up to 30% of energy saving [4].

The performance of ESS depends not only on equipment of storage device, but also on the energy management strategy (EMS). Therefore, the control and EMS have a significant role on increasing energy saving on railway systems. Various control algorithms have been proposed for the control of on board ESS, namely, linear PI controller [7], whose parameters are adjusted using Ziegler-Nichols method and Fuzzy self-tuning PI controller [8], the FSTPI enables an online adaptation of the PI gains using fuzzy rules. These controllers do not take into account the nonlinear behavior of the studied system. Therefore, a nonlinear controller must be introduced. Indeed, a nonlinear SMC has been designed for controlling the ESS, in order to improve dynamic and tracking performance of the system.

The reminder of the paper is structured as follows, Sect. 2 sheds light on the overall mathematical model of railway system, Sect. 3 presents the EMS and SMC. The results of the proposed controller are presented and discussed in Sect. 4. The last section is reserved to conclusions.

2 Railway Traction System

A DC electric rail system is powered by substations distributed along the railway line. As illustrated in Fig. 1. The electric energy flows from substations to electric motors via the pantograph and the catenary. The overhead line is supplied by substations in AC or DC in different voltage levels.

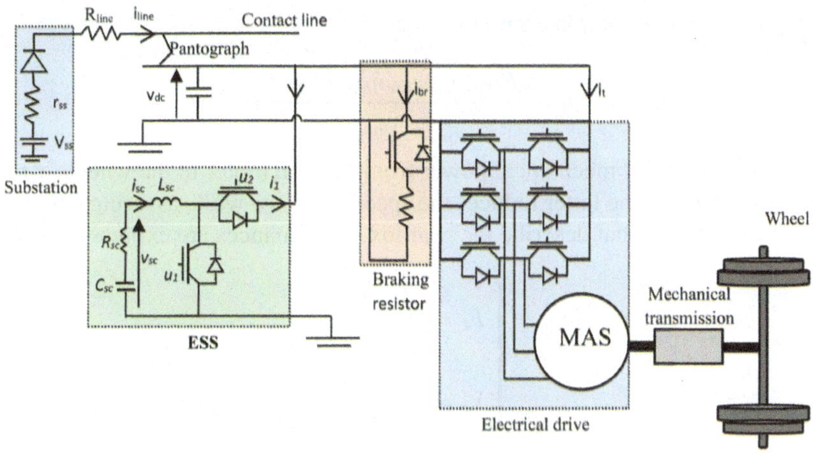

Fig. 1. Railway traction system configuration

2.1 Distribution Network Modeling

The railway distribution network is made by substations, catenary and rails, the schematic of substations is shown in Fig. 1, it is made by transformers and diode rectifiers. Therefore, it is considered as a constant voltage source. The voltage drop at the terminals of diode rectifiers is due to the non-ideality of semiconductors and it is modeled by a series resistance.

In order to simulate the movement of the train on the rails, the catenary is modeled as a variable resistor R_{line}, whose value depends on train position from substation, and it is given by Eq. 1.

$$R_{line} = x_t * R_0 \tag{1}$$

where x_t refers to train position, and R_0 is the resistance per unit length of the catenary.

2.2 Vehicle Modeling

In this study, the approach adopted for modeling the train is backward modeling, known as effect-cause modeling [9]. Therefore, the speed is considered as input, and the required power by the train is calculated. The train absorbs energy from catenary when accelerating and generate energy when decelerating. Hence, it can be considered as a reversible current source which is controlled according to the actual traction power given by Eq. 2:

$$P_t = v_{dc} i_t = F_t v_t (\eta_{mec} \eta_{inv} \eta_{em})^{-\mathrm{sgn}(F_t)} \tag{2}$$

where v_{dc}, i_t, F_t and v_t, are the overhead line voltage, traction current, traction effort and velocity, respectively.

Using Eq. (2), the traction current is given by:

$$i_t = \frac{F_t v_t (\eta_{em}\eta_{inv}\eta_{mec})^{-\text{sgn}(F_t)}}{v_{dc}} \tag{3}$$

The transmission of mechanical power from traction motor to the wheels is realized by means of gearbox, the latter reduces the speed of motor while assuring a high output torque. The equations that describe the gearbox performances are expressed as follows:

$$\begin{cases} F_t = \sum_{i=1}^{N} \frac{T_{e,i}\tau}{R_{wheel}} \\ V_t = \frac{R_{wheel}\omega_m}{\tau} \end{cases} \tag{4}$$

where $T_{e,i}$ and N are torque and number of traction motors. R_{wheel} and ω are radius and speed of wheels. τ is the gear ratio.

2.3 Energy Storage System Modeling

The ESS is made by SC as energy storage device connected to a bidirectional DC-DC converter, the latter can act as boost converter or a buck converter according to the implemented EMS. Indeed, in charging mode of ESS ($I_{sc,ref} < 0$) it acts as a boost converter, although in discharging mode ($I_{sc,ref} > 0$) it acts as Boost converter [10]. Consequently, the binary variable k is introduced.

$$k = \begin{cases} 1 \;\; i_{sc,ref} > 0 \;\; (Boost\ Mode) \\ 0 \;\; i_{sc,ref} < 0 \;\; (Buck\ Mode) \end{cases} \tag{5}$$

Boost Mode $k = 1$
In this mode, the control signal u_2 is set to zero ($u_2 = 0$) and u_1 is a variable PWM input. Therefore, using Fig. 1, the ESS nonlinear model is represented in Eq. (6):

$$\begin{cases} \frac{di_{sc}}{dt} = -(1-u_1) * \frac{v_{dc}}{L_{sc}} - \frac{R_{sc}}{L_{sc}} i_{sc} + \frac{u_{sc}}{L_{sc}} \\ i_1 = (1-u_1) i_{sc} \end{cases} \tag{6}$$

SC internal voltage u_{sc} is estimated using luenberger observer [7], it is given by Eq. (7):

$$\begin{cases} \frac{d\hat{u}_{sc}}{dt} = -\hat{u}_{sc} \frac{1}{R_{sc}C_{sc}(\hat{u}_{sc})} + v_{sc} \frac{1}{R_{sc}C_{sc}(\hat{u}_{sc})} + k_0(i_{sc} - \hat{i}_{sc}) \\ \hat{i}_{sc} = \frac{\hat{u}_{sc} - v_{sc}}{R_{sc}} \end{cases} \tag{7}$$

Buck Mode $k = 0$

In this mode, the control signal u_1 is fixed to zero ($u_1 = 0$), u_2 is a variable PWM input. Also, investigating Fig. 1, the ESS nonlinear model is:

$$\begin{cases} \frac{di_{sc}}{dt} = -u_2 * \frac{v_{dc}}{L_{sc}} - \frac{R_{sc}}{L_{sc}} i_{sc} + \frac{u_{sc}}{L_{sc}} \\ i_1 = u_2 * i_{sc} \end{cases} \tag{8}$$

Rearranging Eq. (6) and Eq. (8) the nonlinear model of Buck-Boost converter is represented by Eq. (9):

$$\begin{cases} \frac{di_{sc}}{dt} = -\alpha * \frac{v_{dc}}{L_{sc}} - \frac{R_{sc}}{L_{sc}} i_{sc} + \frac{u_{sc}}{L_{sc}} \\ \alpha = k(1 - u_1) + (1 - k)u_2 \end{cases} \tag{9}$$

The ESS is designed as to absorb the amount of energy available during braking phase. The maximum braking energy $E_{sc,\max}$ is expressed by Eq. (10):

$$E_{sc,\max} = \eta_{mec}\eta_{inv}\eta_{em}\eta_{dc}\eta_{sc} * \frac{1}{2}(m_{sc} + m_t)v_{t,\max}^2 \tag{10}$$

where m_{sc} is the SC pack mass, m_t train mass, η_{sc}, η_{dc}, η_{em}, η_{inv}, and η_{mec} are SC, buck-boost, electromagnetic, inverter and mechanical parts efficiencies, respectively; $v_{t,max}$ is the maximum speed. The SC mass is related to $E_{sc,\max}$ by Eq. (11), where α is SC energy density.

$$E_{sc} = \alpha * m_{sc} \tag{11}$$

Substituting Eq. (11) in Eq. (10), the braking energy is calculated as follows (Table 1):

$$E_{sc,\max} = \frac{\eta_{mec}\eta_{inv}\eta_{em}\eta_{dc}\eta_{sc} m_t v_{t,\max}^2}{2\alpha - \eta_{mec}\eta_{inv}\eta_{em}\eta_{dc}\eta_{sc} v_{t,\max}^2} \tag{12}$$

using Kirchhoff laws, the overall mathematical model of the system is readily derived from Fig. 1:

$$\begin{cases} V_{ss} - (r_{line} + r_{ss})i_{line} = v_{dc} \\ i_{line} = i_t - i_{sc} - \frac{v_{inv}}{r_{br}} \\ F_t - F_r = (m_t + m_{sc})\frac{dv_t}{dt} \\ F_r = a + bv_t + cv_t^2 \end{cases} \tag{13}$$

where F_r is resistance force; *a*, *b* and *c* parameters refer to friction forces due to the air and rail-wheel contact. Reference current $i_{sc,\,ref}$ and F_t are inputs of the system.

3 Energy Management Strategy

3.1 Regenerative Braking Energy Recovery

The objective of the installation of ESS is recycling RBE, i.e. save energy when the train is decelerating and reuse it in the next acceleration, according to the EMS portrayed by

Table 1. Specifications of supercapacitor module (Maxwell BMOD0063 P125)

Parameters	Value
Rated voltage $(V_{sc,max})$	125 V
Rated Capacitance $C_{sc,m}$	63 F
Series resistance $r_{sc,m}$	18 mΩ
Max current $I_{sc,max}$	750 A
Stored Energy	140 Wh
Mass	58 kg
α	1.75 Wh/kg
Dimensions (mm)	762 × 425 × 265 mm^3

Fig. 2. The conversion of kinetic energy into electrical energy and vice-versa is managed by the balance described by Eq. (14) [8]:

$$\frac{1}{2}C_{sc}(V_{sc,\max}^{2} - u_{sc,ref}^{2}) = \frac{1}{2}k(m_t + m_{sc})v_t^2 \tag{14}$$

where $k = (\eta_{em} \times \eta_{inv} \times \eta_{mec} \times \eta_{dc} \times \eta_{sc}) \times sgn(F_t)$ reflects electrical and mechanical efficiencies.

Using (14), $u_{sc,\ ref}$ SC reference voltage is then expressed as in (15):

$$u_{sc,ref} = f(v_t) \simeq V_{sc,\max}^2\sqrt{1 - k\frac{(m_t + m_{sc})v_t^2}{C_{sc}V_{sc,\max}^2}} \tag{15}$$

The power of the SC is obtained as follows:

$$P_{sc} = i_{sc}u_{sc} = -C_{sc}u_{sc}\frac{du_{sc}}{dt} \tag{16}$$

Substituting Eq. (15) in Eq. (16), leads to Eq. (17) that describes the SC reference power.

$$P_{sc,ref} = kv_t(m_t + m_{sc})\frac{dv_t}{dt} \tag{17}$$

Finally, the reference current is calculated according to Eq. (18). $\tilde{P}_{sc}$ is a compensation component which refers to incertitude in the determination of SC capacitance, it is realized by means of a control loop of SC estimated and reference voltages (Fig. 3).

$$i_{sc,ref} = \frac{P_{sc,ref} + \tilde{P}_{sc}}{u_{sc,ref}} \tag{18}$$

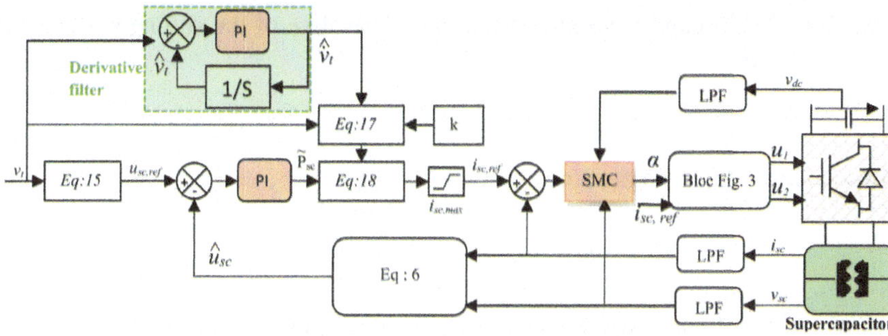

Fig. 2. Energy management strategy

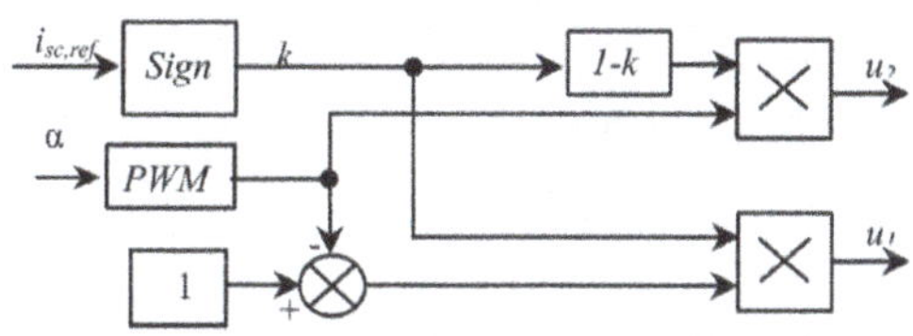

Fig. 3. Control signals u_1 and u_2 generation

3.2 Analysis and Design of Sliding Mode Controller

The control objective is to make the SC current track its desired value provided by the EMS. Accordingly, the SMC approach is adopted to design a suitable controller for the nonlinear model described by Eq. (9). Indeed, let us introduce the tracking error:

$$e = i_{sc} - i_{ref} \tag{19}$$

The time derivative of the error is given as:

$$\dot{e} = \frac{di_{sc}}{dt} - \frac{di_{ref}}{dt} \tag{20}$$

Substituting Eq. (9) in (20), leads to the following expression of the error derivative:

$$\dot{e} = -\alpha * \frac{v_{dc}}{L_{sc}} - \frac{R_{sc}}{L_{sc}} i_{sc} + \frac{u_{sc}}{L_{sc}} - \frac{di_{ref}}{dt} \tag{21}$$

The sliding surface is defined as follows:

$$\sigma = \left(\frac{d}{dt} + \lambda\right)^{n-1} * e \tag{22}$$

where, n is relative degree and λ is a positive constant.

According to Eq. (9), it can be seen that the relative degree is 1. Indeed, the sliding surface is given as follows:

$$\sigma = e \tag{23}$$

From Eq. (23), it can be noted that the time derivative of the sliding surface can be expressed as follows:

$$\dot{\sigma} = \dot{e} \tag{24}$$

Consequently, Eq. (24) becomes:

$$\dot{\sigma} = -\alpha * \frac{v_{dc}}{L_{sc}} - \frac{R_{sc}}{L_{sc}} i_{sc} + \frac{u_{sc}}{L_{sc}} - \frac{di_{ref}}{dt} \tag{25}$$

The condition of attractiveness necessary for the sliding surface to converge to zero in a finite time is:

$$V(\sigma) = \frac{1}{2}\sigma^2 \tag{26}$$

The time derivative of V is given as follows:

$$\dot{V} = \dot{\sigma}\sigma < 0 \tag{27}$$

In order to guarantee that condition in Eq. (27) is true, i.e. $\dot{V}$ is absolutely negative, $\dot{\sigma}$ is chosen according to the following speed arrival law, where K is a positive constant:

$$\dot{\sigma} = -Ksign(\sigma) \tag{28}$$

Therefore, using Eq. (21) and Eq. (28), the control law becomes:

$$\alpha = \frac{L_{sc}}{v_{dc}} \left(Ksign(\sigma) - \frac{R_{sc}}{L_{sc}} i_{sc} + \frac{u_{sc}}{L_{sc}} - \frac{di_{ref}}{dt} \right) \tag{29}$$

4 Numerical Simulations

In this section, the effectiveness of the proposed controller is investigated by simulations. The developed mathematical model has been implemented in MATLAB/SIMULINK software. The speed profile and tractive effort are illustrated in Fig. 4, the corresponding simulation parameters are presented in Table 2.

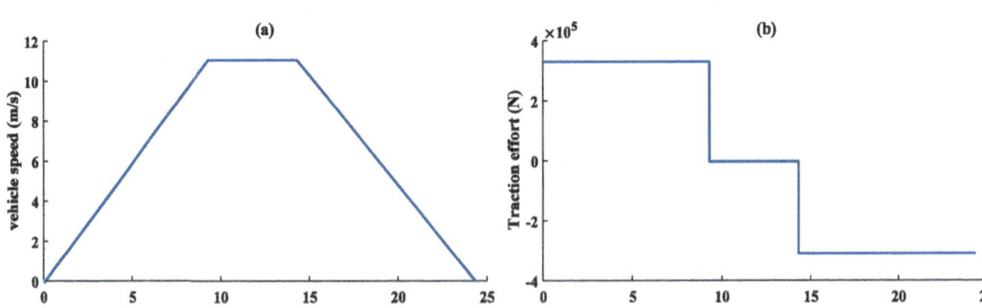

Fig. 4. (a) Speed profile, (b) Traction effort

Table 2. Parameters of simulation

Parameters	Value
Substation rated power	3500 kVA
Substation no load voltage	1500 V
Overhead line resistivity	27.6 mΩ/km
Overhead line section	680 mm^2
Mass of train in full load	277800 kg
Wheels diameter	0.84 m
Substation resistance	10 mΩ
Rated power of Induction motors	250 kW
Gearbox ratio	6
SC parallel cells	3
SC series cells	8
SC capacitance	26.3 F
SC resistance	48 mΩ

4.1 Reference Tracking Performance

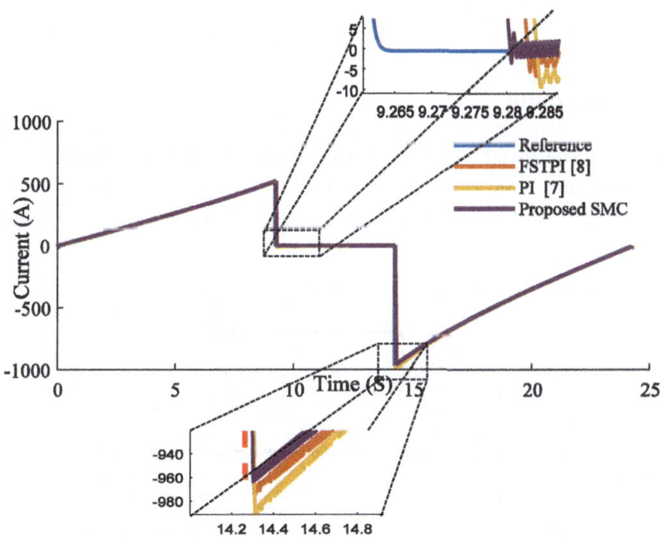

Fig. 5. SC currents

Figure 5 shows the simulation results of SC current. To evaluate the performance of the proposed nonlinear controller, results are compared with the PI and the FSTPI. Indeed, it is clearly seen that using the proposed controller the measured SC current tracks accurately it's reference. Furthermore, to ensure that the tracking performance comparison are fair, the ITAE is measured using ITAE block in Matlab/Simulink library, applying the proposed strategy the ITAE is reduced by 14.39% and 21.81% compared to FSTPI and PI, respectively.

Figure 6 displays the SC state of charge (SOC), which is related to its internal voltage. It can be clearly seen that the SC SOC is decreasing when the train is accelerating, which means that the ESS releases energy. When the train is braking the SOC is increasing, i.e. the ESS absorbs the available RBE. At the cruise phase no energy is supplied by the ESS.

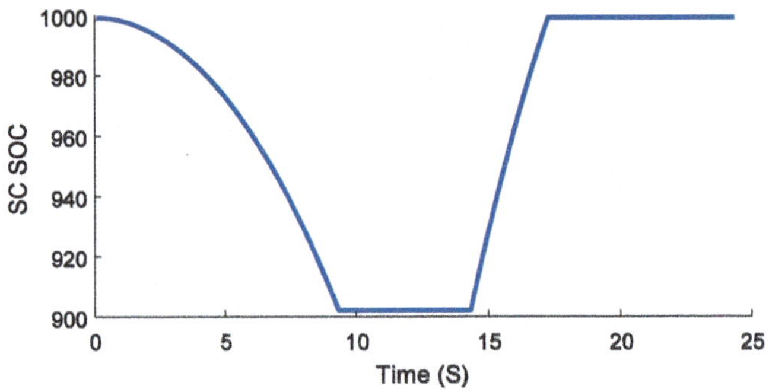

Fig. 6. SC state of charge (SOC)

4.2 Dynamic Performances

So as to assess the dynamic performance of the investigated controller, a step response simulation is performed. The obtained results are portrayed in Fig. 7, it can be clearly perceived that the settling time and rise time are improved using the proposed method.

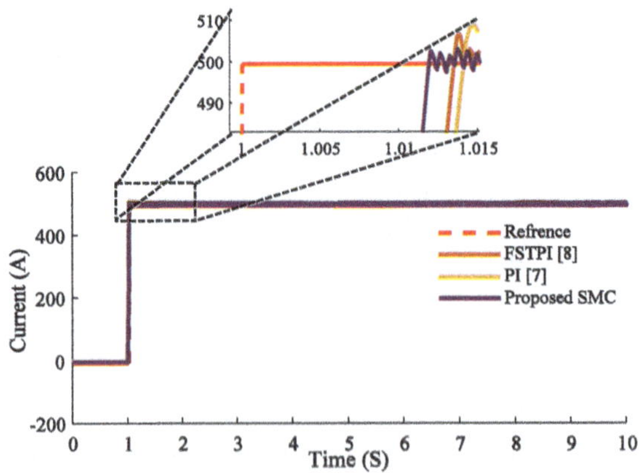

Fig. 7. Step response of SC currents

Also, the overshoot is reduced, the step response characteristics are extracted using the *stepinfo()* Matlab command, it returns a structure containing dynamic performance indicators. Table 3 summarizes results regarding rise time, settling time and overshoot of the three controllers.

Table 3. Rise time, settling time and overshoot

	SMC	FSTPI	PI
Rise time (s)	0.0094	0.0107	0.0112
Settling time (s) (5%)	1.0116	1.0132	11.78
Overshoot (A)	1.1346%	1.3998%	1.8999%

5 Conclusions

In this paper, a sliding mode controller for onboard energy storage system for railway braking energy recovery was developed. The efficiency of the proposed controller is validated in Matlab/Simulink environment. Consequently, significant improvements are recorded by the proposed controller versus the FSTPI and the PI controllers. Indeed, reduction of settling time, rise time and overshoot are realized. Furthermore, a high precision is achieved, it is illustrated by the ability of the system to track the desired value and the minimization of the ITAE. Therefore, it can be guaranteed that the SMC can effectively improve the EMS, and leads to increase energy saving in railway systems. The future work will shed the light on the significant improvement in terms of energy efficiency.

Acknowledgment. This work was supported by the Moroccan National Railways Office "ONCF" in the framework of the Railway Energy Efficiency Research project.

References

1. Novak, H., Lesic, V., Vasak, M.: Hierarchical model predictive control for coordinated electric railway traction system energy management. IEEE Trans. Intell. Transp. Syst. **20**, 2715–2727 (2019)
2. Xu, Y., Jia, B., Li, X., Li, M., Ghiasi, A.: An integrated micro-macro approach for high-speed railway energy-efficient timetabling problem. Transp. Res. Part C Emerg. Technol. **112**, 88–115 (2020)
3. Li, Z., Chen, L., Roberts, C., Zhao, N.: Dynamic trajectory optimization design for railway driver advisory system. IEEE Intell. Transp. Syst. Mag. **10**, 121–132 (2018)
4. Khodaparastan, M., Mohamed, A.A., Brandauer, W.: Recuperation of regenerative braking energy in electric rail transit systems. IEEE Trans. Intell. Transp. Syst. **20**, 2831–2847 (2019)

5. Roch-Dupré, D., Cucala, A.P., Pecharromán, R.R., López-López, Á.J., Fernández-Cardador, A.: Simulation-based assessment of the installation of a Reversible Substation in a railway line, including a realistic model of large traffic perturbations. Int. J. Electr. Power. Energy Syst. **115**, 105476 (2020)
6. Ceraolo, M., Lutzemberger, G., Meli, E., Pugi, L., Rindi, A., Pancari, G.: Energy storage systems to exploit regenerative braking in DC railway systems: Different approaches to improve efficiency of modern high-speed trains. J. Energy Storage **16**, 269–279 (2018)
7. Ciccarelli, F., Iannuzzi, D., Tricoli, P.: Control of metro-trains equipped with onboard supercapacitors for energy saving and reduction of power peak demand. Transp. Res. Part C: Emerg. Technol. **24**, 36–49 (2012)
8. Eziani, S., Ouassaid, M.: Adaptive fuzzy self-tuning PI controller of onboard energy storage system for railway braking energy recovery. In: 2019 International Symposium on Advanced Electrical and Communication Technologies, ISAECT 2019, pp. 1–6 (2019)
9. Khodaparastan, M., Dutta, O., Saleh, M., Mohamed, A.A.: Modeling and simulation of dc electric rail transit systems with wayside energy storage. IEEE Trans. Veh. Technol. **68**, 2218–2228 (2019)
10. El Fadil, H., Giri, F., Guerrero, J.M., Tahri, A.: Modeling and nonlinear control of a fuel cell/supercapacitor hybrid energy storage system for electric vehicles. IEEE Trans. Veh. Technol. **63**, 3011–3018 (2014)

Design of Inductor-Capacitor Circuits for Wireless Power Transfer for Biomedical Applications

Josephine Gloria Ling Ling Goh[1], Marwan Nafea[1(✉)], and Mohamed Sultan Mohamed Ali[2]

[1] University of Nottingham Malaysia, 43500 Semenyih, Selangor, Malaysia
marwan.nafea@nottingham.edu.my
[2] Universiti Teknologi Malaysia, 81310 UTM Johor Bahru, Johor, Malaysia

Abstract. This paper presents an analytical design approach for planar inductor-capacitor (LC) circuits for biomedical wireless power transfer (WPT) applications. This research makes use of the resonant inductive coupling between a transmitter and receiver coil in a series-parallel topology. The micro-electromechanical systems (MEMS)-based LC circuits are operated within a frequency range of 10–300 MHz. Several design cases are realized by varying the values of the number of turns, line width, and spacing width of the coil, while maintaining resonant frequency ranges circuits sizes that are compatible with biomedical applications and MEMS fabrication standards. In addition, the effects of such variations on the resonant frequency and quality factor are investigated. The findings of this paper present a simple approach to achieve different design requirements of planar LC circuits in WPT applications.

Keywords: MEMS · Quality factor · Resonant frequency · Wireless power transfer

1 Introduction

The concept of wireless power transfer (WPT) offers the ability to transmit electrical power wirelessly from a transmitter to a receiving device. Ever since its discovery in the 1890s by Nikola Tesla, WPT has transformed the way electricity is understood and used. Many industries have integrated this technology into their devices for convenient usage as well as to save on space by eliminating battery cells and wiring [1, 2]. There are two main categories in WPT, namely, radiative techniques (far-field) and non-radiative techniques (near field) [3]. Radiative techniques allow for long-range transmission of power. The power is usually propagated in the form of electromagnetic waves. These waves include lasers, microwaves, and radio waves. In radiative techniques, the power leaves the antenna disregarding the presence of a receiver or not. When transmitted, the radiation will leave the transmitter in all directions, and loss will occur, especially when the receiver is small. The radiation decays slowly compared to non-radiative techniques

J. A. Mat Jizat et al. (Eds.): iCITES 2020, AISC 1350, pp. 81–90, 2021.
https://doi.org/10.1007/978-3-030-70917-4_9

(to the squared of the distance), and prove to be very useful for long-distance applications [4].

Non-radiative techniques, also known as near-field, make use of magnetic field coupling to transfer power. This is usually used when the transmitter and the receiver are around one wavelength apart [5]. As this does not make use of radiation, there is no power transmitted when a receiver is absent. The power decreases at an exponential rate in relation to distance, thus being able to operate only in short distances as its name suggests. Near field occurs either through inductive coupling or capacitive coupling, which make use of magnetic fields and electric fields, respectively [6].

Capacitive coupling makes use of electric fields acting across the dielectric between capacitors, which are made up of two metal plates or two metal electrodes. Each electrode acts as a transmitter and receiver, where the transmitting electrode generates an alternating voltage, which in turn creates an alternating current flow due to the electrostatic induction [7]. As frequency increases, so do the capacitance of the plates as well as the power transmitted. This effect can be created either intentionally or can be an unwanted result. As long as the two conductors are close to each other, capacitive coupling can be achieved. When this effect is unwanted, it is known as parasitic capacitance [8]. Inductive coupling uses induction, which creates magnetic fields in a coil of wire. The transmitter coil is supplied with an alternating current (AC), and a magnetic field is generated around it. Placing another coil of wire into the magnetic field can generate (or rather, induce) current in it. This current can then be used directly or can be converted into direct current (DC) via a rectifier. Inductive coupling is generally preferred over capacitive coupling as it allows for higher power transmission in a higher density as compared to the latter [9]. This is called mutual inductance, which is related to the coupling coefficient that in turn measures the passing magnetic effect in between the coils.

As WPT is becoming more well known, more devices are starting to integrate this technology into their systems. One such field is the area of biomedical devices, which are widely used [10]. Extensive research is also being conducted for new innovations and improvements, and as of current, there are several technologies that integrate WPT as a mean of power transmission. One such instance is implantable biomedical devices, which are embedded in the body. Portable biomedical devices also benefit greatly from this technology, making the process of charging devices easier [11]. As biomedical devices are usually small (especially for those which are implantable), they make use of micro-electromechanical systems (MEMS) [12]. MEMS are small machines that consist of both electronic and mechanical components. They vary in size, ranging from a few millimeters to the minuscule micrometers. The purpose of MEMS is to control and sense stimuli, which can, in turn, be used to actuate effects on a bigger scale [13]. This proves to be very important in biomedical devices, as they are expected to trigger something with their small size. MEMS play a significant role in the development of the circuits as it senses changes in its surroundings and uses that stimuli as a signal to start the processes which follow. Biomedical implants have been an increasingly popular technology over the past few years, with different applications and uses across the medical field. These devices are typically attached to or placed inside the body, and they require a power source to function. Batteries have been the typical source of energy for these devices, as it is the most reliable. However, there are physical limitations both in the supply it can store and in size, which may prove to be a hindrance to biomedical implants [14]. Size plays an important role in the structure of an implant, and how small it can get would be especially limited by the presence of a battery. The size of the implants would increase with the installation of batteries inside, which may be a problem as this increases the risk of cell damage, inflammation, and even cell death [15]. Thus, the size of an implant must be taken into careful consideration when it is being manufactured. One possible way to reduce the size of the biomedical implants is to incorporate WPT as an electrical power source. This could prove to be useful in reducing the size of the implant while ensuring that it can still carry out its functions [16].

This study focuses on investigating the design process of inductor-capacitor (LC) circuits for wireless power transfer applications in wearable and implantable biomedical devices. Furthermore, the research analyzes the wireless performance of the system in these applications, making use of mathematical equations analyze the results. As the performance of the circuit is based largely on the receiver planar coil, the performance of the coil will be determined on varying certain factors and observing the output. Relationships and correlations between variables will be analyzed to look at the effects of changing one factor on another, and these can be used to change the design in such a way as to optimize its performance in its respective purpose.

2 Design and Working Principle

The circuit configuration used in this work is a series-parallel topology. This is because it allows for a better power transfer efficiency as compared to other topologies. Also, it can support greater loads at the wanted specific resonant frequency (f_r) and has a larger range of power transfer. Moreover, it can support any additional rectifiers if DC is desired as they work better in parallel. In order to determine the design of the WPT circuit, f_r (in Hertz) can be calculated as follows [17]:

$$f_r = \frac{1}{2\pi\sqrt{LC}} \tag{1}$$

where L and C are the inductance and capacitance of the coil (in Henry and Farad), respectively. This resonant frequency value should in such a way so that it is not too high to pose as a threat to the cells surrounding the implant [18], neither is it too low, thus being able to sufficiently transmit power as its large value also allows for a greater quality factor (Q) [19]. Using this, the desired L and C value can be determined using a suitable inductance and capacitance values for the circuit. John Archibald Wheeler presented many formulae to calculate planar inductance. A simplified formula has been made in order to achieve the same purpose:

$$L = \frac{K_1 \mu_0 n^2 d_{avg}}{1 + K_2 \rho} \tag{2}$$

where K_1 and K_2 are constants, ρ is the fill ratio and d_{avg} is the average diameter of the coil, which is varied based on the inner diameter (d_{in}) and outer diameter (d_{out}). The fill ratio, ρ_{fr}, depicts how spaced out the inductor coil is [16, 20]. The values of ρ_{fr} and d_{avg} can be calculated using the following equations [21]:

$$\rho_{fr} = \frac{(d_{out} - d_{in})}{(d_{out} + d_{in})} \tag{3}$$

$$d_{avg} = 0.5(d_{out} + d_{in}) \tag{4}$$

The values of the constants K_1 and K_2 are 2.34 and 2.75, respectively, according to the shape of the planar coil. The proposed design of the receiver coil is shown in Fig. 1. As this design makes use of a double-sided copper-clad polyimide flexible PCB, the square coil is connected to the capacitor plates in parallel via track connection on the top layer while using an under-bridge at the bottom layer. A detailed fabrication process of this circuit is presented in [22].

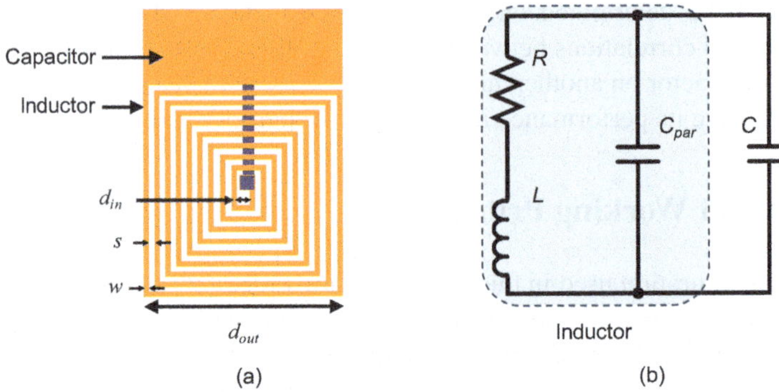

Fig. 1. (a) Sample layout of the LC circuit design (not to scale). (b) Equivalent circuit of the LC circuit.

A double-sided PCB was chosen as it allows for the capacitor plates to be etched into the same PCB as the receiver circuit coil, with one plate on top and the other at the bottom. The self-inductance, *L*, is fundamental in setting the value of the resonance of the coil. In order to further investigate *L*, the effect of track width spacing *s* and track width *w* on *L* and f_r are investigated. The Modified Wheeler's equation is used for the calculation of *L*. As the values of *s* and *w* change, *L* value changes as well. The change in resonant frequency can be observed through the variation of *L*, which may provide further insight on how to achieve the required resonance. Equation (5) is used to obtain the coil resonance:

$$R = \frac{\rho l}{A_{co}} \tag{5}$$

where ρ represents the material resistivity (16.78 nΩ·m for copper), while l is the length of the coil, and A_{co} is the cross-sectional area of the track [23]. The resistance of the coil affects the quality factor negatively and cause energy loss in a form of Joule heat [24].

This heat is desired in certain applications, where the LC circuit is used as a wireless heater [25].

At high frequency values, parasitic elements can affect the operation of the circuit when they come into play and are inevitable. Parasitic capacitance plays a major role in the calculation of the inductor and resonant frequency, as seen from Eq. (6). It is present within the circuit or the electronics component itself due to its proximity with one another, causing unwanted storage of charge within the circuit [26]. In planar coils, parasitic capacitance is mainly caused by the proximity of the spiral copper tracks from each other. The total capacitance, including the desired and parasitic capacitance, found in a circuit which uses capacitor plates can be calculated by [27]:

$$C = \frac{\varepsilon_0 \varepsilon_r A_{pl}}{d_c} + \varepsilon_0 (0.9 + 0.1 \times \varepsilon_r) \frac{d_C}{s} [4(d_{out} - wn)(n-1) - 4sn(n+1)] \quad (6)$$

where ε_0 is the vacuum permittivity (8.854×10^{-12}), A_{pl} is the area of parallel plates (8.4 mm^2), ε_r is the dielectric constant of the material (3.4), and d_c is the separation gap (50 μm). The first part of Eq. 6 represents the capacitance of the parallel plates (~5 pF), and the second part represents the parasitic capacitance of the coil, which varies based on the dimensions of the coil. This can then be implemented into the Eqs. (3) and (4) to find the values of d_{avg} and ρ_{fr}. Then, the value of Q can be found using the following relationship [28]:

$$Q = \frac{1}{R}\sqrt{\frac{L}{C}} \quad (7)$$

3 Results and Discussion

In this work, two different sets of LC circuits were investigated. In the first set, the value of n was varied in the range of 5–13 turns with an increment of 2 turns, while maintaining constant values of w, s, and d_{in} at 100, 150, and 800 μm, respectively. In the second set, to study the effect of the track width and spacing on the inductance, which consequently change the fill ratio, w was varied in the range of 50–175 μm with an increment step of 25 μm, and s was varied in the range of 200–75 μm with a decrement step of 25 μm to vary the fill ratio between 0.15 and 0.55. This was done while fixing the values of n and d_{out} to be 10 and 5.6 mm, respectively.

The effect of changing the values of the aforementioned variables on the resistance of the coil can be seen in Fig. 2 and Fig. 3. From Fig. 2, it can be observed that the increasing the operating frequency causes the value of the of the resistance of the coil to increase proportionally, especially for low n values $\leq$9. At higher n values, the relationship is between the frequency and resistance is almost linear at frequency values $\leq$150 MHz. Beyond that, the relationship exhibits a semi-exponential behavior. Furthermore, it can be seen that increasing the number of turns causes an increment in the resistance of the coil, which is caused by increasing the length of the coil (see Eq. 5). In Fig. 3, it can be noticed that increasing the line width of the coil causes the value of the resistance to drop, which is caused by increasing the cross-sectional area of the coil (see Eq. 5). A similar linear behavior to the one in Fig. 2 is observed at frequency values ≤150 MHz. Such behavior can be explained as a result of the Eddy currents and skin effect, which are more dominant at higher frequencies.

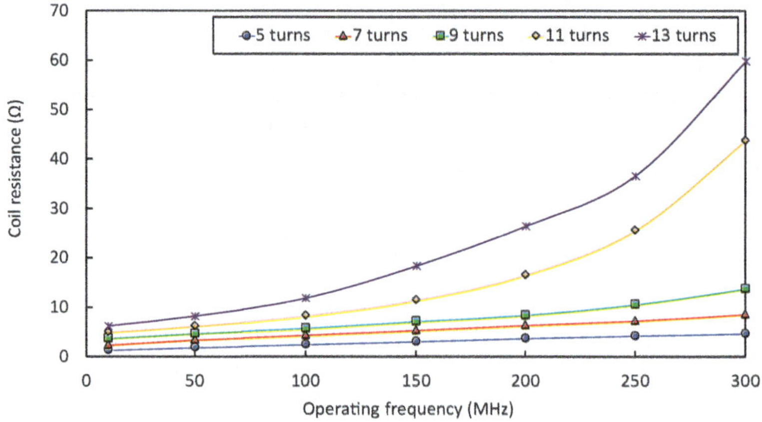

Fig. 2. Frequency-dependent resistance of different LC circuits with a fixed inner diameter of 800 μm and different number of turns.

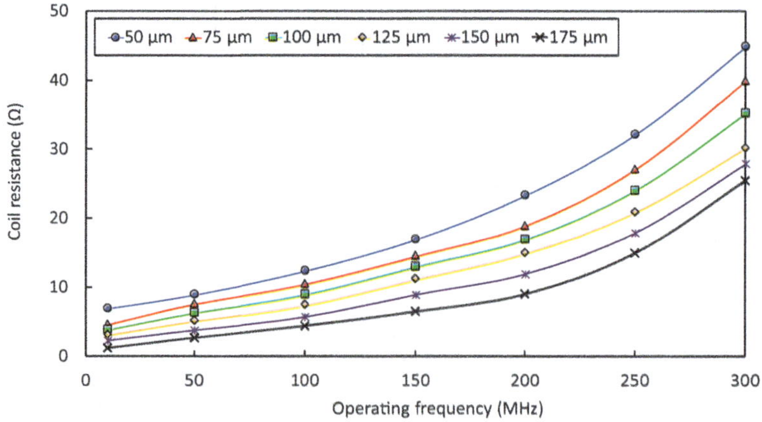

Fig. 3. Frequency-dependent resistance of different LC circuits with 10 turns, fixed outer diameter, and different line widths.

In Fig. 4, it can be seen that the number of turns has a negative effect on the resonant frequency and a positive effect on the quality factor. This can be explained as a result of increasing the inductance value when increasing the number of turns, which inversely proportional to the resonant frequency (see Eq. (1)) and directly proportional to the quality factor (see Eq. (7)), despite the fact that increasing the number of turns increases the resistance of the coil. However, the change in the resistance is less compared to the one in the inductance. On the other hand, in Fig. 5, it can be indicated that increasing the line width has a positive effect on the resonant frequency and quality factor. This is due to the fact that increasing the line width while reducing the spacing width leads to lower inductance values, since the inner turns of the coil are closer to the center of the spiral and thus contribute more negative and less positive mutual inductance.

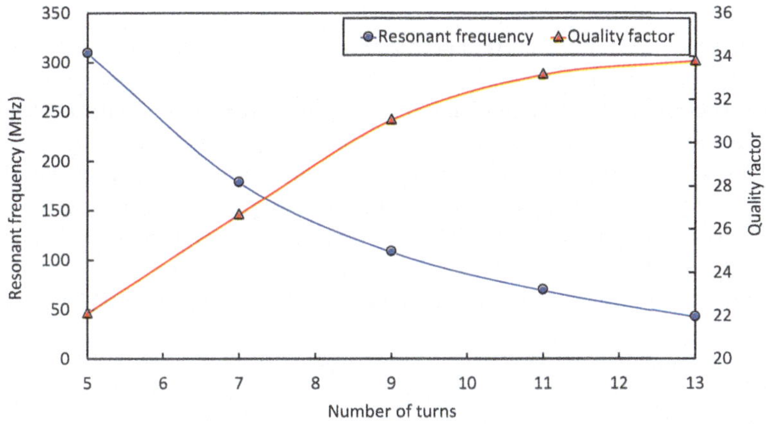

Fig. 4. Resonant frequency and quality factor values of different LC circuits with a fixed inner diameter of 800 μm and different number of turns.

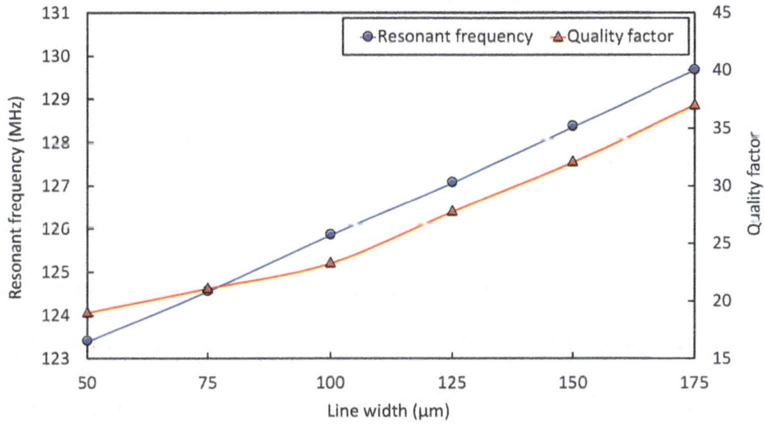

Fig. 5. Resonant frequency and quality factor values of different LC circuits with 10 turns, fixed outer diameter, and different line widths.

The aforementioned relationships and results agree well with previously reported finite element analysis and experimental results [18, 22]. Thus, this approach can be used to design WPT systems for various biomedical applications. In such applications, the size of the device is one of the important factors that should be taken into consideration. Thus, such designs involve a trade-off between different factors in order to achieve the desired performance. In other applications, it is required to use the receiver LC circuit as a hater based on Joule heating concept, which requires increasing the resistance of the coil. Such design tasks can be performed using optimization methods, such as particle swarm optimization, and finite element analysis [29, 30].

4 Conclusion

In this paper, an LC circuit is used in both primary and secondary circuits, consisting of both an inductor and capacitor which generates the transmission signal at the resonant frequency. The transmitter circuit makes use of an air-core solenoid with a tuning capacitor while the receiver circuit uses a square planar inductor coil with capacitor plates. To optimize the WPT system for power supply and Joule heating, the relationships between different factors and variables were investigated. For direct power supply, a high Q value in the receiver coil is required as this indicates that less power is lost during transmission. The results showed that the Q value of the circuit was affected by the inductance of the coil; as inductance increases, Q increases as well. This in turn was determined by the number of turns, track width, and the track width spacing. The total resistance of the circuit also accounts for the Q value, as Q reduces while total resistance increases. The total resistance is again affected by the number of turns, track width, and the track width spacing. On the other hand, Joule heaters require high resistance in order to convert the transmitted power to heat. The total resistance in the receiver circuit causes Joule heating to take place, and the more resistance there is the more power is converted to heat. Future work will involve utilizing optimization methods and finite element analysis to address further design requirements and achieve an optimal performance.

Acknowledgment. The authors acknowledge financial support from the University of Nottingham Malaysia.

References

1. Hojjati-Firoozabadi, A., Azarfar, A., Shahabadi, M.: Compact wireless power transfer system with microstrip-driven coupled dielectric resonators. AEU – Int. J. Electron. Commun. **127**, 153445 (2020)
2. Chee, P.S., Minjal, M.N., Leow, P.L., Ali, M.S.M.: Wireless powered thermo-pneumatic micropump using frequency-controlled heater. Sens. Actuat. A: Phys. **233**, 1–8 (2015)
3. Eteng, A.A., Rahim, S.K.A., Leow, C.Y., Jayaprakasam, S., Chew, B.W.: Low-power near-field magnetic wireless energy transfer links: a review of architectures and design approaches. Renew. Sustain. Energy Rev. **77**, 486–505 (2017)
4. Sun, L., Ma, D., Tang, H.: A review of recent trends in wireless power transfer technology and its applications in electric vehicle wireless charging. Renew. Sustain. Energy Rev. **91**, 490–503 (2018)
5. Chhawchharia, S., Sahoo, S.K., Balamurugan, M., Sukchai, S., Yanine, F.: Investigation of wireless power transfer applications with a focus on renewable energy. Renew. Sustain. Energy Rev. **91**, 888–902 (2018)
6. Chee, P.S., Nafea, M., Leow, P.L., Ali, M.S.M.: Thermal analysis of wirelessly powered thermo-pneumatic micropump based on planar LC circuit. J. Mech. Sci. Technol. **30**(6), 2659–2665 (2016)

7. Li, L., Wang, Z., Gao, F., Wang, S., Deng, J.: A family of compensation topologies for capacitive power transfer converters for wireless electric vehicle charger. Appl. Energy **260**, 114156 (2020)
8. Frechter, Y., Kuperman, A.: Analysis and design of inductive wireless power transfer link for feedback-less power delivery to enclosed compartment. Appl. Energy **278**, 115743 (2020)
9. Ustun, D., Balci, S., Sabanci, K.: A parametric simulation of the wireless power transfer with inductive coupling for electric vehicles, and modelling with artificial bee colony algorithm. Meas.: J. Int. Meas. Confed. **150**, 107082 (2020)
10. Iwasaki, W., Ishida, S., Kondo, D., Ito, Y., Tateno, J., Tomioka, M.: Monitoring of the core body temperature of cows using implantable wireless thermometers. Comput. Electron. Agric. **163**, 104849 (2019)
11. Zhang, B., Carlson, R.B., Smart, J.G., Dufek, E.J., Liaw, B.: Challenges of future high power wireless power transfer for light-duty electric vehicles—technology and risk management. eTransportation **2**, 100012 (2019)
12. Mohd Ghazali, F.A., Hasan, M.N., Rehman, T., Nafea, M., Mohamed Ali, M.S., Takahata, K.: MEMS actuators for biomedical applications: a review. J. Micromech. Microeng. **30**(7), 073001 (2020)
13. Rehman, T., Faudzi, A.A., Nafea, M., Mohamed Ali, M.S.: PDMS-based dual-channel pneumatic microactuator using sacrificial molding fabrication technique. In: 2019 20th International Conference on Solid-State Sensors, Actuators and Microsystems and Eurosensors XXXIII, TRANSDUCERS 2019 and EUROSENSORS XXXIII, pp. 1788–1791 (2019)
14. Truong, B.D., Andersen, E., Casados, C., Roundy, S.: Magnetoelectric wireless power transfer for biomedical implants: effects of non-uniform magnetic field, alignment and orientation. Sens. Actuat. A **316**, 112269 (2020)
15. Dinis, H., Mendes, P.M.: A comprehensive review of powering methods used in state-of-the-art miniaturized implantable electronic devices. Biosens. Bioelectron. **172**, 112781 (2020)
16. Nafea, M., Nawabjan, A., Mohamed Ali, M.S.: A wirelessly-controlled piezoelectric microvalve for regulated drug delivery. Sens. Actuat. A: Phys. **279**, 191–203 (2018)
17. Salem, M., Ramachandaramurthy, V.K., Jusoh, A., Padmanaban, S., Kamarol, M., Teh, J., Ishak, D.: Three-phase series resonant DC-DC boost converter with double LLC resonant tanks and variable frequency control. IEEE Access **8**, 22386–22399 (2020)
18. Nafea, M., Ahmad, N., Wahap, A.R., Ali, M.S.M.: Modeling and simulation of a wireless passive thermopneumatic micromixer. In: Communications in Computer and Information Science, pp. 312–322 (2017)
19. Barman, S.D., Reza, A.W., Kumar, N., Karim, M.E., Munir, A.B.: Wireless powering by magnetic resonant coupling: Recent trends in wireless power transfer system and its applications. Renew. Sustain. Energy Rev. **51**, 1525–1552 (2015)
20. Luo, Z., Wei, X.: Analysis of square and circular planar spiral coils in wireless power transfer system for electric vehicles. IEEE Trans. Industr. Electron. **65**(1), 331–341 (2018)
21. Duan, Z., Guo, Y.X., Kwong, D.L.: Rectangular coils optimization for wireless power transmission. Radio Sci. **47**(3), 1–10 (2012)
22. Nafea, M., Abuzaiter, A., Kazi, S., Mohamed Ali, M.S.: Frequency-controlled wireless passive thermopneumatic micromixer. J. Microelectromech. Syst. **26**(3), 691–703 (2017)
23. Raju, S., Wu, R., Chan, M., Yue, C.P.: Modeling of mutual coupling between planar inductors in wireless power applications. IEEE Trans. Power Electron. **29**(1), 481–490 (2014)
24. Sekiya, N., Monjugawa, Y.: A novel REBCO wire structure that improves coil quality factor in mhz range and its effect on wireless power transfer systems. IEEE Trans. Appl. Supercond. **27**(4), 1–5 (2017)
25. Nafea, M., Abuziater, A., Faris, O., Kazi, S., Ali, M.S.M.: Selective wireless control of a passive thermopneumatic micromixer. In: 2016 IEEE International Conference on Micro Electro Mechanical Systems (MEMS), pp. 792–795 (2016)

26. Sinha, S., Kumar, A., Regensburger, B., Afridi, K.K.: A new design approach to mitigating the effect of parasitics in capacitive wireless power transfer systems for electric vehicle charging. IEEE Trans. Transp. Electrif. **5**(4), 1040–1059 (2019)
27. Nafea, M., Baliah, J., Ali, MSM.: Modeling and simulation of a wirelessly-powered thermopneumatic micropump for drug delivery applications. Indon. J. Electr. Eng. Inf. **7**(2), 182–189 (2019).
28. Zhang, Y., Zhao, Z., Chen, K.: Frequency decrease analysis of resonant wireless power transfer. IEEE Trans. Power Electron. **29**(3), 1058–1063 (2014)
29. Rifai, S.A.M., Nafea, M., Debnath, SK., Bagchi, S.: Hybrid hysteresis-inversion and PSO-tuned PID control for piezoelectric micropositioning stages. In: 2020 IEEE Student Conference on Research and Development (SCOReD), pp. 206–210. IEEE (2020).
30. Kim, D., Abu-Siada, A., Sutinjo, A.: State-of-the-art literature review of WPT: Current limitations and solutions on IPT. Electr. Power Syst. Res. **154**, 493–502 (2018)

Perceived Risk and Benefits of Online Health Information Among Parents in Malaysia

Mira Kartiwi[1](✉), Teddy Surya Gunawan[2], and Jamalludin Ab Rahman[3]

[1] Department of Information Systems, Faculty of Information and Communication Technology, International Islamic University Malaysia, 53100 Kuala Lumpur, Malaysia
mira@iium.edu.my

[2] Department of Electrical and Computer Engineering, Faculty of Engineering, International Islamic University Malaysia, 53100 Kuala Lumpur, Malaysia

[3] Department of Community Medicine, Faculty of Medicine, International Islamic University Malaysia, 53100 Kuala Lumpur, Malaysia

Abstract. In the last decade, studies show that the use of the Internet to access health-related information is rapidly increasing, despite the questionable quality of health information provided on various online sources. Searching for health-related information is often aimed at self-healing or self-improvement of health quality and improving the health quality of other people in our care, such as our children or other family members. Despite the increased use of the Internet to access health information, there is limited research examining how perceived risk and self-efficacy may affect how one perceived the benefits they may obtain from searching health information online, especially among parents in developing countries such as Malaysia. The study shows the significant association between parents' perceived risk and perceived self-efficacy on the benefits of searching for health information online. The data was analyzed using Partial Least Square (PLS) technique. The study also found a positive association between perceived benefits of seeking health information online and their satisfaction with the information retrieved. The Goodness of Fit (GoF) of the model developed in this study is 0.809, which exceeds the cut-off value of 0.36 and shows that the model performs well.

Keywords: Parenting · Risk · Benefit · Self-efficacy · Satisfaction · Health Information · Online

1 Introduction

The Internet has improved patient involvement in health care decisions by allowing the patient to seek relevant information online. It includes searching and retrieving information on available treatment options, awareness of certain medications' side effects, and choosing the suitable physicians. As parents, the Internet has also become comprehensive information to improve children's upbringing and health conditions [1, 2].

Many previous studies on online health information-seeking behavior mainly focused on identifying the benefits and barriers to search for online health information [3]. Little attention has been given to factors that may influence how one perceived the

J. A. Mat Jizat et al. (Eds.): iCITES 2020, AISC 1350, pp. 91–97, 2021.
https://doi.org/10.1007/978-3-030-70917-4_10

benefits of retrieving online information to improve the wellbeing, especially among parents. Perceived self-efficacy and risk of searching online health information have also been identified as critical health information seeking issues. Therefore, this paper aims to examine how parents' perceived self-efficacy and perceived risk of online health information may influence the perceived benefits and, consequently, their decision to continue searching for health information online.

2 Literature Review

The Internet has shown its remarkable impacts on how information has been distributed and shared in a global context. It creates the explosion of information, and the rate of information growth over the Internet has been increasing drastically. However, many researchers agreed that posting and sharing any information on the Internet triggered some issues. Among others, it triggered the issue of the assessment of information accuracy and validity [4].

Amante et al. [5] and Hirvonen et al. [6] reported that health information is one of the most investigated topics online. Recent research also shows that parents, especially mothers of young children, preferred to seek health information online for their children [7, 8]. Among the drivers for seeking online information is the perception that their child health care provider did not provide sufficient detail to their queries. They obtained more information about the disease, social support, treatment, and other medical alternatives through the Internet with search engines [9].

The use of the Internet has eliminated boundaries between patients and health information; thus, the identifiable effect this has on patients is a sense of empowerment over their health and healthcare interactions [8]. Gage and Panagakis [8] also found that the Internet also offers the information that strengthens patients' abilities to question their physician's opinion, look for other treatment options, or even searching for second opinions online [10].

Previous research also reported that online health literacy helps citizens in developing economies have more control over their health and family members [11]. However, the extent of benefits often affected by education level, particularly on assessing the information quality [11]. The information-seeking behavior in developing countries reveals that citizens' information behavior is divided between those who are educated and those who are not educated. The information behavior of educated citizens is similar to the behavior of information users in developed countries. However, the citizens who are not educated perceived the information they obtained online can be especially crucial since information access is often limited.

Other frequently reported critical factors in information-seeking behavior are perceived self-efficacy and perceived risk. Perceived self-efficacy is one of the main concepts of social cognitive theory, and it refers to one's confidence in his/her ability to execute processes and produce results [12]. A previous study also found that positive efficacy expectations reduce anxiety, which is very important in the initial stages of information seeking [13].

On the other hand, perceived risk is conceptualized as the reason or cause for self-protect behavior [14, 15]. These studies reported that a common strategy of dealing with heightened risk levels is by seeking more information about the topic issues. Hence, individuals may consider by having sufficient information about the topic; they will be better able to manage risk and make the requisite behavioral changes.

3 Methodology

3.1 Sample

The study was conducted in the Klang Valley region, Malaysia. A random sampling method was used. The enumerators visited several schools and public places (such as shopping complexes) to get parents' responses. The data used in this study was collected from respondents using a structured questionnaire administered by the enumerators. The enumerators provide the respondents' options to fill the questionnaire personally or administered it as an interview approach to ensure quality and genuine responses. The respondents were also offered to fill the questionnaire or conduct the interview in the language that they are comfortable with (such as in Malay or English).

3.2 Instruments

The instrument used in this study began with a demographic section and then questions on the general use of the Internet and the list of the health-related websites that the parents frequently visited (if any). The instruments also collected a brief assessment of the parent's general health behaviors, knowledge, attitudes, and perceived risk of consuming health information online. The final survey section asked the parents about their satisfaction with the information obtained from the Internet. Besides, to ensure the quality of information collected, we used multiple-choice questions, for which participants could easily indicate their responses.

3.3 Data Analysis

In this study, the researchers adopted one method of structural equation modeling (SEM), namely partial least square (PLS) to analyze the models. Hair et al. [16] suggested that PLS is more suitable for a theory development study. Hence, due to the exploratory nature of this study, the researchers used the SmartPLS 3.0 to run the main model and test the proposed framework developed. Bootstrapping was used to improve the results generated due to the small sample sizes. The proposed framework in this study uses reflective constructs.

4 Results and Findings

A total of 519 responses were collected. Table 1 shows the descriptive statistics of the respondents' profile.

Previous studies conducted in developed countries have identified several main reasons why health information seekers are going online. Among the frequently reported drivers for going online are: to find the information on the side effect of prescribed medication, searching for supplementary information to understand the illness better, and verify the credentials of the healthcare providers [5, 11, 17, 18].

Figure 1 shows the extent and motivations for seeking health information online among parents in Malaysia. As can be seen, most of the parents searching for health information online were to understand the symptoms before meeting the doctor. In addition to that, seeking information on the medication's side effects is another reason parent search for information online to improve their children's health. Finding information online to understand better the specific illness is another primary motivation for parents to supplement the information they received from the medical practitioners.

Table 1. Summary of respondents' profile

Gender	Male	240	46.2%
	Female	279	53.8%
Marital status	Married	456	87.9%
	Divorced	63	12.1%
Education	Primary School or Informal Education	50	9.6%
	High school	132	25.4%
	Diploma	86	16.6%
	Bachelor	197	38.0%
	Master	54	10.4%
Employment status	Employed	358	69.0%
	Unemployed	78	15.0%
	Retired	6	1.2%
	Self Employed	77	14.8%

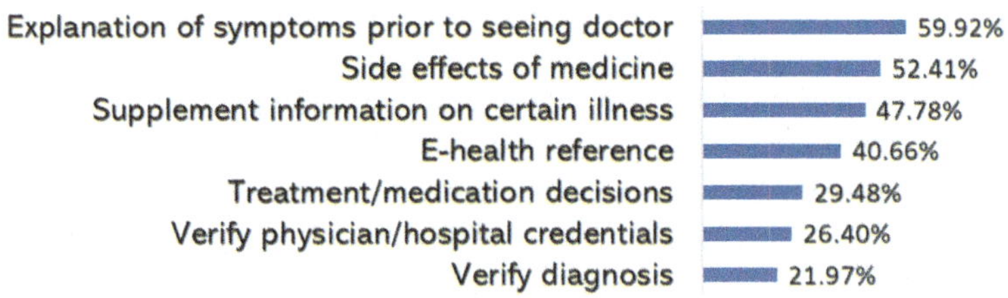

Fig. 1. The extent of health information seeking among parents

4.1 Results of PLS Path Modeling

To test the proposed framework developed in this study, the researcher adopted Partial Least Square (PLS) for both the measurement and structural models for data analysis. The proposed framework in this study uses reflective constructs.

Table 2 shows the matrix of path analysis results for the constructs included in the instruments, i.e., Perceived Self-Efficacy (PSE), Perceived Risk for Health (PRH), Perceived Benefits (PB), and Information Satisfaction (IS).

Table 2. Result of PLS Path Analysis

	Sample mean (M)	Standard deviation (STDEV)	T Statistics (|O/STDEV|)	P values
PB -> IS	0.482	0.057	8.406	0
PRH -> PB	0.331	0.049	6.766	0
PSE -> PB	0.476	0.048	9.871	0

As can be seen, Table 2 shows that all paths of the constructs included in the analysis generated acceptable effect sizes. As expected, perceived self-efficacy positively associated with the perceived benefits of online health information seeking (Path Coefficient = 0.482, p = 0.000). This finding supports a previous study conducted by Rains [13], who reported the critical role of efficacy in information seeking, particularly to reduce anxiety. The result also shows the significant association between perceived risk for health and perceived benefits (Path Coefficient = 0.332, p = 0.000). This finding suggested that parents who perceived they have sufficient information to manage the risk of having incorrect information are more likely to perceived more significant benefits of searching health information online.

The PLS assessment result also shows a significant association between perceived benefits and information satisfaction (Path Coefficient = 0.476, p = 0.000). As expected, parents who perceived more benefits from searching health information online are more likely to be satisfied with the quality of information retrieved from online sources.

Figure 2 shows the result of the PLS assessment of the proposed framework.

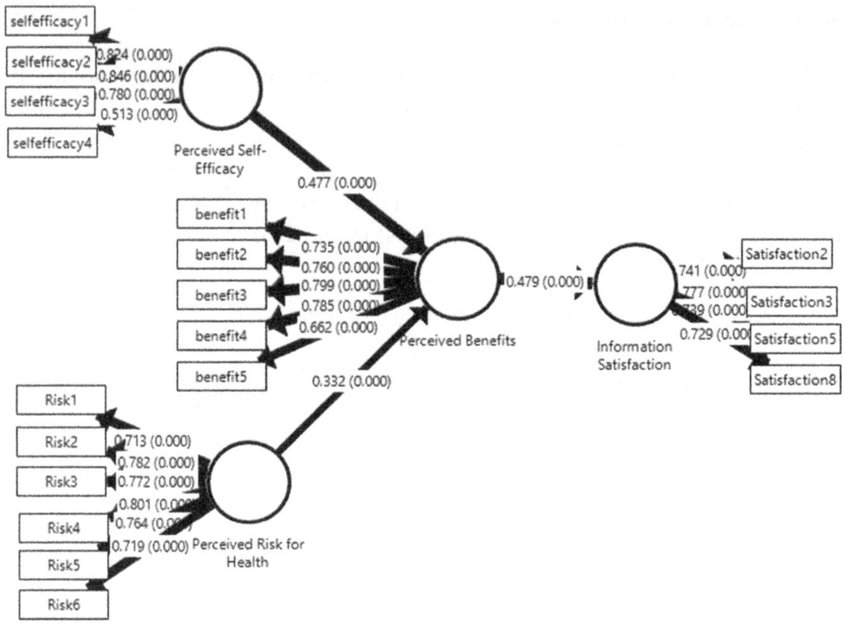

Fig. 2. Path analysis result on the proposed constructs

It is interesting to note that all paths are in the model's expectation-confirmation process, which is significant at the 0.05 level. To obtain the model's global validity, Tenenhaus et al. [19] suggested a Goodness of Fit (GoF) measure for the PLS path modeling. The calculated GoF value is 0.809, which way exceeds the cut-off point of 0.36. This result shows that the model performs well.

5 Conclusion

This research aims to assess the influence of parents' perceived self-efficacy and perceived risk to health, the perceived benefits of seeking health information online, and the satisfaction of the information retrieved. The findings in this research corroborate research which found the central role of self-efficacy in information seeking. The results of the model (with the Goodness of Fit (GoF) of 0.809) indicate that parents with adequate efficacy to search for information online and better control the risk of obtaining incorrect information for health, are more likely to perceive more benefits for searching online information. Consequently, this turns to greater satisfaction with the quality of health information they obtained online for their children.

Hence, this concludes that, in general, the strategy of heightening parents' risk perceptions should also be accompanied by efforts to enhance self-efficacy. It is when both factors are high that we see improvement in positive online health information-seeking behaviors.

Further research is needed to identify other critical factors that influence parents' motivation to seek health information online and their ability to assess the information they retrieved. Such understanding may help governments, public health experts and medical practitioners improve their service and formulate the best supports in helping the parents make an informed decision on improving the wellbeing of their children.

Acknowledgment. The researchers in this study would like to acknowledge the Malaysian Ministry of Higher Education (MOHE) for the financial funding of this research through the Fundamental Research Grant Scheme (FRGS) - FRGS17–012-0578.

References

1. Jaks, R., et al.: Parental digital health information seeking behavior in Switzerland: a cross-sectional study. BMC Public Health **19**(1), 1–11 (2019)
2. Sebelefsky, C., et al.: Internet health seeking behaviour of parents attending a general paediatric outpatient clinic: a cross-sectional observational study. J. Telemed. Telecare **21**(7), 400–407 (2015)
3. Ren, C., et al.: Health information in the digital age: an empirical study of the perceived benefits and costs of seeking and using health information from online sources. Health Inf. Libr. J. **36**(2), 153–167 (2019)
4. Manyika, J., Roxburgh, C.: The great transformer: the impact of the Internet on economic growth and prosperity. McKinsey Glob. Inst. **1**, 0360–8581 (2011)
5. Amante, D.J., et al.: Access to care and use of the Internet to search for health information: results from the US national health interview survey. J. Med. Internet Res. **17**(4), e106 (2015)
6. Hirvonen, N., et al.: Information behavior in stages of exercise behavior change. J. Am. Soc. Inform. Sci. Technol. **63**(9), 1804–1819 (2012)
7. Khoo, K., et al.: Health information seeking by parents in the Internet age. J. Paediatr. Child Health **44**(7–8), 419–423 (2008)
8. Gage, E.A., Panagakis, C.: The devil you know: parents seeking information online for paediatric cancer. Sociol. Health Illn. **34**(3), 444–458 (2012)
9. Harmsen, I.A., et al.: Parental information-seeking behaviour in childhood vaccinations. BMC Public Health **13**(1), 1219 (2013)
10. Zucco, R., et al.: Parents seeking health-related information on the Internet: cross-sectional study (2013)
11. Dutta, R.: Information needs and information-seeking behavior in developing countries: a review of the research. Int. Inf. Libr. Rev. **41**(1), 44–51 (2009)
12. Bandura, A.: Social cognitive theory of personality. Handb. Pers. **2**, 154–196 (1999)
13. Rains, S.A.: Seeking health information in the information age: the role of internet self-efficacy. West. J. Commun. **72**(1), 1–18 (2008)
14. Rimal, R.N.: Perceived risk and self-efficacy as motivators: understanding individuals' long-term use of health information. J. Commun. **51**(4), 633–654 (2001)
15. Rimal, R.N., Real, K.: Perceived risk and efficacy beliefs as motivators of change use of the risk perception attitude (RPA) framework to understand health behaviors (2003)
16. Hair, J.F., Ringle, C.M., Sarstedt, M.: PLS-SEM: indeed a silver bullet. J. Mark. Theory Pract. **19**(2), 139–152 (2011)
17. Li, N., et al.: Reasons for and predictors of patients' online health information seeking following a medical appointment. Fam. Pract. **31**(5), 550–556 (2014)
18. Bernhardt, J.M., Felter, E.M.: Online pediatric information seeking among mothers of young children: results from a qualitative study using focus groups. J. Med. Internet Res. **6**(1), e7 (2004)
19. Tenenhaus, M., et al.: PLS path modeling. Comput. Stat. Data Anal. **48**(1), 159–205 (2005)

Wearable Textile Antenna Using Thermal-Print of Clothes Iron for the Indoor Wireless Remote Monitoring

Kishen Pulanthran, Keerrthenan Yoorththeran, and Noorlindawaty Md. Jizat(✉)

Faculty of Engineering, Multimedia University, Persiaran Multimedia, 63100 Cyberjaya, Selangor, Malaysia
noorlindawaty.jizat@mmu.edu.my

Abstract. Wearable technology has grown tremendously throughout the time and is well known in transmitting or receiving the data for the remote monitoring in the telehealth applications. However, the existing wearable technology including the textile-antennas have limitation due to the high cost material, challenging fabrication, metal fatigue, porous and produce high loss. To address this issue, in this research, a low cost method using thermal-print of clothes iron is proposed. A low cost material of conductive carbon is used where no sintering or fabrication process is needed while provide the affordable fabrication cost. The textile antenna had return loss, S_{11} of −31.6 dB, the gain of 4.1 dB, the bandwidth of 6.9%, while the directivity was 3.812 dB at 5.8 GHz. The antenna prototype provides a convenient and unobtrusive communication subject to flexing and bending. In the prototype module testing, the transceiver antenna can transmit and receive the signal effectively up till 3 m' coverage. Moreover, after repeated flexing, the surface and RF performance of the textile antenna did not degrade, so it can be a suitable device in the future generations, especially in indoor wireless remote monitoring.

Keywords: Wearable · Thermal-print · Textile antenna · Clothes iron · Remote monitoring

1 Introduction

Remote monitoring surveillance requires that the monitoring does not impact the daily behaviors of the elderly as this arises as a solution to the complexities of ubiquitous monitoring of individuals in applications such as telehealth care communication. The combination of non-invasive wearable sensors and IT allows the elderly to obtain the assistance they need, while still living in their own homes, rather than in impersonal and costly nursing homes [1–5]. For these factors, with the low cost and low latency of signal transmission, the need for a communication solution that can be implemented has increased.

Textile-antennas in the wearable sensors have opened possibilities in integrating the wireless functions to the garment which can provide the comfort to the user. In a

J. A. Mat Jizat et al. (Eds.): iCITES 2020, AISC 1350, pp. 98–107, 2021.
https://doi.org/10.1007/978-3-030-70917-4_11

non-intrusive way, the latest clothing generation will be able to sense, communicate data and harvest energy [6–9], where the textile- antenna connects the fabric into the communication system [10–12]. The properties of the textile antenna must be small, lightweight, inexpensive, readily integrated and have good electrical and mechanical properties to serve the purpose. Some recent research of the textile antenna includes the embroidered, ShieldId textile, woven textile, metal-textile fabrics and conductive yarn-based fabric [13–17].

Although the textile antenna is thin and lightweight, the existing designs experience limitation in terms of high cost of fabrication, constantly changing the water molecules, difficult sintering, metal fatigue, porous and high loss. Textile fabrics are distinguished by good elasticity and durability, making them curvature- adaptive. The structure, however, is typically deformed after adjusting to the surface. In order to overcome these limitation, thermal-print method is proposed where no sintering process is involved and relatively high conductivity and uniform metallic element can be achieved without affecting the antenna performance.

The proposed work introduces the textile antenna which is designed using a low cost technique of thermal-print method using the clothes iron. The textile antenna has the return loss of −13.92 dB, the gain of 4.1 dB and the bandwidth of 6.9% at 5.8 GHz. The prototype antenna is tested as the transceiver module for both transmitter and receiver parts, where the antenna works effectively until 3 m which make it suitable for the indoor wireless monitoring applications. Moreover, the remarkable surface and the RF performance of the textile antenna do not degrade and after repeated flexing and are capable of maintaining robustness.

2 Textile Antenna Design

The textile antenna substrate is the cotton material with the dielectric permittivity, ε_r = 1.7 and the loss tangent of 0.04. The antenna design is printed using the conductive carbon ink from Ugent Tech Sdn. Bhd on the absorbent heat paper. The printed design is then transferred onto the cotton substrate using the clothes iron, where the wax is heated at 120 °C and transferred for 10 min.

2.1 Geometry and Antenna Design

The design structure of the textile antenna being proposed is illustrated in Fig. 1 in the first layer, which consists of the antenna patch and the CPW ground, while the cotton substrate is the second layer. The dimensions of the stepped impedance are tabulated in Table 1 respectively. Based on the mathematical model, the design is optimized according to the (1)–(2) where w represents the width, l represents the length, f_0 is the operating frequency and c is the speed of light in the vacuum with the antenna size of 0.12 λo × 0.14 λo.

$$w = \frac{c}{2f_0}\sqrt{\frac{\varepsilon_r + 1}{2}} \tag{1}$$

$$l = \frac{c}{2f_0\sqrt{\varepsilon_r}} - 2\Delta l \tag{2}$$

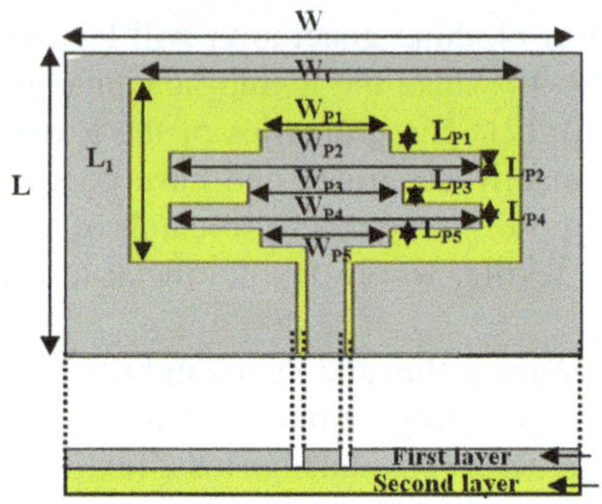

Fig. 1. Antenna design geometry.

Table 1. Dimensions of the textile antenna.

Parameter	Size (mm)	Parameter	Size (mm)
W	40	L	50
W_1	30	L_1	30
W_{P1}	10	L_{P1}	3.0
W_{P2}	24	L_{P2}	3.0
W_{P3}	12	L_{P3}	3.5
W_{P4}	24	L_{P4}	3.5
W_{P5}	10	L_{P5}	3.4
W_F	3.028	L_F	17.5
H1	1	H2	0.035

2.2 Thermal-Print Antenna Using Clothes Iron

The development of the textile antenna prototype includes the steps of the printing process and transferring the design from the absorbent heat to the cotton by the thermal-print process. Ink cartridge is modified to the print head of the DCP J100 office printer by replacing the conventional ink with the conductive carbon ink. The antenna design is printed on the ink absorbent heat transfer paper, where this design is printed for ten times and on the same coordinate to get the quality printing with an accurate dimension. This is then followed by the drying process to eliminate the excess solvents on the top part of the carbon conductive ink, by curing the samples.

A clothes iron with the 204 °C/200 °F is used to transfer the design from the ink absorbent heat transfer paper to the cotton substrate and subsequently the sample is pressed for 30 s. The detailed process flow of the textile antenna sample is illustrated in Fig. 2. The continuity is tested before the design is being transferred to the cotton substrate as shown in Fig. 3. After the bending and stretching test, the efficiency of the antenna is shown in Fig. 4, where, after repeated bending (more than 30 times) and after the stretching test, the remarkable surface and the RF output of the textile antenna will not degrade.

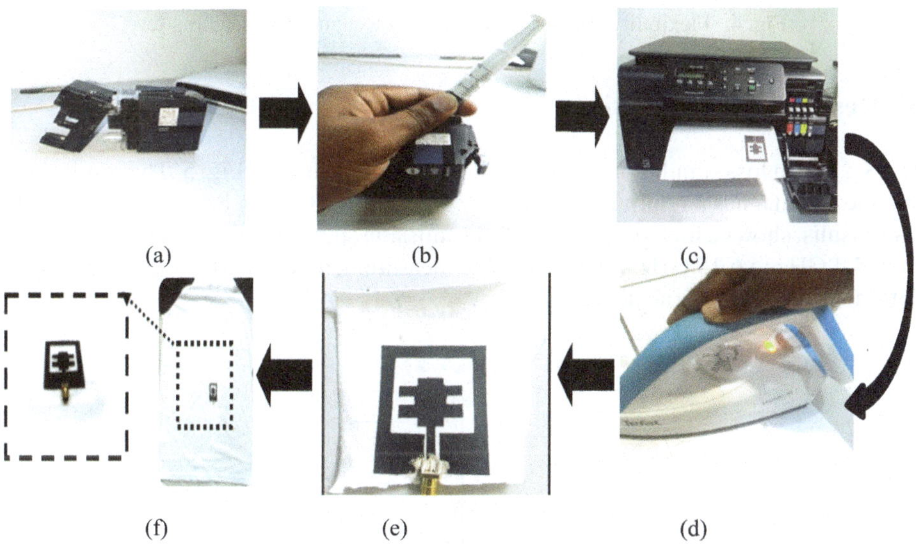

Fig. 2. Antenna development process flow: a) Ink cartridge modification b) Conductive ink transfer c) Printing on the heat paper d) Manual thermal using iron clothes e) Textile antenna sample with the SMA connector f) Final design of the textile antenna

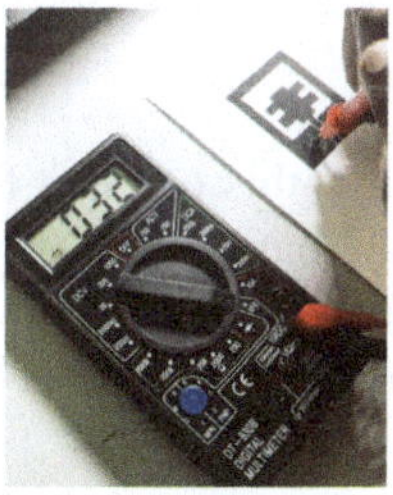

Fig. 3. Continuity test on the antenna.

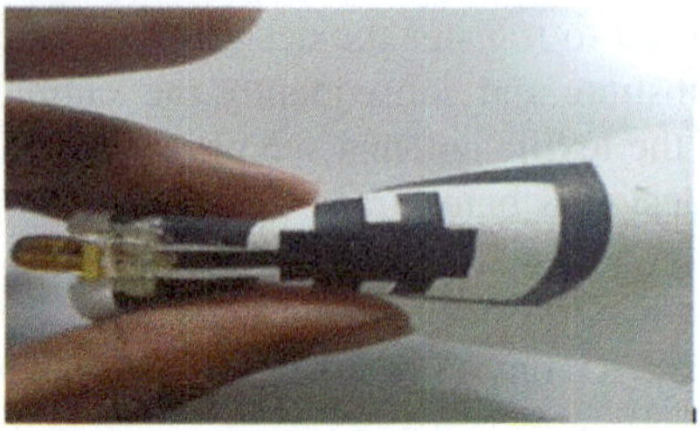

(a)

(b)

Fig. 4. Flexibility test due to (a) bending test and (b) stretching test

3 Results and Discussion

The S_{11} return loss simulation for the textile antenna is shown in Fig. 5. The result shows that the antenna has a return loss value of less than 10-dB return loss, $S_{11} = -31.6$ dB. This results, shows a high percentage of transmission signal and the bandwidth coverage from 5.4 GHz to 6.15 GHz at 5.8 GHz. The measurement results show that the range of 750 MHz bandwidth is achieved.

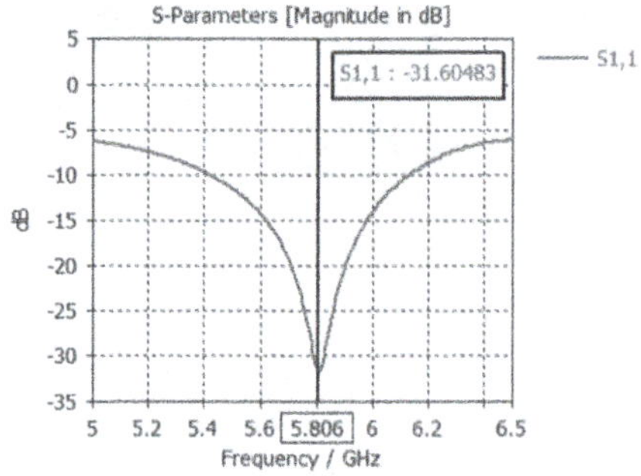

Fig. 5. The simulation return loss, S_{11} of the textile antenna

Bandwidth can be determined for this antenna from the values obtained in Fig. 6 and Eq. (3) are used to calculate the bandwidth, where BW represents Bandwidth, f_2 represents the maximum frequency at -10 dB, f_1 is the minimum frequency at -10 dB and f_0 is the resonant frequency. The textile antenna bandwidth is 13% with a 750 MHz range as used in Eq. (3).

$$BW = \frac{f_2 - f_1}{\sqrt{f_1 f_2}} \times 100\% \tag{3}$$

The simulated radiation pattern of the proposed antenna is shown in the two principal planes, E (y-z) and H (x-z) plane. The suggested characteristics of antenna radiation are investigated at, Phi = 90°. The x-z plane is the H-plane and the y-z plane is the E-plane, according to the coordinate scheme. Figure 6(a) shows the E-plane and H-plane radiation patterns of the proposed textile antenna, while the gain of 3.81dBi is shown in Fig. 6(b). The result shows that in the y-z plane HPBWs the 3-dB half power beam width (HPBW) is 48.9° and the x-z plane HPBW is 41.4°, respectively.

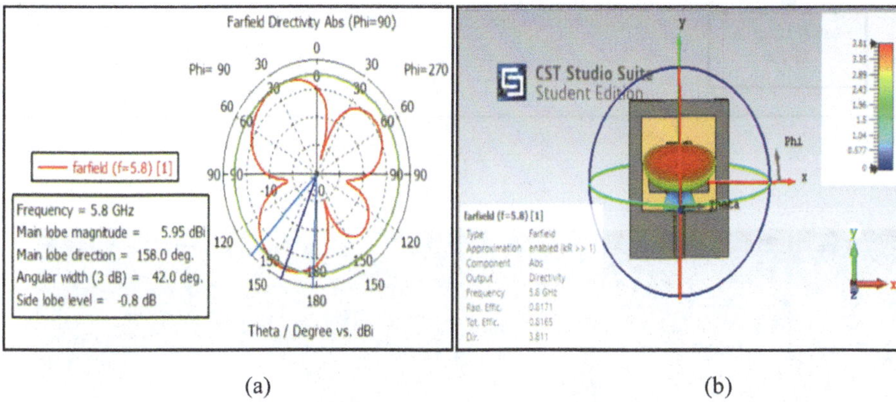

Fig. 6. Simulated (a) radiation pattern for E (red) and H (green) plane (b) gain in the scalar form.

4 Prototype Testing Using Transceiver Module

The designed textile antenna is tested using the AV transceiver module where the textile antenna is used as the transmitter in Fig. 7 while tested as the receiver in Fig. 8 test configuration. In the transmitter testing, the AV receiver is fixed to the television display and receives the AV signal from the textile antenna transmitter. While in the receiver testing, TS 5828 module of 5.8G model of transmitter is used to transmit the signal to the textile antenna. The module has 48 channels which range from 5.362 GHz to 5.945 GHz. A drone camera has been connected to the TS5828 to capture the video signal for both testing. The camera and the transmitter are powered using the Lipo 3S battery which has 11.1 V. RC832 5.8 G AV receivers is then used to receive the AV signal from the TS5828 transmitter. This receiver that has been connected to television is powered by the DC of 12 V.

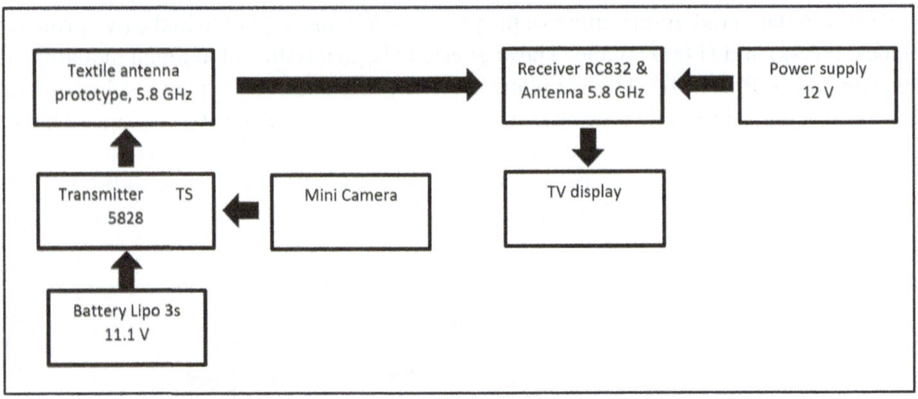

(a)

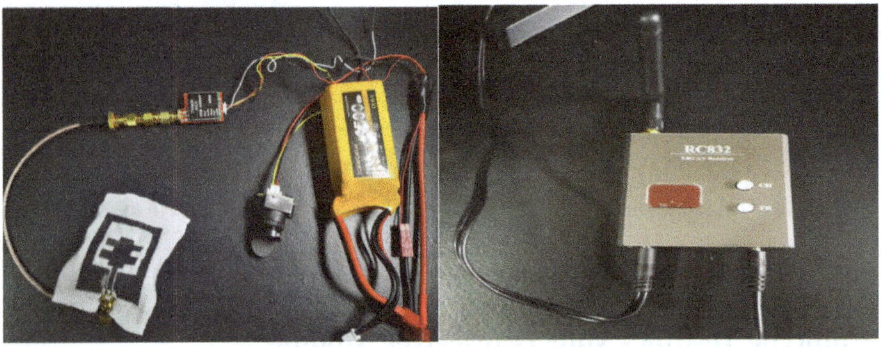

(b)

Fig. 7. Textile antenna transmitter module (a) test connection flow (b) prototype

From the AV transceiver testing, the textile antenna manages to transmits and receives the AV signal at the frequency of 5.8 GHz. In this testing modules, the prototype antenna is placed to the transmitter and receiver module respectively. As shown in Fig. 9(a), the textile antenna is used as a transmitter and is connected to the mini camera. This module is then tested up till 3 m from the receiver module, RC 832 at the TV display, while in Fig. 9(b), the textile antenna is connected to the RC 832 at the TV as the receiver. Both testing show that the textile antenna able to transmit and receive the AV signal by furthering it to the 3-m distance away from the AV modules without any signal lost. The operative and stable signal transmitted and received from the textile antenna prototype shows that this device can be used as the indoor wireless monitoring applications.

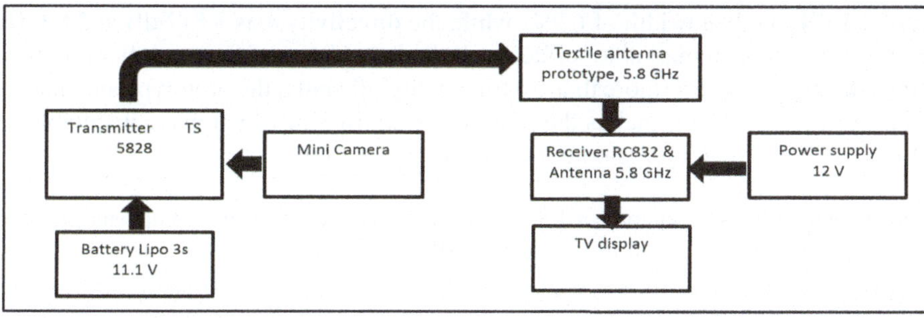

(a)

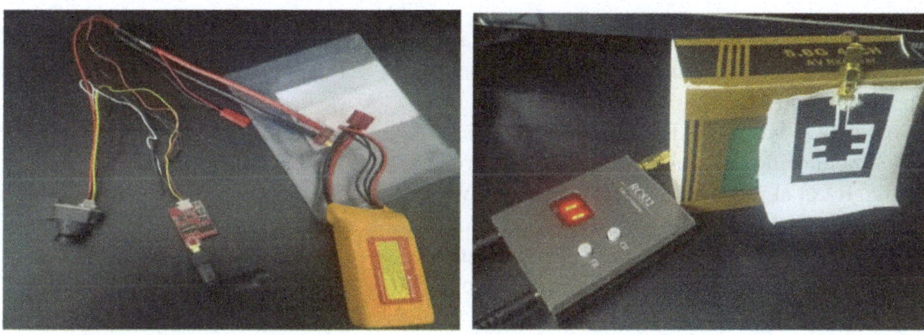

(b)

Fig. 8. Textile antenna receiver module (a) test connection flow (b) prototype

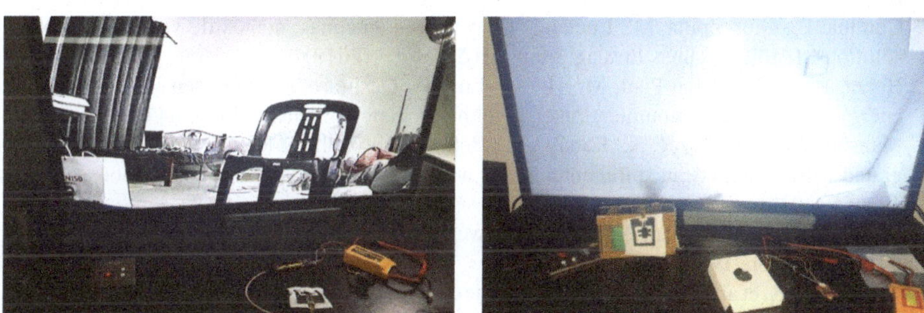

(a) (b)

Fig. 9. Visualization of the textile antenna as (a) transmitter and (b) receiver from the television image.

5 Conclusion

In this research, by using the thermal-transfer printing of clothes iron at 5.8 GHz and the conductive carbon ink as the metallic part, a low cost textile antenna has been designed on the cotton substrate. The textile antenna has the return loss, S_{11} of −31.6 dB, the gain of 4.1 dB, the bandwidth of 6.9%, while the directivity was 3.812 dB at 5.8 GHz. The prototype shows remarkable surface and consistent RF performance after repetitive flexing and stretching test (more than 30 times). Significantly, the prototype antenna can transmit and receive the signal as the transceiver up till 3 m which make it suitable for the low cost indoor wireless remote monitoring applications.

Acknowledgements. The authors wish to acknowledge the facilities and equipment given for testing purposes at the Faculty of Engineering, MMU Cyberjaya.

References

1. Kishen, P., et al.: A low-cost textile antenna using thermal-transfer printing. In: 16th IEEE International Colloquium on Signal Processing & Its Applications (CSPA), Langkawi, pp. 162–165 (2020)
2. Corchia, L., Monti, G., Benedetto, E., and Tarricone, L.:Wearable antennas for remote health care monitoring systems. Int. J. Antennas Propag. 1–11 (2017)
3. Fang, R., et al.: Compact and low-profile UWB antenna based on graphene-assembled films for wearable applications. Sensors **20**(9), 2552 (2020)
4. Basir, A., Yoo, H.: A stable impedance-matched ultrawideband antenna system mitigating detuning effects for multiple biotelemetric applications. IEEE Trans. Antennas Propag. **67**(5), 3416–3421 (2019)
5. Bouazizi, A., et al.: A dual-band case-printed planar Inverted-F antenna design with independent resonance control for wearable short range telemetric systems. Int. J. RF Microw. Comput.-Aided Eng. **29**(8), e21781 (2019)
6. Qiu, W., Hu, Z., Xiao, L.: Design of a wearable textile broadband RF ambient power harvester. Preprints. 2019090041, 1–8 (2019)
7. Corchia, L., Giuseppina, M., Luciano, T.: Wearable antennas: nontextile versus fully textile solutions. IEEE Antennas Propag. Mag. **61**(2), 71–83 (2019)
8. Borza, P.N., Machedon-Pisu, M., Hamza-Lup, F.: Design of wireless sensors for IoT with energy storage and communication channel heterogeneity. Sensors **19**, 3364 (2019)
9. Mansour, M., Le Polozec, X., Kanaya, H.: Enhanced broadband RF differential rectifier integrated with archimedean spiral antenna for wireless energy harvesting applications. Sensors **19**, 655 (2019)
10. Haerinia, M., Noghanian, S.: A printed wearable dual-band antenna for wireless power transfer. Sensors **19**, 1732 (2019)
11. Kennedy, T.F., et al.: Body-worn e-textile antennas: the good, the low-mass, and the conformal. IEEE Trans. Antennas Propag. **57**(4), 910–918 (2019)
12. Letavin, D.A., Knyazev, N.S., Malkin A.I.: Measurement of electrophysical properties of fabric materials and their use in the implementation of antennas. In: 2019 Ural Symposium on Biomedical Engineering, Radioelectronics and Information Technology (USBEREIT), pp. 436–438 (2019)
13. Liu, Y., Xu, L., Li, Y., Ye, T.: Textile based embroidery-friendly RFID antenna design techniques. In: Proceedings of the 2019 IEEE International Conference on RFID (RFID), Pisa, Italy, pp. 1–6 (2019).
14. Ibanez, I., Alomainy, A.: Dielectric characterization of non-conductive fabrics for temperature sensing through resonating antenna structures. Materials **13**, 1271 (2020)
15. Abbas, B., et al.: Full embroidery designed electro-textile wearable tag antenna for WBAN application. Sensors **19**(11), 2470 (2019)

16. Liu, Y., et al.: Textile based embroidery-friendly RFID antenna design techniques. In: 2019 IEEE International Conference on RFID, pp. 1–6 (2019)
17. Song, Y., Le Goff, D., Riondet, G., Mouthaan, K.: Polymer-based 4.2 GHz patch antenna. In: Proceedings of the 2020 International Workshop on Antenna Technology, Bucharest, Romania (2020)

Smart Calling Doorbell Using GSM Module

N. Y. N. Shahrom and Nor Azlinah Md Lazam(✉)

University Malaysia of Computer Science and Engineering, Cyberjaya, Malaysia
nisha.yasmin@gmail.com, norazlinah@unimy.edu.my

Abstract. Smart Calling Doorbell using GSM module help the owner of the house to identify the visitor through a phone call without having to go the door or main entrance. The main component used for this system is SIM900 GSM GPRS Shield and Arduino UNO R3. The GSM module provides a solution to the visitor and house owner communicate to each other by using the AT command. When the doorbell is pressed, it will generate a voice call from the GSM shield module to the house owner's phone number that has been registered and entered earlier through the Arduino serial monitor. Visitors can communicate with the house owner through the speaker and microphone attached to the doorbell panel. An LCD is also mounted to the doorbell panel to display the message send by owner to the visitor whenever the phone call made to the house owner is failed or unanswered. The LCD will also display any message send by the owner through SMS by sending the SMS first to the GSM module. In the situation where the owner is not at home, he/she can leave a message and the message will be displayed on the LCD. The visitor will be acknowledged on the owner's availability by reading the SMS on the LCD. Since most of the answering machine is placed at the door of the main entrance, this project is an ideal solution for those who have mobility problem or have difficulty either to reach the door or to the answering machine.

Keywords: GSM GPRS module · Smart calling doorbell · SMS · Visitor

1 Introduction

In the early days, a simple doorknocker was sufficient to acknowledge someone's arrival. Now, installing a smart doorbell is not just a necessity to notify the presence of someone but also due to safety purposes. The doorbell development and design started a long time ago by using the door knockers which are still found in medieval castles attached to the doors and they are progressing with time until the last decades. Safety issue has triggered the invention of smart automation the system of doorbell. In 1831, the first doorbell that could be rung at a distance via an electric wire was invented by Joseph Henry [1]. Since its invention, doorbells have played a significant part in ensuring the safety of the modern houses. Depending on the wall wiring specification, there are two types of doorbells which are wired doorbells and wireless doorbells [2, 3]. Wireless doorbells nowadays are gaining popularity as they feature an array of varied options and owners are relieved of the hassle of the complicated wiring.

J. A. Mat Jizat et al. (Eds.): iCITES 2020, AISC 1350, pp. 108–117, 2021.
https://doi.org/10.1007/978-3-030-70917-4_12

Usually, modern buildings are fitted with a wireless doorbell system that uses radio technology to draw signal from the doorbell and respond to the door remotely [4, 5]. Although the wireless doorbells are much more user-oriented than the wired ones, they do not always meet the needs of modern houses for the following three arguments.

First scenario, a doorbell with only sound effects. People with physical disabilities will have difficulties to reach the front door to find out who is coming to the house. This goes same to the house owner who does not have a doorbell answering machine. Second, a house owner who have a doorbell with answering machine. Those answering machines are usually fixed near the front door (main entrance). So, if the users want to answer to the doorbell, they must walk to the answering machine. Third scenario, the owner is not at home so there is no way to answer to the answering machine nor admitting the guests. The guest will need to wait for some time to find out if the owner is not at home. In view to these, a smart calling doorbell system for home use is proposed.

This project is the ideal solution for those who have a mobility problem and cannot answer either at the door or answering machine. These people can easily know the presence of the visitor by taking up a phone call or texting to the doorbell to give the response message which will be displayed to the visitors at the LCD screen of the doorbell.

This project aims to build a smart calling doorbell that can make a phone call to the house owner using a GSM module and to receive a message from the house owner and to be displayed on the LCD. A wireless doorbell is developed to allow two-way communication between the house owner and visitor through a speaker and microphone attached at the doorbell by using technology of GSM module. The system will make a voice call using SIM900 GSM GPRS Shield and Arduino UNO R3 board by using the AT Command and the Arduino serial communication. Through LCD, communication via SMS is made possible by using the AT Command and the Arduino serial communication.

Since the doorbell technology has been around for quite sometimes research studies and literature reviews on three existing products similar to this project development have been made to choose the most suitable components for Smart Calling Doorbell. Based on the comparison, a SIM900 GSM GPRS Module is chosen to make a phone call function in this project. The proposed product is mainly focused on the calling function where the existing products do not have. Unlike other existing products, the proposed product does not require Wi-Fi function. This is due to avoid the constant use of mobile data and internet. The proposed product avoids malfunction of system whenever there is no internet connection available.

Global System for Mobile Communication (GSM) and General Packet Radio Service (GPRS) are the basic mode of communication technology that support multiuser, multi-session and two-way network management. GSM GPRS Module is one of the most commonly used communication modules in embedded systems. The GSM GPRS module is used to allow communication between the microcontroller and the GSM GPRS network. GSM GPRS Module allows microcontrollers to communicate wirelessly with other devices and instruments. This wireless connectivity of the microcontroller opens up to a wide range of applications The studies on Quectel M95 GSM Module, IOT-GA6 GPRS GSM Module and SIM900 GSM GPRS Shield Module [6–10] lead to a decision on choosing GSM GPRS Arduino Shield based on SIM900 Quad Band Module for the proposed system because the module comes with an embedded microphone and speaker ports that will be used in this project. Plus, the module is compatible with Arduino and works on frequencies EGSM 900MHz and DCS 1800MHz. Therefore, it provides a way to communicate using the GSM cell phone network and allows to send SMS.

Although there are various types of microcontroller available such as 8051, PIC, AVR and ARM, the AVR microcontroller is preferred because it has more programming flexibility and more numbers of I/O ports. Microcontroller comparative study are made between Raspberry Pi and Arduino board. Arduino UNO R3 Board is selected for it cost, availability and can be programmed by Java, VB.NET and mainly C language. For wireless medium, Arduino sends data via bandwidth channel through carriers such as Wi-Fi, Wi-Max, GSM, GPS, GPRS, and Satellite. Therefore, in this project, the SIM900 GSM GPRS Arduino Shield is used for a wireless data transmission system for data acquisition.

Reviewed on similar existing products highlighted that the proposed system is mainly focused on the calling function where the existing products do not have. Unlike other existing products, the proposed product does not require Wi-Fi function. This is due to avoid the constant use of mobile data and internet. The proposed product avoids malfunction of system whenever there is no internet connection available.

2 System Overview

Three objectives of the project have been determined to cater the problem statements mentioned. The first objective is to develop a wireless doorbell that can allow two-way communication between the house owner and visitor through a speaker and microphone attached at the doorbell by using technology of GSM module. The second objective is to develop a wireless doorbell that can make a voice call from the SIM900 GSM GPRS Shield and Arduino UNO R3 board by using the AT Command and the Arduino serial communication. The third objective of this project is to develop a wireless doorbell that can allow communication via SMS through the LCD by using the AT Command and the Arduino serial communication. To meet the objective, the development of the system

Fig. 1. List of components

hardware is divided into four modules. Module A – configuration between Arduino UNO & breadboard, Module B – Installation of push button and LCD into Module A, Module C – Integration of Module B with the SIM 900 GSM Module and Module D – Installation of microphone and speaker into Module C. Fig. 1 shows the component list for the system.

Arduino is an open-source physical computing platform which implements the Processing Language on a simple input/output (I/O) board. This technology was initially introduced in Interaction Design Institute, Italy on 2005 [3, 7]. The Smart Calling Doorbell using GSM Module is an Arduino-based system. This system is controlled by Arduino Uno R3, a small microcontroller that turns out to be the control unit. AT commands are used to make the call function and send SMS. The command is needed to write into Arduino Serial and the command is sent to Software Serial of GSM Module. A single code is set into the push button through Arduino to trigger the call function of the GSM Module (Fig. 2).

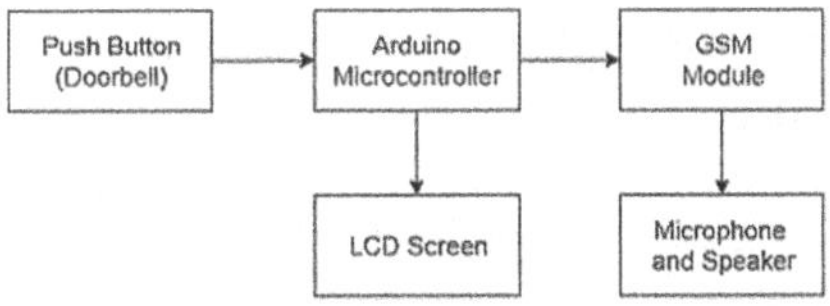

Fig. 2. Block diagram of smart calling doorbell using GSM module

The configuration between Arduino UNO and small breadboard is the main connection of this system where it is needed to extend the power voltage and grounding connection from the Arduino UNO R3. The execution of the project continues with the module integration between the push button and the Arduino board. After the connection is done, the module circuit B is connected to the GSM Module. This part is crucial since the main function of this project depends on it. The baud rate of these two boards is made sure the same. After the hardware part is done, the speaker and microphone are plugged in and that completed the system configuration.

Arduino serial monitor is used for sending AT commands and communicating with the SIM900 GSM Module. The first command sent to serial monitor is AT which is used to initialize auto-bauder. When the GSM replies with OK it means the handshaking process between Arduino and GSM Module is successful. AT + CSQ command is used to check the signal strength. The higher the dB, the better the connection. After all, the connection will depend on the antenna and the location. AT + CCID command is used to get the SIM card information and AT + CREG? command is used to check where the SIM Card is registered on the network (Fig. 3).

The SIM900 GSM module could not be powered using USB nor the 5V Arduino Vout because it cannot provide enough current to power up the SIM900 GSM module. One of the most important parts of getting the SIM900 shield working is supplying it with enough power.

The maximum current draw of the chip is around 2A during transmission burst and it may require around 216mA during phone calls or 80mA during network transmissions. Two different kind of power supplies have been tested to power up the GSM module.

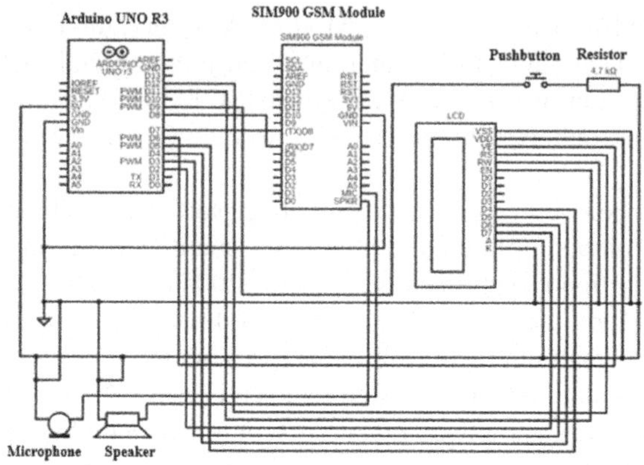

Fig. 3. Schematic diagram of smart calling doorbell using GSM module

The result whether the GSM Module is powered up or not can be observed through the LED Status Indicator [9, 10].

The Arduino hardware has built-in support for serial communication on pins 0 and 1. The native serial support happens via a piece of hardware that is built into the chip called a UART. The Software Serial library has been developed to allow serial communication on other digital pins of the Arduino, using software to replicate the functionality. The used of Arduino controller has been studied and used previously in the development of Child Safety Car Alert System Using Arduino and GSM Module [11].

The SIM900 GSM GPRS shield uses UART protocol to communicate with the Arduino. The chip supports baud rate from 1200 bps to 115200 bps with Auto-Baud detection. In this project, 9600 bps baud rate is used with both New-Line (NL) and Carriage-Return (CR) setting enabled in the Arduino serial monitor.

This system is started when the GSM Module is connected to the Arduino UNO R3 board. To make sure the connection of GSM Module is established, AT Command testing is conducted, or the LED Indicator can be observed Once the system is ready, "Press Doorbell" will be displayed on the LCD Screen (Fig. 4).

Once the pushbutton is pressed, the GSM will call the house owner's phone number that is registered priorly and stored in the Arduino Serial Monitor. Multiple button presses would not affect the system because once the push button status becomes HIGH, it will run the next function called in the code. On the other hand, the LCD will keep on displaying "Press Button" until the push button is pressed. Once the house owner answered the call, the LCD will keep on displaying "Calling" even though the phone call is missed. For the first case when house owner answered the call, microphone and speaker will be activated and it allows the communication between house owner and guest. There is no time limit set for the call. But, after the phone call ends, house owner is needed to send an SMS to the doorbell in order to reset the LCD screen to "Press Doorbell".

The second case of output is when the phone call is not answered, or the house owner is unable to answer the call. As soon as he/she received the missed call he/she must send an SMS to the GSM Module as a feedback message and reset back the doorbell. The

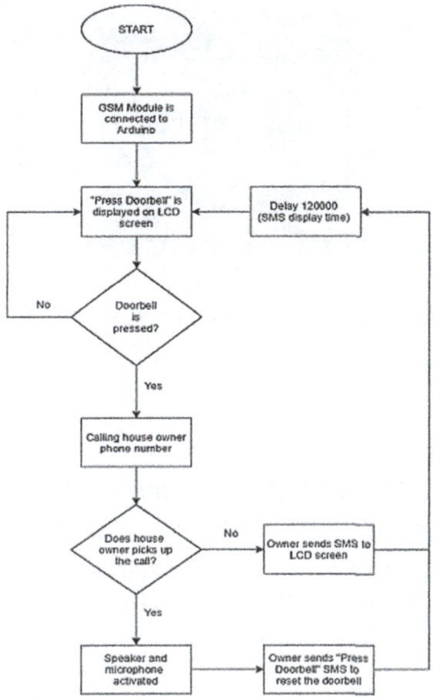

Fig. 4. Flowchart of smart calling doorbell

SMS will have a time limit of 2 min to be displayed on the LCD and after that time, the system will be reset to initial condition.

3 Final Product

Figure 5 shows the final prototype product of smart calling doorbell. The top surface features four main items; the mic, press button, LCD and speaker. The outer box is made of softboard and the dimensions of the box. Fig. 6 shows the actual wiring of the system.

Fig. 5. Front-view of smart calling doorbell prototype and 3D-view of smart calling doorbell prototype with measurements

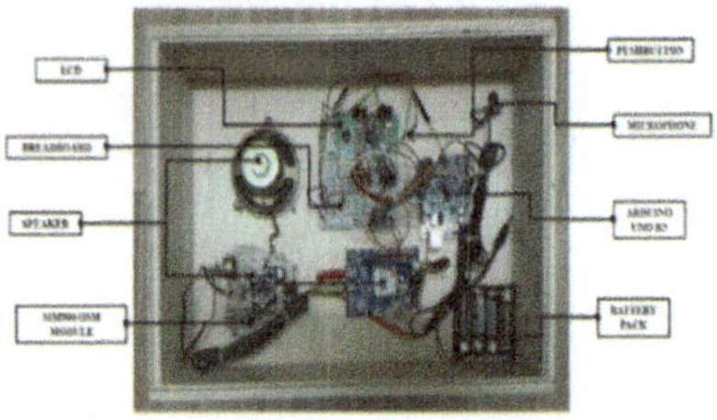

Fig. 6. Inside-view of smart calling doorbell prototype

4 Results

This system started when the GSM Module is connected to the Arduino UNO R3 board. To make sure the connection of GSM Module is established, AT Command testing is conducted, or the LED Indicator can be observed. Once the system is ready, "Press Doorbell" will be displayed on the LCD Screen (Figs. 7 and 8).

Fig. 7. Initial display of the system

Fig. 8. LCD output indicates calling function

Once the pushbutton is pressed, the GSM will call the owner's phone number that has been registered earlier and stored in the Arduino Serial Monitor. Once press the call is activated and the status becomes HIGH, it will run the next function called in the code. On the other hand, the LCD will keep on displaying "Press Button" until the push button is pressed.

For the first case scenario, when the owner answered the call, the LCD will keep on displaying "Calling". Microphone and speaker will be activated and allow communication between house owner and guest. There is no time limit set for the call. When the call ends, the owner then need to send an SMS to the doorbell in order to reset the LCD screen to initial status "Press Doorbell". The second case scenario is when the phone call is not answered, or the owner is unable to answer the call. As soon as he/she received the missed call he/she must send an SMS to the GSM Module as a feedback message and reset the doorbell to its initial status. The SMS will have a time limit of 2 min to be displayed on the LCD and after that time, the system will be reset automatically to its initial condition.

Figure 9 shows the situation of the owner received a call. The guest would be able to communicate with the owner through the microphone and speaker attached on the doorbell (Fig. 10).

Fig. 9. Owner received the phone call

Fig. 10. SMS is displayed on LCD

In the situation when the owner was not able to answer the call, the owner will send an SMS example "not at home' and the message will be displayed on the LCD screen. Unfortunately, the SMS has characters limit. The message displayed on the LCD must be equal to 16 characters or less. The characters after the 16th character will be truncated. Not just that, the LCD also has the time limit to display the SMS. House owner user may set the time up to five minutes or more by editing the code. But for testing purposes, the time limit has been set to 5 s and it has been proved that after 5 s, the LCD Screen will be reset to the main screen which is "Press Doorbell".

5 Discussion

5.1 Power Supply for GSM Module

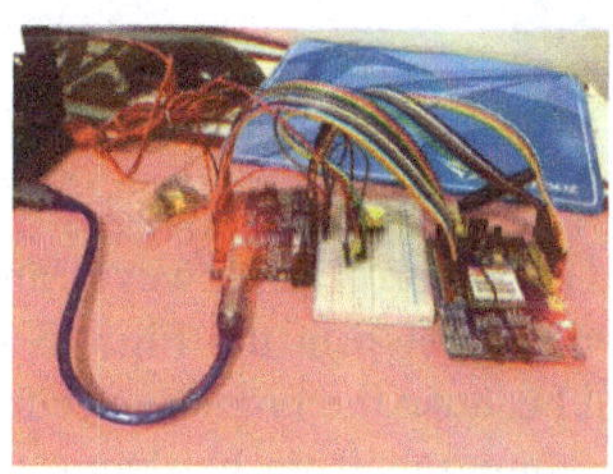

Fig. 11. Power supply for GSM module

Figure 11, the GSM module is being powered up with the 5V Arduino Vout. However, the status of the connection was failed. The failed connections are displayed by on the Arduino serial monitor by showing the garbage value. The SIM900 GSM module could not be powered using USB nor the 5V Arduino Vout because it cannot provide enough current to power up the SIM900 GSM module. Fig. 12 and 13 below shows the output obtained from the serial monitor when the GSM is connected to Arduino power.

5.2 SIM900 UART Connectivity

The Arduino hardware has built-in support for serial communication on pins 0 and 1. The native serial support happens via a piece of hardware that is built into the chip called a UART. The Software Serial library has been developed to allow serial communication

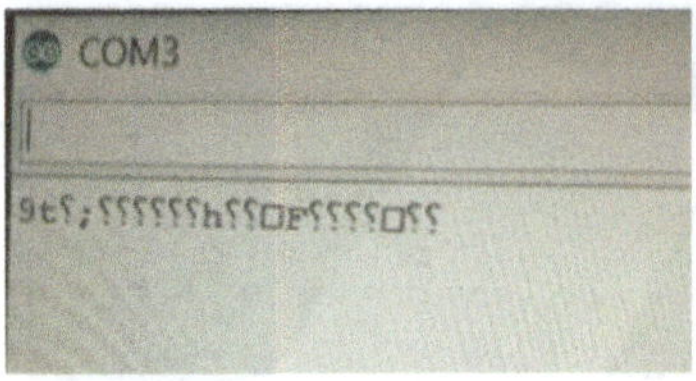

Fig. 12. Garbage value

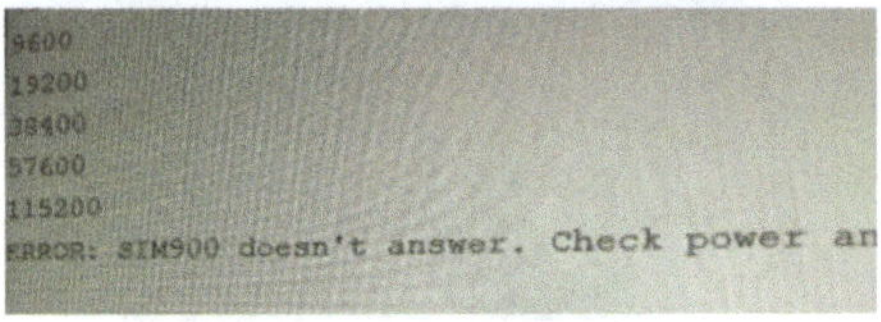

Fig. 13. Error message

on other digital pins of the Arduino, using software to replicate the functionality. The SIM900 GSM GPRS shield uses UART protocol to communicate with the Arduino. The chip supports baud rate from 1200bps to 115200bps with Autobaud detection. In this project, 9600bps baud rate is used with both New-Line (NL) and Carriage-Return (CR) setting enabled in the Arduino serial monitor.The jumpers, (RX, TX) of the shield can either be connected to Software Serial (D8, D7) or Hardware Serial (D1, D0) of the Arduino as shown in Fig. 14.

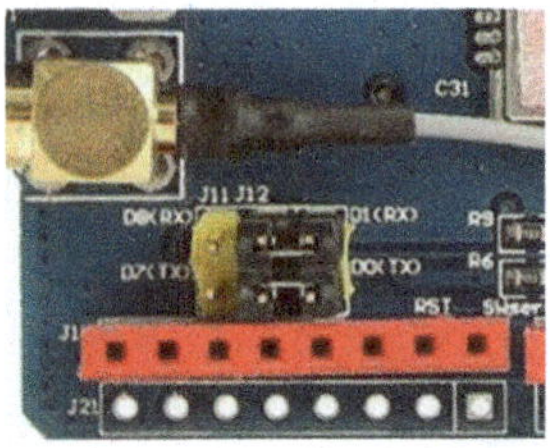

Fig. 14. UART selection jumper

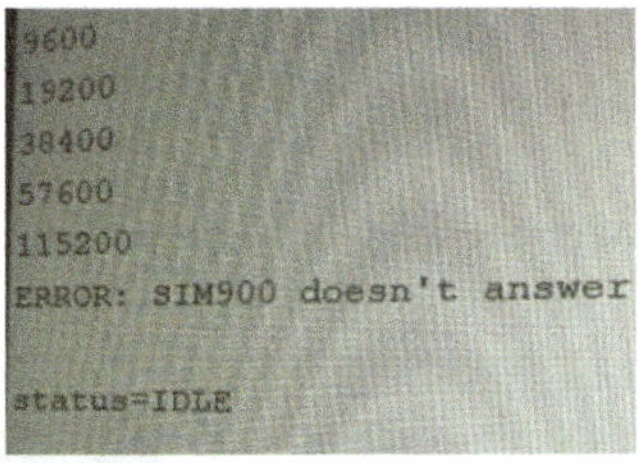

Fig. 15. IDLE status in Arduino monitor

Hardware serial is first used for the project. However, an error of IDLE status occurred in the Arduino Serial Monitor as shown in the Fig. 15 below. The connection using hardware serial is unsuccessful because the board is needed to connect through PL2303 USB to TTL converter. After changing from hardware serial to software serial using the jumpers, the SIM900 GSM module can be finally connected to Arduino and the connection is successful.

6 Conclusion

This project proposed a Smart Calling Doorbell using the GSM Module that help house owner to answer the doorbell anywhere and anytime. The main component used for this system is SIM900 GPRS GSM Shield. The GPRS Shield is based on the SIM900 module from SIMCOM and compatible with Arduino that provides a way to communicate using the AT command. This smart calling doorbell allows user to identify the visitor, through a mobile phone call without having to go to the door. When the pushbutton is pressed, it makes a voice call from the GSM shield module to the house owner's phone number that is stored in the system earlier. Visitors can communicate with the house owner through the speaker and microphone attached to the doorbell panel. An LCD is also attached to the doorbell to display messages and status. This project is an ideal solution for those who have mobility problem and cannot answer either at the door or answering machine. They can easily know the presence of the visitor by taking up the phone call or response through text messages which will be displayed at the LCD screen of the doorbell.

References

1. Aguilar-Gonzalez, R.: The tolling of bells: a traditional communications system in our time. IEEE Consum. Electron. Mag. **8**(4), 67–69 (2019)
2. Quadros, B. Kadam, R., Saxena, K., Shen, W., Kobsa, A.: Dashbell: a low-cost smart doorbell system for home use. DBLP computer science bibliography. CoRR abs/1706.09269 (2017)
3. Muktar, A., Ahmed, A., Salad, A.: A framework for ultrasonic doorbell system with object detection. Electron. Telecommun. Eng. Somali J. Eng. Sci. **2**(1), 1–6 (2017)
4. Tanpure, P., Sandbhor, R., Sarjine, A., Hande, R.: Wireless doorbell. Int. J. Adv. Eng. Res. Dev. (IJAERD) Technophilia-2018, **5**(04) (2018)
5. Ferdous, M.J., Datta, T., Farhana, U., Akanda, O.: Design and development of SMS - based prepaid electricity billing system using GSM and Arduino. Int. J. Comput. Eng. Inf. Technol. **11**(6), 112–118 (2019)
6. Zaghloul, M.S.: Modern architecture tracking system using modern GPS module. Int. J. Sci. Eng. Res. **5**(2), 127–131 (2014)
7. Zaghloul, M.S.: GSM-GPRS Arduino Shield (GS-001) with SIM 900 chip module in wireless data transmission system for data acquisition and control of power induction furnace. Int. J Sci. Eng. Res. **5**(4), 776–780 (2014)
8. Kenneth, D.T., Hintz, J.: Microcontrollers: Architecture, Implementation, and Programming. McGrawHill Professional, Book News, Inc., Portland (1992)
9. Jain, S., Vaibhav, A., Goyal, L.: Raspberry pi based interactive home automation system through E-mail LED (switch). In: International Conference on Reliability, Optimization and Information Technology, India (2014)
10. El Fiorenza, C., Bompelli, S., Sampath, M., Ghogre, M.: Iot smart doorbell surveillance. Int. J. Adv. Res. Innov. Ideas Educ. **2**, 2701–2706 (2018)
11. Hazizan, A., Lazam, N.A.M., Hassan, N.I.: Development of child safety car alert system using arduino and GSM module. In: IOP Conference Series: Materials Science and Engineering, vol. 834, no. 1, p. 012071 (2019)

Development of Smart Home Door Lock System

Hazeem Ahmad Taslim, Nor Azlinah Md Lazam, and Nor Akmar Mohd Yahya(✉)

University Malaysia of Computer Science and Engineering (UNIMY), Cyberjaya, Selangor, Malaysia
akmar.yahya@unimy.edu.my

Abstract. Home automation system is a technological solution. It is the action of applying technology to solve a problem at home. It uses hardware and software technologies that enables control and management over appliances and devices within a home. In this paper, we have developed a low-cost automated door lock system so users can operate systems from anywhere all over the world with internet access. IoT concept is applied in the system development including integration of software and hardware for communication and execution tasks. By using a WIFI enabled microcontroller and connecting devices and user's apps via MQTT protocol, a smart door lock using IoT application has successfully built. The door lock is able to lock and unlock remotely. The users do not need to be on the same network to execute the operation and instruction is conducted via voice command or manually operated on dashboard. The users also will receive feedback as verification for the status of the door when execution take place.

Keywords: IoT · Automation · Microcontroller · MQTT

1 Introduction

Home Automation has been around for a while, but it is viewed as something luxurious due to the high cost and scale of the technology. It remains as a product for high end consumers today. Appliances today can be controlled via an application or from websites. The main objective of home automation is to ease day to day activities by enabling them to control home appliances with their mobile devices. When a user is away from home, they can change the status of appliances like gates and doors from their mobile device. IOT enables this in which the user can manipulate devices in on a secured network. The user remains in control and does not have to rely on another party (neighbors or maid) to unlock doors. This paper will discuss the approach on controlling using IOT. Imagine how useful it is not to have your neighbors collect your parcels and how business owners can remotely unlock their doors for customers. This is made possible today with IOT and MQTT protocol. In the coming years, the IoT is expected to bridge diverse technologies to enable new applications by connecting physical objects together in support of intelligent decision making [1, 3].

According to Lampkin, MQTT is lightweight enough to be supported by the smallest devices, yet robust enough that all important messages get to their destinations every time [2]. With MQTT, devices such as satellite receivers, smart energy meters, sensors

J. A. Mat Jizat et al. (Eds.): iCITES 2020, AISC 1350, pp. 118–126, 2021.
https://doi.org/10.1007/978-3-030-70917-4_13

in automotive vehicles, and personal health care devices can communicate with each other and with other systems or applications. Yokotani states that MQTT provides three transfer modes based on the required reliability: QoS0 (Non assured transmission), QoS1 (Assured transmission), and QoS2 (Assured service on applications). QoS1 is similar to HTTP from a reliability point of view [3]. While HTTP is a symmetric protocol, MQTT has an asymmetric architecture so it is lightweight. Since, in most of the communication for IoT, non-intelligent distributed devices communicate with a server with intelligent ability, asymmetric communication is provided. Because of this point, MQTT is more suitable compared to HTTP. Figure 1 shows MQTT Protocol flow. MQTT consists of two message sets on a connection, "Publish" and "Subscribe". Data blocks are sent by Publish message and are received by Subscribe message. These data blocks are identified by "topic". Receiving data blocks are identified by topics registered by Subscribe message, in advance [4].

There are several ways that can be used which uses different protocols and connectivity technology such as WIFI, Bluetooth and using live analytics with a camera module. There are commercially available Smart Door Solutions however it remains very pricey. Ismail et al. has developed home door lock system using Arduino Uno and Bluetooth connection. The system only can operate when the user is within Bluetooth range distance otherwise, the system is disconnected. This has limited the usage as the user must be nearer to the device to perform the action [5]. There are other systems that use Raspberry Pi that act as microcontroller to perform door lock mechanism as reported in [6, 7] as IoT technology is utilized to execute remotely. This project aims to produce an Arduino based door lock with a notification system to alert the home administrator which can be controlled via a dashboard and voice commands [8, 9].

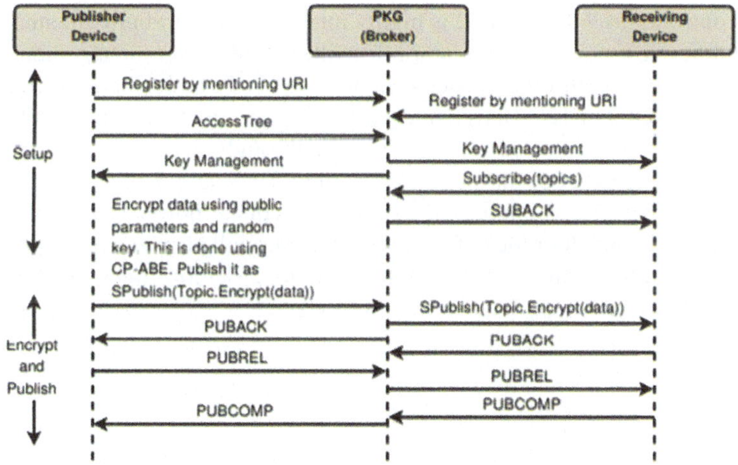

Fig. 1. MQTT protocol flow

By linking a Google account to the system. The door lock can be manipulated to lock and unlock with specific customizable commands. The system will reply commands in real time feedback, users are prompted with a notification of assurance when a command is executed. At the end of each successful executed command, the system is designed to send out a notification into a Telegram channel to keep as a recording to the users when the status of the door lock changes.

2 System Development

2.1 Development Process

To develop this system prototype, Rapid Application Development (RAD) method was selected. This development method has become one of the commonly used and powerful development methods that falls under agile development techniques. It emphasizes heavily on rapid prototyping and iterative delivery as a method of software development. It is suitable for the components and functions that developed simultaneously alongside the development of the whole project. Throughout the process, no detail preplanning required therefore it can be altered or changed accordingly from the feedback provided by the users [10]. RAD Model has changed over the years, but the 4 basic steps still provide the same continuity. The steps are as follows: Defining Requirements, Prototyping, Testing and Finalize/Deploy.

2.2 System Overview

Arduino IDE 1.8.10 [11] is an open-source programming software for writing code called for the integrated Arduino circuit board such as Arduino Uno R3 [12]. For this work, the software will be used to create the program for the door lock that enable the circuit board to communicate with the actuator and the broker. Arduino IDE will be used to configure the microcontroller to WIFI. ESP32 is highly integrated with in-built antenna switches, power amplifier, low noise receives amplifier, filters, and power management modules. ESP32 can interface with other systems to provide Wi-Fi and Bluetooth functionality through its SPI/SDIO or I2C/UART interfaces. The door lock used is a 12 V Solenoid Door Lock which is connected to 12 V adapter through an ACDC Switching Relay. As the operating voltage out on the ESP32 is only 3.3 V and 5 V, an external AC DC 12 V Switching Adapter is connected to supply adequate power to pull and push the slug inside the solenoid door lock. The supplied power will also be used to power up the switching relay in the Arduino-based system. Figure 2 demonstrates the configuration of assemble hardware components of the prototype.

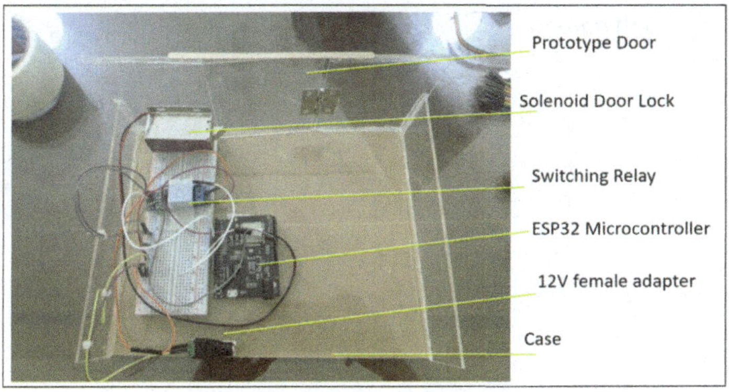

Fig. 2. Assembly hardware components

Upon executing the task, it will trigger the microcontroller to manipulate the status of the door lock. This is done from the user's dashboard. As the door locks, a high (1) is sent to relay switch to pull the slug in. When it receives no signal or low (0) the slug pushes out and locks the door. The microcontroller then sends a signal back to Telegram Messenger as a backlog [9, 13]. Figure 3 illustrates the software and hardware integration of the developed system.

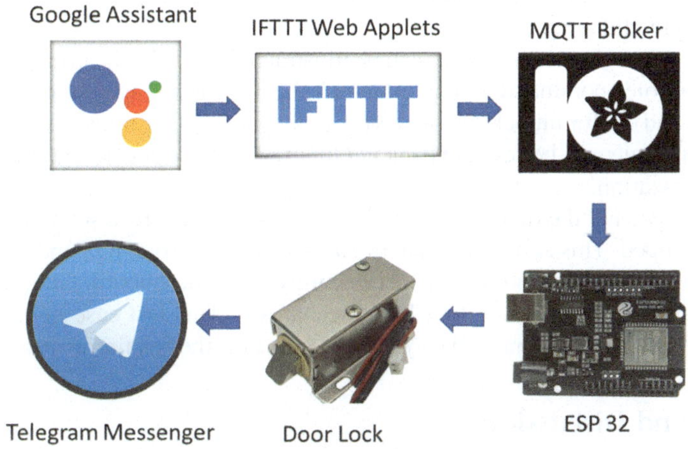

Fig. 3. Hardware and software integration

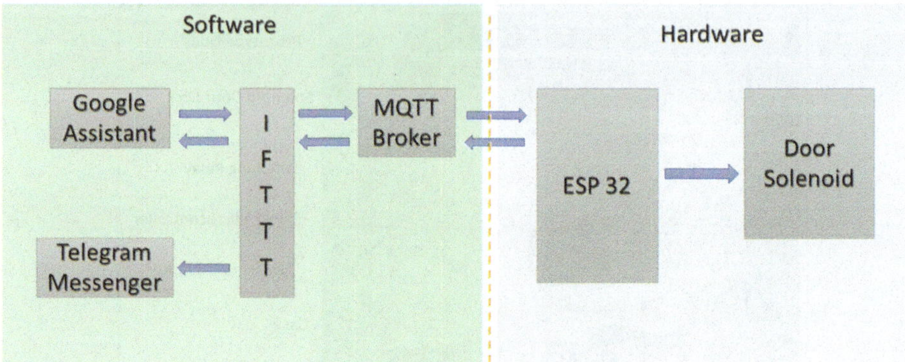

Fig. 4. Block diagram of door lock system

Figure 4 demonstrates the block diagram of the developed door system. The software part comprises of user's apps (Google Assistant and Telegram Messenger) and IFTTT Applets as an interface between users' apps and MQTT Broker. The hardware part covers on microcontroller ESP 32 and the door solenoid itself. The system integrated with Google Assistant apps as a medium to receive voice command for action execution. The MQTT protocol is used for secured network to execute the operation that connected to microcontroller attached to the door solenoid. All activity log will be notified in the Telegram chat bot and Google Assistant apps (most available apps to the users) for users monitoring and validation.

Workflow of the door lock system can be illustrated in Fig. 5. The process starts with receiving user voice command to lock or unlock the door. Locking door carries input of '1' in MQTT and '0' for unlocking the door. Upon updating the status of door in MQTT Broker, the message will be sent out back to Google Assistant and Telegram Messenger for user's notification.

To validate each of the functionality of hardware and software, a set of tests strategic has been developed. This consists of functionality test on android, iOS and desktop. The test area covers on testing notification, accessibility to the dashboard, Google Assistant voice command instruction and door lock mechanism. This test case helps in finding issues during the development of the system and to meet the requirements set.

3 Result and Discussion

All the test case has been successively carried out. This has showed that the developed system and prototype is well functioned including on testing notification, accessibility to the dashboard, Google Assistant voice command instruction and door lock mechanism.

The mechanism of locking and unlocking the door for this prototype can be explained as follows. Generally, the client (user) sends information directly to the endpoint (door lock). In MQTT protocol, broker filters all messages from client and distribute them correctly to subscriber (door lock). The filtering activity of the broker makes it possible to control which client and subscriber receives which message. The input from clients are filtered based on topics which is tagged in each message. The receiving client

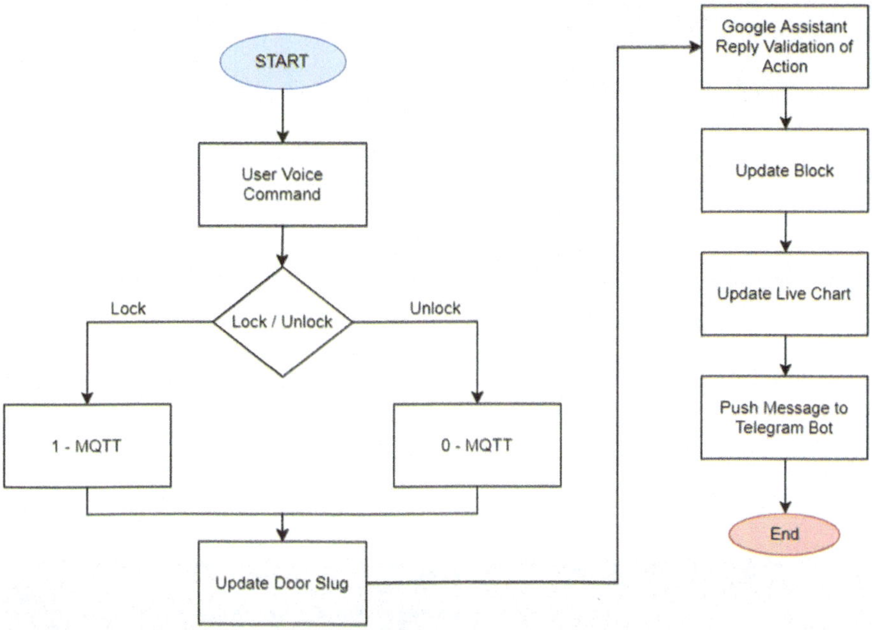

Fig. 5. Workflow of lock/unlock door

(dashboard, Telegram and Google Assistant) subscribes to the broker for topics of interest. From that point on, the broker ensures that the receiving client gets all message published into subscribed topics. MQTT works asynchronously as most client libraries work asynchronously and are based on call-backs. MQTT systems have the logic on the broker-side. Topics are published as a string or unsigned bits depending on the need of the client. In the meantime, the instruction executed in dashboard is linked with the microcontroller embedded at the door solenoid for triggering lock and unlock action.

Upon completion of the prototype development, it was proven that the system able to execute the door lock mechanism through Google Assistant voice command. The Google Assistant will provide feedback on the status of the executed instruction. The user will input a pre-determined voice command which is specific to lock or unlock the door lock. The action is then replied by Google Assistant if the command is successfully executed. The action executed will also be logged into a Telegram chat bot so the user can look back on the current door status. Figure 6 demonstrates the Google Assistant that display the commands and feedbacks for door lock system. Figure 7 shows the notifications of locking and unlocking of door lock in Telegram notification. Notification is sent out as verification that the door lock has changed its state. HIGH to LOW or LOW to HIGH (1–0, 0–1).

Fig. 6. Google assistant command and validation action reply

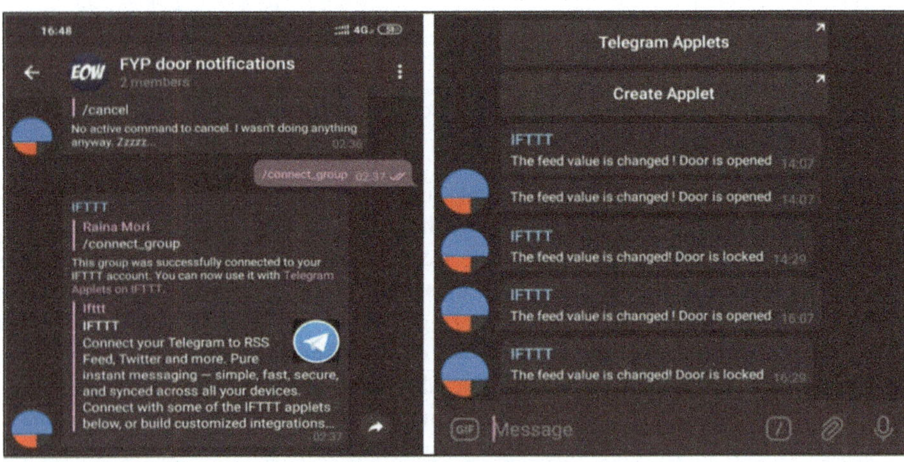

Fig. 7. Telegram door log

Figure 8 shows the dashboard design for the prototype. The dashboard is divided into three section. The first section displays the execution logs of live data and the second one is to illustrate the logs in graphical mode. Meanwhile, the third section is to enable users to manually lock/unlock the door by sliding the slider button to '0' or '1'. The simple design of dashboard and user-friendly interface made easy for anyone to operate it.

From the results, the system developed can control remotely the activities of lock and unlock door using MQTT protocol (IoT Technology). The system also able to receive voice command to execute the activities (lock/unlock door) while integrating it with Google Assistant and Telegram to notify the user on the execution action.

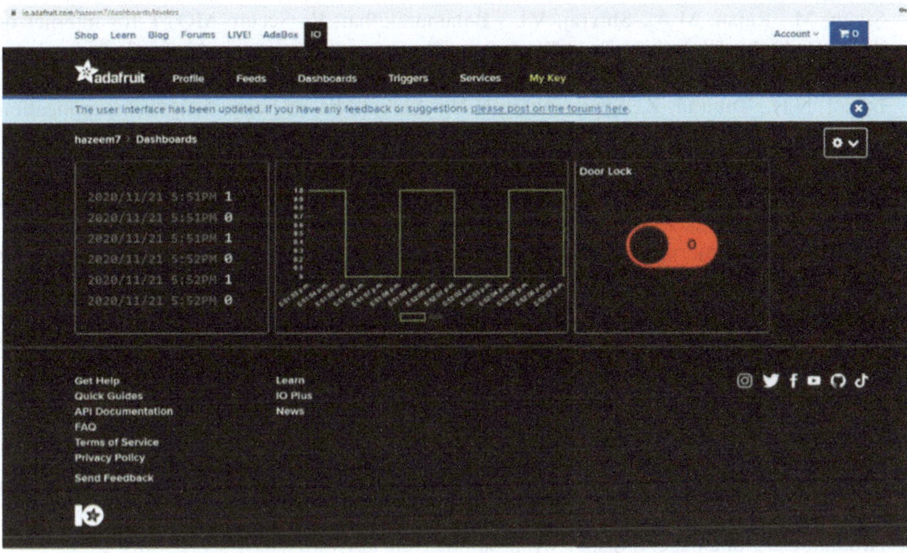

Fig. 8. Dashboard design for desktop

4 Conclusion

An IoT enabled door lock that can be remotely controlled and monitored has been successfully developed. The microcontroller used for this work is an ESP32 and the door lock used is a digital solenoid lock. By using MQTT protocol, it provides a secure channel for the user to communicate directly to the door lock via the middleware broker. Adafruit is used as the middleware as it provides very clean interface and post analytics, hence users can see actuator data in real time. With Google Assistant, the state of door lock and its status can be viewed and monitored. Every execution is then recorded in Adafruit and push to Telegram channel for notification. This system is accessible via mobile devices with internet connectivity. Users also able to track the status of their door locks from anywhere. The interface used is also user friendly and feasible for people who are on-the-go to check the status of door locks at home.

References

1. Al-Fuqaha, A., Guizani, M., Mohammadi, M., Aledhari, M., Ayyash, M.: Internet of things: a survey on enabling technologies, protocols, and applications. IEEE Commun. Surv. Tutor. **17**(4), 2347–2376 (2015)
2. Yokotani T., Sasaki, Y.: Comparison with HTTP and MQTT on required network resources for IoT. In: International Conference Electron Renewable Energy Communication, pp. 1–6 (2017)
3. Lampkin, V., Leong, W.T., Olivera, L., Rawat, S., Subrahmanyam, N., Xiang, R.: Building Smarter Planet Solutions with MQTT and IBM WebSphere MQ Telemetry. IBM Redbooks (2012)

4. Singh, M., Rajan, M.A., Shivraj, V.L., Balamuralidhar, P.: Secure MQTT for internet of things (IoT). In: 5th International Conference Communication System Network Technology, pp. 746–751 (2015)
5. Ismail, N.H., Tukiran, Z., Shamsuddin, N.N.: Android-based home door locks application via bluetooth for disabled people. In: IEEE International Conference on Control System, Computing and Engineering, Penang, Malaysia (2014)
6. Kamat, M.N., Shinde, D.: Smart door security control system using raspberry Pi. Int. J. Innov. Adv. Comput. Sci. **6**, 499–503 (2017)
7. Gupta, R.K., Balamurugan, S., Aroul, K., Marimuthu, R.: IoT based door entry system. Indian J. Sci. Technol. **9**(37), 1–5 (2016)
8. Han, D., Kim, H., Jang, J.: Blockchain based smart door lock system. In: International Conference on Information and Communication Technology Convergence: ICT Convergence Technologies Leading the Fourth Industrial Revolution (2017)
9. Chowdhury, M., Nooman, M., Sarker, S.: Access control of door and home security by raspberry Pi through internet. Int. J. Sci. Eng. Res. **4**, 550–558 (2013)
10. Do, H., Use, Y.: Rapid application development (RAD): what is it and how do you use it? A few RADical steps user interfacing, early and often pros of rapid application development cons of rapid application development, vol. 2016, pp. 12–15 (2019)
11. Arduino CC Homepage. https://www.arduino.cc/en/software. Accessed 25 July 2020
12. Hazizan, A., Md Lazam, N.A., Hassan, N.I.: Development of child safety car alert system using arduino and GSM module. In: IOP Conference Series: Material Science and Engineering, vol. 834 (2020)
13. August Smart Lock Pro Review: The Best Smart Lock Homepage. https://www.tomsguide.com/us/august-smart-lock-pro,review-4970.html. Accessed 26 June 2020

Development of Microwave Antenna for Cancer Treatment

Nurfarhana Mustafa[1], Nur Hazimah Syazana Abdul Razak[1], Nurhafizah Abu Talip Yusof[2], and Mohamad Shaiful Abdul Karim[1](✉)

[1] College of Engineering, Universiti Malaysia Pahang, 26300 Gambang, Pahang, Malaysia
mshaiful@ump.edu.my
[2] Faculty of Electrical Engineering Technology, Universiti Malaysia Pahang, 26600 Pekan, Pahang, Malaysia

Abstract. For cancer treatment, microwave thermotherapy method is widely used especially for liver cancer treatment by using a coaxial biomedical applicator device. This device is used to transfer heat from microwave generator to liver tissue in order to kill cancerous tissue. Neglecting the impedance matching between the applicator and the liver tissue can lead power reflection to the power supply or loss into the surrounding healthy tissue. The impedance of the applicator is hugely dependent on the structure and dielectric material used as the insulator for the biomedical applicator. In this work, we aim to study the best structure of the applicator, which can minimize the impedance mismatch. There are two phases required in order to achieve the aim. In first phase, parameter of the applicator has been estimated using the characteristic of impedance for coaxial structure. In phase two, the biomedical coaxial applicator is designed using the electromagnetic simulator software and the results were analyzed to identify the best parameter used for the design. The result of return loss obtained at 2.45 GHz is −25.42 dB with impedance of 21.23 Ω. From the value of return loss, 0.29% of power will be reflected to the generator or surrounding tissue while another 99.71% of power will be transmitted into the cancer cell. Based on the finding, most power is transmitted efficiently to the liver and less power is reflected to the surrounding which means it is harmless to the surrounding tissue.

Keywords: Ablation · Thermotherapy · Cancer treatment · Impedance matching

1 Introduction

Cancer begins when healthy cell transforms and develops out of control, forming a mass called a tumor. A cancerous tumor is malignant, which means it can grow and spread to other body parts. Cancer that begins in the liver usually called as primary liver cancer. There are three types of primary liver cancer happen in adults. The first one is hepatocellular carcinoma (HCC). HCC is the most common type of primary liver cancer in adults that is the most common cause of death in people. Roughly, 75% of liver cancer are HCC. The second one is Cholangiocarcinoma. Cholangiocarcinoma is a type of cancer that forms in the bile ducts. Approximately around 10% to 20% of liver

J. A. Mat Jizat et al. (Eds.): iCITES 2020, AISC 1350, pp. 127–136, 2021.
https://doi.org/10.1007/978-3-030-70917-4_14

cancer develop in the bile duct of the liver [1]. The last type of primary liver cancer are Angiosarcomas. Angiosarcomas is a cancer of the blood vessel's inner lining and it can occur in any part of the body. The disease occurs mostly in the skin, breast, liver, and deep tissue. Around 1% of liver cancer begins in the blood vessels of the liver and it can grow very quickly. Besides types of primary liver, there are four main stages in liver cancer [1]. The stages are determined by the size of the cancer (tumors) and how the cancer spread into the body. On the first stage of liver cancer, a single tumor grows around 2 cm or less than 2 cm and the cancer has grown or has not grown into the blood vessels. As the liver cancer reach the second stage, the single tumor reach 2 cm and has spread into the liver's blood vessels or there are numbers of tumor in the liver but all of them is less than 5 cm. There are several tumors in the liver but one of them must be larger than 5 cm in third stage of liver cancer. Onto the final stage of liver cancer, the cancer already grown into the blood vessel or has spread to organs around the liver. Besides that, the cancer also grown through the lining that wrap around the abdomen's internal organs.

Traditionally to treat cancer treatment, majority choose to use a major surgical operation and this technique is widely used in medical institutions especially in Malaysia. In this modern technology, minor surgery techniques have been proposed to kill the cancer cells by heating them within temperature range of 42 °C until 45 °C. This is because this range of temperature, which is 42 °C until 45 °C, is good enough to kill the cancer cells. Currently, this technology has been explored using a Magnetic Nano Particle (MNP) [2, 6, 7]. This MNP induces into human body and cancer cell is killed by using heat from magnetic field vibration. This process is called self-regulating. However, there are several problems that need to be solved when using this technique. One of them is to control the temperature so that it can maintain in suitable range during operation. Another problem is to maintain the focus point for the heating process when using magnetic particles. Therefore, to overcome these problems the thermotherapy technique is one of the best options since the heating can be transferred using an antenna applicator. In this technique, a needle-shaped antenna is used to kill the cancer tissue [3, 8].

In conventional design, the design neglects impedance matching between antenna and cancer cell [3, 11, 12]. The return loss, S_{11} drop at frequency of 2.45 GHz for conventional design is −9.82 dB. This situation can conclude that 90% power will transmit into the cancer tissue but there will be another 10% of power will loss. This can be harmful if the power loss is lost to the surrounding of the liver tissue, which can harm normal tissue [10]. This situation happens because the impedance of the applicator mismatch with the impedance of the liver as in theory of impedance matching. The impedance between applicator and liver should be similar to minimize the mismatch between the applicator and the liver. Thus, adjusting the impedance of the applicator is the only option to match the impedance, since the impedance of the liver is unchangeable. In this work, a new design of biomedical applicator is proposed by using microwave coaxial antenna in order to treat liver cancer by minimizing the impedance mismatch so that more power will be supplied to the cancer liver. The coaxial structure is used because of the needle-shaped structure which will ease the surgical procedure in addition to the simplicity in derivation of its impedance based on impedance-line microwave theory.

2 Design of Antenna

2.1 Impedance Theory and Dimension Determination for the Applicator

The difference between the impedance of the biomedical applicator and liver tissue can lead power reflection to the power supply or loss to the surrounding healthy tissue. Thus, the impedance matching is very crucial in this work. In this work, the coaxial biomedical applicator used is shown in Fig. 1. When using structure of coaxial cable, there is a specific formula on how to calculate the impedance. The equation of the impedance of coaxial structure is:

$$Z_o = \frac{138\, log_{10}\left(\frac{2 \times R_2}{2 \times R_1}\right)}{\sqrt{\varepsilon r}} \tag{1}$$

From the equation, the factor that affects impedance, Z_o are the value of inner, R_1 and outer radius, R_2 and, the value of dielectric property, ε_r. Teflon with permittivity (dielectric property) of 2.08 and tan $\delta = 0.00048$ is chosen in this work because this material has very low loss properties. Figure 2 shows the cross-section of coaxial cable where, D_i is the diameter of inner conductor while D_o is the diameter from inner conductor. The value of radii is one of the main factor that should be considered before choosing the suitable dimension for the applicator, in order to have an appropriate size of needle especially for minor surgery. From Fig. 3, the value of impedance is chosen as nearly equal to impedance of liver, which is 21.23 Ω at 2.45 GHz. The impedance of liver was estimated from the simulation based on conventional design of coaxial cable with phantom liver. When the return loss is low the impedance of the coaxial cable can be considered same as the liver. After the impedance value is known, the ratio of R_2/R_1 can be obtained in order to calculate the specific size of the parameter used. From the graph the value of R_2/R_1 is 1.67. The different of R_1 and R_2 is represent the thickness of dielectric material, so the appropriate thickness for fabrication need to be considered during the selection of R_1 and R_2. Thus, value of R_1 is chosen as 0.48 mm and the value for R_2 is 0.8 mm.

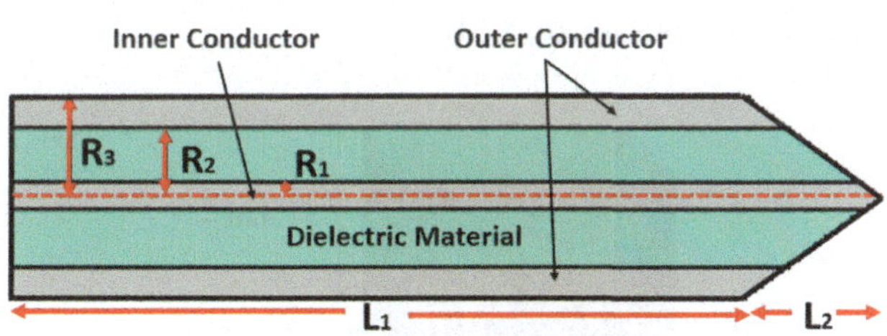

Fig. 1. Structure of needle-shaped antenna.

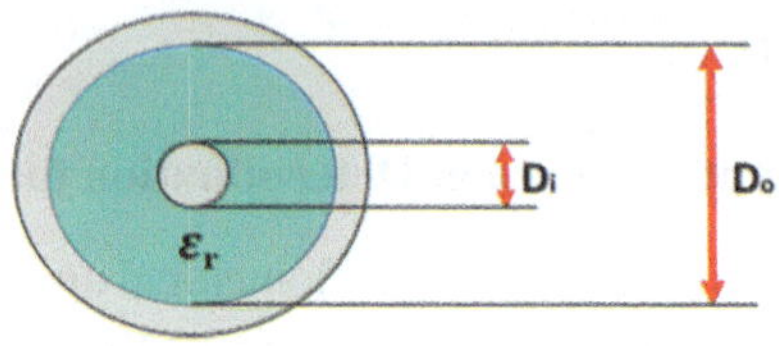

Fig. 2. Coaxial cable cross-section.

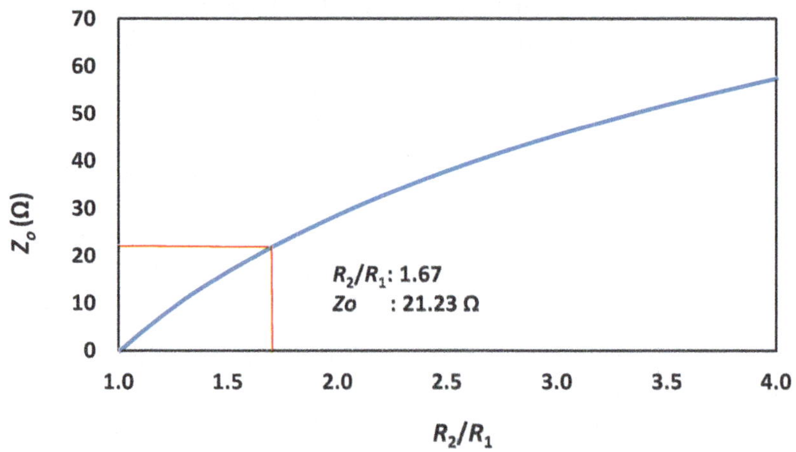

Fig. 3. Impedance characteristic towards ration of R_2 and R_1.

2.2 Phantom of Liver

Liver which affected with cancer cell is considered having the same electrical properties as the healthy liver [5]. Thus, the properties of liver is imported from voxel data of human biological models that having permittivity of 46.83. Since the size of cancer cell for stage one to three is around 10 to 50 mm, a 40 × 40 × 31 mm liver phantom model is modelled as a cuboid in simulation software as shown in Fig. 4. It is necessary in order to monitor the specific absorption rate, SAR.

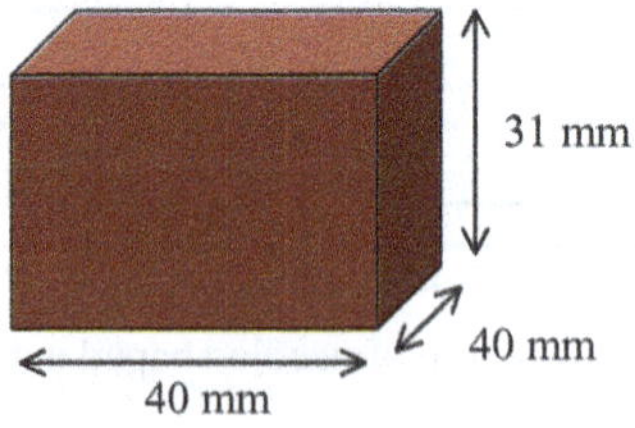

Fig. 4. Model of liver phantom for microwave ablation.

3 Result and Discussion

In this work, the design of the applicator is designed based on design in Fig. 1 but with different parameters. This way is important in order to analyze the effect of different parameters, which are the radius of R_1, R_2, R_3, and the length of L_1, L_2. The analysis of the parameter is needed in order to estimate the impact towards body tissue if the fabrication is imperfect. The parameters of return loss, S_{11} and specific absorption rate, SAR have been simulated. S_{11} is simulated in order to getting the performance of the antenna where S_{11} should lower than −10 dB, while SAR will indicate the energy absorb by human body when exposing to electromagnetic field. In this work, the value for S_{11} and SAR are simulated at frequency of 2.45 GHz which is the common frequency for microwave ablation in liver cancer treatment using CST Studio Suite.

3.1 Parametric Analysis Towards Return Loss

Effect of R_1 Toward Return Loss. From the S_{11} values in Fig. 5, the deepest S_{11} is obtained when $R_1 = 0.48$ mm which indicated the loss is at minimum level. Other than that, the S_{11} does not met the requirement as a good antenna at 2.45 GHz, for example, when $R_1 = 0.25$ mm the value of S_{11} is −7.79 dB which is greater than −10 dB. This value of S_{11} can lead to almost 17% of power reflected back to the surrounding and only 83% power will be transmitted compare to our chosen R_1 which is $R_1 = 0.48$ mm. This R_1 has value of $S_{11} = -25.42$ dB. When value of S_{11} is −25.42 dB, only 0.29% of power will be reflected back while the remaining power will transmit into the cancer tissue.

Effect of R_2 Toward Return Loss. From Fig. 6, the $S11$ values did not suit as the good antenna's condition while using $R2 = 0.75$ mm, 0.83 mm and 0.88 mm. Even though the value of $S11$ for these $R2$ are good and have a good depth which is −18.36 dB, −27.49 dB and −19.46 dB but the $S11$ do not drop at frequency of 2.45 GHz which cannot be accepted. Meanwhile, $R2 = 0.80$ mm was accepted as the good parameter value of $R2$ because the value of $S11$ drop exactly at 2.45 GHz with the value of −25.42 dB.

Effect of R_3 Toward Return Loss. From Fig. 7, almost all five parameters of R_3 has the value of S_{11} that drop at the required frequency. So, the best value for R_3 is at 0.90 mm (red line) which has the depth of $S_{11} = -25.42$ dB. Besides that, at $R_3 = 0.82$ mm and $R_3 = 0.87$ mm, the value of S_{11} at both R_3 have more depth compared to the value of S_{11} at $R_3 = 0.90$ mm. However, those R_3 is not good enough to be selected because the values were too close to the R_2 value which will affect the thickness of the outer conductor. Thus, this design is not practical and difficult for fabrication process. For $R_3 = 1.00$ mm and 1.50 mm the value of S_{11} recorded too close to −10 dB which will degrade during real fabrication.

Effect of L_1 Toward Return Loss. From Fig. 8, all S_{11} are below −10 dB, which indicates that L_1 has less effect on the return loss value. Therefore, L_1 is chosen as 73.00 mm to avoid a bulky design but at the same time the performance is preserved.

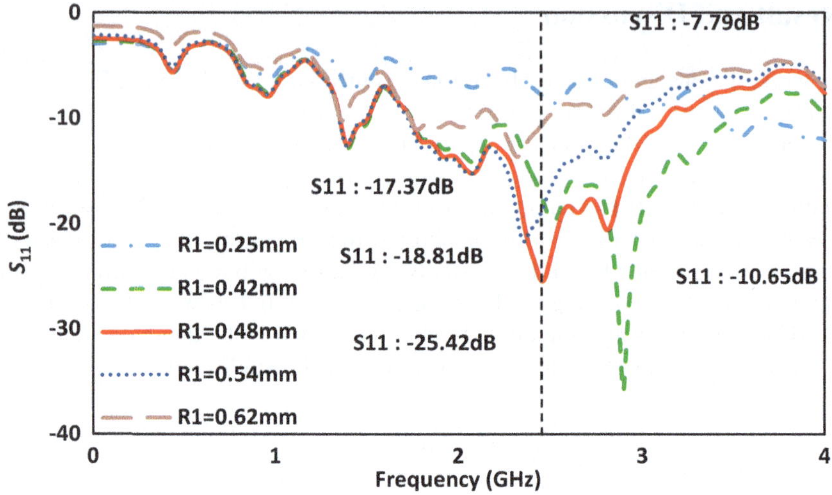

Fig. 5. Effect of inner radius towards return loss.

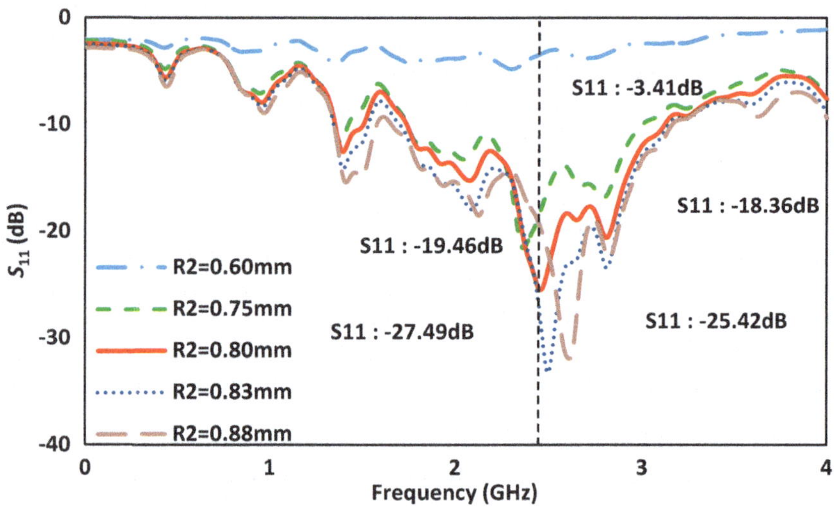

Fig. 6. Effect of dielectric radius towards return loss.

Effect of L_2 Toward Return Loss. From Fig. 9, the value of S_{11} for $L_2 = 4.5$ mm, $L_2 = 5.0$ mm, $L_2 = 8.0$ mm and $L_2 = 10.0$ mm were −7.37 dB, −10.43 dB, −15.79 dB and −8.62 dB, respectively. These parameters cannot be selected because the value of S_{11} are either close to or less than −10 dB. Therefore, the red line is the only line lies at 2.45 GHz and has S_{11} of −25.42 dB.

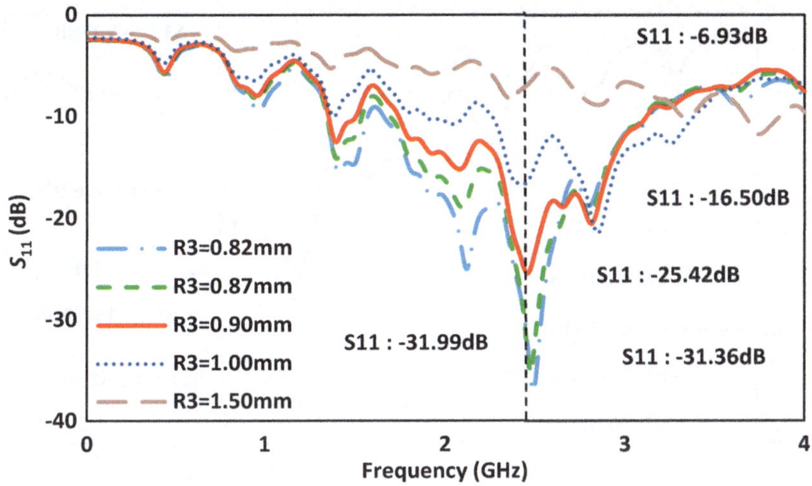

Fig. 7. Effect of outer radius towards return loss.

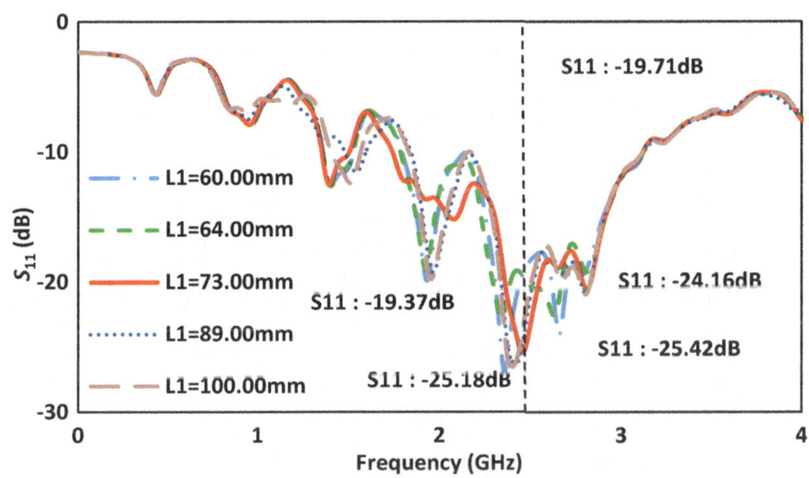

Fig. 8. Effect of L_1 towards return loss.

3.2 Simulation of Return Loss and SAR for the Final Design

From the analysis above, the final parameter for final design are $R_1 = 0.48$ mm, $R_2 = 0.80$ mm, $R_3 = 0.90$ mm, $L_1 = 73.00$ mm and $L_2 = 6.69$ mm. From the parameter value, the impedance for the applicator can be calculated using formula (1) and the value of impedance obtained is $Z_o = 21.23\ \Omega$. By using concept of impedance matching, when the impedance of the applicator match with the impedance of the liver, the reflected heat power will decrease and more heat power will transmit into the liver.

Figure 10 shows the return loss for the final design of the applicator. It is clearly shown that the design can afford a depth of return loss at 2.45 GHz. When 100% power

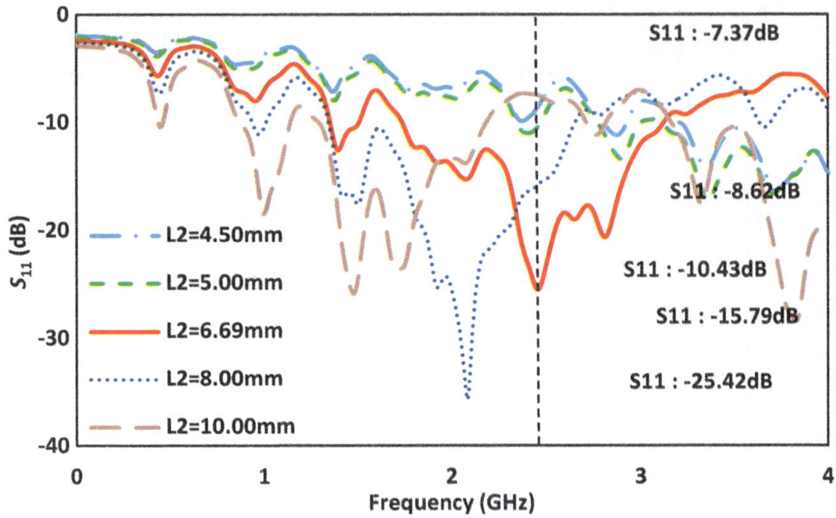

Fig. 9. Effect of L_2 towards return loss.

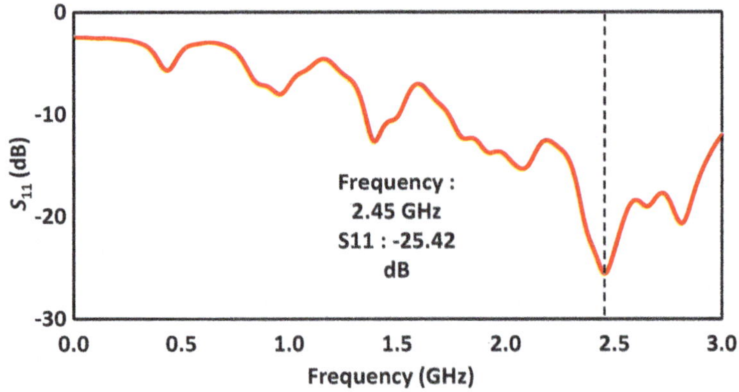

Fig. 10. Return loss characteristics of final design.

is assuming incident the applicator, 0.29% power will be reflected back to the generator or to the surrounding tissue and the remaining power that is 99.71% power will be transmitted into the liver cancer and heat the cancer cell.

Figure 11 shows the simulated SAR at 2.45 GHz. The peak SAR is 104.54 W/kg, which is sufficient for microwave ablation method. In this figure, the heating has a target because the heating region is around the tip of the applicator. The heating is in red zone which maximum heat is apply at the tip of the applicator and along the length of the applicator, the zone is in light blue and almost too dark blue region that has less heat or no heat around the applicator which can reduce the potential to harm normal tissue. From this, we can conclude that, if the applicator match with the targeted tissue, the reflected power that can harm the normal tissue can be minimized and more heating power that can be used to kill the cancer cell.

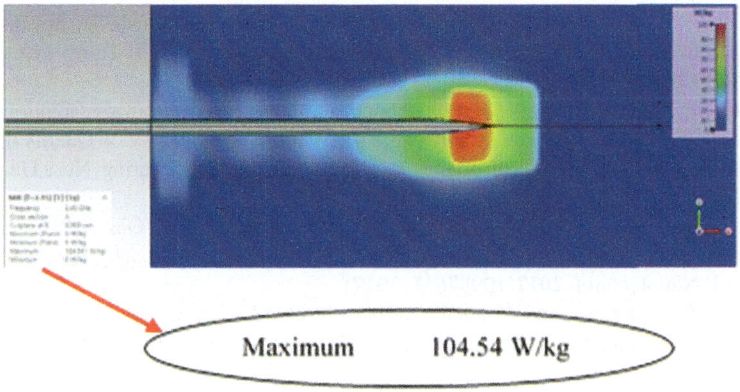

Fig. 11. Simulated SAR distribution at 2.45 GHz.

4 Conclusion

In this paper, a new coaxial biomedical applicator consist of three layer was designed by using microwave thermotherapy procedure. A method that leads to accurately S_{11} value at 2.45 GHz is needed for microwave thermotherapy method in order to guarantee destruction of the cancer tissue and minimizing damage to surrounding healthy tissue. This research was carried out to observe the effects of S_{11} on the destruction/death of cancer cell and input microwave power on the SAR during microwave thermotherapy. The result obtained for S_{11} is less than -10 dB, which indicates the proposed design is applicable to use for microwave thermotherapy. This reduction of S_{11}, makes it possible to supply more power into the cancer cell and reflect less power without heating and harmful towards normal tissue.

Acknowledgment. Authors would like to extend their gratitude to Universiti Malaysia Pahang and Ministry of Education (MOE) under Fundamental Research Grant Scheme (FRGS) with reference no. FRGS/1/2018/STG02/UMP/02/1 or RDU190140. This work also was partly supported by the Research and Innovation Department, Universiti Malaysia Pahang under Postgraduate Research Grants Scheme (PGRS) with reference no. PGRS200314.

References

1. https://www.cancerresearchuk.org/about-cancer/liver-cancer/stages/stage-4. Accessed 12 Dec 2020
2. https://www.express.co.uk/life-style/health/870557/cancer-symptoms-treatment-cells-cure-thermotherapy. Accessed 23 May 2020

3. Urata, M., Kimura, S., Wakino, K., Kitazawa, T.: Optimization of three layer tapered coaxial line applicator for cancer thermothrapy by impedance matching of transition layer sections. In: 2011 IEEE International Symposium on Antennas and Propagation (APSURSI), Spokane, WA, pp. 120–123 (2011)
4. Nakata, T., Yoshitake, H., Wakino, K., Lin, Y.D., Tani, T., Kitazawa, T.: Spectral domain approach to the scattering analysis of coaxial discontinuities. IEICE Trans. Electron. **E90–C**(2), 275–281 (2007)
5. Narpati, F., Basari: Design of ultrawideband applicator for microwave ablation aimed at thermal therapy in liver cancer. In: 2017 15th International Conference on Quality in Research (QiR): International Symposium on Electrical and Computer Engineering, Nusa Dua, pp. 150–153 (2017)
6. Malik, A., Butt, T.T., Zahid, S., Zahid, F., Waquar, S., Rasool, M., Qazi, M.H., Qazi, A.M.: Use of magnetic nanoparticles as targeted therapy: theranostic approach to treat and diagnose cancer. J. Nanotechnol. **2017**(1098765) (2017)
7. Hayek, S.S.: Synthesis and characterization of CeGdZn-ferrite nanoparticles as magnetic hyperthermia application agents. Adv. Mater. Sci. Eng. **2019**(4868506) (2019)
8. Liu, W., Lin, X., Yang, F., Zeng, J., Song, H., Cui, Y.: Microwave ablation antenna with the ablation pattern being a figure-of-eight. In: 2016 IEEE MTT-S International Microwave Workshop Series on Advanced Materials and Processes for RF and THz Applications (IMWS-AMP), Chengdu, pp. 1–3 (2016)
9. Labonte, S., Blais, A., Legault, S.R., Ali, H.O., Roy, L.: Monopole antennas for microwave catheter ablation. IEEE Trans. Microwave Theory Tech. **44**(10), 1832–1840 (1996)
10. Neira, L.M., Sawicki, J., Van Veen, B.D., Hagness, S.C.: Characterization and analysis of wideband temperature-dependent dielectric properties of liver tissue for next-generation minimally invasive microwave tumour ablation technology. In: 2018 IEEE/MTT-S International Microwave Symposium, Philadelphia, PA, pp. 911–914 (2018)
11. Keangin, P., Rattanadecho, P.: A numerical investigation of microwave ablation on porous liver tissue. Adv. Mech. Eng. (2018)
12. Luyen, H., Gao, F., Hagness, S.C., Behdad, N.: High frequency microwave ablation for targeted minimally invasive cancer treatment. In: The 8th European Conference on Antennas and Propagation (EuCAP 2014), The Hague, pp. 1478–1482 (2014)

Review on Motor Imagery Based EEG Signal Classification for BCI Using Deep Learning Techniques

K. Venu(✉) and P. Natesan

Kongu Engineering College, Perundurai, India
venu.cse@kongu.edu, natesanp@kongu.ac.in

Abstract. Motor Imagery (MI) based Electroencephalography (EEG) Signals analysis plays an important role in Brain- Computer Interface (BCI) which is used in several applications such as controlling the wheelchair by motor impaired patients, as recovery gadgets for stroke patients, as an input apparatus for several gaming devices. EEG signals are used in several application since, classification process of the signal is made automatic and more accurate using deep learning techniques. In recent days several deep learning architectures were used for analyzing the EEG signal effectively with the availability of large amounts of data. Deep learning techniques for MI based EEG analysis is surveyed from 2015 to 2020 to give a detailed description of various newly designed classification techniques. How the EEG signals are analyzed in each and every phase of its processing is also explained along with its accuracy. This paper reviewed various approaches, pre-processing techniques, the dataset used for classification of MI task from the EEG Signal.

Keywords: Deep learning · EEG signal · Motor imagery · Classification · CNN · RNN

1 Introduction

The activity of the brain is measured by using Electroencephalography (EEG). Brain activity patterns can be captured by using EEG and converted into information for an application using a system called Brain Computer Interface (BCI) [1]. For example, by imagining the movement of the hands, cursor of the computer can be made to move by using BCI [2] without involving any physical activity of the user. Here when the user imagines a particular movement, then Motor Imagery (MI) based EEG Signal will be generated in the brain. This MI-EEG signal is processed and converted into commands which is understandable by the peripheral device and then the cursor movement is done. MI-EEG Based BCI can be used in many applications such as controlling the wheelchair by motor impaired patients [3], as recovery gadgets for stroke patients [4], as input apparatus for several gaming devices [5], to design an flexible interface that can behave according to the mental state of the user [6] and many more [7, 8].

J. A. Mat Jizat et al. (Eds.): iCITES 2020, AISC 1350, pp. 137–154, 2021.
https://doi.org/10.1007/978-3-030-70917-4_15

There are two phases in BCI. In the first phase, the system is calibrated and it is called as a offline training phase. In the second phase, only the EEG patterns are recognized and translated into commands. It is known as the operational online phase [9]. In the BCI system first EEG patterns were taken and pre-processed using several filters which are spectral and spatial [10]. Then the high level and low level features were gathered from EEG Signal. At last, these extracted features were used for classifying the signal [11] and to provide the appropriate command with respect to the signal for the given application [12]. Finally, the user will be provided with the information whether a command has been detected from the EEG or not.

There were several key elements such as pre-processing and classification tasks are involved in the MI-BCI system. The pre-processing is done in terms of filtering and transformation of EEG Signal into images etc., Variety of classifiers are used to perform classification task [13–17]. There are various challenges while analyzing the EEG Signal such as signals non-stationary, non-linear, week, multivariate time series data, presence of artifacts, high dimensionality, low signal-to – noise ratio, low spatial resolution, Complex, differs from subject to subject and even differs between different trails of the same subject, suffer from channel crosstalk.

In this paper different filters, transforms, classifiers used in various works for BCI are explained along with the dataset description. This paper is organized as follows. Section 2 provides information about the usage of various transforms and filters for signal pre-processing. Then Sect. 3 presents a description about the dataset used in various works followed by the summary of different types of classifiers. Section 4 provides the conclusion.

2 EEG Signal Preprocessing Methods

Removing of abnormal samples, Each signal is converted as a zero mean signal by subtracting the original signal by average amplitude. In [18] The signal – to – noise ratio is improved by doing baseline correction. The background noise is removed from the EEG Signal by using the spectral subtraction technique.

EEG signals are classified into bands of various frequency ranges using Band Pass Filter. Each frequency band reveals information about various activities of the brain corresponding to different parts of the body. MI related activities are recorded in the frequency band of range 7 Hz to 30 Hz. Hence it is extracted from the input EEG Signal using band pass filtering approach in [19–21]. The frequency ranges from 6 to 13 Hz is called mu band where as the frequency range from 13 to 30 Hz is known as as beta band. During the motor imagery task the power spectrum in mu band decreases and in beta band it is increased. It is called as Event Related Desynchronization [ERD] and Event Related Synchronization[ERS] respectively. This can be determined using CSP techniques [22]. In [23] Butterworth Band-Pass filter s used for filter the required signal range from the given EEG. The original EEG Data is converted into set of vectors without compromising its original characteristics and values by using several feature extraction algorithms which are explained below.

Riemannian Geometry [24] is one of the feature extraction techniques in which the covariance matrix is used to provide the frequency and spatial information of the given input EEG Signal. This covariance matrix is combined with the power spectrum density features and then used for classification.

2.1 Various Wavelet Transformation Techniques

Continuous Wavelet Transform. Time – Frequency computation methods like wavelet transform is used to extract time and frequency related features by decomposing the raw EEG signals. These features can be used to classify the signals accurately. In [19, 25–27] the Continuous Wavelet Transform (CWT) is used to represent the original EEG signal into wavelets. In the CWT technique, given EEG signals are converted into two-dimensional signals which provide the time-frequency representation of the given signal. The output of CWT are Time- frequency images that provides efficient time and frequency localization which will be helpful during classification. Time-frequency images in the specified range of mu and beta rhythms are used for training the classifiers. Amplitude, scale and time related information are present in the image which is obtained using CWT.

BSS algorithm is an unsupervised algorithm that can also be combined with CWT. It is used to extract the independent components from the given EEG Signal in [25]. The components are sorted based on mu and beta frequencies. The components at the starting time are the components which are active during the MI activities. Then the artefacts components will be moved to last. Then CWT is used to generate image with temporal, spatial, frequency information. By continuing these steps training data will be generated.

Short Time Fourier Transform [STFT] and Stockwell Transform. STFT is used to construct spectral image from the given EEG Signal in [22, 30–32]. It is effective since it will take various frequency features and less number of parameters for optimization. Images of 16 × 32 and 23 × 32 size are extracted from mu and beta band respectively. Images that corresponds to beta frequency bands are resized to 15 × 32using cubic interpolation method so that both band images will have a similar effect. In [18] the EEG Signal is converted into frequency domain from its time domain using stockwell transform.

Wavelet Packet Transform. Wavelet packet transform is a type of multi scale transform which used to decompose the EEG Signal into two separate bands of multiple scales in [26, 40, 41]. The decomposed data contain both the time and frequency related information. The given EEG signal is converted into different form using various transformation methods such as CSP, EMD, Fourier transform and discrete cosine transform. Then triple frame matrix is formed using the representation obtained from different transformation techniques. Then it is passed to CNN for feature extraction in [42]. Several dimensionality reduction techniques were discussed in [43].

2.2 Conditional Empirical Mode Decomposition Algorithm

Empirical Mode Decomposition[EMD] is one of the signal dependent decomposition technique which is used to analysis nonlinear and non-stationary signals. Hence it is used to analysis EEG signals. In this technique, the given EEG Signal is decomposed into several band- limited intrinsic mode function [IMFs] and a residual components.

These IMFs are combined together to create a new EEG in different format in [28] which is used for data augmentation. There are some conditions and drawbacks exist in EMD while selecting the IMFs for further processing. In [29] IMFs are selected by using median frequency measures. This will have high SNR value. In order to select those IMFs effectively Conditional Empirical mode decomposition[CEMD] technique is used. In [20] only the right and left hand movement is going to be classified. The first five IMFs will give the information for hand movement classification. Hence only those five IMFs are taken in this paper. In order to select those five IMFs first correlation coefficient condition and second relative energy occupancy rate condition are used in CEMD, after retrieving the IMFs from the given EEG signal of a channel. Now all the selected IMFs combined to form an another EEG Signal.

In order to extract time series information from the EEG Signal of a channel with strong anterior-posterior dependence 1D convolution is performed. But Event-related synchronization (ERS) and Event-related de-synchronization (ERD) that exist between the EEG signals channels are not captured using 1D convolution. To insert those information an EEG Signal combination method is used in [20]. Thus a new EEG Signal is obtained after several pre-processing steps. Then it is passed on to a classification model for classifying the given task.

2.3 Common Spatial Pattern [CSP]

CSP is used in [23, 29, 33] for the purpose of feature extraction since it is very effective in this task. I is the spatial filtering technique. The eigenvalues are extracted from the EEG data for further processing using CSP technique. In [34] The pre-processing of EEG signals for MI task is done by using a time window. 16 frequency pass bands is created by employing Chebyshev Type II filter by decomposing the EEG signal using a filter bank with 16 filters. CSP technique is used to determine the desynchronization and synchronization events in the EEG Signals. In [21] The similarity among the features extracted using this techniques is measured by using Kullback–Leibler divergence (KL). In [35] The efficiency of CSP algorithm is improved by increasing the number EEG signal channels by using various frequency bands. In this method a set of spatial filters are taken in which variances of one class is high while comparing with other data. It is the idea used in CSP for extracting features with large difference in [36]. In CSP the channel having similar information are combined together.

Filter- Bank Common Spatial Patterns[FBCSP] is the extended version of CSP approach [37]. It is used to find linear combination of channels with spatial filter that contains data from all the classes present in the problem [38]. This steps involved is given as follows. The EEG Signal is divided into nine different frequency bands using filter bank. Then by applying CSP on each frequency band to get spatial filters. Then the index value is provided for each channel in the frequency band of each class. These index are used for extracting the corresponding channels in each frequency bands. Then the analytical form of the given EEG Signal is obtained using signal envelope. A signal matrix is formed by using the signal envelope output. This is used for training the model for classification.

2.4 Complex Morlet Wavelets and Power Spectrum Density[PSD]

Complex Morlet Wavelets [28, 39] is used to convert the EEG data into tensor so that it can be easily processed by neural network model. PSD which is frequency domain analysis technique is used as the feature extraction method [36]. Here the EEG data where the amplitude changes with time is taken and converted into frequency domain by using correlation function and period gram method. Then the analysis is performed in frequency domain.

3 Dataset Description

There are different MI –EEG signal dataset is available for the classification. The Table 1 provides description about various datasets available.

Table 1. Dataset description

Dataset	Subjects/Persons	No. of electrodes	MI task
Dataset 2a from BCI Competition IV	9	22	Left hand, right hand, foot or tongue
Dataset 2b from BCI Competition IV	9	3	Right hand and left hand
Dataset IVa from BCI Competition III	5	118	Right hand and foot
BCI Competition III	25	3	Right hand and left hand
BCI Competition II Dataset III	1	3	Right hand and left hand
High Gamman Dataset	14	128	Left hand, right hand, feet and rest
The Physionet database	109	10	Left fist, right fist, both fists and both feet
BNCI Horizon 2020	15	61	Forearm supination/pronation, elbow movement, hand grasping, and rest

4 Various Deep Learning Techniques used for Classifying MI\Tasks

4.1 Convolution Neural Network

Simplified CNN is used in [19] for the classification of left and right hand movements. A Six layer simplified CNN is used in this work for the purpose of classification. It consists of an input and output layer, two convolutional layers, flatten and Fully connected layers. The input layer receives the input as Time- Frequency image. Then in the first convolutional layer frequency features are extracted by moving along the time axis of the input by using 8 filters. Batch normalization is also done in this layer. Whereas in the second convolutional layer features related time is extracted by moving along the horizontal axis with the help of 16 filters. In the fatten layer, all the frequency and the time related features are combined together. Then at last in the output layer softmax function is used to classify the output. Here the pooling layer is not used in order to simplify the network and because of which all the extracted features are analyzed.

In [20] 1D Multi-scale convolutional neural network is used to classify the given EEG Signal for the motor imagery task. 1D Convolution is used to extract time series data from the given EEG Signal. Multi scale represents different size of kernels used in convolution layer. Holistic features can be extract from the given image using kernel with large size, whereas kernel with smaller size will extract detailed features in an effective manner. In [20] all the features are extracted by using one large and one small kernel in two different convolution layers respectively. Pooling layers are also used to maintain the important features. Then all the features extracted are combined together and used for the classification task. In [20] Relu activation function is used in convolution layers and at the output layer softmax function is used for classification. Loss function used in the model is cross- entropy. Stochastic Gradient Descent is algorithm which is used for training the given the model along with the dropout rate as 0.5.

CNN with convolution, polling, padding, fully connected layers are used for the task of classification in [25, 27, 44]. In [33] an extension of feedforward neural network along with different kinds of kernel such as Gaussian and polynomial kernels were used.

5 Layer CNN model is used for classification in [45]. The five layers consist of one input layer, two hidden layers which is used to extract features and two fully connected layers which is used for classification purpose. Here vectors are used as convolution kernels instead of matrix. The idea behind is not to mix both time and space information as one kernel.

There are 2 convolutional layers present in the CNN model proposed in [46]. The convolution operation is done along the time axis in the first layer. In this layer linear pre- filtering operation is done for each and every channel's EEG signal. Whereas in the second layer only the convolution operation is performed. Here the number of kernel used is equal to the number of channels. Table 2 provides various pre-processing techniques, architecture and accuracy for various papers using BCI Competition IV Dataset 2a. In some of the work the result is compared with their previous work and improvement is shown, for all those reference papers ** is placed in the accuracy field. The model in [32] which uses Short Time Fourier Transform [STFT] as preprocessing technique and Convolutional Neural Network as classification model provides best accuracy as 93% for the classification of BCI Competition IV 2a Dataset.

The given EEG signal is classified into four different classes using the CNN architecture [23]. Two dimensional image is given as input to the CNN model. It is obtained by taking spectrogram of the signal which is received as output of the CSP Filter. Spectrogram is obtained by using STFT and Discrete Fourier Transform. Then the images

Table 2. Parameters used for classifying BCI Competition IV dataset 2a with 4 classes

Reference paper	Pre-processing technique	Architecture used	Accuracy	Size of the kernal	Optimization algorithm
[24]	Common spatial pattern	CNN	82.39%	7 × 7	Adam
[37]		CNN	74.46%	7 × 7, 3 × 3	Adam with Negative log-likelihood
[38]		RNN	**	–	Stochastic gradient descent algorithm
[53]		LSTM	81.50%	–	Back propagation through time algorithm
[34]		CNN+LSTM	83%	3 × 3	Adam
[21]		Transfer Learning	**	–	–
[36]		Transfer Learning	85.70%	–	–
[35]		Support Vector Machine	**	–	–
[29]	Empirical mode decomposition	Support Vector Machine	79.19%	–	–
[32]	STFT	CNN	93%	1 × 1 to 5 × 5	CP Mixed Net
[51]	Nil	3D CNN	52.17%	2 × 2	Adam with Negative log-likelihood
[42]		CNN	75.70%	2 × 2	Adam
[44]		CNN	75%	10 x 1	Mini-batch stochastic gradient descent

are stacked together to form a signal images which is given as input to the CNN model for classification.

Normal convolution operation along with a layer of separable convolution operations are combined together to form a CNN Model which is used for classification task. Separable convolution are nothing but a Depth wise convolution which is followed by point wise convolution operation. It is used in this model since it can reduce the number of parameters [47]. The next advantage is that it can learn the individual feature maps by kernel summarizing because of which the relationship between and inside the feature maps can be decoupled and optimally merge easily. Then finally softmax layer is used to perform classification.

Forearm movement classification is done by using hierarchical flow Convolution neural network with enhanced linear unit as activation function [48]. Filtering based on coincidence is used to extract feature and classification is done by CNN in [49]. CNN is used in [41] with 3-D input to classify the Motor Imagery task. The first two dimension represent the time information and frequency band information is shown in third dimension.

3D input is first created to use it in 3D CNN [51]. Initially 2D representation of the images is generated by keeping the spatial information. Then it is extended with third dimension using temporal information and it is given as input to 3D CNN. In this model 3D filters are used in the convolution layer.

Relu is used in most of the CNN model since it is less likelihood to gradient vanishing problem and it has high rate of convergence. This will increase the process of optimization with the help of steep gradient even when there is a high deviation from the optimal value. CNN may lead to inaccurate. Classification result because it can't correctly extract spatial information between objects just by using pooling operation [50].

The classification model which consist of CNN along with Triplet input. The EEG Signals are classified using triplet network. Triplets of inputs (x, x+, x−) are used for training the network. Here x– anchor, x+ - positive instance are similar to each other. Where as x– is the difference from x. The performance of the classifier is improved by decreasing the background and increasing the spectrogram plot area. The Triplet network will optimize only one embedding difference that could be either between anchor and positive instance or between anchor and negative instance by optimizing the ratio of Euclidean. In this network [18] more number of parameters are shared. Hence the number of parameters to be learnt get decreased. Thus by increasing the training process and by reducing the memory requirements. The loss function is optimized by changing the weights assigned to the CNN encoders in each and every iteration during the training process. The loss function decreases when the Euclidean distance between the embedding of the images belongs same class decreases and different class increases. The CNN Model comprises of 7 × 7 convolutional layer, 2D adaptive average polling layer, 2D adaptive max polling layer, 4 dense layers and 2 fully connected layers are used to extract the embedding from the images. Final softmax layer is used for the purpose of classification.

In [24] the classification is done by following the given steps. First the data augmentation is done by using moving time frame along with the window size. The features are extracted using CSP technique. Feature selection algorithm which is wrapper based technique and it is optimized by using PSO algorithm to provide more effective results. Then CNN architecture is used with fully connected layer and softmax layer is used for classification. Adam optimizer is used as learning algorithm in order to reduce the error rate.

The CNN Model via Variational AutoEncoder and Stacked Autoencoder. Convolution Neural Network [CNN] and Variational AutoEncoder [VAE] are used to classify the signal [30]. The CNN model is trained using backpropagation algorithm and the VAE model is trained using gradient – based methods. An autoencoder is called undercomplete when the dimension of the code is less when compared to the dimension of the input.The training algorithm will extract the most important features from the training data if undercomplete autoencoder is used. CNN followed by VAE is used combinely inorder to classify the data effectively.

Several CNN models has been constructed and trained, the features extracted from all the models has been concatenated and it is known as feature fusion. Multi – layer CNN feature fusion is used since some model may be best in identifying different good features hence by combining all the features a good number of best features will be obtained. Feature fusion process is done by using Multi Layer Perceptron and Autoencoder. MLP is constructed with two hidden layers in [42]. Both the layers are trained using the concatenated feature vector. In order to get the output probability value softmax function is used. Autoencoders are used to reconstruct the input as an output. Here in this paper cross encoding autoencoders are trained in such a way that they automatically reconstruct the input by extracting robust features from the combined feature set. EEG Signal Motor Imagery task is classified by using CNN and stacked autoencoder [SAE] [22]. Here the input is framed in such as that it include time, frequency, location information.

The Classification Model which consist of Convolutional Neural Network [CNN] and Long Short – Term Memory [LSTM] Networks. In [34] 16×16 image is given as input to the CNN model which has 3 hidden layers. In each of the convolution layer 3×3 size filters are used to extract the features. Then in the pooling layer 2×2 size filters are used to reduce the size of the image. In order to reduce overfitting dropout methodology is used in the third layer by using the value as 0.5. However to reduce the feature lose zero padding technique is used in this model. CNN is used to identify the spatial features from the EEG Signal. Since EEG is time series signal LSTM is employed to extract temporal features. A Deep LSTM Model is built by using several LSTM Cells, a input and output layer and three hidden layers. In both CNN and LSTM architecture cross-entropy is used as loss function and adaptive moment estimator is used as training algorithm.

The CNN Model with Wavelet Neural Network and EEG Source Imaging [ESI]. 2 layer convolution neural network is used. In first layer 200 filters with the filter size of 3×1 is used in [28]. Whereas in the second layer 200 filters with the filter size of 2×2 is used. Pooling layer contains pool size of 4×4. In order to extract the time – frequency information Wavelet Neural Network is used. Here instead of convolution layer in CNN wavelets are used for feature extraction in Wavelet Neural Network. Stochastic Gradient decent based learning algorithm is used in both of these models.

ESI consist of two methods to solve the problem of EEG forward and inverse such as boundary element method and Weighted minimum norm estimation method A novel approach of decoding EEG four-class motor imagery tasks via scout ESI and CNN [39]. The region of interest has been identified by creating ten scouts from the motor cortex. Morlet wavelet approach is used extract the features. The extracted features are passed into CNN for 4 different types of classifications.

4.2 Long Short Term Memory [LSTM]

In [53] LSTM is employed for the classification of EEG Signal. Effective signal representation is extracted by removing the interference using 1-D aggregate approximation technique after normalizing the signal. Channel weighting technique is used for useful information extraction by using spatial filters in the hidden layers. The output obtained is fed into the two LSTM which will be executed parallely. Here Softmax layer is used at last in order to display the probability of the prediction labels.

The time related features such as mean, zero crossing, variance, Peak to Peak distance were extracted. Then the frequency related features from four different bands such as delta, theta, alpha, beta is also extracted. All these features extracted are used in LSTM for classification in [54].

2 layer LSTM and 4 layer neural network is used in [55] to classify the EEG Signal. In Neural network the radial basis operation is carried out with the help of 1-D gradient descent activation function. Statistical features were extracted and it is used for classification with the taken LSTM and NN model. The model uses binary cross entropy as the loss function and adam optimizer is used for learning.

The spatial temporal information can't be extracted simultaneously and dynamic correlation can't be captured using traditional deep learning scheme. Hence these problems are addressed by using LSTM and Spatial Convolution network in [52]. Spatial information are extracted from the convolution layer. Discrete Wavelet Transforms are used to extract time- frequency sequence. It can perform well on spectral and time domain. Batch normalization is also used along with the transform.

Table 3 provides various pre-processing techniques, architecture and accuracy for various papers using BCI Competition IV Dataset 2b. In some of the work the result is compared with their previous work and improvement is shown, for all those reference papers ** is placed in the accuracy field. Best accuracy as 92% for two class classification in BCI Competition IV 2b Dataset is provided in [45] which uses Fast Fourier Transformation as preprocessing technique and Convolutional Neural Network as classification model.

4.3 Recurrent Neural Network [RNN]

The Spatial and frequency features were extracted from the given EEG Signal using FBCSP technique. These features extracted were used to train the RNN [38].This architecture consist of gated recurrent unit and LSTM unit along with the input, recurrent hidden and the output layers. RNN can effectively extract the time series sequential pattern.

4.4 Support Vector Machine [SVM]

The features extracted from CSP technique is used to train the support vector machine classifier. In this work there are three SVM classifiers [35]. The final decision is given by combining the output of all the classifiers using posterior probability. The optimal hyperplane is identified by optimizing the objective function using Lagrange multiplier method.

Table 3. Parameters used for classifying BCI competition IV dataset 2b with 2 classes

Reference paper	Pre-processing technique	Architecture used	Accuracy	Size of the kernal	Optimization algorithm
[19]	Continuous wavelet transform	CNN	83.20%	–	–
[27]		CNN	**	3 × 3	Backpropagation with gradient decent
[20]	Conditional empirical mode decomposition	CNN	82.61%	10 × 10, 3 × 3, 1 × 1	Stochastic gradient descent (SGD) algorithm
[30]	Short time fourier transform	CNN+VAE	**	3 × 3	Backpropagation with gradient decent
[22]		CNN+SAE	**	30 × 3	Backpropagation algorithm
[31]		CapsNet-Based Framework	**	1 × 25, 1 × 7	Stochastic gradient descent (SGD) optimizer
[45]	Fast fourier transformation	CNN	92%	28 × 1, 1 × 10	Gradient descent method
[40]	Fast fourier transform and wavelet package decomposition	Restricted Boltzmann Machine (RBM)	**	–	Backpropagation algorithm
[54]	Nil	LSTM	83.2%	–	Stochastic Adam optimizer

A pretrained AlexNet CNN architecture is used as a feature extractor [42]. It consist of 5 convolution layer along with three fully connected layer and polling layer. An input matrix which is obtained using different transformation techniques are given to this CNN and features will be extracted and it is given as output. Then these extracted features were used to train the SVM for classification. Convex optimization technique is used to find an optimal hyperplane for SVM Classification. In [29] SVM and Linear discriminant analysis are used for classification.

4.5 Restricted Boltzmann Machine [RBM]

Restricted Boltzmann Machine is used for the classification of EEG signal in [22]. First the given EEG Signal is decomposed by using fast Fourier transform and wavelet package decomposition. Here a four layer neural network which consist of three layers of RBM and one output layer with softmax function is used. The RBM consist of two layers one is visible to which the input signal is passed. The other layer is the hidden layer which is used to reconstruct the signal. In both of these layers stochastic binary units were used. It is trained by using the output obtained from the transformations. There are two steps involved in the training process. First RBMs were individually trained in an unsupervised manner in order to reconstruct the output. Then the second step is fine tuning of the whole network parameters using supervised learning along with the conjugate gradient method.

Table 4, 5 provides various pre-processing techniques, architecture and accuracy for various papers using Dataset IVa from BCI Competition III and various other datasets respectively. In some of the work the result is compared with their previous work and improvement is shown, for all those reference papers ** is placed in the accuracy field.

Best accuracy for classifying BCI Competition III dataset is provided in [42] as 96.34% by using the combination of Discrete Cosine Transform, Fourier transform, Common Spatial Pattern, and Empirical Mode Decomposition as preprocessing techniques and CNN, SVM model as classification technique.

Table 4. Parameters used for classifying dataset IVa from BCI competition III with 2 classes

Reference paper	Pre-processing technique	Architecture used	Accuracy	No. of convolution layer/Hidden layer	Size of the Kernal
[25]	Continuous wavelet transform	CNN	94.66%	2	$3 \times 1, 1 \times 2$
[42]	Discrete cosine transform, fourier transform, common spatial pattern, and empirical mode decomposition,	CNN + SVM	**	5	10×1
[33]	Common spatial pattern	Extreme Learning Machine	**	–	–
[52]	Discrete wavelet transformation decomposition	LSTM	90%	3	–

4.6 Transfer Learning

The calibration time has been considerable reduced without compromising the classification accuracy using transfer learning technique [21]. In transfer learning, previously learnt features from the generalized data which is similar to the target data will be used for classifying the given target data. Target data represents the current data being considered for classification. In order to make the classification parameter similar between the target and the generalized data regularization parameter is used in the objective function along with different weights being assigned for different features. Here weighted logistic regression classifier is used along with transfer learning inorder to make classification. [36] uses transfer learning concept along with support vector machine for classifying motor imagery task. Pretrained well known architecture like Alexnet, ResNet are used in [26] for classifying using transfer learning.

Table 5. Parameters used for classifying various dataset

Reference paper	Pre-processing technique	Architecture used	Dataset	Accuracy	Size of the Kernal	Optimization algorithm
[28]	Empirical mode decomposition	CNN wavelet neural network	BCI competition II dataset III	90.10%	2 × 2	Stochastic gradient descent optimizer
[28]	Wavelet packet transform	Transfer learning		97.06%	–	–
[39]	Morlet wavelet approach	CNN	Physionet MI dataset	96%	3 × 3	Adam
[46]	Nil	CNN		58.58%	30 × 1, 64 × 1	Adam
[48]	Nil	CNN	BNCI horizon 2020	52%	1 × 36, 36 × 36,	Adam

4.7 Capsule Network

A group of neurons form a capsule in this capsule network [31]. Each capsule consist of activity vector which is used to exhibit different parameters of an individual entity. The length and the orientation of the capsule's activity vector express the probability of the existence of the entity and instantiation parameters respectively. Hence network architecture which is capsule based exhibits different properties such as size and position with the help of activity vectors. In this capsule network inorder to extract the spatial information effectively which is the main drawback of CNN an effective algorithm called "dynamic routing by agreement" is used. It is an iterative algorithm. The output of one capsule is given to appropriate capsule which is at its higher level. Capsule architecture model consist of a convolution layer, primary and digital capsule layer, decoder and a fully connected layer. A capsule layer can perform classification task effectively.

4.8 End to End Framework

The spatial and temporal features were extracted using the first block and convolution block with mixed scal are used to represent the temporal features. Then the classification block is used to classify the given signal. The classification accuracy is not changed by including perturbation to amplitudes of the spectrums. This kind of data augmentation is done in [32].

A collection of discriminative features are formed in [56]. It is formed by combining the spectral – spatial input and the features learnt using CNN. Spectral-spatial input with unique brain signal pattern is designed by using discriminative frequency band that measures the power of features in various classes. Then the covariance between matrix is obtained using the spectral-spatial input. The Covariance matrix is used for classification.

5 Conclusion

In this work a review of MI classification task by using EEG signals is done. Most of the work is proposed by using CNN model with different architectures. Combination of CNN and other techniques is also used. Beyond CNN the next widely used architecture for analyzing EEG is RNN or LSTM. Several pre-processing techniques used in various work is also described in this paper. Commonly used pre-processing technique is Common Spatial Pattern. In Four class classification, CNN model used along with Short Time Fourier Transform and Common Spatial Pattern provided good accuracy when compared with other traditional models. Whereas in Two class classification, Transfer learning based CNN model with some pre- processing techniques are used to provide better accuracy. In most of the papers stochastic gradient descent is used as the optimizer. Hence it is concluded that CNN used along with Common Spatial Pattern and Short Time Fourier as the pre processing technique and stochastic gradient descent as the optimizer provides better accuracy in classifying motor imagery task in EEG signals. Accuracy for classifying the task as either left or right hand movement, hand or foot movement ie., two class classification is good when compared to four classes classification such as left hand, right hand, feet and rest, forearm supination/pronation, elbow movement, hand grasping, and rest.

References

1. Wolpaw, J.R., Wolpaw, E.W.: Brain-Computer Interfaces: Principles and Practice. Oxford University Press, Oxford (2012)
2. Wolpaw, J.R., McFarland, D.J., Neat, G.W., Forneris, C.A.: An EEG-based brain-computer interface for cursor control. Electroencephalogr. Clin. Neurophysiol. **78**(3), 252–259 (1991). https://doi.org/10.1016/0013-4694(91)90040-B
3. Pfurtscheller, G., Müller-Putz, G.R., Scherer, R., Neuper, C.: Rehabilitation with brain-computer interface systems. Comput. (Long. Beach. Calif) **41**(10), 58–65 (2008). https://doi.org/10.1109/MC.2008.432
4. Ang, K.K., Guan, C.: Brain-computer interface in stroke rehabilitation. J. Comput. Sci. Eng. **7**(2), 139–146 (2013). https://doi.org/10.5626/JCSE.2013.7.2.139

5. Coyle, D., Principe, J., Lotte, F., Nijholt, A.: Guest editorial: brain/neuronal-computer game interfaces and interaction. IEEE Trans. Comput. Intell. AI Games **5**(2), 77–81 (2013). https://doi.org/10.1109/TCIAIG.2013.2264736
6. Zander, T.O., Kothe, C.: Towards passive brain-computer interfaces: applying brain-computer interface technology to human-machine systems in general. J. Neural Eng. **8**(2) (2011). doi: https://doi.org/10.1088/1741-2560/8/2/025005.
7. Lotte, F., et al.: A review of classification algorithms for EEG-based brain-computer interfaces: a 10 year update. J. Neural Eng. **15**(3) (2018). https://doi.org/10.1088/1741-2552/aab2f2
8. Ge, S., et al.: Sinusoidal signal assisted multivariate empirical mode decomposition for brain-computer interfaces. IEEE J. Biomed. Heal. Inf. **22**(5), 1373–1384 (2018). https://doi.org/10.1109/JBHI.2017.2775657
9. Lotte, F.: A tutorial on EEG signal processing techniques for mental state recognition in brain-computer interfaces a tutorial on EEG signal processing techniques for mental state recognition in brain-computer interfaces. Eduardo Reck Miranda; Julien Castet. Guide to Brain-Computer Music Interfacing. Springer (2014). https://hal.inria.fr/hal-01055103. Accessed 06 Sept 2020
10. Blankertz, B., Tomioka, R., Lemm, S., Kawanabe, M., Müller, K.-R.: Optimizing spatial filters for robust EEG single-trial analysis (2008)
11. Lotte, F., Congedo, M., Lécuyer, A., Lamarche, F., Arnaldi, B.: A review of classification algorithms for EEG-based brain-computer interfaces. J. Neural Eng. **4**(2) (2007). https://doi.org/10.1088/1741-2560/4/2/R01
12. Brain-Computer Interfaces 2: Technology and Applications | Wiley. https://www.wiley.com/en-in/Brain+Computer+Interfaces+2:+Technology+and+Applications-p-9781848219632. Accessed 06 Sept 2020
13. Hsu, W.Y.: Fuzzy Hopfield neural network clustering for single-trial motor imagery EEG classification. Expert Syst. Appl. **39**(1), 1055–1061 (2012). https://doi.org/10.1016/j.eswa.2011.07.106
14. Ai, Q., et al.: Feature extraction of four-class motor imagery EEG signals based on functional brain network. J. Neural Eng. **16**(2) (2019). https://doi.org/10.1088/1741-2552/ab0328.
15. Chen, M., Fang, Y., Zheng, X.: Phase space reconstruction for improving the classification of single trial EEG. Biomed. Signal Process. Control **11**(1), 10–16 (2014). https://doi.org/10.1016/j.bspc.2014.02.002
16. Hortal, E., et al.: SVM-based brain-machine interface for controlling a robot arm through four mental tasks. Neurocomputing **151**(P1), 116–121 (2015). https://doi.org/10.1016/j.neucom.2014.09.078
17. Xiao, X., Xu, M., Jin, J., Wang, Y., Jung, T.P., Ming, D.: Discriminative canonical pattern matching for single-trial classification of ERP components. IEEE Trans. Biomed. Eng. **67**(8), 2266–2275 (2020). https://doi.org/10.1109/TBME.2019.2958641
18. Alwasiti, H., Yusoff, M.Z., Raza, K.: Motor imagery classification for brain computer interface using deep metric learning. IEEE Access **8**, 109949–109963 (2020). https://doi.org/10.1109/ACCESS.2020.3002459
19. Li, F., He, F., Wang, F., Zhang, D., Xia, Y., Li, X. : A novel simplified convolutional neural network classification algorithm of motor imagery EEG signals based on deep learning. Appl. Sci. **10**(5) (2020). https://doi.org/10.3390/app10051605
20. Tang, X., Li, W., Li, X., Ma, W., Dang, X.: Motor imagery EEG recognition based on conditional optimization empirical mode decomposition and multi-scale convolutional neural network. Expert Syst. Appl. **149** (2020). https://doi.org/10.1016/j.eswa.2020.113285
21. Azab, A.M., Mihaylova, L., Ang, K.K., Arvaneh, M.: Weighted transfer learning for improving motor imagery-based brain-computer interface. IEEE Trans. Neural Syst. Rehabil. Eng. **27**(7), 1352–1359 (2019). https://doi.org/10.1109/TNSRE.2019.2923315

22. Tabar, Y.R., Halici, U.: A novel deep learning approach for classification of EEG motor imagery signals. J. Neural Eng. **14**(1), 16003 (2017). https://doi.org/10.1088/1741-2560/14/1/016003
23. Shajil, N., Mohan, S., Srinivasan, P., Arivudaiyanambi, J., Arasappan Murrugesan, A.: Multiclass classification of spatially filtered motor imagery EEG signals using convolutional neural network for BCI based applications. J. Med. Biol. Eng. (0123456789) (2020). https://doi.org/10.1007/s40846-020-00538-3
24. Majidov, I., Whangbo, T.: Efficient classification of motor imagery electroencephalography signals using deep learning methods. Sensors (Switzerland) **19**(7), 1–3 (2019). https://doi.org/10.3390/s19071736
25. Ortiz-Echeverri, C.J., Salazar-Colores, S., Rodríguez-Reséndiz, J., Gómez-Loenzo, R.A.: A new approach for motor imagery classification based on sorted blind source separation, continuous wavelet transform, and convolutional neural network. Sensors (Switzerland) **19**(20), 1–4 (2019). https://doi.org/10.3390/s19204541
26. Kant, P., Laskar, S.H., Hazarika, J., Mahamune, R.: CWT based transfer learning for motor imagery classification for brain computer interfaces. J. Neurosci. Methods **345**(February), 108886 (2020). https://doi.org/10.1016/j.jneumeth.2020.108886
27. Lee, H.K., Choi, Y.S.: Application of continuous wavelet transform and convolutional neural network in decoding motor imagery brain-computer Interface. Entropy **21**(12), 1–1 (2019). https://doi.org/10.3390/e21121199
28. Zhang, Z., et al.: A novel deep learning approach with data augmentation to classify motor imagery signals. IEEE Access **7**, 15945–15954 (2019). https://doi.org/10.1109/ACCESS.2019.2895133
29. Gaur, P., Pachori, R.B., Wang, H., Prasad, G.: An automatic subject specific intrinsic mode function selection for enhancing two-class EEG-based motor imagery-brain computer interface. IEEE Sens. J. **19**(16), 6938–6947 (2019). https://doi.org/10.1109/JSEN.2019.2912790
30. Dai, M., Zheng, D., Na, R., Wang, S., Zhang, S.: EEG classification of motor imagery using a novel deep learning framework. Sensors (Switzerland) **19**(3), 1–6 (2019). https://doi.org/10.3390/s19030551
31. Ha, K.W., Jeong, J.W.: Motor imagery EEG classification using capsule networks. Sensors (Switzerland) **19**(13) (2019). https://doi.org/10.3390/s19132854
32. Li, Y., Zhang, X.R., Zhang, B., Lei, M.Y., Cui, W.G., Guo, Y.Z.: A channel-projection mixed-scale convolutional neural network for motor imagery EEG decoding. IEEE Trans. Neural Syst. Rehabil. Eng. **27**(6), 1170–1180 (2019). https://doi.org/10.1109/TNSRE.2019.2915621
33. Zhang, Y., et al.: Multi-kernel extreme learning machine for EEG classification in brain-computer interfaces. Expert Syst. Appl. **96**, 302–310 (2018). https://doi.org/10.1016/j.eswa.2017.12.015
34. Search, H., et al.: A novel hybrid deep learning scheme for four-class motor imagery classification (2017)
35. Luo, J., Gao, X., Zhu, X., Wang, B., Lu, N., Wang, J.: Motor imagery EEG classification based on ensemble support vector learning. Comput. Methods Programs Biomed. **193** (2020). https://doi.org/10.1016/j.cmpb.2020.105464.
36. Zheng, M., Yang, B., Xie, Y.: EEG classification across sessions and across subjects through transfer learning in motor imagery-based brain-machine interface system. Med. Biol. Eng. Comput. **58**(7), 1515–1528 (2020). https://doi.org/10.1007/s11517-020-02176-y
37. Sakhavi, S., Guan, C., Yan, S.: Learning temporal information for brain-computer interface using convolutional neural networks. IEEE Trans. Neural Netw. Learn. Syst. **29**(11), 5619–5629 (2018). https://doi.org/10.1109/TNNLS.2018.2789927

38. Luo, T., le Zhou, C., Chao, F.: Exploring spatial-frequency-sequential relationships for motor imagery classification with recurrent neural network. BMC Bioinf. **19**(1), 18 (2018). https://doi.org/10.1186/s12859-018-2365-1
39. Hou, Y., Zhou, L., Jia, S., Lun, X.: A novel approach of decoding EEG four-class motor imagery tasks via scout ESI and CNN. J. Neural Eng. **17**(1) (2020). https://doi.org/10.1088/1741-2552/ab4af6.
40. Lu, N., Li, T., Ren, X., Miao, H.: A deep learning scheme for motor imagery classification based on restricted Boltzmann machines. IEEE Trans. Neural Syst. Rehabil. Eng. **25**(6), 566–576 (2017). https://doi.org/10.1109/TNSRE.2016.2601240
41. Zhang, Y., Zhang, X., Sun, H., Fan, Z., Zhong, X.: Portable brain-computer interface based on novel convolutional neural network. Comput. Biol. Med. **107**(November 2018), 248–256 (2019). https://doi.org/10.1016/j.compbiomed.2019.02.023
42. Amin, S.U., Alsulaiman, M., Muhammad, G., Mekhtiche, M.A., Shamim Hossain, M.: Deep learning for EEG motor imagery classification based on multi-layer CNNs feature fusion. Futur. Gener. Comput. Syst. **101**, 542–554 (2019). https://doi.org/10.1016/j.future.2019.06.027
43. Tan, P., Wang, X., Wang, Y.: Dimensionality reduction in evolutionary algorithms-based feature selection for motor imagery brain-computer interface. Swarm Evol. Comput. **52**, 100597 (2020). https://doi.org/10.1016/j.swevo.2019.100597
44. Amin, S.U., Alsulaiman, M., Muhammad, G., Bencherif, M.A., Hossain, M.S.: multilevel weighted feature fusion using convolutional neural networks for EEG motor imagery classification. IEEE Access **7**, 18940–18950 (2019). https://doi.org/10.1109/ACCESS.2019.2895688
45. Tang, Z., Li, C., Sun, S.: Single-trial EEG classification of motor imagery using deep convolutional neural networks. Optik (Stuttg) **130**, 11–18 (2017). https://doi.org/10.1016/j.ijleo.2016.10.117
46. Dose, H., Møller, J.S., Iversen, H.K., Puthusserypady, S.: An end-to-end deep learning approach to MI-EEG signal classification for BCIs. Expert Syst. Appl. **114**, 532–542 (2018). https://doi.org/10.1016/j.eswa.2018.08.031
47. Lawhern, V.J., Solon, A.J., Waytowich, N.R., Gordon, S.M., Hung, C.P., Lance, B.J.: EEGNet: a compact convolutional network for EEG-based brain-computer interfaces. J. Neural Eng. **15**(5) (2016). https://doi.org/10.1088/1741-2552/aace8c
48. Jeong, J.H., Lee, B.H., Lee, D.H., Yun, Y.D., Lee, S.W.: EEG classification of forearm movement imagery using a hierarchical flow convolutional neural network. IEEE Access **8**(Mi), 66941–66950 (2020). https://doi.org/10.1109/ACCESS.2020.2983182
49. Gao, Z., Li, Y., Yang, Y., Dong, N., Yang, X., Grebogi, C.: A coincidence-filtering-based approach for CNNs in EEG-based recognition. IEEE Trans. Ind. Inf. **16**(11), 7159–7167 (2020). https://doi.org/10.1109/TII.2019.2955447
50. Pfurtscheller, G.: Event-related synchronization (ERS): an electrophysiological correlate of cortical areas at rest. Electroencephalogr. Clin. Neurophysiol. **83**(1), 62–69 (1992). https://doi.org/10.1016/0013-4694(92)90133-3
51. Zhao, X., Zhang, H., Zhu, G., You, F., Kuang, S., Sun, L.: A multi-branch 3D convolutional neural network for eeg-based motor imagery classification. IEEE Trans. Neural Syst. Rehabil. Eng. **27**(10), 2164–2177 (2019). https://doi.org/10.1109/TNSRE.2019.2938295
52. Yang, J., Yao, S., Wang, J.: Deep fusion feature learning network for MI-EEG classification. IEEE Access **6**, 79050–79059 (2018). https://doi.org/10.1109/ACCESS.2018.2877452
53. Wang, P., Jiang, A., Liu, X., Shang, J., Zhang, L.: LSTM-based EEG classification in motor imagery tasks. IEEE Trans. Neural Syst. Rehabil. Eng. **26**(11), 2086–2095 (2018). https://doi.org/10.1109/TNSRE.2018.2876129

54. Zhang, G., Davoodnia, V., Sepas-Moghaddam, A., Zhang, Y., Etemad, A.: Classification of hand movements from EEG using a deep attention-based LSTM network. IEEE Sens. J. **20**(6), 3113–3122 (2020). https://doi.org/10.1109/JSEN.2019.2956998
55. Nagabushanam, P., Thomas George, S., Radha, S.: EEG signal classification using LSTM and improved neural network algorithms. Soft Comput. **24**(13), 9981–10003 (2020). https://doi.org/10.1007/s00500-019-04515-0
56. Kwon, O.-Y., Lee, M.-H., Guan, C., Lee, S.-W.: Subject-independent brain-computer interfaces based on deep convolutional neural networks. IEEE Trans. Neural Networks Learn. Syst. **PP**(D1), 14 (2019). https://doi.org/10.1109/tnnls.2019.2946869

Deep Learning Techniques for Breast Cancer Diagnosis: A Systematic Review

B. Krishnakumar(✉) and K. Kousalya

Kongu Engineering College, Perundurai, Erode 638 060, India
krishnakumar.cse@kongu.edu

Abstract. Breast cancer analysis is carried out by many computer aided diagnosis methodologies. Recently among all methodologies deep learning techniques provides better accuracy. The image segmentation, extraction of features and classification process are done automatically using deep learning techniques. Hence deep learning techniques plays an important role in analysing images. In this paper various models of deep learning used for the classification of different types of cancer such as Benign, Malignant, In-situ, Invasive Ductal Carcinoma is described. The image processing steps are explained for each and every model starting from input layer to output layer. Transfer learning techniques along with fine tuning process is also given. It also survey works which uses images from various modalities such as screen film mammography digital mammography, ultrasound, magnetic resonance imaging, digital breast tomosynthesis. This review provides an overview of all the deep learning models used for breast cancer classification of images from various modalities.

Keywords: Breast cancer classification · Deep learning · Transfer learning

1 Introduction

Breast cancer is one of the most dangerous type of cancer which may leads to death. Therefore breast cancer early detection and diagnosis will improve the survival rate of the patients by providing flexible treatment [1–4]. Screening of the breast cancer is done by utilizing several modalities such as screen film mamaography digital mammography, ultrasound, magnetic resonance imaging, digital breast tomosynthesis. All these modalities provides the internal anatomy of the breast tissues with high spatial resolution images. The challenges lies in finding the breast information fully even though the breast tissues are overlapped each other in order to find the suspicious lesions [5, 6].

There are four different types of breast cancer such as Benign, Malignant, In-situ, Invasive Ductal Carcinoma [IDC] [7, 8]. Cancer cells which are of benign type are not transmitted to any other parts of the body, whereas malignant type will be transferred to new parts of the body easily. In in-situ type of cancer lobules, ducts present in the breast will be expanded since normal cells in the breast is replaced by the abnormal cells with cancer tissues. The next stage is invasive ductal carcinoma where the cancer cells were spread to all the breast tissues surrounding the point of origin by collapsing the

J. A. Mat Jizat et al. (Eds.): iCITES 2020, AISC 1350, pp. 155–171, 2021.
https://doi.org/10.1007/978-3-030-70917-4_16

ducts. Early identification of the type of cancer will be very much useful for providing the appropriate treatment.

Breast cancer diagnosis for classifying the type of cancer is performed over thirty years by computer aided models [9, 10]. By using these models a cost effective diagnosis and treatment is performed with high sensitivity and accuracy. Among all the technique deep learning is widely used because of its effective performances. Deep learning is used in many areas such as medical imaging, analysis of medical images [11–15]. Some review papers has been published in this area. A brief introduction about those review papers are given below. Various methods used for detection of lesion by using various modality images is provided in [16, 17]. The main focuses on the analysis of histology and mammography images are shown in [13]. This paper focus on reviewing of various deep learning architectures and transfer learning techniques used for classifying images from various breast screening modalities.

The paper is organized as follows with the first section which provides overall process involved during the classification of breast cancer types using computer aided models. Then in the next section various deep learning models used were described. Third section provides tabulation based description about various architectures used for different dataset along with its accuracy. The last section is conclusion.

1.1 Overall Process in Classification

The classification process are taken place in five steps [37]. First data is collected and it is pre-processed in order to remove duplicates, strain normalization is also done in this phase. Augmentation process such as rescaling is carried out as the third step. Then the model is trained by using the training data. Fine tuning of the weights are done by using learning algorithm. Finally test set is applied on the trained model in order to get the classification output. Many different models are used for classification. All these models are explained below.

2 Various Models Used for Classification

2.1 Stacked Sparse Autoencoders and Softmax Regression as Classifier

Autoencoder is a type of artificial neural network. In order to produce the given input information in a different form as output. It is trained using unlabelled data which is known unsupervised learning. Autoencoder takes an important role in deep learning platform. Sparse autoencoder is also similar to autoencoder but with some sparseness constraints embedded on each and every hidden neuron present in the neural network. Sparse penalty term is used in this sparse autoencoder. Stacked sparse autoencoders forms a deep learning network which consist of several sparse autoencoders arranged one after another. The result of each layer is passed on to the successive layers. At last output layer consist of softmax function which acts as a classifiers. Stacked sparse autoencoders is one of the most efficient deep learning architectures.

In two sparse autoencoders has been trained to produce two different representations [18]. Initially one sparse autoencoder which is stacked at the first layer is trained using the given dataset which is known as first representation. The output received from the first encoder using the neurons of first hidden layer is passed on to next sparse autoencoder. This will again generates an another output which is known as second representation. These two representations are combined to generate third representation. Since there are three representation three softmax classifier is used correspondingly for classification. To reduce the error and to improve the performance. The fine tuning operation is to carried out for all the input layer, hidden layer which produces different representations and the softmax layer which produces the output.

Gene expression is used along with stacked autoencoder for extracting high level representations for each and every data [19]. High level features of all the images were combined together and to form a complex feature using a hidden layer. By using all these learnt features the given data is classified into different clusters by using K- Means clusters.

GoogleNet. GoogleNet consist of three hidden layers in which convolution, pooling layers and rectified linear units are used. It is followed by two fully connected layers. In [20] the same googleNet architecture is proposed with single filter which is obtained by combining several filters in convolution layer with different sizes. Because if using single filter the number of parameters used and the computational complexity are reduced to some extent.

VGGNet. VGGNet architecture is as like as Alexnet architecture with some additional convolution layers. There are 13 convolution layers along with pooling, rectification unit and 3 fully connected layers. Filters of size 3×3 is used in convolution layer whereas 2×2 is used in pooling layer. VGGNet Performance is high when compared to AlexNet architecture.

ResNet. ResNet is one of the most efficient deep residual network which provides very good accuracy while classifying ImageNet dataset. It is mainly used in large – scale data analysis. Here the degradation problem is rectified by using various different sized convolution filters at different layers. This model takes only less time for training itself. The shortcut links between the layers are provided in ResNet with the help of many basic residual blocks. These shortcuts can create more number of layers and when the number of layers gets increased then performance can also be increased. The shortcut connection also helps resnet to skip some of the entries in the network. Hence vanishing gradient problem during backpropagation and loss of information which is transmitted in the hidden layers are reduced with the help of skip connections. ResNet-50 architecture contains 50 convolution layers along with fully connected layers.

DenseNet. In DenseNet architecture all the nodes in one layer is connected to each and every node in another layer. It is as like as Resnet with all the nodes connected to each other in a feed forwarded manner. Number of connections are very high in these type of network. The number of parameters used in Densenet is less when compared to other architectures. It also solves the vanishing gradient problem. Training Imagenet data over DenseNet model increases the computation cost but provides good performance in terms of accuracy.

Popular CNN architectures such as ResNet, GoogleNet, VGGNet are used to extract all low level features from the given cytology images in [20]. All the extracted features are combined and it is given as input to the fully connected layer for the purpose of classification. Pretraining of these famous architectures were done with the help of generalized images. Transfer learning theory is used to extract features from the given images.

Mammography images were effectively classified using [21]. First the preprocessing steps such as image normalization and data enhancement is done in order to clear the interference caused due to light and to reduce overfitting respectively. Then Dense Net architecture is used by replacing the first layer by using the Inception structure for the classification of the given input images.

2.2 Transfer Learning

Transfer learning is the one in which background knowledge which is used to solve one problem can be used for solving similar problems using the same model. First the model is trained for one particular task using its relavant dataset. Then again the model is trained with actual dataset for the given problem. There are two main steps involved in the process of transfer learning such as choosing of pre-trained model, size and similarity of the problem. When the size of the dataset for the taken problem is less and it is very much similar to the data in which the model is trained then there will be problem of overfitting. Similarly when the size of the dataset for the taken problem is high and it is similar to the data in which the model is trained then the problem of overfitting will be less but the pretrained model should be tuned according to the problem taken. In three pretrained architecture namely GoogLeNet, VGGNet, and ResNet were used to do transfer learning [20]. These pretrained models were initially trained with the help of ImageNet dataset. Hence these model can learn generic features easily without any new training. While using these model for breast cancer classification problem all the features extracted using the models were combined and processed using fully connected layer in order to classify whether the given data belongs to the type malignant or benign.

Large amount of training data is required to train deep learning models. Medical Images with pathology proven disease is very less when compared to Imagenet data because of its demand in data-to-date life. Hence it take more time, expertise and cost for the process of compiling medical related data. Transfer learning can be used in such scenarios where all the basic level features are learned by the model using generic data and higher level features are only learned using given medical dataset. Proper fine tuning of the model is to be done in order to learn the basic features correctly.

Multi-stage transfer learning methodology is proposed in [22]. First transfer learning is applied to DCNN which is already trained with ImageNet data. Now the same DCNN is trained again with the target mammography data by doing so, the performance could be increased. In this training process only the first layer weights are taken from pre-trained model, remaining all layers were trained again with the target dataset. Then two fully connected layers were added in order to perform classification after training with the target data in the second phase DBT training is performed to detect the presence of masses in mammograms without performing any additional fine tuning operation. Several features were extracted in phase two training step. Important features were extracted by using recursive feature elimination methodology. DCNN pruning is also done in stage two with the help of genetic algorithm.

In Residual Network and Densely Connected Convolutional Networks are used to detect ductal carcinoma in the histopathology images [23]. [24] uses deep belief network for classifying breast cancer. Each and every image is transformed into three different scales and then it is used for training the CNN Model for the classification task [25].

2.3 Deep Convolutional Neural Network [DCNN]

DCNN is employed in to detect epithelial and stromal tissues in the given histological images automatically by extracting the features [8]. DCNN uses raw pixel intensity values whereas the older methods uses color and texture information for identifying tissues. The DCNN provides better accuracy when compared with methodologies. The identification of these tissues in the image will be very much useful while analysing images using computerized system.

Deep Learning techniques are applied to detect the presence of mitosis in [26]. The main idea behind this proposed model is utilization of small patches of images for training the model. Various handcrafted features such as intensity, textural, morphological were used in this work by extracting from the training data. The shape of the mitosis is provided by morphological features. The textural and statistical behaviour of the mitosis region is provided by using texture, intensity features respectively.

The classification of mitotic and non-mitotic nuclei is done by using two phase CNN model in [27]. The two phase CNN model will reduce the class skewness by following the steps given below. First binary thresholding, blue ratio are the techniques which are used to segment the images. Then the patched images are used for training the phase I CNN. The augmentation process is carries out by undersampling and oversampling of non-mitotic, mitotic examples respectively. Oversampling is done by using rotation and flipping techniques where as undersampling is done by using k-means clustering using blue ration histogram. In order to reduce the class imbalance problem, the dataset is modified as balanced set before sending it to phase II CNN Model. The main advantage of this proposed technique is reduction of skewness, training and evaluation time.

DCNN is used along with kernals of multiple size and skip connections in order to overcome the drawbacks of hand-crafted features such as change in size and shape which may lead to poor performance [28]. There are two steps involved in the proposed system. First it is checked for the presence of malignant tumors then in the next step solid nodules where identified from the image. In order to combine these two steps effectively class activation maps are used along with the cross training algorithm which is used to train the network.In this work ultrasonography images were taken for classification.

Breast Cancer Cell Nuclei Classification by Using DCNN. The histopathology images is classified as nuclei or non-nuclei by using DCNN [29]. The proposed system is suitable for the dataset where there is less number of labelled data and large number of unlabelled data. As the first step stacked denoising autoencoder is employed to extract high level features from the given raw images. Then these features along with the labelled data is used to train the DCNN model and optimization operation is also performed. Finally the images is classified as nuclei or non-nuclei images.

DCNN and Boosting Tree Classifier. The combination of Deep Convolution Neural Network [DCNN] and Boosting Tree classifier is used to classify the given image into four different classes such as normal tissue, benign, insitu and invasive carcinomas [30]. Initially the common space and the variance is reduced in the image by using stain normalization which is used to enhance the detection performance [30, 31]. DCNN is used to extract the low dimensional features from the high dimensional input images. In the last layer of DCNN one dimensional feature vector is created and it is given as input to tree classifier. Tree Classifiers are used in this work since it can tackle the problem of imbalance in data and also less number of training samples. This provides better accuracy since boosting classifier has the capacity of converting week into strong classifier by adjusting the weights accordingly.

DCNN and SVM. The proposed system consist of two segmentation steps [32]. In the first step the Region of Interest [ROI] is determined manually. In the second step region and threshold based method is used for finding ROI. Transfer learning is used along with AlexNet for feature extraction and classification. Support Vector Machine [SVM] is used after the fully connected layer in the AlexNet model inorder to improve the performance. Image Quality Assessment technique is used along with CNN [33]. First the pixels which are affected by lesion is identified by CNN and for those pixels, large amount of weight values were used. Kernel values for the image is calculated by using Multi-SVM technique to achieve better classification accuracy.

2.4 Convolution Neural Network [CNN]

CNN model is used for classifying the image into benign, malignant, healthy [34]. First data preprocessing is done by splitting the larger images into smaller patches which is used to reduce the processing time in CNN. Then the patches are forwarded as input to CNN model which consist of input, two hidden and an output layer. At last lesion detector is used for detecting objects in a given image by using multibox detector. By using this the classification is done. Thermal breast images were analysed using image processing techniques to detect the sign of cancer cells [35]. The features were extracted and passed to CNN which is optimized using Bayes algorithm for classification.

Nuclei – Guided Feature Extraction Method Based on CNN. The proposed framework consist of three steps namely: detection of nuclei, pre – training of the model and the last step is fine-tuning of the parameters [36]. First nuclei from the image is identified by employing blob detection methodology. By using the nuclei's location information the neural network model is trained and features are identified based on the pattern of the nuclei. The next step is pretraining of auto encoder using the nuclei information which is used to get rid of the redundancies and to extract the discriminative patterns from the unlabelled input data. Three different models are stacked together to form a network using which the high level features can be extracted from the input images. The weights and bias values present in each model is trained and optimized separately. Since all the models were combined together, now weights and bias should be optimized globally using supervised information and this is the last step after this classification is done by using softmax function.

CNN and LSTM. Convolution neural network and Long Short Term Memory are combined together to form a model for classification. The last layer in the model consist of softmax and support vector machine which is used for making decisions. Three different models were used for analysing the data [37]. First only CNN Model is used. Then LSM model and the third model is created by using both CNN and LSTM. Adaboost algorithm is used along with deep learning for tumor classification [38]. Here adaboost is used for selecting appropriate features and those features were passed to CNN and LSTM model for classification.

FCN and Bi-LSTM. Fully connected network along with bidirectional long short term memory is used to classify high dimensional data in [39]. First the input image is fed into FCN to encode the features. Then the feature are flatten and passed on to the next Bi-LSTM model for classification.

Inception Recurrent Residual CNN [IRRCNN]. IRRCNN model is constructed based on Inception Network, RESNET and Recurrent convolution neural network [40]. It consist of input layer, recurrent based convolution layer, inception and residual units, output layer with softmax function.This model uses less number of learning parameters and provide good performance.

Mixture Ensemble CNN. Mixture Ensemble CNN is used for classification [41]. CNN based experts and gating networks are present in the proposed model. Each experts are specially designed for one ROI of the input image. Gating network is used to find the optimal weight which are used to combine the output of the experts. The model is used for MRI images.

2.5 Multiple Instance Learning

Multiple Instance Learning which is a supervised learning framework is used for diagnosing breast cancer patients [42]. Instead of labelling all the instances here the instances are grouped together into different bags and they are labelled. CNN which is a non-parametric approach along with multiple instance learning provides best result in the task of classification.

Multiple Instance and Multi Label Learning. Side level classifiers and ROI level classifiers are using multiple instance and multi –label learning technique [43]. The main difference between traditional learning and multiple instance learning is by using the concept of bags. Many positive as well as negative instance will be placed in a single training bag with its associated class label which is bag level. There are two types of bags available such as positive and negative. Positive bag will contain atleast one positive instance where as in negative bag all the samples will be of type negative. But during the training process itself the labels are not revealed. Here each and every sample will be represented by a bag with multiple instance. Multiple class labels are provided to each and every bag. So only this methodology is proposed as multiple instance and multi label learning and it provides better accuracy.

Deep learning with Enhanced Loss Function [ELF]. In order to reduce the processing time and the error rate in autoencoder, K-Means clustering, deep learning techniques were used along with enhanced loss function [ELF] using linear transformation filter [44]. First the cell level information and tissue level information is gathered by extracting the patches from the given image.Then these images are clustered by using autoencoder along with K-Means clustering technique. Now the image is classified using the deep learning model with enhance loss function which decreased the classification error rate.

2.6 Xception Model

Google has designed a deep learning model which provides good classification accuracy for the large amount of dataset. In this model depthwise separable convolution layer is used. In order to solve the problem of overfitting and to obtain high classification accuracy. Two designs were proposed in [45]. First the model is designed by using layers from the xception model to extract the features from the given image. Second transfer learning is used to tune the model designed.

2.7 Restricted Boltzmann Machine [RBM]

Restricted Boltzmann Machine is a type of artificial neural network which is generative stochastic in nature. RBM neural network learns probability distribution from its input data.

2.8 Deep Belief Network [DBN]

Deep Belief Network is a type of deep neural network with a generative model which consist of many different layers with hidden nodes. The connection is established between each layers. DBN is a combination of many RBM learning modules with many neurons represents units in each layer. Data is processed in each layer.In each layer hidden units are used to extract features from the given data. The neurons or units present in each layer in connected by using matrix with weighted connections. RBM are stacked together to form DBN. Learning task using DBN is very much computational intensive.

Fine tuning can be done to DBN similar to neural network model. DBN Networks belongs to generative pretrained network in which dimensionality reduction of input features can be done with the help of non linear hidden layers. DBN performs well in pattern recognition, Handwritten digits recognition applications. The accuracy of DBN can be improved in unsupervised learning with the help of back propagation pass techniques. DBN [24] provides better accuracy when pretraining of the model is done using unsupervised learning prior to the supervised training of the target dataset.

Various learning algorithms such as RMSProp, adaptive gradient, nesterov, stochastic gradient descent, accelerated gradient, Adam were used for initializing and updating the weights of the convolution neural network [46]. Among all the leaning algorithms stochastic gradient descent provides better accuracy. The given image is classified as benign and malignant using the CNN Model. The performance of the model is improved by using multiple filters in the convolution layer and by using batch normalization.

DNN for Breast Cancer Screening. Global and local information from the image is extracted in [47] by using two-phase neural network. The neural network which is a patch level network with large capacity is used to find pixel level and breast level labels. The proposed method provides heatmaps which pinpoints the suspicious findings location. The proposed model uses architecture which is similar to ResNet inorder process large images by taking less amount of time for classification. In [31] ultrasound images are processed using neural network for classification. Here median filter and adaptive weighted filters are used to improve the accuracy.

BreastNet Model. BreastNet model is built by using attention modules in the residual architecture [48]. Augmentation techniques such as flip, rotation and shift are used to modify the features of the original image. Then the attention modules are used to select and process the ROI in the image. The hypercolumn technique is also used to provide accurate classification. Remaining parts are same compared with other resnet architectures.

HIC-Net. Window based classification is performed with some pre-processing steps in order to provide better prediction and to improve the training rate [49]. In the pre-processing step first the artefacts are removed to provide better learning facility. Then fluctuations are removed from the background of the image inorder to display the cells effectively. As a result of pre-processing, cancer affected region will be larger but light in colour where as non affected region will be dark and small in size. This pre-processing steps will help the CNN model to learn the features easily.

2.9 Classification Using Patches Based on Deep Learning Techniques

Discriminative Patches were chosen by using clustering and CNN methods in [50]. First the cell level and the tissue level features were extracted as patches in unequal size from the given images by using sliding window technique. In the next step the obtained patches are divided into different clusters by applying K-Means clustering algorithm and discriminative patches were obtained. Then the final feature image corresponding to the each and every original image is obtained by combining the features extracted from the large and small patches which is obtained in the previous two steps.

2.10 Mass Lesion Detection

Mass lesions in mammography is classified using CNN in [50]. The lesions are expressed as ROIs with stable size. Then normalization is done by employing global contrast normalization. Here transfer learning is used based on ResNet50 and other state-of-art-architectures for classification. A model which maps the features which are abnormal in mammogram to its histopathological representation in [13]. Here two different model were used one is Mammography model and the other is Histology Model. Each model are trained separately in order to reduce the error. Nottingham Grading System is used to differentiate nuclei, mitoses and tubules. Along with this high level and low level features are also extracted from both the models. These features are combined and analysis is performed to provide classification result.

3 Comparison of Various Models Used for Classification

Two class classification means the data will be classified either as Benign or Malignant where as Four class classification represents the four different classes such as Benign, Malignant, In-situ, Invasive Ductal Carcinoma. Table 1 represents various model used for classifying BreakHis dataset which contain two different class images. Table 2 represent the model used various other datasets. Table 3 provides the model used for classifying four class dataset.

Table 1. Various deep learning models used to perform two class classification of BreaKHis dataset

Reference paper	Pre-processing technique	Classification technique	Training algorithm	Implementation Framework	No. of epoch	Accuracy
[42]	Augumentation which generate images as bag of words	Non parametic multiple instance learning CNN technique	Hyperparameters are optimized using grid search	Keras and Theano framework	100	92.10%
[51]	Stain Normalization, Image resize, Augumentation	CNN	Gradient descent solver	Matlab code	30	97.74%
[37]	K Mean and Mean shift Clustering	DNN	–	–	500	91.00%
[39]	Fully Convolutional network for network feature extraction	Bi-LSTM	Stochastic gradient descent and Adam optimizer	Matlab code	8 to 20	95.69%
[44]	Linear transformation filter	AutoEncoder with Enhanced Loss Function [ELF], K-Means clustering	–	Tensorflow	–	97.00%
[40]	Resizing, Augumentation	Inception recurrent residual CNN	Stochastic gradient decent	Keras and tensorflow framework	150	99.05%
[45]	Augumentation	Xception model	Adam optimizer	Keras and tensorflow framework	60	93.43%
[48]	No preprocessing	BreastNet model	Adam optimizer	Matlab and python	100	98.08%
[23]	–	Transfer learning	Backpropagation	–	30	91.57%

Table 2. Different deep learning techniques used for various datasets for 2 class classification

Reference paper	Classification technique	Preprocessing/Training techniques	Dataset	Number of training and testing data/epoch	Accuracy
[26]	DCNN	Handcrafted feature extraction	Histopathology WSI images	1136 images for training, 991 for testing	88.89%
[49]	CNN	Gaussian filter with sigma value 2, Stochastic gradient Descent		23040 images for training, 5056 for testing	96.21%
[18]	Sparse autoencoders	Gready layer wise approach for training	Wisconsin breast cancer dataset	Epoch range from 50 to 600	98.06%
[24]	DBN	Back propagation technique and Levenberg – Marquardt in neural network		63.84% of dataset for training and 36. 16% for testing	99.68%
[29]	DNN	Stackking denoising autoencoders, Gready layer wise approach	CSB bio-segmentation benchmark	Patches from 29 images for trianing, 3 images for testing	98.27%
[20]	Transfer learning	GoogleNet, VGGNet, ResNet architecture model	Cytology images	6000 images for training and 2000 for testing	97.52%
[34]	CNN	Feature wise data augmentation, Backpropagation for training	Mammography images	25–35 epochs	90.71%

Table 3. Various classification techniques used for four class breast cancer classification

Reference paper	Preprocessing methods	Classification technique	Training algorithm	Number of training and testing samples	Accuracy	Dataset/Number of classes
[30]	Stain normalization, augumentation	DCNN and boosting tree classifier	Gradient descent algorithm	232 and 246	96.04%	Bioimaging 2015 breast histology classification
[40]	Image resize, augumentation	Inception recurrent residual CNN	Stochastic gradient descent algorithm	249 and 229	96.67%	
[25]	Multi Scale image convertion	CNN	Adam Optimizer	400 and 100	91.75%	BACH Dataset
[8]	Generating sub images using superpixel and fixed size square window	DCNN	Gready wise training approach	656 and 720	84.34%	Haematoxylin and eosin (H&E) stained slides of breast biopsies

4 Conclusion

Deep Learning provides better accuracy when compared to other approaches in medical image analyzing process. Recent work proves that deep learning performs better even when it is trained by using small amount of data. In deep learning techniques the process of feature extraction, segmentation of Region of Interest, classification process can done effectively when less human intervention. Just by building appropriate model and training it correctly with dataset the classification process can be done easily. In this review paper various deep learning model used for classifying the type of cancer is described. The model used for various images modalities along with the number of classes it can classify is provided. The accuracy comparison between different models used for the same dataset is also given in this work. It is concluded that better accuracy is provided for 2 class classification process. In future more research work can be contributed towards 4 class classification.

References

1. Zhang, J., Saha, A., Soher, B.J., Mazurowski, M.A.: Automatic deep learning-based normalization of breast dynamic contrast-enhanced magnetic resonance images, July 2018. https://arxiv.org/abs/1807.02152. Accessed 12 Sept 2020

2. Zhang, J., Saha, A., Zhu, Z., Mazurowski, M.A.: Breast tumor segmentation in DCE-MRI using fully convolutional networks with an application in radiogenomics.In: Medical Imaging 2018: Computer-Aided Diagnosis, vol. 10575, p. 29, February 2018. https://doi.org/10.1117/12.2295436
3. Kim, H., Hwang, S. : Scale-invariant feature learning using deconvolutional neural networks for weakly-supervised semantic segmentation. Undefined (2016)
4. Kim, D.H., Kim, S.T., Ro, Y.M.: Latent feature representation with 3-D multi-view deep convolutional neural network for bilateral analysis in digital breast tomosynthesis. In: ICASSP, IEEE International Conference on Acoustics, Speech and Signal Processing - Proceedings, vol. 2016-May, pp. 927–931, May 2016. https://doi.org/10.1109/ICASSP.2016.7471811
5. Yousefi, M., Krzyżak, A., Suen, C.Y.: Mass detection in digital breast tomosynthesis data using convolutional neural networks and multiple instance learning. Comput. Biol. Med. **96**, 283–293 (2018). https://doi.org/10.1016/j.compbiomed.2018.04.004
6. Shin, S.Y., Lee, S., Yun, I.D., Kim, S.M., Lee, K.M.: Joint weakly and semi-supervised deep learning for localization and classification of masses in breast ultrasound images. IEEE Trans. Med. Imaging **38**(3), 762–774 (2017). https://doi.org/10.1109/TMI.2018.2872031
7. Debelee, T.G., Amirian, M., Ibenthal, A., Palm, G., Schwenker, F.: Classification of mammograms using convolutional neural network based feature extraction. In: Lecture Notes of the Institute for Computer Sciences, Social-Informatics and Telecommunications Engineering. LNICST, vol. 244, pp. 89–98, September 2018. https://doi.org/10.1007/978-3-319-95153-9_9
8. Xu, J., Luo, X., Wang, G., Gilmore, H., Madabhushi, A.: A deep convolutional neural network for segmenting and classifying epithelial and stromal regions in histopathological images. Neurocomputing **191**, 214–223 (2016). https://doi.org/10.1016/j.neucom.2016.01.034
9. Tang, J., Rangayyan, R.M., Xu, J., El Naqa, I.E., Yang, Y.: Computer-aided detection and diagnosis of breast cancer with mammography: recent advances. IEEE Trans. Inf. Technol. Biomed. **13**(2), 236–251 (2009). https://doi.org/10.1109/TITB.2008.2009441
10. Yassin, N.I.R., Omran, S., El Houby, E.M.F., Allam, H.: Machine learning techniques for breast cancer computer aided diagnosis using different image modalities: a systematic review. Comput. Methods Program. Biomed. **156**, 25–45 (2018). https://doi.org/10.1016/j.cmpb.2017.12.012
11. Zhang, Z., Liang, X., Dong, X., Xie, Y., Cao, G.: A sparse-view CT reconstruction method based on combination of DenseNet and deconvolution. IEEE Trans. Med. Imaging **37**(6), 1407–1417 (2018). https://doi.org/10.1109/TMI.2018.2823338
12. Wang, R., Liang, X., Zhu, X., Xie, Y.: A feasibility of respiration prediction based on deep Bi-LSTM for real-time tumor tracking. IEEE Access **6**, 51262–51268 (2018). https://doi.org/10.1109/ACCESS.2018.2869780
13. Hamidinekoo, A., Denton, E., Rampun, A., Honnor, K., Zwiggelaar, R.: Deep learning in mammography and breast histology, an overview and future trends. Med. Image Anal. **47**, 45–67 (2018). https://doi.org/10.1016/j.media.2018.03.006
14. Qin, W., et al.: Superpixel-based and boundary-sensitive convolutional neural network for automated liver segmentation. Phys. Med. Biol. **63**(9) (2018). https://doi.org/10.1088/1361-6560/aabd19
15. Xiao, T., Liu, L., Li, K., Qin, W., Yu, S., Li, Z.: Comparison of transferred deep neural networks in ultrasonic breast masses discrimination. Biomed Res. Int. **2018** (2018). https://doi.org/10.1155/2018/4605191
16. Burt, J.R., et al.: Deep learning beyond cats and dogs: recent advances in diagnosing breast cancer with deep neural networks. Br. J. Radiol. **91**(1089) (2018). https://doi.org/10.1259/bjr.20170545

17. Hu, Z., Tang, J., Wang, Z., Zhang, K., Zhang, L., Sun, Q.: Deep learning for image-based cancer detection and diagnosis – a survey. Pattern Recognit. **83**, 134–149 (2018). https://doi.org/10.1016/j.patcog.2018.05.014
18. Kadam, V.J., Jadhav, S.M., Vijayakumar, K. : Breast cancer diagnosis using feature ensemble learning based on stacked sparse autoencoders and softmax regression. J. Med. Syst. **43**(8) (2019). https://doi.org/10.1007/s10916-019-1397-z.
19. Guo, Y., Shang, X., Li, Z.: Identification of cancer subtypes by integrating multiple types of transcriptomics data with deep learning in breast cancer. Neurocomputing **324**, 20–30 (2019). https://doi.org/10.1016/j.neucom.2018.03.072
20. Khan, S.U., Islam, N., Jan, Z., Ud Din, I., Rodrigues, J.J.P.C.: A novel deep learning based framework for the detection and classification of breast cancer using transfer learning. Pattern Recognit. Lett. **125**, 1–6 (2019). https://doi.org/10.1016/j.patrec.2019.03.022
21. Li, H., Zhuang, S., Li, D., Zhao, J., Ma, Y.: Benign and malignant classification of mammogram images based on deep learning. Biomed. Signal Process. Control **51**, 347–354 (2019). https://doi.org/10.1016/j.bspc.2019.02.017
22. Samala, R.K., Chan, H.-P., Hadjiiski, L.M. : Evolutionary pruning of transfer learned deep convolutional neural network for breast cancer diagnosis in digital breast tomosynthesis, no. May (2009)
23. Celik, Y., Talo, M., Yildirim, O., Karabatak, M., Acharya, U.R.: Automated invasive ductal carcinoma detection based using deep transfer learning with whole-slide images. Pattern Recognit. Lett. **133**, 232–239 (2020). https://doi.org/10.1016/j.patrec.2020.03.011
24. Abdel-Zaher, A.M., Eldeib, A.M.: Breast cancer classification using deep belief networks. Expert Syst. Appl. **46**, 139–144 (2016). https://doi.org/10.1016/j.eswa.2015.10.015
25. Yang, Z., Ran, L., Zhang, S., Xia, Y., Zhang, Y.: EMS-net: ensemble of multiscale convolutional neural networks for classification of breast cancer histology images. Neurocomputing **366**, 46–53 (2019). https://doi.org/10.1016/j.neucom.2019.07.080
26. Saha, M., Chakraborty, C., Racoceanu, D.: Efficient deep learning model for mitosis detection using breast histopathology images. Comput. Med. Imaging Graph. **64**, 29–40 (2018). https://doi.org/10.1016/j.compmedimag.2017.12.001
27. Wahab, N., Khan, A., Lee, Y.S.: Two-phase deep convolutional neural network for reducing class skewness in histopathological images based breast cancer detection. Comput. Biol. Med. **85**, 86–97 (2017). https://doi.org/10.1016/j.compbiomed.2017.04.012
28. Qi, X., et al.: Automated diagnosis of breast ultrasonography images using deep neural networks. Med. Image Anal. **52**, 185–198 (2019). https://doi.org/10.1016/j.media.2018.12.006
29. Feng, Y., Zhang, L., Yi, Z.: Breast cancer cell nuclei classification in histopathology images using deep neural networks. Int. J. Comput. Assist. Radiol. Surg. **13**(2), 179–191 (2018). https://doi.org/10.1007/s11548-017-1663-9
30. Vo, D.M., Nguyen, N.Q., Lee, S.W.: Classification of breast cancer histology images using incremental boosting convolution networks. Inf. Sci. (Ny) **482**, 123–138 (2019). https://doi.org/10.1016/j.ins.2018.12.089
31. Mohammed, M.A., Al-Khateeb, B., Rashid, A.N., Ibrahim, D.A., Abd Ghani, M.K., Mostafa, S.A.: Neural network and multi-fractal dimension features for breast cancer classification from ultrasound images. Comput. Electr. Eng. **70**, 871–882 (2018). https://doi.org/10.1016/j.compeleceng.2018.01.033
32. Ragab, D.A., Sharkas, M., Marshall, S., Ren, J.: Breast cancer detection using deep convolutional neural networks and support vector machines. PeerJ **2019**(1), 1–23 (2019). https://doi.org/10.7717/peerj.6201
33. Fang, Y., Zhao, J., Hu, L., Ying, X., Pan, Y., Wang, X.: Image classification toward breast cancer using deeply-learned quality features. J. Vis. Commun. Image Represent. **64**, 102609 (2019). https://doi.org/10.1016/j.jvcir.2019.102609

34. Ting, F.F., Tan, Y.J., Sim, K.S.: Convolutional neural network improvement for breast cancer classification. Expert Syst. Appl. **120**, 103–115 (2019). https://doi.org/10.1016/j.eswa.2018.11.008
35. Ekici, S., Jawzal, H.: Breast cancer diagnosis using thermography and convolutional neural networks. Med. Hypotheses **137**(December 2019), 109542 (2020). https://doi.org/10.1016/j.mehy.2019.109542
36. Zheng, Y., et al.: Feature extraction from histopathological images based on nucleus-guided convolutional neural network for breast lesion classification. Pattern Recognit. **71**, 14–25 (2017). https://doi.org/10.1016/j.patcog.2017.05.010
37. Al Nahid, A., Mehrabi, M.A., Kong, Y.: Histopathological breast cancer image classification by deep neural network techniques guided by local clustering. Biomed. Res. Int. **2018** (2018). https://doi.org/10.1155/2018/2362108
38. Zheng, J., Lin, D., Gao, Z., Wang, S., He, M., Fan, J.: Deep learning assisted efficient AdaBoost algorithm for breast cancer detection and early diagnosis. IEEE Access **8**, 96946–96954 (2020). https://doi.org/10.1109/ACCESS.2020.2993536
39. Budak, Ü., Cömert, Z., Rashid, Z.N., Şengür, A., Çıbuk, M.: Computer-aided diagnosis system combining FCN and Bi-LSTM model for efficient breast cancer detection from histopathological images. Appl. Soft Comput. J. **85**, 105765 (2019). https://doi.org/10.1016/j.asoc.2019.105765
40. Alom, M.Z., Yakopcic, C., Nasrin, M.S., Taha, T.M., Asari, V.K.: Breast cancer classification from histopathological images with inception recurrent residual convolutional neural network. J. Digit. Imaging **32**(4), 605–617 (2019). https://doi.org/10.1007/s10278-019-00182-7
41. Rasti, R., Teshnehlab, M., Phung, S.L.: Breast cancer diagnosis in DCE-MRI using mixture ensemble of convolutional neural networks. Pattern Recognit. **72**, 381–390 (2017). https://doi.org/10.1016/j.patcog.2017.08.004
42. Sudharshan, P.J., Petitjean, C., Spanhol, F., Oliveira, L.E., Heutte, L., Honeine, P.: Multiple instance learning for histopathological breast cancer image classification. Expert Syst. Appl. **117**, 103–111 (2019). https://doi.org/10.1016/j.eswa.2018.09.049
43. Mercan, C., Aksoy, S., Mercan, E., Shapiro, L.G., Weaver, D.L., Elmore, J.G.: Multi-instance multi-label learning for multi-class classification of whole slide breast histopathology images. IEEE Trans. Med. Imaging **37**(1), 316–325 (2018). https://doi.org/10.1109/TMI.2017.2758580
44. Acharya, S., Alsadoon, A., Prasad, P.W.C., Abdullah, S., Deva, A. : Deep convolutional network for breast cancer classification: enhanced loss function (ELF). J. Supercomput. (0123456789) (2020). https://doi.org/10.1007/s11227-020-03157-6
45. Li, L., et al.: Multi-task deep learning for fine-grained classification and grading in breast cancer histopathological images. Multimed. Tools Appl. **79**(21–22), 14509–14528 (2020). https://doi.org/10.1007/s11042-018-6970-9
46. Burçak, K.C., Baykan, Ö.K., Uğuz, H: A new deep convolutional neural network model for classifying breast cancer histopathological images and the hyperparameter optimisation of the proposed model. J. Supercomput. (0123456789) (2020). https://doi.org/10.1007/s11227-020-03321-y
47. Wu, N., et al.: Deep neural networks improve radiologists' performance in breast cancer screening. IEEE Trans. Med. Imaging **39**(4), 1184–1194 (2020). https://doi.org/10.1109/TMI.2019.2945514
48. Toğaçar, M., Özkurt, K.B., Ergen, B., Cömert, Z.: BreastNet: a novel convolutional neural network model through histopathological images for the diagnosis of breast cancer. Phys. A Stat. Mech. its Appl. **545** (2020). https://doi.org/10.1016/j.physa.2019.123592.
49. Öztürk, Ş, Akdemir, B.: HIC-net: a deep convolutional neural network model for classification of histopathological breast images. Comput. Electr. Eng. **76**, 299–310 (2019). https://doi.org/10.1016/j.compeleceng.2019.04.012

50. Li, Y., Wu, J., Wu, Q.: Classification of breast cancer histology images using multi-size and discriminative patches based on deep learning. IEEE Access **7**, 21400–21408 (2019). https://doi.org/10.1109/ACCESS.2019.2898044
51. Murtaza, G., Shuib, L., Wahab, A.W.A., Mujtaba, G., Raza, G.: Ensembled deep convolution neural network-based breast cancer classification with misclassification reduction algorithms. Multimed. Tools Appl. **79**(25–26), 18447–18479 (2020). https://doi.org/10.1007/s11042-020-08692-1

Hybridized Metaheuristic Search Algorithm with Modified Initialization Scheme for Global Optimization

Zhi Chuan Choi, Koon Meng Ang, Wei Hong Lim(✉), Sew Sun Tiang, Chun Kit Ang, Mahmud Iwan Solihin, Mohd Rizon Mohamed Juhari, and Cher En Chow

Faculty of Engineering, Technology and Built Environment, UCSI University, 56000 Kuala Lumpur, Malaysia
limwh@ucsiuniversity.edu.my

Abstract. Metaheuristic search algorithms (MSAs) such as particle swarm optimization (PSO) and differential evolution (DE) have gained significant attentions from practitioners to solve different engineering applications. Nevertheless, these MSAs are only effective to solve certain problem categories due to No Free Lunch Theorem. A modified hybrid algorithm namely chaotic oppositional based hybridized DE with PSO (CO-HDEPSO) is proposed to address the aforementioned drawbacks in order to solve different global optimization problems more effectively. The chaotic oppositional based initialization scheme is first designed into CO-HDEPSO, aiming to produce initial population with better quality. A hybridization framework is introduced in CO-HDEPSO to combine the strengths of PSO and DE simultaneously for further optimization performance enhancement. Comprehensive analyses are performed to compare the performances of CO-HDEPSO with five MS algorithms using CEC 2014 benchmark functions. The proposed CO-HDEPSO is reported to demonstrate more competitive search accuracy than its peer algorithms in most benchmark functions and able to solve 26 out of 30 CEC 2014 test functions with best F_{mean} values.

Keywords: Chaotic system · Differential evolution (DE) · Hybridization · Opposition-based learning · Particle swarm optimization (PSO)

1 Introduction

Metaheuristics search algorithms (MSAs) have recently emerged as a prominent approach to handle various real-world engineering optimization problems such as those in [1–11] by utilizing the working mechanisms motivated by different natural phenomena. Generally, MSAs are divided into two major branches of evolutionary algorithms (EAs) and swarm intelligence (SI) algorithms that are inspired by Darwinian's theory of evolution and social behaviors of animals, respectively. Among all existing MSAs, differential evolution (DE) [12] and particle swarm optimization (PSO) [13] are considered as the popular EA and SI algorithm, respectively, due to their competitive advantages such as strong global search capability, simplicity implementation, fast convergence

J. A. Mat Jizat et al. (Eds.): iCITES 2020, AISC 1350, pp. 172–182, 2021.
https://doi.org/10.1007/978-3-030-70917-4_17

speed and etc. Since the inceptions of DE and PSO, different improved variants of these algorithms have been proposed to further enhance their search performances through various approaches (e.g., parameter adaptation, modified learning strategy or neighborhood structure, hybridization and etc.) such as those reported in [14–18]. Nevertheless, there are some research issues that remain as open-ended challenges of MSAs. For instance, the initial population of MSAs produced using random initialization in general have questionable quality because this method involves random guess without referring to any information of search environment [19]. A poor initial population can either produce particles that are close to local optima or too far away from global optima, hence jeopardizing the overall optimization performance of MSAs. Another popular issue of MSAs is the "No Free Lunch Theorem" [20] that explains the limitations of MSAs in handing various optimization problems. More robust mechanisms are needed to regulate the contradictory requirements of fast convergence and high population diversity for solving different challenging problems effectively.

A modified hybrid algorithm known as chaotic oppositional based hybridized DE with PSO (CO-HDEPSO) is designed to tackle the aforementioned challenges. Some notable novelty of current work are presented as:

- A chaotic oppositional based initialization scheme (COBIS) is incorporated into CO-HDEPSO to produce an initial population with better overall quality.
- A new hybridization scheme is designed to regulate the fast convergence speed and high population diversity of algorithm by leveraging benefits of DE and PSO.
- Comprehensive analyses are conducted to compare the proposed CO-HDEPSO with other peer algorithms by using the CEC 2014 test functions.

As shown in the remaining sections, the related works of current study will be presented in Sect. 2. The working mechanisms of CO-HDEPSO are elaborated in Sect. 3. This is followed by thorough simulation analyses in Sect. 4. The conclusions of current study are provided in Sect. 5.

2 Related Works

2.1 DE

Inspired by the natural evolution process, DE was proposed by [12] to tackle global optimization problems. Initially, a set of DE individuals with population size of N is randomly generated and each n-th individual of $X_n = \left[X_{n,1}, \ldots, X_{n,D}\right]$ is considered as a potential solution of an optimization problem with dimension D, where $n = 1,\ldots, N$. During the evolution process, mutation is first performed to produce a donor vector $W_n = \left[W_{n,1}, \ldots, W_{n,D}\right]$ for each target vector $X_n = \left[X_{n,1}, \ldots, X_{n,D}\right]$. One of the frequently used mutation strategies to produce W_n is "DE/rand/1", i.e.

$$W = X_{r1} + F(X_{r2} - X_{r3}) \tag{1}$$

where F is a scaling factor varies from 0 to 1; X_{r1}, X_{r2} and X_{r3} are three randomly selected solutions, where $n \neq r1 \neq r2 \neq r3$.

A crossover process is subsequently performed to generate an offspring vector of $U_n = [U_{n,1}, \ldots, U_{n,D}]$ for each n-th individual based on the corresponding vectors of $X_n = [X_{n,1}, \ldots, X_{n,D}]$ and $W_n = [W_{n,1}, \ldots, W_{n,D}]$. Let d be the dimension index and Cr refer to the crossover probability with the value between 0 and 1. Then,

$$U_{n,d} = \begin{cases} W_{n,d}, & \text{if } rand_d \leq Cr \\ X_{n,d}, & \text{otherwise} \end{cases} \tag{2}$$

A selection process is further conducted to compare $f(X_n)$ and $f(U_n)$, i.e., the fitness values of X_n and U_n, respectively. For minimization problem, the solution with smaller fitness value is considered better and it will survive in the next generation. The evolution process that involves mutation, crossover and selection is repeated for all N solutions in DE population until the predefined termination conditions are met.

2.2 PSO

The idea of PSO was discussed in [13], as motivated by complex social cooperative behaviors of animals in locating the food sources. Similar with DE, every particle in PSO population is considered as candidate solution of a given problem. The current position of each n-th particle is represented as $X_n = [X_{n,1}, \ldots, X_{n,D}]$ with a velocity vector of $V_n = [V_{n,1}, \ldots, V_{n,D}]$, where $n = 1, \ldots, N$. Each PSO particle is also able to remember the best solution found so far and store in its personal best position of $P_n^{best} = \left[P_{n,1}^{best}, \ldots, P_{n,D}^{best}\right]$. For the best solution found by PSO swarm, it is stored in the global best position $G^{best} = [G_1^{best}, \ldots, G_D^{best}]$. Both of the P_n^{best} and G^{best} represent the self-cognitive and social components to guide each n-th particle moving towards the new locations of solution space throughout the search processes.

Suppose that $V_{n,d}, X_{n,d}, P_{n,d}^{best}$ and G_d^{best} refer to the current velocity, current position, personal best position and global best position of n-th particle in the d-th dimension, respectively. The new velocity and position of each n-th particle in every iteration are updated as:

$$V_{n,d} = \omega V_{n,d} + c_1 r_a \left(P_{n,d}^{best} - X_{n,d}\right) + c_2 r_b \left(G_d^{best} - X_{n,d}\right) \tag{3}$$

$$X_{n,d} = X_{n,d} + V_{n,d} \tag{4}$$

where ω is a time-varying inertia weight; c_1 and c_2 represent the acceleration coefficients; r_a and r_b represent two uniformly distributed random numbers that vary from 0 to 1. For each n-th particle, the fitness value of updated X_n, i.e., $f(X_n)$ is compared with $f\left(P_n^{best}\right)$. Both of the personal best position and fitness are updated if the updated X_n is better than P_n^{best}, i.e., $f(X_n) < f\left(P_n^{best}\right)$ for minimization problems. The same mechanisms are adopted in updating the position vector and fitness value of global best particle. The search processes of particles using Eqs. (3) and (4) are iterated as long as that the predefined termination criteria of PSO are not fulfilled.

3 CO-HDEPSO

A chaotic oppositional based initialization scheme (COBIS) is first designed in this paper to address the drawbacks of random initialization scheme. At the beginning of search process, the ergodicity and non-repetition natures of chaotic map are leveraged to generate the initial chaotic swarm of CO-HDEPSO with better diversity, aiming to address premature convergence issue. Let φ_0 and $\varphi_z \in [0, 1]$ be the initial condition and z-th sequence of chaotic variable, respectively. A modified sine map is then used to generate the chaotic sequence as follows [21]:

$$\varphi_{z+1} = \sin(\nu\varphi_z), z = 1, \ldots, Z \tag{5}$$

where $\nu = \pi$ is a bifurcation coefficient of modified sine map; Z is the maximum chaotic sequences. Suppose that X_d^{min} and X_d^{max} represent the smallest and largest values of design variables, respectively, where $d = 1,\ldots, D$. Let φ_z be the final sequence of chaotic variable obtained by modified sine map. The d-th dimension of each n-th chaotic swarm member can then be initialized as $X_{n,d}^{CS}$ based on φ_z, where $n = 1,\ldots, N$, as follows:

$$X_{n,d}^{CS} = X_d^{min} + \varphi_z\left(X_d^{max} - X_d^{min}\right) \tag{6}$$

Suppose that $X_{n,d}^{OBL}$ refers to the d-th dimension of every n-th opposite swarm member of $X_{n,d}^{CS}$ that is generated using oppositional based learning strategy. Then,

$$X_{n,d}^{OBL} = X_d^{min} + X_d^{max} - X_{n,d}^{CS} \tag{7}$$

The pseudocode used to generate an initial population of CO-HDEPSO is presented in Fig. 1. Denote $\mathbf{P}^{CS} = \left[X_1^{CS}, \ldots, X_N^{CS}\right]$ and $\mathbf{P}^{OBL} = \left[X_1^{OBL}, \ldots, X_N^{OBL}\right]$ as the population sets used to store the chaotic and opposite swarm members, respectively. A new population set $\mathbf{P}^{MRG}$ with population size of $2N$ is obtained by merging both $\mathbf{P}^{CS}$ and $\mathbf{P}^{OBL}$, where $\mathbf{P}^{MRG} = \mathbf{P}^{CS} \cup \mathbf{P}^{OBL}$. After sorting all solution members in $\mathbf{P}^{MRG}$ from best to worst based on their fitness values, the first N members of $\mathbf{P}^{MRG}$ with more promising fitness values are identified as an initial population of CO-HDEPSO, i.e., $\mathbf{P}^{Initial} = [X_1, \ldots, X_N]$.

A hybrid scheme is designed in CO-HDEPSO to further evolve $\mathbf{P}^{Initial} = [X_1, \ldots, X_N]$ by leveraging different search characteristics of DE and PSO. Under this hybridization scheme, DE and PSO are considered as the main and secondary optimizers used to perform searching of new solutions, respectively. During the optimization process of CO-HDEPSO, a mutation scheme of Eq. (1) with more explorative characteristic is selected to produce a donor vector of W_n for each n-th individual solution. The corresponding offspring vector of U_n is then obtained by using Eq. (2). The greedy selection operator is used to compare U_n with the original solution X_n. If U_n has better fitness than X_n, the latter solution and its fitness are replaced by those of former solution. Similar selection scheme is performed to update the values of G^{best} and $f\left(G^{best}\right)$.

If the offspring solution of U_n produced by DE has inferior fitness than X_n, PSO is activated as the secondary optimizer to further evolve X_n. This enables the n-th individual

Algorithm 1: COBIS

Input: D, N, X_d^{min}, X_d^{max}, Z and v

Initialize $\mathbf{P}^{CS} \leftarrow \emptyset$ and $\mathbf{P}^{OBL} \leftarrow \emptyset$
for each n-th particle **do**
    for each d-th dimension **do**
        Randomly generate $\varphi_0 \in [0,1]$ and set $z = 1$;
        while $z \leq Z$ **do**
            Update φ_z using Eq. (5) and update $z \leftarrow z + 1$;
        end while
        Calculate $X_{n,d}^{CS}$ and $X_{n,d}^{OBL}$ using Eqs. (6) and (7) respectively;
    end for
    Update $\mathbf{P}^{CS} \leftarrow \mathbf{P}^{CS} \cup X_n^{CS}$ and $\mathbf{P}^{OBL} \leftarrow \mathbf{P}^{OBL} \cup X_n^{OBL}$;
end for
$\mathbf{P}^{MRG} = \mathbf{P}^{CS} \cup \mathbf{P}^{OBL}$;
Evaluate fitness of all solutions of $\mathbf{P}^{MRG}$ and sort them from best to worst;
Extract the first N solutions of $\mathbf{P}^{MRG}$ with best fitness to form initial population;

Output: $\mathbf{P}^{Initial} = [X_1, \ldots, X_N]$ and the associated fitness values

Fig. 1. Pseudocode of COBIS to generate initial population of CO-HDEPSO.

solution X_n to uncover different regions of search space with alternative trajectory. Let c be an acceleration coefficient, the V_n of each n-th solution is updated as:

$$V_n = \omega V_n + c\left(G^{best} - X_n\right) \tag{8}$$

Unlike original PSO, the velocity update strategy adopted by CO-HDEPSO in Eq. (8) only considers the social component. This is because the solution X_n of every n-th individual in CO-HDEPSO is only updated when better solution is found, hence it is essentially same as the personal best position P_n^{best}. The self-cognitive component can then be discarded from Eq. (3) due to the nullified effect of P_n^{best} and X_n. Based on new V_n obtained from Eq. (8), the new position X_n^{new} of each n-th individual is calculated with Eq. (4). Fitness evaluation of X_n^{new} is performed and then compared with those of X_n and G^{best}. Both of the X_n and G^{best} will be replaced by X_n^{new} if the latter solution is fitter. Otherwise, the inferior X_n^{new} will be discarded.

The complete working mechanisms of CO-HDEPSO are summarized in Fig. 2. COBIS is anticipated to produce an initial population with better quality. The chaotic characteristic of COBIS enables more thorough searching, while its oppositional based learning can facilitate wider coverage of search space during initialization process. The hybrid scheme adopted by CO-HDEPSO is expected to offer better capability in handling the contradict requirements of fast convergence speed and high population diversity of algorithm by leveraging the benefits of DE and PSO. The iterative search process using DE and PSO is repeated until the fitness evaluations counter γ exceeds the predefined maximum fitness evaluation number Υ. The global best position G^{best} obtained by CO-HDEPSO is returned at the end of optimization.

4 Comparative Studies

4.1 Simulation Settings for Comparative Studies

Simulation analyses of CO-HDEPSO are performed using 30 benchmark functions from CEC 2014 [22]. These benchmark functions have different characteristics at their fitness landscapes and they are not easily solved. The more detailed information can be obtained from [22].

```
Main Algorithm: CO-HDEPSO
Inputs: D, N, X_min, X_max , Cr, F, ω, c_1, c_2, γ, Υ
Initialize γ ← 0 and generate initial population using COBIS (Algorithm 1);
γ ← γ + 2N;
G^best ← X_1 , f(G^best) ← f(X_1);      //Initialize global best position and fitness
while γ ≤ Υ do
      for each n-th solution do        //Perform searching using DE
         Randomly select X_r1, X_r2 and X_r3 from population;
         Generate donor vector W_n by mutation using Eq. (1);
         Generate offspring vector U_n by crossover using Eq. (2);
         Ensure U_n is within search boundaries and then evaluate its fitness;
         γ ← γ + 1;
         if f(U_n) < f(X_n) then
            X_n ← U_n, f(X_n) ← f(U_n);
            if f(X_n) < f(G^best)then
               G^best ← P_n^best, f(G^best) ← f(P_n^best);
            end if
         else                                //Perform searching using PSO
            Update velocity V_n using Eq. (8);
            Calculate new position X_n^new using Eq. (4);
            Ensure X_n^new is within search boundaries and then evaluate its fitness;
            γ ← γ + 1;
            if f(X_n^new) < f(X_n) then
               X_n ← X_n^new, f(X_n) ← f(X_n^new);
               if f(X_n^new) < f(G^best) then
                  G^best ← X_n^new, f(G^best) ← f(X_n^new);
               end if
            end if
         end if
      end for
end while
Output: G^best, f(G^best)
```

Fig. 2. Pseudocode for the complete framework of CO-HDEPSO.

Performance metrics used in this study are mean fitness F_{mean} and standard deviation *SD* that implies for the search accuracy and consistency of algorithm, respectively. Smaller *SD* is better as it indicates a compared algorithm has better consistency in solving the given benchmark function with similar global best fitness value.

The performances of CO-HDEPSO in solving all test functions from CEC 2014 are compared with other five metaheuristics known as: PSO [13], DE [23], a hybrid optimizer abbreviated as DE-PSO [24], accelerated PSO (APSO) [25] and unconstrained version of the multi-swarm PSO without velocity (MPSOWV) [26]. Population size and maximum fitness evaluations numbers of all algorithms are set as $N = 100$ and $\Upsilon = 10,000 \times D$, respectively. All compared algorithms are simulated using Matlab 2018a in a personal computer (Intel ® Core i5-7300HQ CPU @ 2.5 GHz) for 30 independent runs to solve all CEC 2014 functions at $D = 10$.

4.2 Performance Comparisons for All Optimizers

The F_{mean} and *SD* values obtained by all compared algorithms to solve CEC 2014 test functions are presented in Table 1. The boldface and underlined fonts are used to indicate the best and second-best F_{mean} results, respectively. The performance comparison of all optimizers can be summarized as *w/t/l* and *#BMF*. *w/t/l* indicates that the proposed CO-HDEPSO deliver better performance in *w* test functions, deliver similar performance in *t* test functions and deliver worse performance in *l* test functions. The number of function that can be solved by each optimizer with best F_{mean} is represented as *#BMF*.

The proposed CO-HDEPSO has produced three best F_{mean} values in solving the three unimodal functions (F1 to F3). It is also able to locate the near global optima of functions F2 and F3. This is followed by DE-PSO and MPSOWV that solve these unimodal functions with two and one second best F_{mean} values, respectively. In contrary, APSO is identified as the most inferior optimizer to deal with unimodal functions because it produces three worst F_{mean} values in all functions F1 to F3. CO-HDEPSO demonstrates the most competitive search accuracy to solve all simple multimodal functions (F4 to F16) by producing nine best F_{mean} values. This is followed by MPSOWV that produces three best and two second best F_{mean} values. DE is the poorest optimizer for producing four worst F_{mean} values, suggesting its high tendency to be misled by the local optima information.

The performance of CO-HDEPSO remains competitive in dealing with hybrid functions of F17 to F22 by solving all these functions with best F_{mean} values. Another hybrid algorithm DE-PSO is the second-best optimizer for being able to solve four hybrid functions with the promising F_{mean} values. Despite showing good performance to solve the unimodal and simple multimodal functions, MPSOWV suffers with serious performance degradation when solving the hybrid functions with inferior search accuracy as indicated by the relatively large F_{mean} values. APSO is the worst optimizer because it produces the poorest F_{mean} values in five out of six hybrid functions. The proposed CO-HDEPSO is able to achieve the best search accuracies in solving the composite functions (F23 to F30). DEPSO and MPSOWV are also reported to solve some of these challenging benchmark functions with premising accuracies. In contrast, APSO produces five worst F_{mean} values in solving the composite functions.

From Table 1, PSO and DE have limitations because these two optimizers only perform well in some selected test function, hence aligning with No Free Lunch Theorem. Both of the DE-PSO and MPSOWV are able to solve CEC 2014 functions with relatively good accuracies because these algorithms leverage the benefits of search characteristics contributed by various metaheuristics, implying the effectiveness of hybrid scheme to solve challenging optimization problems. Despite of their performance gains, both of DE-PSO and MPSOWV are still outperformed by the proposed CO-HDEPSO because the latter algorithm is able to solve 26 out of 30 CEC 2014 test functions with best F_{mean} values. The main factor that leads to the performance differences between CO-HDEPSO and with those DE-PSO and MPSOWV are the modified initialization scheme adopted by CO-HDEPSO in producing the initial population with better quality that can be beneficial to algorithm, i.e., better robustness in handling premature convergence issue and rapid convergence speed.

Table 1. Performance comparison between CO-HDEPSO with five peers in 30 test functions

Function	Metrics	PSO	DE	DE-PSO	APSO	MPSOWV	CO-HDEPSO
F1	F_{mean}	1.31E+05	2.67E+04	8.21E+02	1.33E+06	2.06E+03	**6.93E+02**
	SD	1.52E+05	1.02E+04	9.76E+02	1.13E+06	6.47E+02	8.61E+02
F2	F_{mean}	1.27E+03	2.71E+02	4.56E−04	1.91E+03	5.79E+01	**3.11E−04**
	SD	1.80E+03	2.17E+01	6.41E−04	1.40E+03	8.18E+01	4.37E−05
F3	F_{mean}	2.29E+02	2.15E−02	8.52E−04	2.77E+04	6.81E−05	**9.11E−06**
	SD	1.08E+02	1.08E−02	8.43E−04	8.62E+03	9.09E−05	7.44E−06
F4	F_{mean}	1.98E+01	2.18E+00	8.29E−04	1.96E+01	1.74E+01	**7.45E−04**
	SD	2.12E+01	3.04E+00	4.63E−04	2.15E+01	2.46E+01	9.36E−04
F5	F_{mean}	2.03E+01	2.02E+01	2.03E+01	**2.00E+01**	2.04E+01	2.01E+01
	SD	5.43E−02	3.47E−02	1.16E−01	7.13E−08	4.82E−02	6.33E−04
F6	F_{mean}	9.95E−01	1.40E+00	2.07E+00	9.53E−01	6.43E−01	**5.32E−01**
	SD	1.38E+00	5.59E−03	2.84E+00	1.34E+00	9.06E−02	6.59E−01
F7	F_{mean}	2.33E−01	3.93E−01	1.11E−01	2.19E−01	**4.67E−02**	1.76E−01
	SD	3.31E−02	1.74E−01	1.04E−02	5.58E−02	3.13E−02	8.51E−02
F8	F_{mean}	5.47E+00	1.50E+01	3.98E+00	1.34E+01	6.47E+00	**3.48E+00**
	SD	4.93E+00	1.08E+00	1.41E+00	4.92E+00	7.04E−01	7.04E−01
F9	F_{mean}	7.96E+00	2.94E+01	1.54E+01	2.79E+01	9.95E+00	**6.47E+00**
	SD	1.41E+00	2.31E+00	2.11E+00	1.41E+00	2.81E+00	2.11E+00
F10	F_{mean}	1.98E+02	1.02E+02	4.43E+01	6.86E+02	6.83E+01	**3.87E+01**
	SD	8.84E+01	2.54E+01	3.36E+01	4.41E+02	2.23E+01	2.33E+01
F11	F_{mean}	1.81E+02	9.24E+02	3.06E+02	7.49E+02	3.69E+02	**1.37E+02**
	SD	8.87E+01	2.95E+02	2.60E+02	1.98E+02	2.95E+00	3.20E−01
F12	F_{mean}	2.60E−01	3.20E−01	5.22E−02	2.57E−01	6.94E−01	**1.25E−02**
	SD	2.27E−01	9.83E−02	3.51E−02	3.55E−02	3.66E−01	1.66E−02
F13	F_{mean}	2.63E−01	3.15E−01	1.49E−01	2.85E−01	**9.43E−02**	1.27E−01
	SD	1.68E−02	1.74E−02	1.13E−02	1.22E−01	2.14E−02	8.30E−03
F14	F_{mean}	1.57E−01	2.11E−01	1.74E−01	3.17E−01	3.73E−01	**1.19E−01**
	SD	3.37E−02	2.97E−02	7.60E−03	1.04E−01	1.76E−02	1.34E−02
F15	F_{mean}	9.53E−01	3.45E+00	9.92E−01	1.42E+00	6.25E−01	**4.64E−01**
	SD	5.75E−01	2.25E−01	4.83E−01	8.72E−01	1.81E−01	1.48E−01
F16	F_{mean}	2.82E+00	3.39E+00	2.04E+00	3.98E+00	**5.30E−01**	1.34E+00
	SD	3.43E−01	9.76E−02	4.08E−01	6.87E−01	3.40E−01	9.47E−02
F17	F_{mean}	6.24E+02	1.11E+02	9.05E+01	1.73E+04	7.41E+03	**6.81E+01**
	SD	4.67E+02	2.22E+01	3.04E+01	1.95E+04	7.53E+03	3.90E+01

(*continued*)

Table 1. *(continued)*

Function	Metrics	PSO	DE	DE-PSO	APSO	MPSOWV	CO-HDEPSO
F18	F_{mean}	5.60E+03	9.37E+00	8.81E+00	1.52E+04	6.33E+02	**7.91E+00**
	SD	5.33E+03	4.29E+00	1.06E+00	1.45E+04	8.23E+02	1.24E−01
F19	F_{mean}	9.76E−01	1.38E+00	1.12E+00	1.99E+00	2.34E+00	**8.08E−01**
	SD	1.13E+00	3.42E−01	2.53E−01	2.45E−01	3.25E+00	6.91E−01
F20	F_{mean}	8.49E+02	3.60E+00	3.91E+00	7.45E+03	1.33E+01	**3.16E+00**
	SD	2.09E+01	1.68E+00	4.82E−01	8.37E+03	2.47E−01	7.84E−01
F21	F_{mean}	3.54E+01	1.57E+01	1.19E+01	4.57E+03	6.81E+01	**9.51E+00**
	SD	3.83E+01	1.23E+01	4.37E+00	5.09E+03	7.25E+01	5.80E+00
F22	F_{mean}	8.82E+01	9.74E+00	1.35E+00	2.12E+02	2.08E+01	**4.55E−01**
	SD	9.54E+01	5.58E+00	1.27E+00	2.55E+01	6.63E−01	1.03E−01
F23	F_{mean}	**3.29E+02**	**3.29E+02**	**3.29E+02**	3.30E+02	**3.29E+02**	**3.29E+02**
	SD	3.22E−13	2.07E−07	0.00E+00	8.30E−02	0.00E+00	0.00E+00
F24	F_{mean}	1.26E+02	1.43E+02	1.17E+02	1.28E+02	**1.16E+02**	**1.16E+02**
	SD	4.06E+00	1.27E+00	3.01E+00	6.91E+00	5.81E−01	5.73E+00
F25	F_{mean}	1.81E+02	1.70E+02	1.56E+02	1.83E+02	1.64E+02	**1.42E+02**
	SD	3.15E+01	4.43E+01	2.42E+01	2.83E+01	5.06E+01	1.17E+01
F26	F_{mean}	**1.00E+02**	**1.00E+02**	**1.00E+02**	**1.00E+02**	**1.00E+02**	**1.00E+02**
	SD	1.20E−01	4.21E−02	5.46E−03	3.58E−02	2.92E−02	6.08E−03
F27	F_{mean}	2.01E+02	4.68E+00	3.47E+00	2.01E+02	1.76E+02	**2.48E+00**
	SD	2.81E+02	3.80E−04	2.13E−01	2.82E+02	2.44E+02	8.04E−02
F28	F_{mean}	6.77E+02	3.76E+02	3.76E+02	4.60E+02	3.75E+02	**3.69E+02**
	SD	2.67E+02	2.65E+00	2.65E+00	2.92E+01	7.93E+00	2.43E−03
F29	F_{mean}	2.45E+02	**2.24E+02**	**2.24E+02**	1.34E+03	3.89E+02	**2.24E+02**
	SD	2.49E+01	1.84E−02	1.74E−01	2.19E+02	8.05E+01	1.89E−01
F30	F_{mean}	1.06E+03	5.02E+02	4.87E+02	1.43E+03	1.24E+03	**4.84E+02**
	SD	2.12E+02	9.92E−01	4.24E+00	5.55E+02	1.71E+02	4.84E+00
	w/t/l	28/2/0	27/3/0	26/3/1	28/1/1	24/3/3	–
	#BMF	2	3	3	2	6	26

5 Conclusion

A new hybrid algorithm known as CO-HDEPSO is designed in this paper to address various global optimization problems effectively. COBIS is designed to produce an initial population with better quality to benefit CO-HDEPSO from the perspective of convergence speed and ability to handle premature convergence. A hybrid scheme is also incorporated into CO-HDEPSO to leverage the search characteristics of both DE and PSO, in order to achieve better balancing of rapid convergence speed and high diversity of algorithms. Thorough comparative studies were performed to evaluate CO-HDEPSO using CEC 2014 functions and it is reported to outperform all peer algorithms. As the extended works of current study, the impacts of different chaotic maps on the convergence characteristics of CO-HDEPSO can be further investigated. More thorough theoretical analyses on the algorithm framework of CO-HDEPSO is worth to be investigated as well.

References

1. Natarajan, E., Kaviarasan, V., Lim, W.H., Tiang, S.S., Tan, T.H.: Enhanced multi-objective teaching-learning-based optimization for machining of delrin. IEEE Access **6**, 51528–51546 (2018)
2. Yao, L., Chen, Y.-Q., Lim, W.H.: Internet of things for electric vehicle: an improved decentralized charging scheme. In: 2015 IEEE International Conference on Data Science and Data Intensive Systems (2015)
3. Yao, L., Lim, W.H.: Optimal purchase strategy for demand bidding. IEEE Trans. Power Syst. **33**(3), 2754–2762 (2018)
4. Natarajan, E., Kaviarasan, V., Lim, W.H., Tiang, S.S., Parasuraman, S., Elango, S.: Non-dominated sorting modified teaching–learning-based optimization for multi-objective machining of polytetrafluoroethylene (PTFE). J. Intell. Manuf. **31**(4), 911–935 (2019)
5. Yao, L., Lim, W., Tiang, S., Tan, T., Wong, C., Pang, J.: Demand bidding optimization for an aggregator with a genetic algorithm. Energies **11**(10), 2498 (2018)
6. Tarawneh, M.A., Yu, L.J., Tarawni, M.A., Ahmad, S.H., Al-Banawi, O., Batiha, M.A.: High performance thermoplastic elastomer (TPE) nanocomposite based on graphene nanoplates (GNPs). World J. Eng. **12**(5), 437–442 (2015)
7. Yu, L.J., Ahmad, S.H., Kong, I., Tarawneh, M.A., Flaifel, M.H.: Preparation and characterisation of NiZn ferrite/multiwalled nanotubes thermoplastic natural rubber composite. Int. J. Mater. Eng. Innov. **4**(3/4), 214 (2013)
8. Yu, L.J., Sahrim, A.H., Kong, I., Tarawneh, M.A.: Microwave absorbing properties of nickel-zinc ferrite/multiwalled nanotube thermoplastic natural rubber composites. Adv. Mater. Res. **501**, 24–28 (2012)
9. Yao, L., Damiran, Z., Lim, W.H.: Direct load control of central air conditioning systems using fuzzy optimization. In: 2016 IEEE 16th International Conference on Environment and Electrical Engineering (EEEIC) (2016)
10. Yu, L.J., Ahmad, S., Appadu, S., Kong, I., Tarawneh, M., Flaifel, M.: Comparison of magnetic and microwave absorbing properties between multiwalled carbon nanotubes nanocomposite, nickel zinc ferrite nanocomposite and hybrid nanocomposite. World J. Eng. **11**(4), 317–322 (2014)
11. Muhieldeen, M.W., Kuang, Y.C.: Saving energy costs by combining air-conditioning and air-circulation using CFD to achieve thermal comfort in the building. J. Adv. Res. Fluid Mech. Theremal Sci. **58**, 84–99 (2019)
12. Price, K.V., Storn, R.M., Lampinen, J.A.: The differential evolution algorithm. In: Differential Evolution: A Practical Approach to Global Optimization, pp. 37–134. Springer, Heidelberg (2005)
13. Kennedy, J., Eberhart, R.: Particle swarm optimization. In: IEEE International Conference on Neural Networks, IV, pp. 1942–1948. IEEE Press, Perth (1995)
14. Bilal, Pant, M., Zaheer, H., Garcia-Hernandez, L., Abraham, A.: Differential evolution: a review of more than two decades of research. Eng. Appl. Artif. Intell. **90**, 103479 (2020)
15. Bonyadi, M.R., Michalewicz, Z.: "Particle swarm optimization for single objective continuous space problems: a review. Evol. Comput. **25**(1), 1–54 (2017)
16. Lim, W.H., Isa, N.A.M., Tiang, S.S., Tan, T.H., Natarajan, E., Wong, C.H., Tang, J.R.: A self-adaptive topologically connected-based particle swarm optimization. IEEE Access **6**, 65347–65366 (2018)

17. Lim, W.H., Isa, N.A.M.: Particle swarm optimization with dual-level task allocation. Eng. Appl. Artif. Intell. **38**, 88–110 (2015)
18. Das, S., Mullick, S.S., Suganthan, P.N.: Recent advances in differential evolution – an updated survey. Swarm Evol. Comput. **27**, 1–30 (2016)
19. Yang, X.: Metaheuristic optimization: algorithm analysis and open problems. arXiv.org (2020)
20. Wolpert, D.: No free lunch theorem for optimization. IEEE Trans. Evol. Comput. **1**, 467–482 (1997)
21. Demir, F.B., Tuncer, T., Kocamaz, A.F.: A chaotic optimization method based on logistic-sine map for numerical function optimization. Neural Comput. Appl. **32**(17), 14227–14239 (2020)
22. Liang, J.J., Qu, B.Y., Suganthan, P.N.: Problem definitions and evaluation criteria for the CEC 2014 special session and competition on single objective real-parameter numerical optimization. Zhengzhou University, Zhengzhou China Computational Intelligence Laboratory (2013)
23. Ronkkonen, J., Kukkonen, S., Price, K.V.: Real-parameter optimization with differential evolution. In: 2005 IEEE Congress on Evolutionary Computation (n.d.)
24. Pant, M., Thangaraj, R., Abraham, A.: DE-PSO: A new hybrid meta-heuristic for solving global optimization problems. New Math. Nat. Comput. **07**(03), 363–381 (2011)
25. Yang, X.S., Deb, S., Fong, S.: Accelerated particle swarm optimization and support vector machine for business optimization and applications. In: Communications in Computer and Information Science. CCIS, vol. 136, pp. 53–66 (2011)
26. Ang, K.M., Lim, W.H., Isa, N.A.M., Tiang, S.S., Wong, C.H.: A constrained multi-swarm particle swarm optimization without velocity for constrained optimization problems. Expert Syst. Appl. **140**, 112882 (2020)

A Multi-stage SVM Based Diagnosis Technique for Photovoltaic PV Systems

Yassine Chouay(✉) and Mohammed Ouassaid

Mohammadia School of Engineers, Mohammed V University, Rabat, Morocco
yassinechouay@research.emi.ac.ma, ouassaid@emi.ac.ma

Abstract. This paper proposes an improved method for fault diagnosis on the DC side of PV plant. The proposed technique relies on three-stage classification algorithms to detect and distinguish between eight of the most common faults occurring in the PV generator. The first part of the detection algorithm uses the power loss analysis approach to identify the presence of potential fault from the comparison of the measured and expected generated power. The second part relies on a comparison between the extracted and reference PV characteristic to identify the fault type. The last part is based on Support Vector Machine (SVM) algorithm that interferes to classify faults with the same signature. The simulation results have proven that the proposed method is capable of identifying and distinguishing between eight of the most common DC side faults with a 100% accuracy.

1 Introduction

As it is well known, photovoltaic (PV) integration is having a huge growth as one of the cleanest renewable energy sources. This increase is related also to the sustainability of PV technologies and their environmental and economic benefits. However, the extraction of electrical energy from the PV systems still suffers from efficiency limits and maintenance accumulation. In fact, PV systems may encounter faults occurrences that drastically limit its efficiency, reliability, and even the whole system safety. Therefore, the interest in fault diagnosis methods is nowadays constantly developed in the literature to ensure the optimal functioning of PV systems.

Over the years, many fault detection techniques have been proposed for detecting and classifying faults in PV systems. Some of these methods rely on the comparison of real-time parameters with the predicted ones, known as Model-Based Difference Measurement (MBDM) to identify faults in the system [1, 2]. Other methods compare the PV real-time attributes with their threshold limits to detect faults, without using the predictive model [3, 4]. Other approaches use the direct analysis of the effect of the faults on the voltage and current waveforms called Output Signal Analysis (OSA) [5, 6]. Usually, it applies some adapted analysis on the output signal to discover the anomalies and faults occurrence.

In order to overcome some of the limitations of the conventional methods, other Machine learning (ML) based techniques [7, 8] were also introduced. For example, in [7] A Decision Tree (DT) algorithm was trained to construct an array level PV model

J. A. Mat Jizat et al. (Eds.): iCITES 2020, AISC 1350, pp. 183–193, 2021.
https://doi.org/10.1007/978-3-030-70917-4_18

with validation of up to 99% accuracy in detecting line-line faults, shading faults, and ground faults. In [8], Probabilistic Neural Network (PNN) was used to detect open circuit faults and line-line faults in a PV array.

In other cases, more than one approach is combined in a single algorithm as a hybrid diagnosis technique [9, 10]. As in [9], the Output signal analysis method was combined with Support Vector Machine (SVM) to identify line-line faults. Alternatively, in [10] a model-based difference measurement technique is combined with an ANN approach to detect and identify several faults occurring in the PV string.

In this paper, an accurate and robust fault detection and diagnosis technique is presented. It allows the identification of eight of the common faults occurring in the PV system at the level of cells, modules or PV strings. The proposed method is based on a hybrid detection algorithm combining the simplicity of the model-based difference measurement and the accuracy of ML techniques. The first part is based on power losses analysis, then the evaluation of a number of I–V characteristic parameters against ideal characteristic under the same conditions. The last part extends the accuracy and allows the distinction between similar faults using a Support Vector Machine (SVM) model.

The rest of this paper is organized as follows. In Sect. 2, the targeted faults and their effect on the PV characteristic are presented. The proposed fault diagnosis technique is introduced in Sect. 3. Section 4, presents and discusses the simulation results of testing the adopted diagnosis system under various faults. Finally, the conclusions are drawn in Sect. 5.

2 PV System Under Faulty Operation

Generally, faults can occur in different parts of the PV system such as the PV array, the storage system, the inverter, and the grid etc… Since the AC side and storage system usually have their diagnosis system, this paper intends to detect only the faults occurring in the PV array (DC side). In order to identify and detect the presence of any fault, its effect must be noticed. The work presented in this paper deals with the eight faults types indicated in Table 1. The effects of these faults on the I–V characteristic are illustrated in Fig. 1.

From Fig. 1, one can conclude that each fault has a different effect on the I–V characteristic called also the fault signature. These fault signatures vary in three main regions, including the short circuit, the open circuit and the Maximum Power Point (MPP) region.

Table 1. Different fault types identified by the proposed method.

Location	Fault name	Notation
Module	Open circuit fault in any cell or module	F1
	Bypass diode, cells or modules short circuited	F2
	Bypass diode, cells or modules shunted	F3
String	Shor circuit at the terminal of the string	F4
	Partial shading	F5
	Partial shading combined with an open-circuited bypass diode	F6
	Partial shading in a combined with a connection fault	F7
Cables	Connection resistance between PV modules	F8

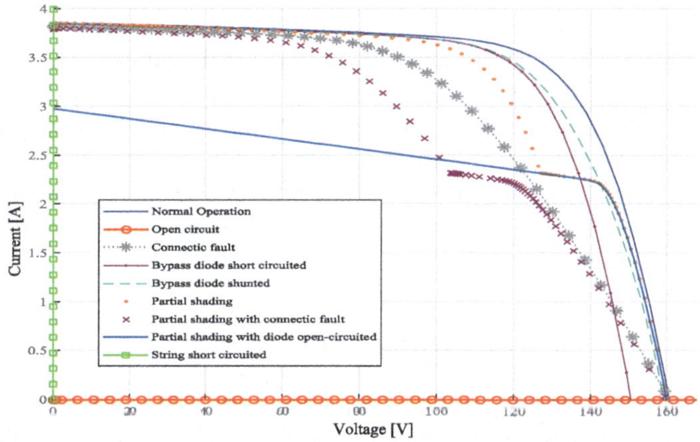

Fig. 1. I–V characteristics of a PV string in normal and faulty operation.

3 The Proposed Fault Diagnosis System

3.1 The Proposed Diagnosis Configuration

Since the aim of this work is to detect fault occurrence in the DC side of a PV system, a simple string connected to a DC load via a DC-DC converter as illustrated in Fig. 2 is proposed. The diagnosis system relies on the continuous measurements and comparison of the MPP output power with a reference power, under the same operating climatic conditions. The difference between the measured MPP and the reference estimated power is compared with a predefined threshold, taking into consideration measurement and estimation errors. If the difference exceeds the power threshold, the system is considered under faulty operation, triggering then the characteristic tracing operation. The extracted I–V characteristic is then stored and compared to the reference characteristic to identify the fault type.

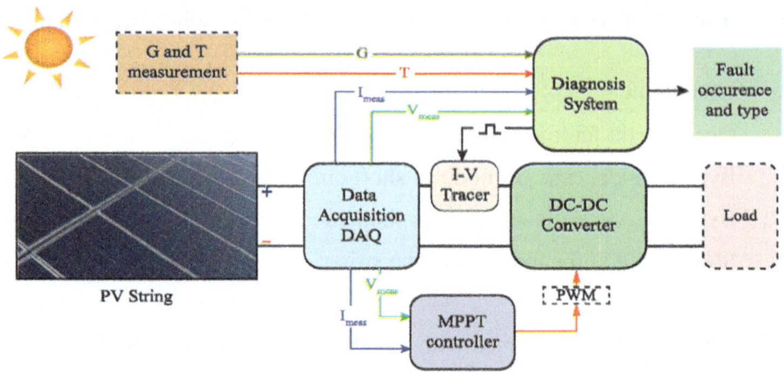

Fig. 2. The proposed fault diagnosis system implementation.

3.2 Fault Occurrence Detection (Algorithm 1)

In order to identify fault occurrence in real-time, this paper proposes a continuous monitoring of the string produced DC power with reference to the estimated power (P_{Ref}). The reference power indicates power produced by the system in normal operation (fault-free operation). It is calculated using the one diode PV model [11], by taking into consideration the measurement of real operating climatic conditions (irradiance and temperatures). To avoid false alarms and mismatch measurement, an appropriate threshold must be defined. In this case, the power threshold T_P is calculated using a model estimation error of 5.9% [12], and a measurement error of 3%. The T_P threshold is then given in Eq. (1):

$$T_P = 0.059 + 0.03 = 0.089 \tag{1}$$

The flowchart of the first algorithm is depicted in Fig. 3. The difference between the reference and the measured PV system output power (ΔP) is compared with a threshold T_P to conclude the state of the system under observation.

3.3 Model-Based Difference Measurement Approach (Algorithm 2)

If the system is considered in faulty operation, a set of ratios is extracted from the measured I–V characteristic and compared with their thresholds in a second algorithm. The thresholds for current and voltage are calculated the same way as the power threshold. In this case, the model estimation error for current and voltage are set to 3.0% and 2.9%, respectively [12]. Equation (2) and (3) gives the final formula for the current and voltage thresholds T_C and T_V:

$$T_C = 0.03 + 0.02 = 0.05 \tag{2}$$

$$T_V = 0.029 + 0.01 = 0.039 \tag{3}$$

These thresholds are compared to the relative ratios between reference and measured PV string attributes. The ratios can be summarized into five main characteristic ratios including:

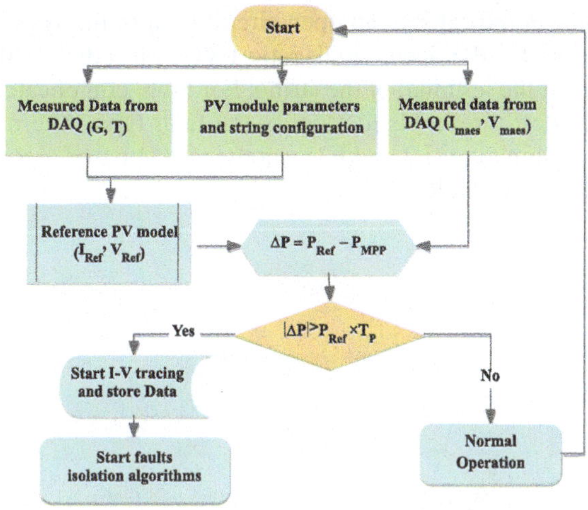

Fig. 3. Flowchart of the fault occurrence detection algorithm.

The ratio between the number of measured and reference MPP peeks given by:

$$R_N = \frac{N_{meas}}{N_{Ref}} \tag{4}$$

The ratio between the measured and reference short circuit current expressed as:

$$R_{I_{SC}} = \frac{\left|I_{SC_{Ref}} - I_{SCmeas}\right|}{I_{SC_{Ref}}} \tag{5}$$

The ratio between the measured and reference open circuit voltage written as:

$$R_{V_{OC}} = \frac{\left|V_{OC_{Ref}} - V_{OCmeas}\right|}{V_{OC_{Ref}}} \tag{6}$$

The ratio between the measured and reference current at the MPP given by:

$$R_{I_M} = \frac{\left|I_{M_{Ref}} - I_{Mmeas}\right|}{I_{M_{Ref}}} \tag{7}$$

The ratio between the measured and reference voltage at the MPP expressed as:

$$R_{V_M} = \frac{\left|V_{M_{Ref}} - V_{Mmeas}\right|}{V_{M_{Ref}}} \tag{8}$$

The results of the comparison, lead to the identification of faults having a different combination of attributes. The calculated ratios and thresholds are compared according to the flowchart depicted in Fig. 4.

As indicated in Fig. 4 and from the results obtained in Fig. 1, faults could be grouped at first by the number of MPP peeks. If RN is less than unity, it means that the fault is related to a partial shading. On the other hand, if the short circuit current ratio RISC is equal to unity, it's an indicator to an open circuit fault in the system. Else, when the same condition is true for the open circuit ratio RVOC, it indicates that a short circuit fault is taking place at the terminal of the string. For more complicated cases, the ratios are compared with their thresholds to identify the fault type. Alternatively, in case of faults with the same signature (F2 and F3), another SVM based algorithm is trained to distinguish between theses faults.

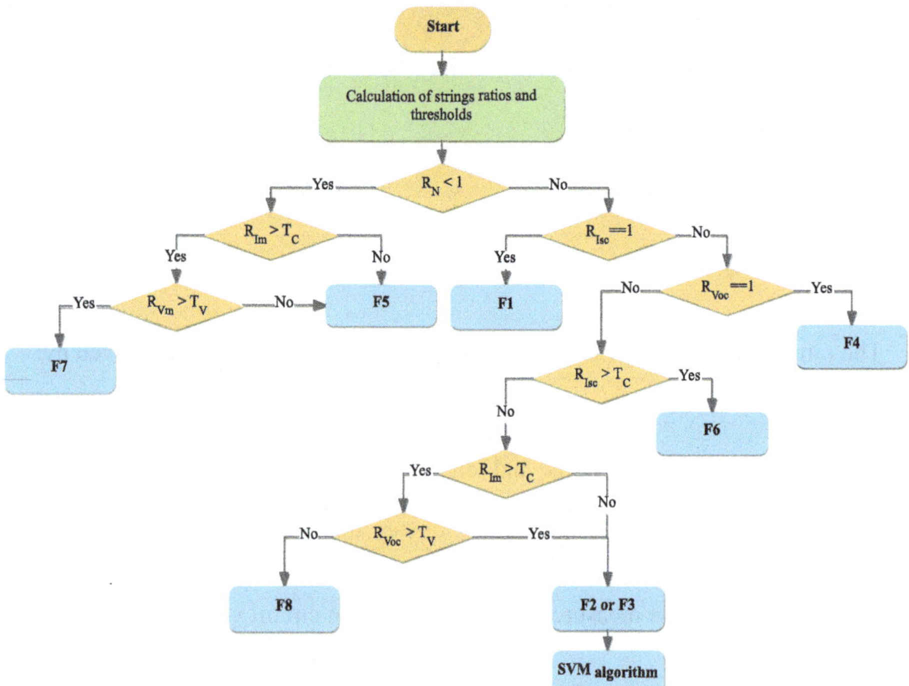

Fig. 4. The flowchart of the fault identification algorithm (Algorithm 2).

3.4 Machine Learning Based Algorithm (Algorithm 3)

As it can be seen in the last section, the detection algorithm (Algorithm 2) is unable to distinguish between two faults that have similar signature. Although, there are some slight differences in the I–V characteristic symptoms under these faults. Therefore, in order to classify this type of faults, an SVM model with more classification capabilities has been developed.

SVM Model. Support Vector Machine is considered as a machine learning tool with powerful and fast classification capabilities for signal processing areas, and particularly, for resolving binary classification problems [13]. One of the main advantages of SVM is the kernel trick that allows the extraction of the optimal separation hyper plan. Even if the data set is not linearly separable, a non-linear mapping could be applied via a kernel function from the original data to a higher dimensional separable space.

Feature Analysis and Extraction. The main criteria for features extraction for the SVM algorithm is characterizing the faults and differentiate it from other confusing situations. In that context, and with reference to Fig. 1, the group of faults F2 and F3 have some differences in the MPP point and in the open circuit point. Therefore, three new features are extracted from the I–V characteristic regarding these points:

The first feature is the ratio between the measured and reference value of the MPP current:

$$F_{\mathrm{I}_m} = \frac{I_{mMeas}}{I_{m\mathrm{Ref}}} \tag{9}$$

The second feature is the ratio between the measured and reference value of the MPP Voltage:

$$F_{V_m} = \frac{V_{mMeas}}{V_{m\mathrm{Ref}}} \tag{10}$$

The last feature is the ratio between the measured and reference value of the open circuit voltage:

$$F_{V_{OC}} = \frac{V_{OC_{Meas}}}{V_{OC_{\mathrm{Ref}}}} \tag{11}$$

4 Simulation Results and Discussion

In order to verify the performance of the proposed fault detection technique, a PV string of four PV modules was simulated under MATLAB/Simulink environment. The testing approach is based on testing the performances of each algorithm separately under several situations of fault presence.

4.1 Performance of the Fault Occurrence Detection Algorithm (Algorithm 1)

The first detection algorithm uses the real-time measurement of the MPP power and compares it with the reference power. This late is estimated under the same environmental conditions to emulate the normal functioning of the system. In order to avoid false alarms due mainly to the response time of the MPPT, the detection algorithm starts only when a static MPP is reached. Fig. 5(a) shows the difference between extracted and reference power and the first algorithm decision under faulty and normal operation. On the other hand, Fig. 5(b) presents the evaluation of the power difference and the relative ratio under the same conditions.

As it is shown in Fig. 5(a), in normal operation, the reference power is similar to the measured power in terms of amplitude and evolution. This is seen also in Fig. 5(b), as the power difference does not exceed the relative threshold. However, after the fault occurrence (t = 0.5s and t = 1.3s), the difference between the measured and estimated power exceeds the relative threshold under these conditions. As result, the first algorithm initiates an instantaneous I–V characteristic tracing and store data to be used by the next classification algorithms. The characteristic tracing procedure takes only 0.1s, indicated in yellow in Fig. 5(a) and (b). On the other hand, in normal operation, the power threshold is not stationary and follows the climatic conditions avoiding false classifications. This allows permanent fault detection regarding environmental conditions.

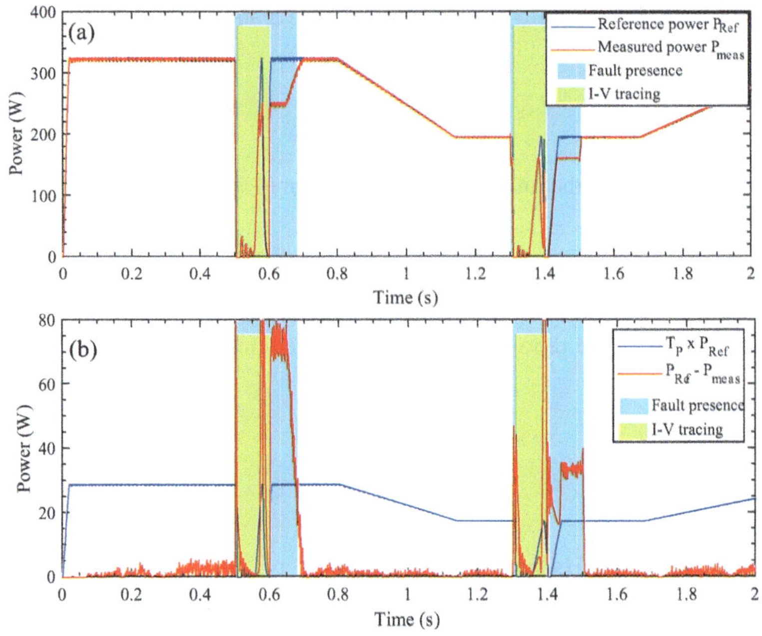

Fig. 5. Evaluation of PV reference power, measured power, power difference and threshold.

4.2 Performance of the Fault Classification Algorithm (Algorithm 2)

In order to verify the classification capabilities of algorithm 2, the same PV system has been considered, and different complicated fault scenarios have been examined under STC:

- **Case 1:** Connection fault by increasing in the series resistance by 10 Ω **(F8)**.
- **Case 2:** Partial shading in PV modules: 60% in the first and 40% in the second **(F5)**.
- **Case 3:** Partial sanding of 50% in one module with a disconnected by-pass diode **(F6)**.
- **Case 4:** Partial sanding of 30% in one module with connection fault **(F7)**.
- **Case 5:** one module short-circuited in PV string **(F2)**.

The decision of Algorithm 2 as well as the calculated thresholds and ratios are reported in Table 2. The ratios exceeding their thresholds are indicated in bold.

The simulation results have shown that the second algorithm is able to identify and classify the faults in the examined PV string correctly. For instance, in case 1, both RIM and RVM exceed their defined thresholds. In this case, the decision algorithm indicates a correct classification of connection fault F8. On the other hand, for case 2, only RIM exceeds its corresponding threshold indicating also a correct decision of a partial shading fault F5.

Although, for faults with the same signature (F2 and F3), this algorithm still unable to distinguish between them as the ratios exceed their thresholds in the same way.

4.3 Performance of the SVM Based Algorithm (Algorithm 3)

As it has been mentioned before, the SVM model uses three features for the classification between two faults F2 and F3. In order for the SVM model to find the optimal separation hyperplane, it should be trained through a set of data including the features and their known fault classes. In this work, a set of 50 samples including 25 samples for each fault class (F2 and F3) was prepared. On the other hand, several kernels were tested for the SVM model in order to find the most accurate one. After training the model using 25% of the data for validation, the Fine Gaussian kernel was found to be the most accurate kernel with a 100% correct classification rate. A comparison between all the SVM kernels in terms of accuracy is presented in Fig. 6.

The validation classification confusion matrix of the trained SVM model is presented in Fig. 7. As it can be noticed for the training data the SVM model indeed reached a 100% accuracy. Another set of data was used to test the accuracy of the trained SVM model with 20 fault case including 10 samples for each fault. As depicted in classification in Fig. 8, the SVM model reaches a 100% classification rate under the new test data.

Table 2. Algorithm 2 Ratios, thresholds, and decision under different fault cases.

Cases	Attributes					Thresholds		Decision
	N	R_{ISC}	R_{VOC}	R_{Im}	R_{Vm}	T_C	T_V	
1 (F8)	1	0.0084	0	**0.3221**	**0.3526**	0.05	0.039	F8
2 (F5)	0.33	0.0009	0.0200	**0.3745**	0.0210	0.05	0.039	F5
3 (F6)	1	**0.4213**	0.0094	**0.4755**	**0.0964**	0.05	0.039	F6
4 (F7)	0.5	0.0044	0.0094	**0.4803**	**0.0823**	0.05	0.039	F7
5 (F2)	1	0	**0.2500**	0.0001	**0.2499**	0.05	0.039	F2 or F3

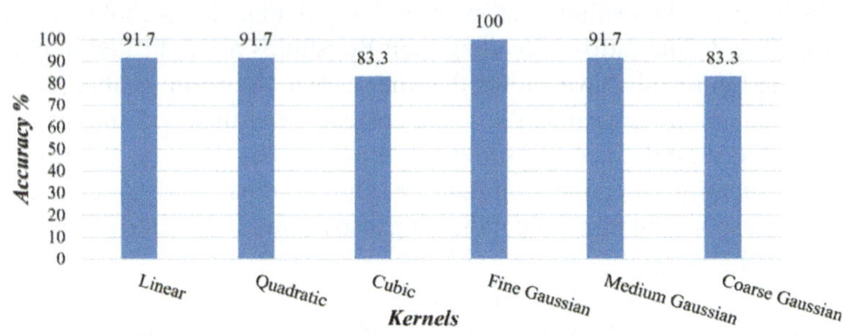

Fig. 6. Different SVM model training kernels accuracy.

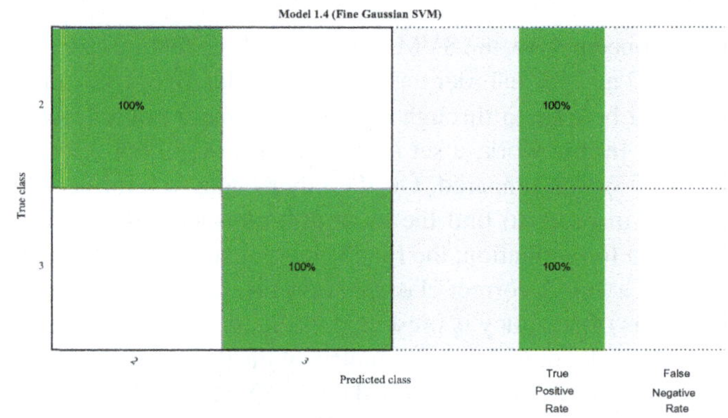

Fig. 7. Validation confusion matrix for the trained SVM model.

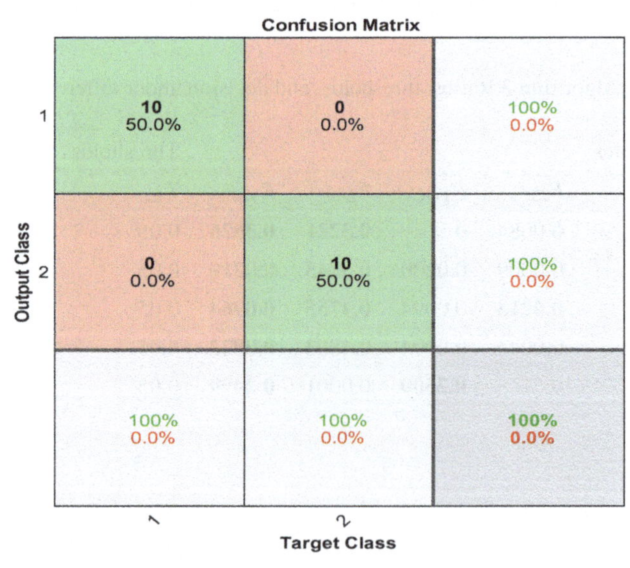

Fig. 8. Test confusion matrix for the trained SVM model.

5 Conclusion

This paper has presented a fault diagnosis technique capable of detecting and isolating eight of the common faults in a PV array. The adopted method is a hybrid approach, based on three algorithms allowing the identification and classification of the addressed faults. The first algorithm is based on power analysis approach, that differentiates between normal and faulty operation. The second is a model difference-based algorithm that detects faults having a different combination of ratios. The last is an SVM based approach, that classifies even faults with similar signatures. The simulation results have confirmed the accuracy of the proposed technique by evaluating the performances of each algorithm separately. The results have also shown that the proposed method detects and classifies the investigated faults with up to 100% correct classification rate.

References

1. Dhimish, M., Holmes, V., Mehrdadi, B., Dales, M.: Simultaneous fault detection algorithm for grid-connected photovoltaic plants. IET Renew. Power **12**, 1565–1575 (2017)
2. Harrou, F., Sun, Y., Taghezouit, B., Saidi, A., Hamlati, M.E.: Reliable fault detection and diagnosis of photovoltaic systems based on statistical monitoring approaches. Renew. Energy **116**, 22–37 (2018)
3. Hariharan, R., Chakkarapani, M., Ilango, G.S., Nagamani, C.: A method to detect photovoltaic array faults and partial shading in PV systems. IEEE J. Photovolt. **6**, 1278–1285 (2016)
4. Dhoke, A., Sharma, R., Saha, T.K.: PV module degradation analysis and impact on settings of overcurrent protection devices. Solar Energy **160**, 360–367 (2018)
5. Roy, S., Alam, M.K., Khan, F., Johnson, J., Flicker, J.: An irradiance independent, robust ground fault detection scheme for PV arrays based on spread spectrum time domain reflectometry (SSTDR). IEEE Trans. Power Electron. **33**, 7046–7057 (2017)
6. Kumar, B., Ilango, G., Reddy, M., Chilakapati, N.: Online fault detection and diagnosis in photovoltaic systems using wavelet packets. IEEE J. Photovolt. **8**, 257–265 (2018)
7. Zhao, Y., Yang, L., Lehman, B., de Palma, J., Mosesian, J., Lyons, R.: Decision tree-based fault detection and classification in solar Photovoltaic arrays. In: Proceedings of Applied Power Electronics Conference (APEC), Orlando, pp 93–99. IEEE (2012)
8. Akram, M., Lotfifard, S.: Modeling and health monitoring of DC side of photovoltaic array. IEEE Trans. Sustain. Energy **6**, 1245–1253 (2015)
9. Yi, Z., Etemadi, A.: Line-to-line fault detection for photovoltaic arrays based on multiresolution signal decomposition and two-stage support vector machine. IEEE Trans. Ind. Electron **64**, 8546–8556 (2017)
10. Chouay, Y., Ouassaid, M.: An intelligent method for fault diagnosis in photovoltaic systems. In: International Conference on Electrical and Information Technologies (ICEIT), Morroco, pp. 1–5. IEEE (2017)
11. Villalva, M., Gazoli, J., Filho, E.: Comprehensive approach to modeling and simulation of photovoltaic arrays. IEEE Trans. Power Electron. **24**, 1198–1204 (2009)
12. Bun, L: Détection et localisation de défauts pour un système PV. Ph.D. thesis in University of Grenoble (2011)
13. Begg, R., Palaniswami, M., Owen, B.: Support vector machines for automated gait classification . IEEE Trans. Biomed. Eng. **52**, 828–838 (2005)

A Framework of IoT-Enabled Vehicular Noise Intensity Monitoring System for Smart City

Md. Abdur Rahim[1], M. M. Rahman[1,2](✉), Md Arafatur Rahman[3], Abu Jafar Md Muzahid[3], and Syafiq Fauzi Kamarulzaman[3]

[1] Department of Mechanical Engineering, College of Engineering, Universiti Malaysia Pahang, 26300 Gambang, Kuantan, Pahang, Malaysia
mustafizur@ump.edu.my

[2] Automotive Engineering Centre, Universiti Malaysia Pahang, 26600 Pekan, Pahang, Malaysia

[3] Faculty of Computing, Universiti Malaysia Pahang, 26600 Pekan, Pahang, Malaysia

Abstract. The noise, sound pollution, and harshness are steadily increasing alarmingly from various vehicles such as lorries, vans, cars, and buses. It is considered a significant concern in our modern life due to the long-term harmful effect on human health. The excessive vehicular noise and sound pollution mainly affecting in specific areas of smart cities such as hospitals, educational institutes, various private/public organizations. As a result, people are suffering from neurocognitive problems. Therefore, controlling the sound pollution from vehicles are incredibly essential to mitigate health issues globally. However, minimal research was conducted by monitoring vehicular noise intensity in a smart city area using smart, reliable, and sophisticated technologies such as the internet of things (IoT) to mitigate these issues. This paper presents an enhanced intelligent IoT-enabled vehicular noise intensity monitoring system to protect the city dweller's health by reducing sound pollution in the smart city. We are planned to propose an architectural framework using noise intensity measuring sensors, 360° Camera, 360° LIDAR, GPS (Global Positioning System) in the vehicle with IoT technology for monitoring individual vehicles and their nearby vehicles over the smart city. The proposed system can monitor real-timely vehicular noises and notify designated stakeholders (i.e., vehicle owner, city authority) promptly and stored in the cloud with valid proofs. The proposed system helps to develop an ideal vehicular noise monitoring system over the smart cities and the widespread significance of mitigating health problems that relate to sound pollution and increase public awareness.

Keywords: Vehicular noise monitoring · Sound pollution · IoT · Smart city · Machine learning algorithm

The authors would like to thank Universiti Malaysia Pahang for providing the laboratory facilities and financial support under the University FLAGSHIP Research Grants (Project number RDU192203), Postgraduate Research Scheme (No. PGRS200325), and International Matching grant RDU192704-UIC191515.

J. A. Mat Jizat et al. (Eds.): iCITES 2020, AISC 1350, pp. 194–205, 2021.
https://doi.org/10.1007/978-3-030-70917-4_19

1 Introduction

The automotive sector has exceptionally improved its service by incorporating advanced and emerging technologies. Self-driving facilities, artificial intelligence for predictive maintenance, IoT technology for remote monitoring, and many others are typically mentionable technologies incorporated into the progress of this sector by minimizing the problems [1]. Presently, among these, IoT technology in automobiles is increasing significantly and enhances the safest and comfortable journey for vehicle users. Similarly, scientists are incorporating IoT technology in various ways to gain maximum efficiency for reducing environmental pollution caused by vehicles [2]. In this case, sound pollution is a key concern of environmental pollution to the environmentalists and environmental organizations.

The various severe human health problems such as sleeplessness, cardiovascular disease [3], neurocognitive impairment of both children and adults, annoyance [2] are affected by excessive noise. In this case, the World Health Organization (WHO) identified that the traffic noise is significantly responsible [4]. Based on the European Union publication, about 40% of the EU country's inhabitants are exposed to at least 55dB [5]. The researcher apprehensions that countries in Asia and the lower developed areas may face more severe vehicular noise than EU as well as relatively developed countries [3, 5]. Therefore, the importance of carrying out a profound research on vehicular sound pollution mitigation at any cost. Since in urban areas, the density of inhabitants is very high due to various private/public organizations, industries, educational/training institutes, hospitals, and so forth. As a result, the importance of different vehicles in this area is conspicuous for the inhabitant's movement and essential things for moving from one place to another place in urban areas, and it is increasing considerably. Besides this, some necessary facilities and advanced services are required for the city dwellers to make their activity faster, more accessible, reliable, safe, and secure in smart cities than conventional city/urban areas with the incorporation of IoT [6]. However, recently this concept more enhanced by ensuring the reliability, security, and productivity of city residents by introducing an emerging IoT technology. Because smart cities are equipped with lots of smart features and intelligent devices to ensure the city dwellers easy, enjoyable, secure life as well as increase their work efficiency [7]. Nonetheless, city inhabitants' productivity is hindered due to their physical problems by the traffic noise, and it is rising to owe to the increasing population and vehicles with changing times [8].

However, the vehicular noise mitigation work is challenging work besides ensuring city dwellers' other necessary facilities in smart city management [9]. IoT technology shows an outstanding possibility [10] to control vehicular noise severity by employing the remote monitoring features using advanced instruments with an efficient procedure. Because, its profound successful practice appeared in vehicles for remote vehicle tracking [2], performance monitoring, emission monitoring [10], remote anti-theft features, driver health condition observation, and ITS (intelligent transportation system) [2] in the smart city area. Hence, this technology is a promising option in vehicular noise pollution prevention for smart cities. It has not yet been used to a large extent by employing sophisticated instruments and upgraded techniques.

This paper aims to provide a distinct architectural concept on vehicular noise intensity monitoring systems using recently emerged IoT technology to enhance the smart city concept and prevent city dwellers' health problems by improving their productivity. Moreover, raise public awareness about vehicular noise issues, their effect, and create a future research scope in the automotive sector for mitigating sound pollution. Also, the proposed framework with advanced and sophisticated instruments for vehicular noise observation systems for the smart city makes it unique a prominent data analytic technique to get the real-time result as much as faster and accurate through data analysis eloquently in the architecture section. On the other hand, this article organized constructively to give a clear idea about previous related research works and make evident the uniqueness of this article. Then in the architecture and working principle sections constructively present instrument details and their connection with the working procedure of the proposed framework to get real-time outcomes.

2 Relevant Research Work

Recent technological advancements escalating the dependability on the internet due to its a bunch of benefits in our daily activity. As a result, the strong influence of it appearing in the automotive sector, and its application scope broadening considerably as IoT technology due to outstanding benefits [2]. However, this technology was made by Kevin Ashton in 1999 [11], and later in 2012, it evolved to self-contained for use in the automotive sector through the various transformation [2]. Again, in 1994 the smart city concept was introduced and gradually modernized by incorporating sophisticated instruments and updated technology to facilitate the quality of life for inhabitants and improve their workability [12]. For smart city management, as a promising technology, IoT capable of ensuring secure, efficient, and economic activities in different aspects to save energy [13], reduce pollution, increase productivity and reliability [14]. Arasteh et al. [15] investigated the benefit and application of IoT technology in smart city management as an indication of its significance. Such as smart house systems to save energy, intelligent surveillance systems for security, environment pollution monitoring system to mitigate health problems, ITS for a comfortable journey, intelligent parking system to utilize free space, utility service systems as a basic needs, waste management to prevent pollution and so forth [7]. Hence, the pollution is a significant issue against city dwellers' health to reduce the workability and ultimately affect the national economy. For leading various unwanted physical irregularities, noise pollution is a significant issue among different pollution problems and Modern Plague that described lucidly by Goines et al. [16].

Major roads, including highways are constructing beside the cities due to urbanization as a result gradually increasing the resident and tendency of using motor vehicles, which emits a massive amount of noise due to horn honking, accelerating vehicles, illegal modification of tailpipes, and old vehicles [15]. The researchers proposed various techniques using contemporary devices and instruments to mitigate the sound pollution intelligently over the city. Jin et al. [17] applied a noise mapping method for monitoring the noise from vehicles used IoT technology with WSN (wireless sensor network) among installed sensors on the traffic lights, street lights, and building facades of the city area. In this system, they also take into consideration the public complaint system. However, complaints sometimes are not authentic due to some complainers may complain intentionally to harass someone though the accused may not be responsible for noise emission. Almehmadi [18] developed a model using Support Vector Machine (SVM) and Random Forest classifiers to predict the probable time of noise based on previous data and to measure current noise over the city and inform respected city authority.

Table 1. Recently devised vehicular noise monitoring system with different techniques and devices for the smart city.

Devices	Technique	Purposes	References
SEN-12642 sensor, raspberry pi, GSM module	Own developed method	Noise monitoring	[8]
Built-in smartphone's microphones	Mobile Crowd Sensing (MCS) platform	Low-cost noise monitoring system development	[9]
NoivelCam, highly directional shotgun microphones	Time difference of arrival (TDOA) based analysis	Noise monitoring in the busy highway	[15]
Noise sensor	Random Forest classifiers and Support Vector Machine	Noise Monitoring of a smart city	[18]
Sound level meter	Geographic Information System (GIS) and spatial analysis tool	Noise Monitoring in a city	[19]
Wireless Acoustic Sensor, Arduino	Frequency domain-based algorithm	Acoustic noise monitoring	[20]

Similarly, Malche et al. [21] was proposed the architecture of the environmental pollution monitoring system using the sensing layer in the zone wise over the city and measure sound level besides other environmental parameters and sent the notification to the city authority using IoT. Patil [8] proposed a model using sound detection, carbon monoxide, gas sensors, GSM (Global System for Mobile Communications) Module, and Raspberry pi using IoT technology to monitor older and heavier vehicles emitted noise as well as emitted other pollutants in the city. Laxmi et al. [19] proposed an innovative approach for identifying mostly noise polluted points using a cycle as a noiseless medium with a sound measuring meter to develop spatial and strategic noise maps of a city by Geographic Information System (GIS) and spatial analysis tool. On the other hand, a cloud-based data analysis system was proposed to monitor and analyze the gathered information for mitigating noise issues in the city by taking necessary preventive activity [22]. Velásquez et al. [23] proposed a low-cost noise pollution monitoring system to deliver the geotagged information and state of pollution to the smart city inhabitants using IoT technology. Fernandez-Prieto et al. [20] proposed a novel approach by designing a complete wireless acoustic sensor network (WASN) to monitor noise levels and prepare precise noise maps over smart cities. Again, for the smart city, a surveillance system was designed using NoivelCam (Noisy vehicle surveillance Camera) system across the city.

This system can keep a continuous video record of drivers inconsiderate behaviours for creating noise by unnecessary acceleration, racing, illegal tailpipe modification, driving aged vehicles besides noise level. For the smart city, the recently devised few vehicular noise monitoring systems are presented in Table 1.

However, the description above used their contemporary instruments with available techniques. Currently, the severity of noise in the city areas are increasing alarmingly and affect the city dwellers' productivity, hindering the national economic growth drastically, and that needs to mitigate effectively. Until now, previously used instruments, applied methods, and techniques are insufficient to overcome the confronted difficulties. Because during the literature study, we found that some methods used for an individual vehicle, some proposed systems used obsolete instruments and techniques, some proposals are not feasible to measure the information accurately with necessary proofs, and so forth. Also, the number of vehicles is increasing tremendously besides the rising number of city residents. As a result, we have to devise something new that can collect data accurately, analyze fastly, deliver real-timely with strong evidence of individual vehicle and nearby source for demotivating people from vehicular sound noise emission by employing sophisticated instruments and advanced techniques. Again, recently, available some devices and techniques will capably gather, analyze, store, and deliver the noise-related data efficiently and real-timely by ensuring the required security. Thus, the IoT technology incorporation with noise intensity measuring sensors, 360° Camera, 360° LIDAR, GPS features in individual vehicles using machine learning (ML) algorithm technique has a considerable possibility for reducing noise in the smart city area by analyzing fastly, accurately and sharing real-time. The conceptualized framework and machine learning model architecture are proposed in this study to give a clear understanding of vehicular noise monitoring.

3 Architecture and Modeling of Proposed System

3.1 Architecture

Figure 1 presents the vehicular noise monitoring system in a smart city is comprised of three layers such as sensing layer, gateway, network, security and service layer, and application layer. The sensing layer consists of a sensor, camera, and LIDAR for monitoring vehicular noise intensity with strong evidence like captured video and object distance from the vehicle. Raspberry Pi 3 Model B (IoT development board), controls the sensing nodes as well as connects with the second layer. For the second layer case, the gateway is designated for node authentication and transmission of collected data using the internet. The 4G-LTE Module is proposed for establishing communication with the available 2G/3G/4G-LTE network to transfer the data at a higher rate. These data store in the cloud and categorize the data for future jurisdiction using a web server as a service option of this layer. Finally, the interface facility to the proposed system ensures through the application layer. As a result, the users can monitor in real- timely as well as generate a monthly report to take necessary action for violating the noise pollution rules of the city using a smartphone, computers, and so on.

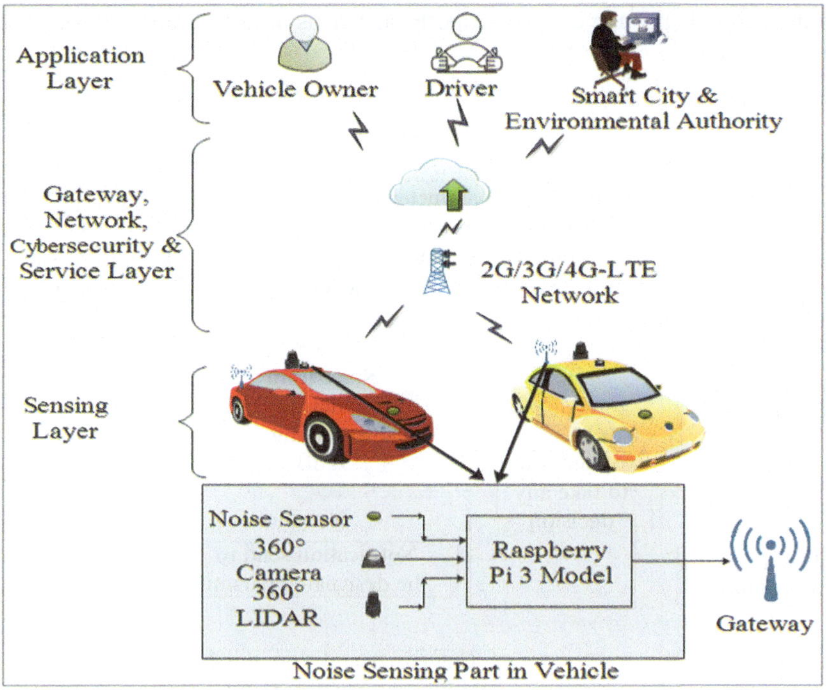

Fig. 1. Architecture of the noise monitoring system for smart city management

3.2 Machine Learning Algorithm Modelling

Fig. 2 presents the machine learning algorithm modeling. The steps related to the modeling eloquently described to understand straightforwardly as follows:

Data Acquisition. This step is responsible for gathering the data to train the developed model. For data acquisition, usually installed sensing devices done this work. In this stage, noise pattern for horn hocking or engine also is needed to define beside the noise level to classify data whether from vehicular source or other sources in surroundings.

Data Preprocessing. This phase is designated to clean up raw data by eliminating null values, unnecessary data, changing one function to another, and so on.

Input Feature Engineering. It is defined as the process of employing domain knowledge for creating new features from raw data that improve the predictivity of the ML algorithm.

Model Training and Trained Model. In this case, mostly Support Vector Machine (SVM), Decision Trees, Nearest Neighbour Algorithms, and Random Forest Classifiers are preferable to use as the featured ML algorithm. However, model training is defined as measuring good value for all data using the provided ML algorithm with training data. Therefore, it is conducted by feeding processed data using the featured algorithm and enhance the estimation performance as a trained model through tuning hyper-parameters.

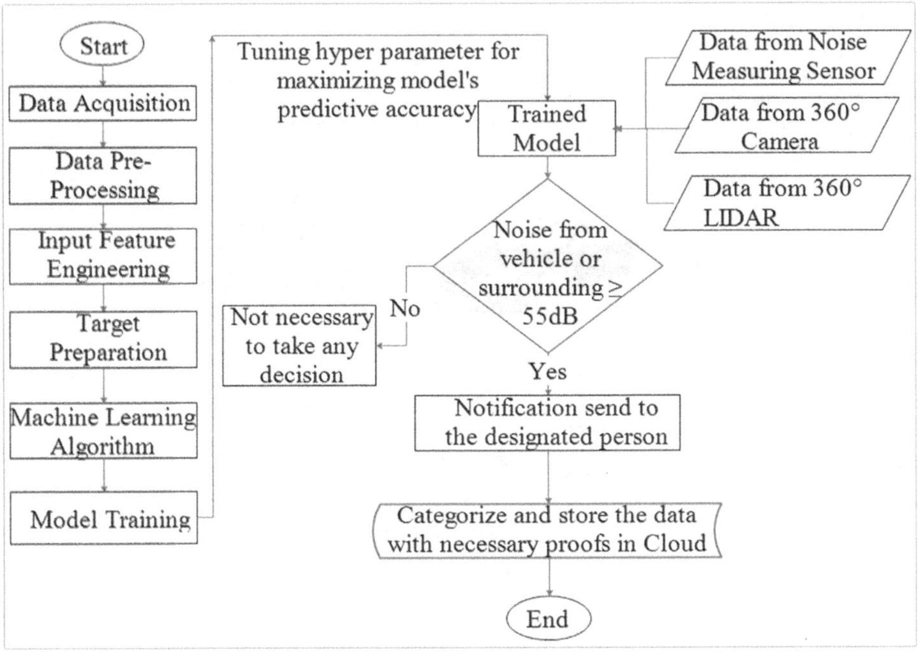

Fig. 2. Machine learning algorithm of the noise monitoring system for smart city management

Finally, the proposed algorithm also classify the noise source, like emitted by the horn system or the engine, or other vehicle parts based on predefined information besides categorizing based on the sound level. Also, it determines whether that noise is created by a vehicle or comes from surroundings.

4 Working Principle of Proposed Prototype System and Required Components

The vehicular noise monitoring system is designed by proposing upgraded instruments and advanced techniques capable of accurately analyzing the gathered data and sending real-time sound pollution notification quickly to enhance smart city management activity. Moreover, the working principle and instrument are details discussed as follows.

4.1 Sensing Layer

The sensing layer is designed so that the noise intensity sensor remains active to sense the driver's raised noise and surrounding noise. Since maximum cases, the drivers are responsible for creating the noise by using an existing horn system. In this framework, we planned to synchronize the horn system to distinguish whether the noise is created by the vehicle or outside the vehicle like surroundings. Besides that, the camera and LIDAR also work simultaneously where the 360° camera captures the video in excessive noise intense producing point (more than 55dB) based on the measured distance of any objects from the vehicle by 360° LIDAR. Again, in some cases, the driver may create higher intensity noise intensionally or unintentionally in the busiest location to alarm the nearest people, vehicles, or other objects. The 360° LIDAR of the proposed framework measures the minimum distance of vehicles' from any objects of surroundings. Then the proposed Machine Learning Algorithm analyzes the collected information and determines whether the noise is created unnecessarily by the driver or not; if yes, then the 360° camera will instantly capture a video of surroundings with other evidence for further jurisdiction. After scrutinizing, the authority may use the measured distance as references to dig out the primary purpose of horn hocking during the jurisdiction process. Similarly, if the nearest object emits noise unnecessarily and more than the permissible limit, then that also be measured by the system whether their intention is valid or not based on the distance from the vehicle. After that, if the system found any invalid intension, then the proposed system will notify the city councilor to take further action with the captured video and other proofs.

Moreover, these sensing nodes connect with the Raspberry Pi 3 Model B, and it acts as a host to operate the overall system. In addition, the 4G-LTE Module node connects with that to act as a gateway for transferring collected data to the cloud using an available network. However, the details of the required instruments for the sensing layer are described as follows:

Noise Intensity Sensor. To measure the vehicles' emitted noise in the smart city area, we propose model TA120 of the CESVA brand noise measuring sensor because it is suitable for smart city requirements and has higher noise level measuring accuracy with minimal error. However, the configuration of this sensor is described as follows [24].

- Class 1 precision sensor based on IEC 61672–1
- Lightweight and small size
- IP65 protection
- Required power12V DC
- Ethernet, GPRS, and Wi-Fi communication facility
- HTTP, HTTPS protocol
- Noise measuring level 28 to 120 dBA360°

Camera. 360° camera vigorously requires capturing the video around the vehicle as proof if noise level found above the tolerable limit. For this, we proposed a 360° WAYLENS brand camera with Model: 99-A0000044–01 to capture the comprehensive range video when required.

360° LIDAR. We proposed 360° LIDAR to measure the object's accurate distance from the vehicles honking the horn and the distance to the noise source in the vehicle's surroundings. It is provided 360° mapping with high-resolution for short to long-distance accurately. As a result, it is measured the distance of the objects that are moving around the vehicles. Thus, the distance measurement is essential that needs to distinguish for further jurisdiction to find the actual reason behind the emission of noise vehicle itself or any other objects from surroundings. In this case, the measured distance is easily calculated by Eq. (1) [25] using LIDAR light traveling time for both emitter and photodetection sensors after reflecting the targeted object.

$$d = v * t/2 \tag{1}$$

Where d, t, and v are signified distance, the time between journey, and light velocity, respectively. However, LIDAR has numerous advantages, including wide and long-distance coverage, accurate object identification, and so forth [25].

IoT Development Board. We propose Raspberry Pi 3 Model B as an IoT development board due to a small-sized single-board computer developed by Raspberry Pi Foundation, UK [26]. The configuration of the proposed development board is provided as follows:

- Quad-Core (4x ARM Cortex-A53, 1.2 GHz),
- 1 GB RAM, Wi-Fi,
- Bluetooth 4.1/BLE
- 802.11n WLAN
- micro SD card storage
- 4 USB ports,
- HDMI port, and 40 GPIO pins
- Two 5V pins & Two 3.3V pins.

4.2 Gateway, Network, Cybersecurity and Service Layer

The selected 4G-LTE Module establishes communication with the available 2G/3G/4G-LTE easily using the internet. On the other hand, this layer's service section involved three matters such as an analytic server, a cloud storage server, and the webserver. The analytic server performs according to the proposed ML algorithm model. We proposed that the system notify and store the data with the required evidence in the cloud when the vehicle emits noise more than 55 dB. Because in the urban area, the maximum tolerable noise level is 55 dB, and the maximum people's health is affected by the noise level from 75dB [27]. We proposed categorizing the measured noise in the smart city area by 55 to $<$ 75 dB, 75 to 85 dB, and $>$ 85dB as Moderate, High, and Critical noise, respectively. Besides that, the analytic system also performs necessary action to store video evidence, vehicle location co-originates, and the object's distance from the vehicle by following the preliminary discussion procedures in this section. Furthermore, it ensures the availability of these data in the webserver to help future jurisdiction activity, whether the driver/surrounding object creates noise intentionally or unexpectedly. Then the city authority can make a decision based on noise pollution rules. The used instruments and techniques for this layer is presented as follows:

Intelligent Data Analytics for Noise Monitoring in Smart City. The data used to make a fruitful decision by processing through data acquisition, data analytics (processing and management), and data interpretation [17]. Among these, data analytics is a significant concern to get fast and accurate outcomes from massive amounts of gathered data. However, machine learning is a powerful choice to facilitate the rapid processing of large-volume data for IoT technology. The integration of the ML technique with IoT paves the way for potential improvement in efficiency, productivity, accuracy, and overall cost-savings for IoT devices. Because it is characterized as a family of data analysis techniques, where the developing model process is automated on training data, i.e., requires negligible human participation. Therefore, the method of categorizing data into different groups and finding reasons for producing noise based on object distance, capturing video, and co-ordinates are entirely conducted in automated ways. Therefore, incorporating the ML technique in IoT-enabled noise monitoring systems in smart cities helps improve computation and communication capability, precise decision-making, and exceptional controllability intelligently [28].

Privacy and Cybersecurity. Currently, cybersecurity and privacy is the most prevalent and discussed issue among the researcher and service receiver because intruder or hacker may abuse the shared information and spoil the purpose of the promising system instead of getting a better outcome. In this case, biometric and two-factor authentication system with location co-ordinates considered, which will help to ensure the security about collected data and privacy protection of users [2].

4.3 Application Layer

The involved stakeholders are vehicle owners, vehicle drivers, city authorities, and environmental protection agents in the application layer. The owner and vehicle driver became aware of their activity and tried to avoid emitting noise unnecessarily due to the real-time alert system. On the other hand, the city admin and environmentalists may take legislative action against the vehicle users, the owner using the webserver, and the cloud server by following stated city rules. Also, using co-origin information, the proposed frameworks can quickly identify most noise-producing locations and may take necessary action.

Cloud Server. For vehicular noise monitoring in the smart city, the cloud server is proposed as a common platform for vehicle users, vehicle owners, city councilor's offices, and the pollution control department. In this case, the specific website to get the required information using any available web browser. The proposed system has good capabilities to send a notification of required information to the vehicle owner, the driver, or others through SMS or Email using IoT technology. Before that, useful information extracts and stores in the designated cloud as per the categorization mentioned above through the cloud computing platform [28]. Moreover, the analytic system also facilitates to present in web server for stakeholders.

5 Conclusion

Due to gradual modernization and urbanization, the human dependency on transportation is substantially increasing for its immense beneficial aspects that lead to the excessive vehicular noise emission in addition to other pollutants. As a result, the intensity of noise rises significantly in the urban area, and alarmingly affects human health. In the city area, this problem arises continuously and reducing the work productivity of city dwellers. We proposed a vehicular noise monitoring framework to reduce noise pollution and increase vehicle user's awareness. The proposed framework architecture elaborately presented ML model development procedure, instrument details, and working principle to include sophisticated instruments and recognized techniques to get fast, reliable, and accurate outcomes in real-time. Therefore, it helps mitigate the noise problem and help future researchers develop a more advanced noise monitoring system for smart city residents.

References

1. Rahman, M.A.: Reliability analysis of ZigBee based intra-vehicle wireless sensor networks. In: Lecture Notes in Computer Science (including subseries Lecture Notes in Artificial Intelligence and Lecture Notes in Bioinformatics), pp 103–112 (2014)
2. Abdur Rahim, M., Arafatur Rahman, M., Rahman M.M., et al.: Evolution of IoT-enabled connectivity and applications in automotive industry: a review. Veh. Commun. 100285 (2020). https://doi.org/10.1016/j.vehcom.2020.100285
3. Münzel, T., Sørensen, M., Schmidt, F., et al.: The adverse effects of environmental noise exposure on oxidative stress and cardiovascular risk. Antioxid. Redox Signal **28**, 873–908 (2018)
4. Fritschi, L., Brown, A.L., Kim, R., et al.: Burden of disease from environmental noise: quantification of healthy life years lost in Europe (2011)
5. Berglund, B., Lindvall, T., Schwela, D.H.: New WHO Guidelines for Community Noise. Noise Vib. Worldw. (2000). https://doi.org/10.1260/0957456001497535
6. Albino, V., Berardi, U., Dangelico, R.M.: Smart cities: definitions, dimensions, performance, and initiatives. J. Urban Technol. (2015). https://doi.org/10.1080/10630732.2014.942092
7. Arasteh, H., Hosseinnezhad, V., Loia, V., et al.: IoT-based smart cities: a survey. In: 2016 IEEE 16th International Conference on Environment and Electrical Engineering (EEEIC), pp. 1–6. IEEE (2016)
8. Patil, P.: Smart IoT based system for vehicle noise and pollution monitoring. In: Proceedings - International Conference on Trends in Electronics and Informatics, ICEI 2017 (2018)
9. Zappatore, M., Longo, A., Bochicchio, M.A.: Using mobile crowd sensing for noise monitoring in smart cities. In: 2016 International Multidisciplinary Conference on Computer and Energy Science, SpliTech 2016 (2016)
10. Manna, S., Bhunia, S.S., Mukherjee, N.: Vehicular pollution monitoring using IoT. In: International Conference on Recent Advances Innovations in Engineering, ICRAIE 2014, pp. 1–5. https://doi.org/10.1109/ICRAIE.2014.6909157
11. Ashton, K.: That "internet of things" thing. RFiD J. **22**, 97–114 (2009)
12. Dameri, R.P., Cocchia, A.: Smart city and digital city : twenty years of terminology evolution. In: X Conference of the Italian Chapter of AIS, ITAIS 2013, pp. 1–8 (2013)
13. Rahman, M.A., Mukta, M.Y., Yousuf, A., et al.: IoT based hybrid green energy driven highway lighting system. In: 2019 IEEE International Conference on Dependable, Autonomic and Secure Computing, International Conference on Pervasive Intelligence and Computing, International Conference on Cloud and Big Data Computing, International Conference on Cyber Science and Technology Congress (DASC/PiCom/CBDCom/CyberSciTech), pp 587–594. IEEE (2019)

14. Rahman, M.A., Kabir, M.N., Azad, S., Ali, J.: On mitigating hop-to-hop congestion problem in IoT enabled Intra-Vehicular communication. In: 2015 4th International Conference on Software Engineering and Computer Systems (ICSECS), pp 213–217. IEEE (2015)
15. Agha, A., Ranjan, R., Gan, W.-S.: Noisy vehicle surveillance camera: a system to deter noisy vehicle in smart city. Appl. Acoust. **117**, 236–245 (2017). https://doi.org/10.1016/j.apacoust.2016.05.025
16. Goines, L., Hagler, L.: Noise pollution: a modern plague. South Med. J. **100**, 287–294 (2007). https://doi.org/10.1097/SMJ.0b013e3180318be5
17. Jin, J., Gubbi, J., Marusic, S., Palaniswami, M.: An information framework for creating a smart city through internet of things. IEEE Internet Things J. **1**, 112–121 (2014). https://doi.org/10.1109/JIOT.2013.2296516
18. Almehmadi, A.: Smart city architecture for noise pollution mitigation through the internet of things. Int. J. Comput. Sci. Netw. Secur. **18**, 128–133 (2018)
19. Laxmi, V., Dey, J., Kalawapudi, K., et al.: An innovative approach of urban noise monitoring using cycle in Nagpur, India. Environ. Sci. Pollut. Res. (2019). https://doi.org/10.1007/s11356-019-06817-0
20. Fernandez-Prieto, J.-A., Cañada-Bago, J., Gadeo-Martos, M.-A.: Wireless acoustic sensor nodes for noise monitoring in the city of Linares (Jaén). Sensors **20**, 124 (2019). https://doi.org/10.3390/s20010124
21. Malche, T., Maheshwary, P., Kumar, R.: Environmental monitoring system for smart city based on secure internet of things (IoT) architecture. Wirel. Pers. Commun. **107**, 2143–2172 (2019). https://doi.org/10.1007/s11277-019-06376-0
22. Deshinta, A.D., Lai, E.P.Y., Ponkoodalingan, K.: The air and sound pollution monitoring system using internet of things and cloud based data analysis. INTI J. **1**, 1–7 (2018)
23. Velásquez, P., Vásquez, L., Correa, C., Rivera, D.: A low-cost IoT based environmental monitoring system. A citizen approach to pollution awareness. In: 2017 CHILEAN conference on Electrical, Electronics Engineering, Information and Communication Technologies (CHILECON), pp 1–6. IEEE (2017)
24. CESVA: TA120 - Noise measuring sensor for Smart Solutions - CESVA instruments (2020). https://www.cesva.com/en/products/sensors-terminals/TA120/. Accessed 14 Sep 2020
25. Yu, X., Marinov, M.: A study on recent developments and issues with obstacle detection systems for automated vehicles. Sustainability **12**, 3281 (2020)
26. Toma, A., Popa, Z.: IoT solution for smart cities' pollution monitoring and the security challenges. Sensors **19**, 3401 (2019). https://doi.org/10.3390/s19153401
27. King, G., Roland-Mieszkowski, M., Jason, T., Rainham, D.G.: Noise levels associated with urban land use. J. Urban Health **89**, 1017–1030 (2012). https://doi.org/10.1007/s11524-012-9721-7
28. Adi, E., Anwar, A., Baig, Z., Zeadally, S.: Machine learning and data analytics for the IoT. Neural Comput. Appl. (2020). https://doi.org/10.1007/s00521-020-04874-y

16 nm FinFET Based Radiation Hardened Standard Cell Library Analysis Using Visual TCAD Tool

Jessy Grace[1], Sphoorthy Bhushan[1], Chinnam S. V. Maruthi Rao[2], and Ameet Chavan[1](✉)

[1] ECE Department, Sreenidhi Institute of Science and Technology, Hyderabad, India
ameetchavan@sreendihi.edu.in

[2] ECE Department, Sreyas Institute of Engineering and Technology, Hyderabad, India
maruthirao.chinnaman@sreyas.ac.in

Abstract. The proposed work analyzes the Single Event Transients (SETs) in combinational circuits and Single Event Upsets (SEUs) in sequential elements for 16 nm FinFET device structures implemented using Visual TCAD tool. The FinFET CMOS circuits of traditional and Dual-port Radhard designs are compared for power consumption and particle strike immunity. As the radhard circuits are inherently redundant this causes proportional increase in power consumption which is ~2 times as compared to traditional design. In the presented work, the radiation resilience of the implemented radhard circuits, measured in terms of critical charge (Q_C), has drastically improved by greater than three order as compared to the traditional designs. With this improvement the radhard designs can in turn withstand an accelerated heavy ion particle strike with Linear Energy Transfer (LET) up to 50 MeV-cm^2/μm for SETs and LETs upto 100 MeV-cm^2/μm for SEUs. Finally, the Visual TCAD particle strike simulation showed that the radhard FinFET circuits were resilient and adjacent nodes were not affected for increased energy levels.

Keywords: FinFETs · Radiation hardening · Standard cell · Total Ionizing Dose (TID) · Single Event Effects (SEEs) · Single Event Upsets (SEUs) · LET

1 Introduction

Integrated circuits are vulnerable to Single Event Effects (SEEs) due to ionization caused by high energy particle strike [1–4]. Aggressive technology scaling to Nano level has further created challenges in the design reliability of the circuits to withstand radiation particle strike in the harsh environment like Space [5, 6]. Soft Error Rate (SER) increases by an order with each major technology scaling [7]. At nanoscale, due to increase in clock frequency SETs have become more prominent, they increase the probability of latching incorrect value. The logic gates in the critical path with high probability of failure can be selectively replaced with the new library cells as presented in this work. The ionized particles in the Space environment comprises of heavy ions, high-energy

J. A. Mat Jizat et al. (Eds.): iCITES 2020, AISC 1350, pp. 206–213, 2021.
https://doi.org/10.1007/978-3-030-70917-4_20

electrons and protons. These highly-charged particles create electronic-hole pairs under the active region of the semiconductor device. This results in either functional failure or degradation of the device properties due to parasitic effect with long term exposure [8]. Direct particle strike produces glitches that may result in erroneous output. These glitches are referred to as Single Event Transient (SET) and the circuit is said to be prone to Single Event Upset (SEU) [9, 10]. Since there are many digital components, the device may also become susceptible to Single Event Effects (SEE) that can cause the entire device to become non-functional. Exposure to radiation may also result in loss of sensitivity and reliability of the semiconductor devices [11]. To address the basic issues like bulk scaling, shrinking size and short channel effects that play a major role in radiation environment, FINFET's are preferred over MOSFET devices. FINFET also known as a tri-gate transistor, i.e. is a vertical (non-planar) double gate transistor built on an SOI body (substrate) that looks like the back fin of a fish [12, 13].

The objective of the presented work is two folds - *1) Implement nanoscale SOI FinFET device structure to mitigate the effects of radiation and 2) Design and Analyze radhard standard cell circuits based on these devices for SET and SEU effects using VTCAD tool.* The research paper is organized as follows: Sect. 2 gives the motivation and contribution for the research work conducted. Section 3 explains the Tools and design flow of implementation. Section 4 discusses the design of FinFET circuits Sect. 5 presents the simulation results and Sect. 6 gives the conclusion. The findings of this research will be useful in understanding the robustness and applicability of FinFET designs to radiation impacts.

2 SOI FinFET Device for Radhard Design

The degrading short channel behavior of Planar MOSFET's has led to the necessity of FinFET's [12]. The silicon body of FinFET is rotated on its edge with a vertical orientation, while source and drain are horizontally placed over its body, like in a normal FET as shown in Fig. 1(a). An additional biasing in DG part enables multiple threshold

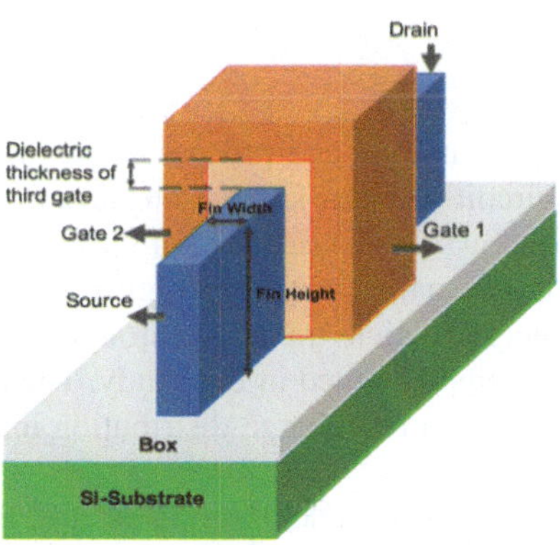

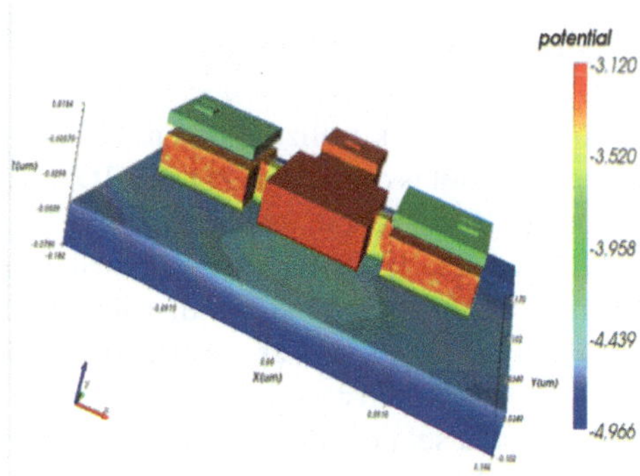

Fig. 1. 3D Structure of SOI FinFET

voltages. The gate is fabricated over fin and due to its vertical structure the FinFET width is reduced [11, 12]. The width is given by

$$W_{min} = 2 * H_{FIN} + T_{FIN} \quad (1)$$

Where,

W_{min} - minimum transistor width, H_{FIN} - Fin Height (fixed), T_{FIN} - Silicon body thickness.

As H_{FIN} is fixed, by placing multiple fins in a parallel device over all can be width be increased. For 'n' parallel fins, the aggregate physical width (W_{total}) of a tied-gate FinFET is given as:

$$W_{total} = n * W_{min} = n * (2 * H_{FIN} + T_{FIN}) \quad (2)$$

Basically, FinFET's are classified as Shorted Gate (SG or 3T) and Independent Gate (IG or 4T) based on their operation. Compared to SG, the IG FinFET's offers more flexibility in applying different levels of signals to each of the two gates that will offer back gate bias to modulate the threshold voltage 'Vth' of the device front gate. With the insertion of the buried oxide layer, nearby fins are also isolated. The electrical layout of the SOI-FinFET is comparable to that of a conventional SOI-MOSFET except that the channel is fabricated as a 3-D bar on top of the substrate (fin) and is enclosed within the gate. Multiple such electrodes can be formed on each side of the device so that current leakage and a drive current are controlled. A simple procedure in manufacturing a bulk multi gate transistor starts with the selection of a substrate. The substrate of a FinFET is a lightly doped p-type with silicon nitrate masking on top of it with a patterned layer for resistance. Channel is formed as fin with anisotropic etching that is time based. The fins are isolated through oxide deposition with a high aspect ratio. To obtain the planar structure the oxide is mechanically polished by a chemical such that the mask acts as a stop layer. One more etching is performed to perform lateral isolation of fins with the oxide. The gate electrode is insulated from the channel by the thermal oxidation process [14].

3 Simulation Tool Setup

The proposed work utilizes Visual TCAD tool to construct and process SOI FinFET device. The tool is capable of 2D and 3D device/circuit structure simulation and can also perform SPICE level circuit simulation. Fig. 1(b) shows the 3D structure of the designed FinFET characterizing its potential distribution. The Visual TCAD tool is highly effective in analyzing the radiation dose distribution of energy particle transport through Monte Carlo particle simulation as described in [15], and implemented in this study. The tool further assists in illustrating the dose distribution to trapped charge distribution inside insulators and semiconductors interfaces.

This tool estimates the problem of charge carriers drift under definite electric field and evaluates the relation of trapped charge distribution to the increased leakage current. Parasitic transistors generated due to TID effect are simulated to different dose rates. In this work, we present 2D and 3D structures of radhard and traditional cells that have been designed with 16 nm SOI FinFET through genius tool by integration of Visual TCAD tool with Cadence. To emulate CMOS devices exposure to moderate amount of dose, a relatively simple model that describes the generation of fixed charge in the oxide is adequate in most cases. This model calculates the carrier generation, recombination, transport and trapping in insulator regions. A user can directly define the total ionizing dose and the dose rate, Genius will calculate the oxide charge density produced in the insulators [16]. More detailed model is required to describe the generation of interface traps due to TID, and its dependence on dose-rate. This effect is critical to BJT devices, but is the subject of a separate application example from Cogenda. The simulation process produces a device with meshed structure. Then simulation is performed using appropriate transport models on the meshed structure. Instead of conventional drift-diffusion model, a hydrodynamic model with quantum corrections (density-gradient model) is used in nanometer Fin-FET.

4 FinFET Based Standard Cell Design

The proposed work implements radhard standard cells at 16 nm technology using SOI FinFET devices. The standard cells considered for this study comprises of – (i) Combinational circuits – Inverter (Fig. 2(a)), 2-input NAND (Fig. 2(c) and NOR (Fig. 2(d)), 2:1 Multiplexer (Fig. 2(e) circuits and Clock buffer inverter chain (Fig. 2(b)) and (ii) Radhard FF with robust latch having cross-coupled redundant nodes (Fig. 2(f)). Traditionally, a component or design is considered Radiation Hardened (Radhard) when it is capable of mitigating and withstanding the effects due to radiation particle strike such as TID, DD and SEE (SEU, SEL & SET) for higher level of exposure and Liner Energy Transfer (LET > 100 MeV-cm^2/mg) [17]. As discussed in Section I, a radiation particle strike generates electron-hole pairs that induces external charge which collects under the pn junction through the diffusion regions [18]. If we consider for example of a conventional inverter gate with output set at logic 0 and assume a particle strike at PMOS diffusion region. The accumulation of external charge will cause the output node voltage to flip temporary to logic 1causing a glitch which will result in a SET [19]. This SET may propagate through stages of logic gate before latching in the flipped value in a storage element and cause erroneous computation [20], as shown in Fig. 2(a) and 3(a). To mitigate this SET effect, radhard versions of the above-mentioned combinational cells are implemented with Dual Port configuration, in case of inverter there are two-ports at the input & output, refer Fig. 2(a).

Under normal operation both input ports (IN_N and IN_P) would have same logic value and output ports (OUT_N and OUT_P) would carry same inverted values. To better understand the SET mitigation method of Dual ported circuits, consider the setup shown in Fig. 2(b). A SET at A_P does not affect the output of a succeeding inverter, because A_P is connected to PMOS of the succeeding inverter. This setup is highly effective for implementing robust clock distribution network under the SET effect, towards the end of the chain a correction inverter is place. The mitigation of SET is shown in Fig. 2(b) of the setup simulation. Similar to the Dual-ported inverter other logic gates such as 2-input NAND, NOR and Mux gates are implemented along with a dual-port Tri-state inverter.

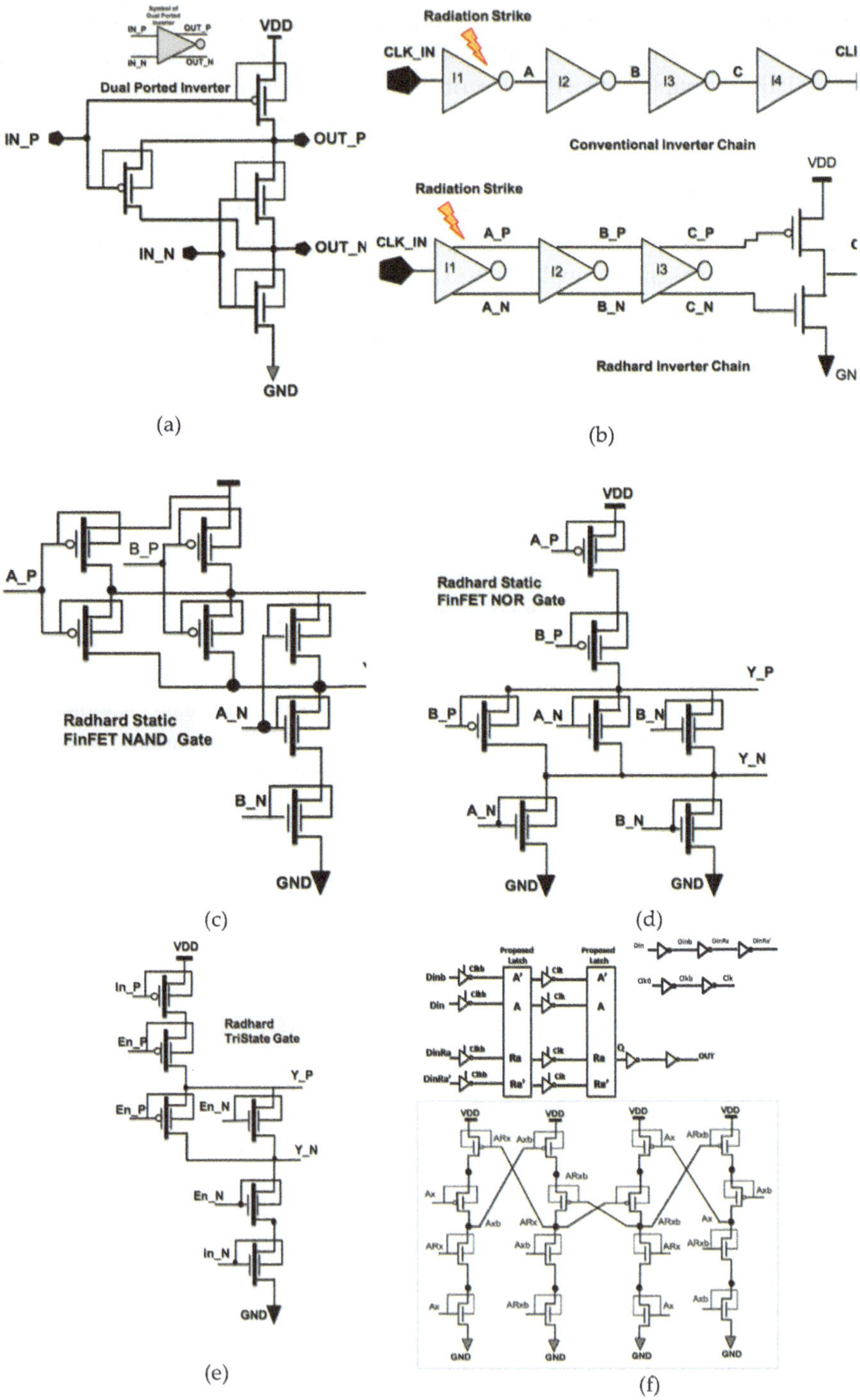

Fig. 2. 16 nm SOI FinFET Radiation Hardened Standard Cell Designs – (a) Dual Ported Inverter Design, (b) Inverter Chain for conventional and dual ported Inverter, (c) Dual Port 2-input Nand, (d) Dual Port 2-input Nor, (e) Dual Port Tri-state Inverter and (f) Dual Port Radhard Latch cell

5 Standard Cell SET and SEU Simulation Results

The radhard standard cell designs presented in the previous section are implemented with the 16 nm FinFET devices. The circuits were simulated using the Visual TCAD to verify their functionality and mitigating effects after a high-energy radiation particle strike. A simple setup of inverter chain comprising of conventional and dual port version of inverter were evaluated for radiation particle strike. As shown in Fig. 3(a), a radiation particle that produced 4fC of critical charge (Qcrit) at node results in a SET. For the traditional inverter chain setup considered the SET propagates through multiple stages and is observed at the output stage. However, a dual ported radhard inverter was able to withstand higher energy particle strike that produced 800fC of critical charge (Qcrit) without producing SET at the following stages and thereby at the output stage, as shown in Fig. 3(b). Similar analysis was performed to verify the Qcrit levels of other standard cells discussed earlier and their comparison with traditional designs is presented in Fig. 4(a). For the radhard designs, as redundant transistors are employed power consumption

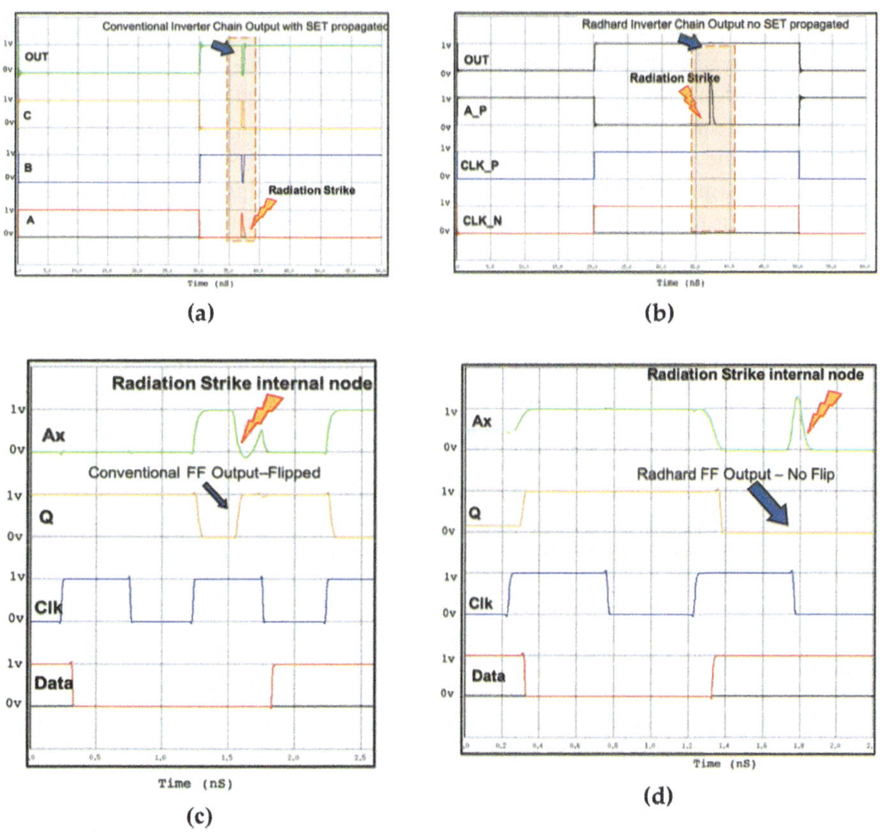

Fig. 3. Particle Strike simulation for SET and SEU analysis of Conventional and Radhard designs. (a) Conventional Inverter chain, (b) Radhard Two Port Inverter, (c) Traditional Flip-flop Design and (d) Dual port Radhard Flip-flop design

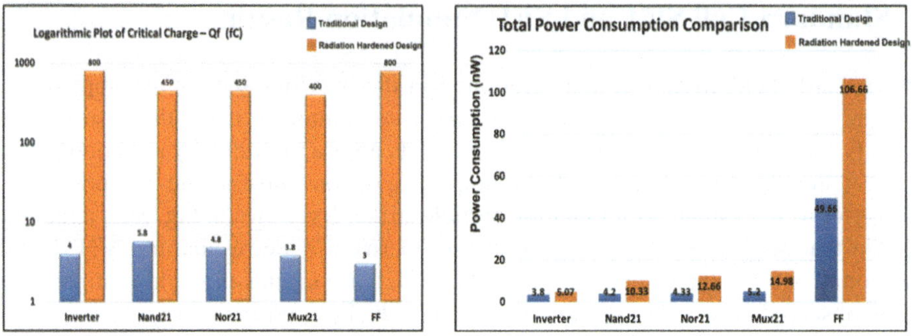

(a) (b)

Fig. 4. (a) Logarithmic scale plot of Critical charge (Qcrit) for SET and SEU comparison Traditional Vs Dual port radhard designs comparison- * For Silicon based devices, an LET of 97 meV-cm^2/mg corresponds to charge disposition of 1 pC/μm [21], (b) Power Consumption comparison between Traditional Vs Dual port radhard designs

metric increases proportionally as shown in Fig. 4(b). The radhard dual port latch was simulated within a Master-Slave Flip-Flop (FF) configuration. The logic value preserve under the radiation particle strike for dual port configuration was verified as against the conventional simple Master-Slave FF. Fig. 3(c) shows that the output logic flipped due to the particle strike, Single Event Upset (SEU) was observed for Qcrit 3fC for conventional design. However, for the dual port radhard design FF the logic value did not flip the output for 800fC of external charge injected as shown in Fig. 3(d).

6 Conclusion

The research work has implemented 16 nm SOI FinFET device structure using Visual TCAD tool. Robust Radhard combinational and sequential standard cell are designed and analyzed for SET (>50 MeV-cm^2/μm) and SEU (>100 MeV-cm^2/μm) tolerance over conventional designs (~1 MeV-cm^2/μm) implemented with the same device. The radhard standard cell circuits showed higher level of resilience to the SEEs with several order higher values of critical charge Qc. For the particle strike analysis, the FinFET based radhard circuits demonstrated reduced external charge distribution due to ionization and reduction of vulnerable nodes as compared to their Bulk CMOS counterpart.

References

1. Li, Y., et al.: A robust hardened latch featuring tolerance to double-node-upset in 28nm CMOS for spaceborne application. IEEE Trans. Circuits Syst. II Exp. Briefs **67**(9), 1619–1623 (2020). https://doi.org/10.1109/TCSII.2020.3013338
2. Dodd, P.E., Shaneyfelt, M.R., Schwank, J.R., Felix, J.A.: Current and future challenges in radiation effects on CMOS electronics. IEEE Trans. Nucl. Sci. **57**(4), 1747–1763 (2010). https://doi.org/10.1109/TNS.2010.2042613
3. Nicolaidis, M. (ed.): Soft Errors in Modern Electronic Systems, vol. 41. Springer, Heidelberg (2010)

4. Fetahović, I., Pejović, M., Vujisić, M.: Radiation damage in electronic memory devices. Int. J. Photoenergy **2013**, 1–5 (2013). https://doi.org/10.1155/2013/170269
5. Anashin, V.S., Koziukov, A.E., Gorchichko, M.E., Zemtsov, K.S., Galimov, A.M., Zebrev, G.I.: Compact modeling of soft error rate in space environment. In: 2016 16th European Conference on Radiation and Its Effects on Components and Systems (RADECS), Bremen, pp. 1–5 (2016). https://doi.org/10.1109/RADECS.2016.8093126.11
6. Oliveira, R.N.M., Lüdke, A.D., Meinhardt, C.: Radiation effects in XOR logic gates at 16nm CMOS and FinFET technology. In: 26th IEEE International Conference on Electronics, Circuits and Systems (ICECS), Genoa, Italy, pp. 590–593 (2019). https://doi.org/10.1109/ICECS46596.2019.8964801
7. Uemura, T., Lee, S., Monga, U., Choi, J., Lee, S., Pae, S.: Technology scaling trend of soft error rate in flip-flops in 1X nm bulk FinFET technology. IEEE Trans. Nucl. Sci. **65**(6), 1255–1263 (2018)
8. Andjelkovic, M., et al.: Design of radiation hardened RADFET readout system for space applications. In: 2020 23rd Euromicro Conference on Digital System Design (DSD), Kranj, Slovenia, pp. 484–488 (2020). https://doi.org/10.1109/DSD51259.2020.00082
9. Edmonds, L.D., Barnes, C.E., Scheick, L.Z.: An introduction to space radiation effects on microelectronics. Jet Propulsion Laboratory, National Aeronautics and Space Administration (2000)
10. Yan, A., Ling, Y., Yang, K., Chen, Z., Yi, M.: Aging-temperature-and-propagation induced pulse-broadening aware soft error rate estimation for nano-scale CMOS. In: 2018 IEEE 27th Asian Test Symposium (ATS), Hefei, pp. 86–91 (2018). https://doi.org/10.1109/ATS.2018.00027
11. Chavan, A., Palakurthi, P., MacDonald, E., Neff, J., Bozeman, E.: Heavy ion characterization of a radiation hardened flip-flop optimized for subthreshold operation. J. Low Power Electron. Appl. **2**, 168–179 (2012). https://doi.org/10.3390/jlpea2020168
12. Razavieh, A., Zeitzoff, P., Nowak, E.J.: Challenges and limitations of CMOS scaling for FinFET and beyond architectures. IEEE Trans. Nanotechnol. **18**, 999–1004 (2019). https://doi.org/10.1109/TNANO.2019.2942456
13. Bttacharya, D., Jha, N.K.: FinFETs: from devices to architectures. In: Advances in Electronics (2014)
14. BhaHadia, S.K., Patel, R.R., Kosta, Y.P.: FinFET architecture analysis and fabrication mechanism. Int. J. Comput. Sci. Issues (IJCSI) **8**(5), 235 (2011)
15. Uznanski, S., Gasiot, G., Roche, P., Autran, J.L., Ferlet-Cavrois, V.: Monte-Carlo based charge sharing investigations on a bulk 65 nm RHBD flip-flop. IEEE Trans. Nucl. Sci. **57**(6), 3267–3272 (2010)
16. Cogenda Pt Ltd.: Visual TCAD – Semiconductor Device Simulator - User's Guide Version 1.72 (2011)
17. Holmes-Siedle, A., Adams, L.: Handbook of Radiation Effects, 2nd edn. Oxford Univ. Press, Oxford (2002)
18. Label, K.A.: Single event effect criticality analysis. NASA Headquarters/Code QW (1996)
19. Singar, S., Joshi, N.K., Ghosh, P.K.: A glitch-free novel DET-FF in 22 nm CMOS for low-power application. J. Nanotechnol. (2018)
20. Hosseinabady, M., Lotfi-Kamran, P., Mathew, J., Mohanty, S., Pradhan, D.: Single-event transient analysis in high speed circuits. In: IEEE International Symposium on Electronic System Design, pp. 112–117 (2011)
21. Schwank, J.R., Shaneyfelt, M.R., Dodd, P.E.: Radiation hardness assurance testing of microelectronic devices and integrated circuits: test guideline for proton and heavy ion single-event effects. Sandia National Laboratories Document 2008-6983

Vehicles Trajectories Analysis Using Piecewise-Segment Dynamic Time Warping (PSDTW)

Muhammad Syarafi Mahmood[1], Uswah Khairuddin[1(✉)], and Anis Salwa Mohd Khairuddin[2]

[1] Malaysia-Japan International Institute of Technology, Universiti Teknologi Malaysia, Jalan Sultan Yahya Petra, 54100 Kuala Lumpur, Malaysia
uswah.kl@utm.my

[2] Department of Electrical Engineering, Faculty of Engineering, Universiti Malaya, Jalan Universiti, 50603 Kuala Lumpur, Malaysia

Abstract. The number of vehicles increases every year in big cities around Malaysia, causing a higher flow of traffic on the road, prompting authorities to increase traffic monitoring to ensure smooth traffic conditions. Traffic surveillance normally conducted using cameras and observed manually by the authorities before they can make decisions on controlling the traffic flow. Continuous monitoring is tedious and prone to error especially during rush hour where traffic volume drastically increased. Intelligent traffic monitoring is possible via trajectory analysis and prediction where the artificial intelligent algorithms learns and cluster the trajectories of vehicles movement. Similarity measure based on distance-based such as Euclidean-distance, Dynamic Time Warping (DTW) and Longest Common Subsequence (LCSS) are less accurate and computationally costly. This paper proposed a combined modified DTW method which merged the piecewise and segmentation measurement in DTW, and the proposed method will be known as the Piecewise-Segment Dynamic Time Warping Distance (PSDTW). PSDTW algorithm is tested on CROSS dataset and compared the time taken and accuracy of the algorithm with previous studied algorithm. The proposed method improves the execution time by average of factor of 4 compared to the SDTW, DTW and LCSS with good results accuracy which have less than 0.01 error rate.

Keywords: Vehicle trajectory analysis · DTW · SDTW · PDTW

1 Introduction

Transportation has become a necessity for people to go about in their daily lives. Over the past 8 years in Malaysia, the number of registered vehicles has a significant increase of 44.44% which accounts to addition of approximately 8 million vehicles, according to the vehicle population report [15] from Malaysia's Department of Road Transport. The increment of vehicle population will affect the environment in terms of air pollution but most importantly it may disrupt emergency services especially during rush hour. In a

J. A. Mat Jizat et al. (Eds.): iCITES 2020, AISC 1350, pp. 214–223, 2021.
https://doi.org/10.1007/978-3-030-70917-4_21

2016 statistical analysis provided by TomTom [16], the urban Malaysian living in Kuala Lumpur spend on average an extra 41 min per day stuck in the traffic which ultimately accumulates to 158 h wasted per year.

With the advancement of surveillance technology and tracking algorithm, vehicle trajectories can be easily obtained with high accuracy. A trajectory is the motion of a moving object with a sequence of the locations which show the route of an object over time [3, 12]. Trajectory data can be described as several object's activities such as the position, speed, or the driver behavior. Measuring the similarity of two objects is a fundamental research problem which is the foundation of many trajectory analysis applications.

Most of the trajectories are sequences with varying sizes and the task of measuring the similarity between two trajectories is a challenging task [13]. This problem has been addressed via proposed methods for measuring the similarity between two trajectories such as Dynamic Time Warping (DTW) [4, 8, 10], Longest Common Sub-Sequences (LCSS) [9, 11] and Edit Distance on Real Sequences (EDR) [14] which mostly used to find the best alignment between trajectories. However, the methodologies have not achieved desirable results, which are to have both good accuracy and fast execution time [1, 2].

This paper will present a new method for trajectories similarity finding which combines two modified Dynamic Time Warping (DTW) methods which are i) the Piecewise-DTW (PDTW), and ii) the Segment-DTW (SDTW). It has been proven in the previous studies that the PDTW can improve the execution time [6] and the SDTW can improve the accuracy of similarity finding between trajectories [5]. Therefore, the combination of both algorithms which can be called as the Piecewise-Segment Dynamic Time Warping Distance (PSDTW) can achieve high accuracy in less execution time.

The paper is organized as follows. In the next section, Sect. 2, the detail of the proposed method will be discussed. In Sect. 3, the paper will discuss experimental results and discussion. Conclusion and future work will be discussed in Sect. 4.

2 Piecewise-Segment Dynamic Time Warping Distance (PSDTW)

The proposed method is inspired by previous research from Keogh [5] and Mao [6] on PDTW and SDTW, respectively.

The DTW algorithm [8] lead to low accuracy reading and sensitive to noise [6] due to not considering the overall trajectory structure. Shape factor in trajectory will be able to improve the trajectory measurement. The sequence of trajectory is regarded as many continuous trajectory segments, and the trajectory shape is obtained by measuring the angle differences between trajectory segments [7].

2.1 Modified Dynamic Time Warping (DTW) Method: The SDTW Algorithm

STDW utilizes prediction distance, segment-segment distance as well as point-point distance to calculate the increasing distance of pair trajectories which can enhance the accuracy measure.

Point Distance. The point-segment distance is the spatial distance of the pair of two trajectory points. Assume P_i (x_i, y_i) is a trajectory point *i* on the trajectory sequence *R* and SP_j (x_j, y_j) is a trajectory *j* on the trajectory sequence *S*. As for the point-segment distance of P_i and SP_j, $dist_{ps}(P_i,SP_j)$ and point-segment distance of SP_j and P_i, $dist_{ps}(SP_j,P_i)$ and $dist_{ps}(SP_j,P_i) \neq dist_{ps}(P_i, SP_j)$. Figure 1 illustrates the point-segment distance computation.

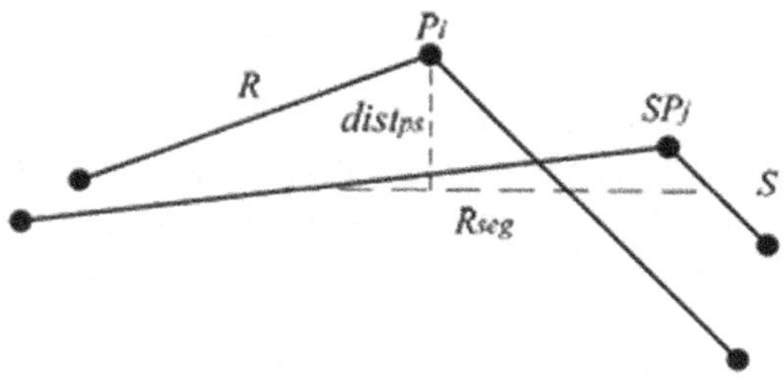

Fig. 1. Point-segment distance

To compute point-segment distance, first need to find the midpoint of $P_{mid1}(x_{mid1}, y_{mid1})$ *of* SP_j*and* SP_{j-1} and the midpoint *of* $P_{mid2}(x_{mid2}, y_{mid2})$ of SP_j*and* SP_{j+1}. The Eq. (1) of the midpoint is:

$$\begin{cases} P_{mid1}(x_{mid1}, y_{mid1}) = ((x_{j-1} + x_j)/2, (y_{j-1} + y_j)/2) \\ P_{mid2}(x_{mid2}, y_{mid2}) = ((x_j + x_{j+1})/2, (y_j + y_{j+1})/2) \end{cases} \tag{1}$$

From the midpoint, then compute the point-distance by measure the shortest distance between P_i and R_{seg}, $dist_{ps}(P_i, R_{seg})$ using Eq. (2).

$$dist_{ps}(P_i, R_{seg}) = \begin{cases} \sqrt{(x_i - x_{mid1})^2 + (y_i - y_{mid1})^2}, & if\ r \leq 0 \\ \sqrt{(x_i - x_{mid2})^2 + (y_i - y_{mid2})^2}, & if\ r \leq L_{seg} \\ \sqrt{(x_i - dx)^2 + (y_{mid2} - dy)^2}, & otherwise \end{cases} \tag{2}$$

Where $r = (x_{mid2} - x_{mid1})$ x $(x_i - x_{mid1}) + (y_{mid2} - y_{mid1})$ x $(y_i - y_{mid1})$ and the L_{seg} is the length of the derivation segment, $dx = (x_{mid1} + (x_{mid2} - x_{mid1})$ x $(r/L_{seg}))$ and $dy = (y_{mid1} + (y_{mid2} - y_{mid1})$ x (r/L_{seg}). The point-segment formula can be summarized by Eq. (3).

$$dist_p(P_i, SP_j) = dist_{sp}(P_i, SP_j) + dist_{ps}(SP_j, P_i) \tag{3}$$

Prediction Distance. After computing the point-segment distance as spatial distance, the algorithm will calculate prediction-distance as a temporal distance. The temporal distance between P_i and SP_j is transformed into the spatial distance between the prediction location of P'_j and SP_j. This method can efficiently improve the precision of similarity measurement. Figure 2 shown the time distance between P_i on trajectory *R* and SP_j on trajectory S is computed.

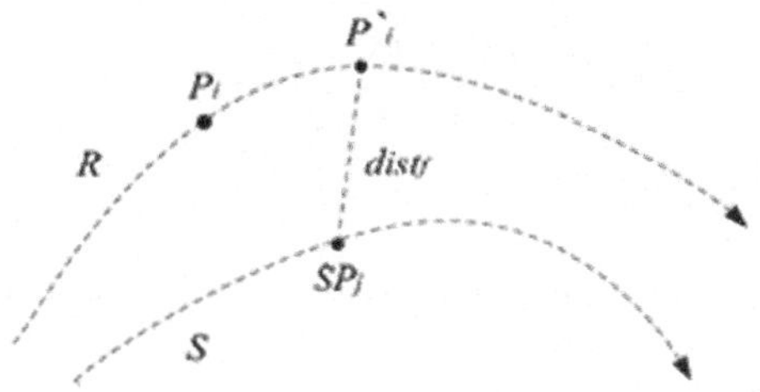

Fig. 2. Prediction distance

The initial reference point with a much earlier timestamp is set as A, whereas the reference point with a later timestamp is referred as B. Their difference is $\Delta t = t_A - t_B$. The spatial coordinates $(X_{B'}, Y_{B'})$ of the prediction position B′ can be computed as Eq. (4).

$$\begin{cases} x_{B'} = x_{i-1} + v_i^x X\,(t_B + \Delta t - t_{i-1}) \\ y_{B'} = y_{i-1} + v_i^y X\,(t_B + \Delta t - t_{i-1}) \end{cases} \tag{4}$$

Where v_i^x and v_i^y is a velocity between two points which can compute from Eq. (5).

$$\begin{cases} v_i^x = (x_i - x_{i-1})/\Delta t_i \\ v_i^y = (y_i - y_{i-1})/\Delta t_i \end{cases} \tag{5}$$

From this parameter, prediction-distance between trajectory can be measure by using Eq. (6).

$$dist_t(\mathrm{A}, \mathrm{B}) = \mathrm{dist}(\mathrm{A}, \mathrm{B}') \tag{6}$$

Where dist(A, B′) is the Euclidean distance between A and B′ in the coordination. The point-segment distance between point A and segment BB′ will also be represent-ed through the prediction distance between point A and B.

Segment-Segment Distance. The segment-segment distance computation is integrated computation from point-distance and prediction-distance algorithm. The point-point distance includes the spatial distance and the temporal distance. The segment distance can be measure using Eq. (7). Figure 3 shown the illustration of measuring segment distance between two trajectory sequences.

$$dist_{st}(P_i, \mathrm{SP}_j) = dist_p\left(P_i, SP_j\right) + tX\ dist_t(P_i, \mathrm{SP}_j) \tag{7}$$

Where t is a time-sensitivity parameter, thus the time dimension cannot be overlooked when $t = 0$. To calculate the segment-segment distance, $dist_{st}(S_i, \mathrm{SS}_j)$ and the angle distance need to be combined. The angle between S_i and SS_j can be computed in Eq. (8) and (9).

$$\theta = \left|arctan2(y_{i+1} - y_i, x_{i+1} - x_i) - arctan2\left(y_{j+1} - y_j, x_{j+1} - x_j\right)\right| \tag{8}$$

$$f(\theta) = \frac{dist_{smid}(S_i, SS_j)}{dist_{max}(R, S)} x(\omega + \theta) \tag{9}$$

Where ω is a variable parameter and the negative shape factor in the equation. The bigger ω, the less sensitive the distance to said factor. If no requirement, let $\omega = 1$. Spatial-temporal distance between midpoints S_i and SS_j is provided in $dist_{smid}(S_i, SS_j)$. While $dist_{max}(R, S)$ is the maximum temporal distance between two points of trajectory sequences R and S. $\frac{dist_{smid}(S_i, SS_j)}{dist_{max}(R, S)}$ is used to dynamically adjust the weight of the shape factor only.

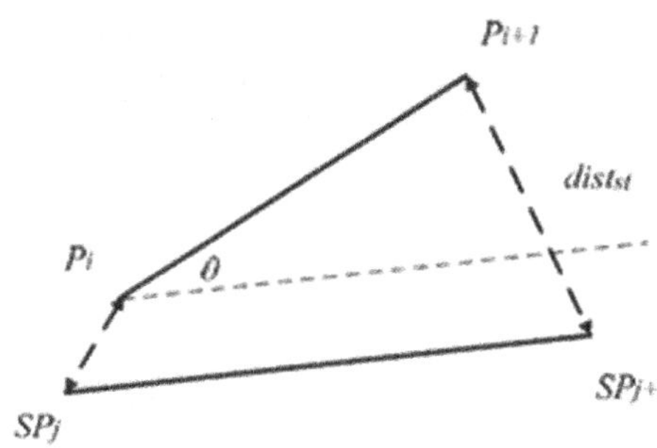

Fig. 3. Segment-segment distance

Segment-Distance DTW. Finally, to compute Segment-DTW, the accumulative is drawn from the theory of DTW algorithm. As previously done for DTW, the similarity measurement of SDTW can be calculated by can be computed using Eq. (10).

$$SDTW(R, S) = \begin{cases} 0, \ if \ n = 0 \ and \ m = 0 \\ \infty, \ if \ n = 0 \ or \ m = 0 \\ dist_s(Head(R), Head(S)) + \min \begin{cases} SDTW(T, \ Rest(S)) \\ SDTW(S, \ Rest(T)), \ otherwise \\ SDTW(S, \ Rest(T)) \end{cases} \end{cases} \tag{10}$$

Where n represents the number of line segment of trajectory R, while m stands for the number of lines segment of trajectory S. $dist_s(Head(R), Head(S))$ represents segment-segment distance between $Head(R)$ and $Head(S)$.

2.2 Modified Dynamic Time Warping (DTW) Method: The PDTW Algorithm

We represent a time series of $X = x_1, \ldots x_n$. Let N be the dimensionality of the transformed time series with $(1 < N < n)$. A time series of X of the extent of n is represented in N space by a vector $X = x_1 \ldots x_N$. The ith element of the X is computed by the following equation.

$$\bar{x}_i = \frac{N}{n}, \sum\nolimits_{j=\frac{n}{N}(i-1)+1}^{\frac{n}{N}i} x_j \tag{11}$$

The data is apportioned into N equivalent sizes frames as for reducing the data from *n-to-N* dimensions. A vector of average value of the data falling within frame is develops to data reduced representation. Figure 4 below show the example of this notation.

Time series consist of *(n)* points are outlined into *(N)* dimensions. The time series is divided into *(N)* frames and the mean of each frame is measured. The measurement is becoming the data reduction representation. Compression rate, c is a proportion of the length of the original time series to the length of reduction data.

$$c = \frac{n}{N} \tag{12}$$

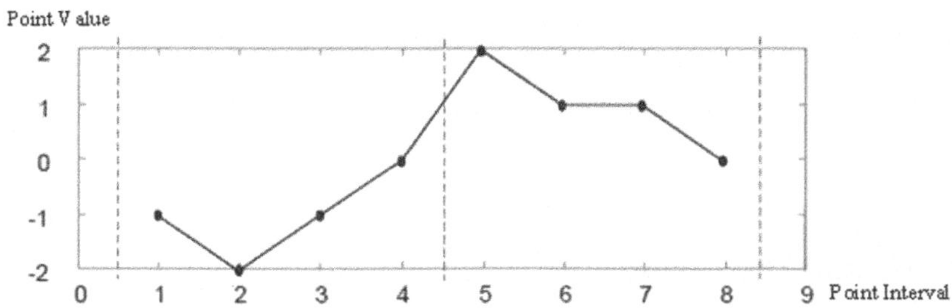

Fig. 4. Illustration of data reduction from n dimensions to N dimensions.

To determine the value of *c*, there is a typical adjustment between memory saving and fidelity. The best compression rate is depending on the structure of the data itself and the task at hand. In this research, we set the *c* value to 2. Figure 5 below a shown example of normal time series and reduction data time series.

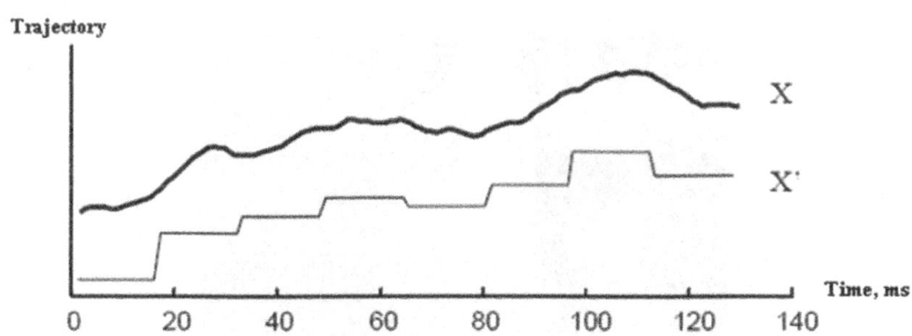

Fig. 5. Time series X and time series X′ (reduction data)

2.3 Modified Dynamic Time Warping (DTW) Method: The PSDTW Algorithm

PSDTW algorithm is the combination of PDTW and SDTW algorithm. The trajectory is been reduced data with equivalent frame as mentioned in PDTW algorithm that called

Piecewise Aggregate Approximation (PAA) [5]. All the trajectories go through PAA process before proceeding with similarity measure. After that, the pre-processes trajectories are comparing with the respective cluster trajectory to find the similarity. SDTW method is using to find the similarity measure. Block diagram below shown the overall PSDTW process (Fig. 6).

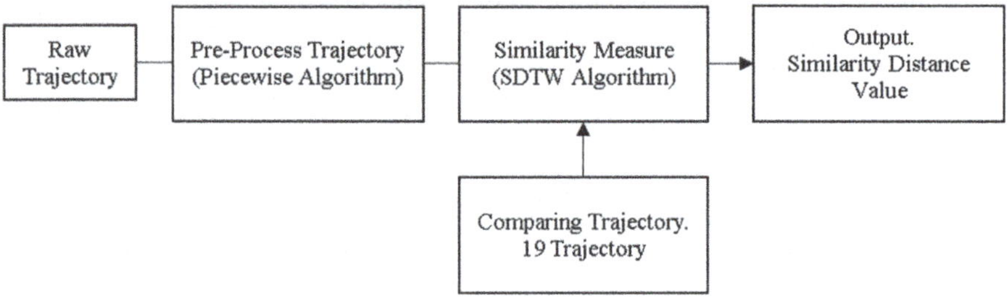

Fig. 6. PSDTW block diagram

3 Result and Discussion

The research is interested in the execution time and accuracy of the PSDTW algorithm over the previous study. The algorithm is been tested using CROSS dataset, the synthetic trajectories of vehicles driving behaviour at crossroads [17]. The dataset is structured into test subset and train subset. Train subset contains of 1,900 trajectories and test subset is 9,700 trajectories. All of the ground truth activity is provided and labelled which contains 19 of vehicle cluster. As in Fig. 7 shown that every colour label in the dataset represent of each vehicle cluster provide from the dataset.

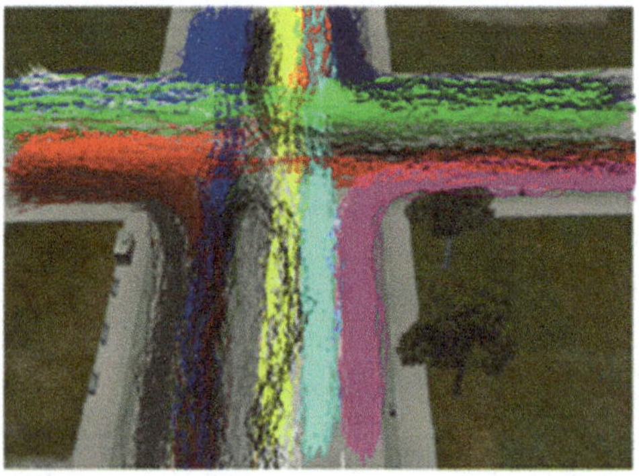

Fig. 7. CROSS dataset

We are comparing the proposed method to the other similarity measure method: original DTW, SDTW, and Longest Common Subsequence (LCSS) in terms of similarity measure accuracy and execution time for per similarity measure. The result is comparing from previous study [1, 2, 5, 6] which are using the CROSS dataset in the experiment. The experiment is carried out on a computer equipped with an AMD Ryzen 7 3750 processor, an NVIDIA GeForce GTX 1660 Ti GPU and 8-GB memory. All the similarity measure and execution time calculation is using MATLAB software.

3.1 Execution Time

The execution time taken is based on the average value of the time. In this research, we ran the algorithm five (5) times to measure the average time of the trajectory similarity measure. The proposed method obtains all similarity measures per cluster in 89 s while the other is more than 5 min to finish the measurement. It means that, for PSDTW, average time taken per similarity measure is around 9.18 ms only, four (4) times faster than other algorithms. Table 1 shows the summarized result of this experiment.

Table 1. Average execution time is taken per similarity measure for others similarity method

Method	PSDTW	SDTW [5]	DTW [1, 5]	LCSS [1, 5]
Time [ms]	9.18	45.10	37.37	33.16

3.2 Accuracy Measurement

In defining the accuracy of the proposed method, error rate (*ER*) will be used and compared with other trajectory similarity measure methods. *ER* is a measure of wrongly clustered trajectory sequences compared to the correct clustering rate (*CCR*). The smaller the amount of *ER*, the better the precision of the method. Equation 13 shown the *ER*.

$$ER = 1 - \frac{1}{N}\sum_{c=1}^{k} p_c \tag{13}$$

N is the number of known trajectories; k is the entire number of groups and p_c is the exact number of the order fitting to the class c. Error rate is finding the ratio between number wrong cluster total number of ground truth, the different between ground truth and the ratio is called Error Rate. As shown in Fig. 7, the accuracy of the proposed method PSTDW is slightly lower compared to the SDTW, although still in the same *ER* percentile range. Therefore, the *ER* is still at the same desired rate which confirms the good accuracy of the method. The slight increase in *ER* is due to the proposed method transforms first the trajectory into a piecewise constant approximation of the original sequence. As example, before the trajectory is being piecewise, the number of points is 12 and after the trajectory been through the piecewise process, the number of points of the trajectory is reduced to 6. Thus, the comparing trajectory's reference point to the in-process trajectory point is reduced process and causes the return value is not similar to the original sequence trajectory. On the other hand, when compared to the DTW and LCSS method, the proposed PSTDW method has lower *ER* hence has better accuracy (Fig. 8).

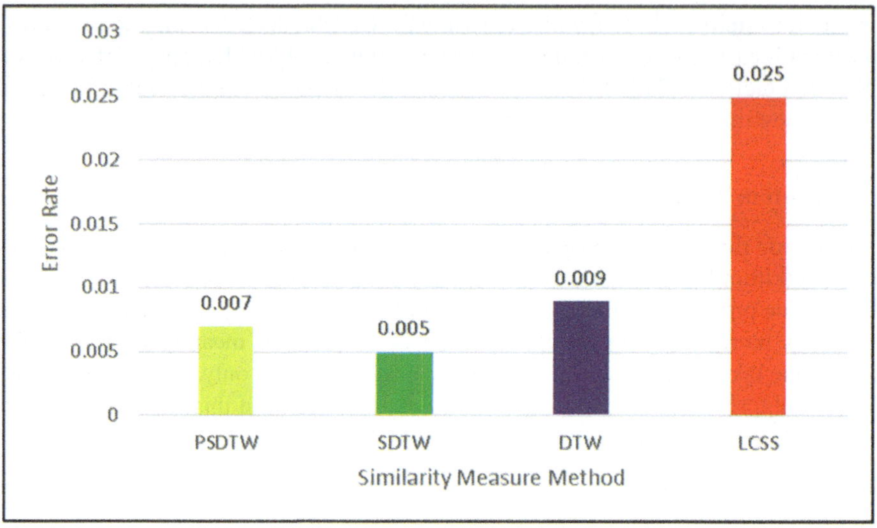

Fig. 8. Comparison of Error rate between different trajectory similarity measure methods

The proposed PSDTW method is almost as accurate as the SDTW method but better than the original DTW method. However, when looking at execution time, the proposed PSDTW is four (4) times faster if compared to other methods. It is very important to have a minimal execution time and this makes the proposed PSDTW method a powerful algorithm that can analyse huge datasets with high accuracy in just a short time. This advantage will also be useful for live trajectory analysis which will make real-time traffic management and control possible.

4 Conclusion

This paper proposed a new trajectory similarity method, the Piecewise-Segment Dynamic Time Warping Distance (PSDTW) for fast and accurate trajectory analysis. The proposed method is a combination of the modified Dynamic Time Warping (DTW) methods which are i) the Piecewise-DTW (PDTW), and ii) the Segment-DTW (SDTW). The combined method inherits both benefits from its predecessors; PDTW's fast execution time and SDTW's high accuracy. Experiment using CROSS dataset shows that the proposed PSDTW method is able to maintain the high accuracy similar to the SDTW method with average error rate around 0.007 and can finish the trajectory similarity analysis around four (4) times faster than other available trajectory analysis method. The short running time will enable analysis of huge datasets and live trajectory analysis hence will allow real-time traffic management and control by the authorities.

Acknowledgement. The authors would like to thank Ministry of Higher Education (MoHE), Malaysia for funding this research paper through a Fundamental Research Grant Scheme (FRGS) entitled "A Study of Vehicle Flow Trajectories Modelling and Prediction for Optimized Traffic Management in Transportation Industry" (vote number 5F265).

References

1. Arfa, R., Yusof, R., Shabanzadeh, P.: Deep dissimilarity measure for trajectory analysis. Commun. Comput. Inf. Sci. **946**, 129–139 (2018). https://doi.org/10.1007/978-981-13-2853-4_11
2. Arfa, R., Yusof, R., Shabanzadeh, P.: Novel trajectory clustering method based on distance dependent Chinese restaurant process. PeerJ Comput. Sci. **2019**(8), 1–5 (2019). https://doi.org/10.7717/peerj-cs.206
3. Atev, S., Miller, G., Papanikolopoulos, N.P.: Clustering of vehicle trajectories. IEEE Trans. Intell. Transp. Syst. **11**(3), 647–657 (2010). https://doi.org/10.1109/TITS.2010.2048101
4. Feng, L., Zhao, X., Liu, Y., Yao, Y., Jin, B.: A similarity measure of jumping dynamic time warping. In: Proceedings of 2010 7th International Conference Fuzzy System Knowledge Discovery, FSKD 2010, vol. 4, no. Fskd, pp. 1677–1681 (2010). https://doi.org/10.1109/FSKD.2010.5569383
5. Keogh, E.J., Pazzani, M.J.: Scaling up dynamic time warping for datamining applications. In: Proceedings of the Sixth ACM SIGKDD International Conference on Knowledge Discovery and Data Mining, pp. 285–289 (2000). https://doi.org/10.1145/347090.347153
6. Mao, Y., Zhong, H., Xiao, X., Li, X.: A segment-based trajectory similarity measure in the urban transportation systems. Sensors (Switzerland), **17**(3), (2017). https://doi.org/10.3390/s17030524
7. Sankoff, D.: Practice of Sequence (1983)
8. Senin, P.: Dynamic Time Warping Algorithm Review. Science (80-), vol. 2007, pp. 1–23 (2008). https://doi.org/10.1109/IEMBS.2007.4353810
9. Soto, D., Soto, W.: The constrained longest common subsequence: theory and experiments, pp. 1–6 (2020). https://doi.org/10.1109/concapanxxxix47272.2019.8977105
10. Van Der Vlist, R., Taal, C., Heusdens, R.: Tracking recurring patterns in time series using dynamic time warping. In: European Signal Processing Conference, vol. 2019, pp. 1–5 (2019). https://doi.org/10.23919/EUSIPCO.2019.8903102
11. Vlachos, M., Kollios, G., Gunopulos, D.: Discovering similar multidimensional trajectories. In: Proceedings of International Conference on Data Engineering, pp. 673–684 (2002). https://doi.org/10.1109/ICDE.2002.994784
12. Yuan, G., Sun, P., Zhao, J., Li, D., Wang, C.: A review of moving object trajectory clustering algorithms. Artif. Intell. Rev. **47**(1), 123–144 (2017). https://doi.org/10.1007/s10462-016-9477-7
13. Zhang, Z., Huang, K., Tan, T.: Comparison of similarity measures for trajectory clustering in outdoor surveillance scenes. In: Proceedings International Conference Pattern Recognition, vol. 3, pp. 1135–1138 (2006). https://doi.org/10.1109/ICPR.2006.392
14. Chen, L., Özsu, M.T., Oria, V.: Robust and fast similarity search for moving object trajectories. In: Proceedings of the 2005 ACM SIGMOD International Conference on Management of Data, pp. 491–502 (2005). https://doi.org/10.1145/1066157.1066213
15. Jabatan Pengangkutan Jalan Raya (JPJ) Malaysia. Laporan bilangan kenderaan di atas jalan raya mengikut negeri 2008–2015. https://www.data.gov.my/data/dataset/bilangan-kenderaan-di-atas-jalan-raya-mengikut-negeri/resource/f0dffdea-354b-416a-90b2-b6ca471e603c?view_id=fa5d1250-a717-4d72-b655-ca852a52da6c
16. TomTom Traffic Index Report. https://www.tomtom.com/en_gb/traffic-index/kuala-lumpur-traffic/
17. Morris, B.T., Trivedi, M.M.: Trajectory learning for activity understanding : adaptive approach. IEEE Trans. Pattern Anal. Mach. Intell. **33**(11), 1–7 (2011)

Real-Time KenalKayu System with YOLOv3

Nenny Ruthfalydia Rosli[1], Uswah Khairuddin[1(✉)], Muhammad Faris Nor Fathi[2], Anis Salwa Mohd Khairuddin[3], and Azlin Ahmad[4]

[1] Universiti Teknologi Malaysia, 54100 Kuala Lumpur, Malaysia
uswah.kl@utm.my
[2] International Islamic University Malaysia, 53100 Kuala Lumpur, Malaysia
[3] Universiti Malaya, 50603 Kuala Lumpur, Malaysia
[4] Universiti Teknologi MARA, 40450 Shah Alam, Selangor, Malaysia

Abstract. An automated tropical wood species recognition system known as KenalKayu has been developed by the Centre for Artificial Intelligence & Robotics (CAIRO) to identify the tropical wood species. The system works very well in offline mode with an accuracy rate of up to 98%. But when it comes to real-time testing, the accuracy rate dropped by about 62%, partly due to low image quality. The system was trained by using ideal quality of wood images that are stored in the database. However, during real-time testing, the quality of wood image captured might be degraded due to motion blur, out of focus and illumination. Therefore, it is challenging to perform accurate recognition via real-time approach. This research proposed an improved KenalKayu prototype by using You Only Look Once version 3 (YOLOv3) algorithm to detect and classify tropical wood species via real-time approach. 60 images from 10 tropical wood species have been trained while another 60 images have been captured and tested during real-time testing. The preliminary test shows promising results where the system is now able to classify tropical wood species in real-time mode with accuracy rate for both training set and testing set are 100%. The average accuracy rate for output probability generated by YOLOv3 is 95.63% .

Keywords: Deep learning · Image analysis · Pattern recognition · YOLOv3

1 Introduction

Malaysia exports logs, sawn timber, plywood, veneer, chipboard, wooden frame, builder joinery, wood furniture and rattan furniture which brings in up to RM 22.5 billion [1]. Currently, wood identification is mainly performed by certified personnel who have dedicated themselves to the field. The difficult process of upgrading experienced staff members is due to the vast number of different types of wood trees found in Malaysian forests. This means that proper techniques need to be in place so that species identification can be more accurate and quicker.

A wood recognition system called KenalKayu developed [2] by the Center for Artificial Intelligence and Robotics (CAIRO) was proposed. The research project which is based on wood microscopic features, has been in place since 2002 and has been significantly improved [2–15] to ensure that it can recognize wood species in both offline and

J. A. Mat Jizat et al. (Eds.): iCITES 2020, AISC 1350, pp. 224–232, 2021.
https://doi.org/10.1007/978-3-030-70917-4_22

real-time mode. This system would help make wood identification more accurate and less susceptible to human error and bias.

KenalKayu can identify a wood type within seconds. This may allow for faster, less expensive and more efficient wood identification. Moreover, a Wood Database Management System is integrated with the recognition system to ensure the whole recognition process to be more organized and systematic. In addition to that, KenalKayu has a web-based version that can be accessed anywhere such as on the internet since its online [6]. A prototype of the KenalKayu program for data collection, perform experiments and data analysis, and for training and testing was provided for demonstration [7]. The prototype includes a microscopic camera equipped with lighting system where it is attached with a conveyor belt system to aid the data collection process. The distance between the camera and the wood sample is the same throughout the experiment. This system can be applied both to online and offline mode. The passing rate for the offline testing was 98%, but when the tested on real-time mode, the passing rate dropped to 62%. The upgraded version of KenalKayu is described in [14] where the system is used without the conveyor system. The images are captured directly using the same camera with previous version of KenalKayu [7] without fixed distance between the wood sample and the camera. The user is required to adjust the camera's position manually in order to capture an ideal image. The accuracy rate of the real-time experiment for this version of KenalKayu is 53%. Thus, there is an urgent need to overcome the limitation of the current real-time wood species recognition system.

There are several researchers have been worked with deep learning method to identify the local wood species. Hafemann et al. [16] have applied Convolutional Neural Networks (CNN) on both macroscopic and microscopic images to classify Brazil's forest species. Another researchers from Korea [17] applied CNN on macroscopic Korean softwood images while researchers from North America [18] also applied CNN on their local macroscopic wood images. Their results show that CNN give a high accuracy rate for their local wood species range from 92% to 99%. This inspired us to apply CNN by using You Only Look Once version 3 (YOLOv3) to detect and classify tropical wood species. The main contribution of this paper is to apply deep learning on KenalKayu system. The proposed methodology and experimental results are presented in Sect. 2 and 3.

2 Methodology

There are numerous techniques used for features extraction based on the characteristics of wood, such as gravity, weight, density, texture, smell and hardness [16]. Besides CAIRO, other researchers reported in [19–26] also work on improving the effectiveness of wood identification system based on the wood texture.

KenalKayu is a wood species recognition system prototype that incorporates pattern recognition and artificial intelligence techniques as shown Fig. 1.

2.1 Data Acquisition and Image Database

Sample was collected from Forest Research Institute of Malaysia (FRIM) and cut into roughly one inch cubes. Fig. 2 shows the sample of a wood cube with 3 surfaces which

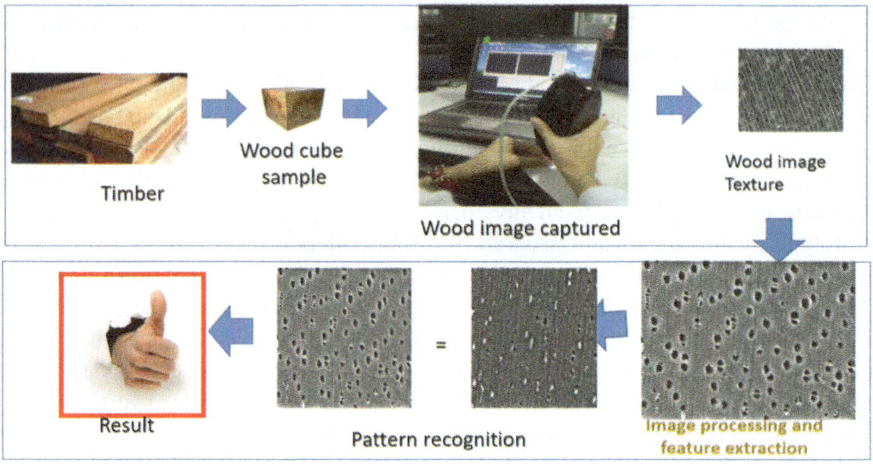

Fig. 1. Wood identification by using KenalKayu.

are the radials section, tangential section and cross section. The cross section is the best surface to capture the macroscopic anatomy of the tropical wood. This task has been done by using a microscopic camera. The size of the images is 768 × 576 pixels. In this research, 10 different tropical wood species were used to test the proposed YOLOv3 method.

Fig. 2. Three surfaces of wood; (X) cross-section surface, (T) tangential surface and (R) radial surface.

For training process, wood images of 10 species such as Bintangor, Chengal, Durian, Gerutu, Jelutong, Keranji, Melunak, Perupok, Ramin and Sesendok were used. Fig. 3 shows the wood anatomy image for the selected 10 wood species. For each species, 60 images were captured from 6 different wood cubes. Hence, the total images trained in YOLOv3 is 600 images. Next, the wood dataset is labelled into a YOLO format.

2.2 Feature Extraction and Classification

Figure 4 shows the previous work of KenalKayu developed in [2]. In the original KenalKayu prototype, wood images are captured by using a microscopic camera which

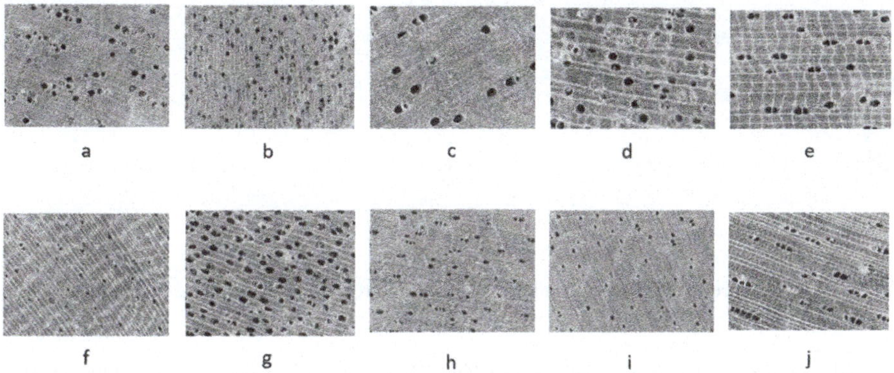

Fig. 3. Wood anatomy image; (a) Bintangor, (b) Chengal, (c) Durian, (d) Gerutu, (e) Jelutong, (f) Keranji, (g) Melunak, (h) Perupok, (i) Ramin and (j) Sesenduk.

will then undergo two main module of the system: feature extraction module and pattern recognition module. The system adopted Gray-Level Co-occurrence Matrix (GLCM) feature extraction technique and Artificial Neural Network (ANN) to classify the wood species.

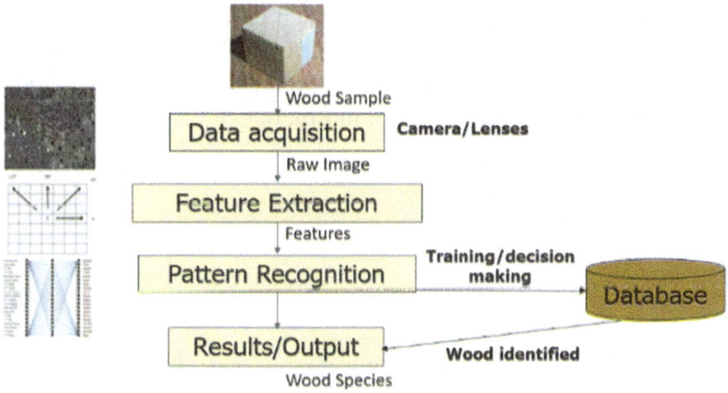

Fig. 4. Process flow of wood recognition system with KenalKayu first prototype.

Another technique that can be consider in detection and classify wood species is YOLO. YOLO is a well-known object detection and deep learning algorithm that was released in May of 2016 [27]. It uses deep learning for object recognition. YOLO first computes a single convolutional neural network over the entire images in order to predict bounding boxes and probabilities, respectively. It will decide the area of the image with the highest probability. Some of the frameworks that are compatible with YOLO are Darknet, Darkflow, Darknet-OpenCV and Darkflow-OpenCV. The CNN architecture being used in YOLOv3 network that is shown in Fig. 5. We used the Darknet-53 network as the backbone that has three separate scale predictions.

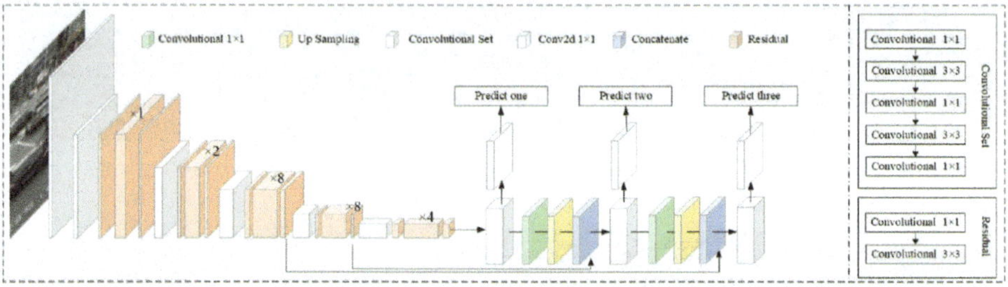

Fig. 5. Architecture of YOLOv3 that been used in this research. The image is taken from [28].

In this paper, an improved version of KenalKayu prototype is developed by adopting YOLOv3 technique to detect and classify tropical wood species. YOLO is a unified model for object detection that has been developed by Redmond et al. [27]. The model is simple to apply and can be trained directly on full images and real-time video. Unlike the ordinary classifier-based methods, YOLO is trained on a loss function that straight away corresponds to detection performance. And the most interesting part is the entire model is trained together. This makes YOLO the fastest general-purpose detector ever and the same time it can identify object even in real-time mode. It is shown that YOLO technique is faster than other detection systems which can detect different datasets. Moreover, it can be tested on multiple of image sizes to deliver a smooth tradeoff between speed and accuracy [29]. In addition, the advantages of implementing YOLOv3 algorithm to solve classification problems are also reported in [30]. Hence, YOLOv3 is adopted to improve the KenalKayu prototype with the aim to improve the accuracy rate in real-time mode. The process flow for this research is shown in Fig. 6.

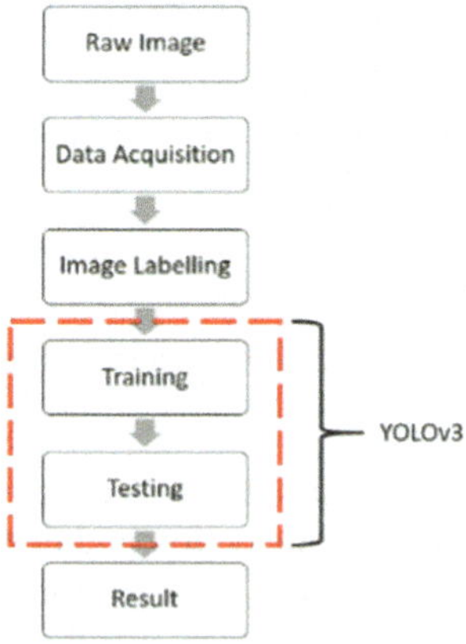

Fig. 6. Process flow of wood recognition system with KenalKayu improved prototype using YOLOv3

3 Experimental and Results

The training process was been done by using Google Colab while the testing process was been done by using PyCharm. The main objective of this experiment is to test the wood samples in real-time mode. Firstly, we use the training set data for testing (we called this set a training set). Then, we used the same cube samples that have been captured for the training set to be the testing set. Each species has 6 cubes. Hence, in total, we have 60 cubes for testing data. The proposed system is tested by using 60 images of each wood species for training data and using 1 test image from each cube for each wood species. Altogether there are 600 images for the training set and 60 images for the testing set. Table 1 shows the result of this experiment while Fig. 7 show its' chart. Figure 8 shows the confusion matrix for both training set and data set.

Table 1. Classification result of 10 tropical wood species.

Class	Training set (%)	Testing set (%)	Average of output probability YOLOv3 (%)
Bintangor	100	100	96.23333
Chengal	100	100	90.86667
Durian	100	100	97.71667
Gerutu	100	100	98.75000
Jelutong	100	100	93.83333
Keranji	100	100	96.88333
Melunak	100	100	95.38333
Perupok	100	100	97.25000
Ramin	100	100	94.61667
Sesenduk	100	100	94.75000
Average (%)	**100**	**100**	**95.62833**

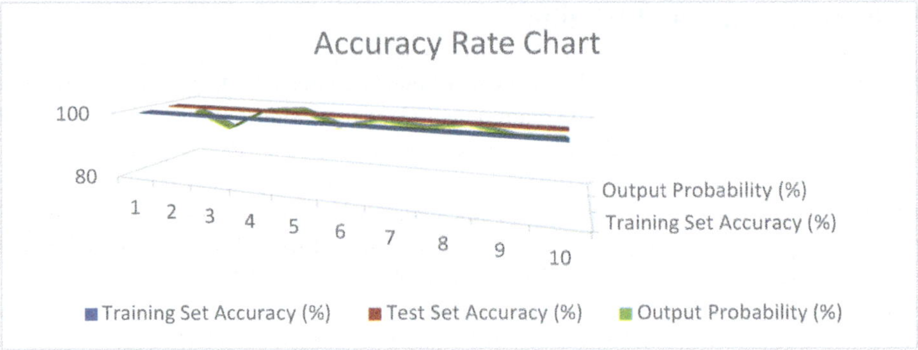

Fig. 7. Accuracy rate chart

Class	Bintangor	Chengal	Durian	Gerutu	Jelutong	Keranji	Melunak	Perupol	Ramin	Sesenduk	Total
Bintangor	60	0	0	0	0	0	0	0	0	0	60
Chengal	0	60	0	0	0	0	0	0	0	0	60
Durian	0	0	60	0	0	0	0	0	0	0	60
Gerutu	0	0	0	60	0	0	0	0	0	0	60
Jelutong	0	0	0	0	60	0	0	0	0	0	60
Keranji	0	0	0	0	0	60	0	0	0	0	60
Melunak	0	0	0	0	0	0	60	0	0	0	60
Perupol	0	0	0	0	0	0	0	60	0	0	60
Ramin	0	0	0	0	0	0	0	0	60	0	60
Sesenduk	0	0	0	0	0	0	0	0	0	60	60

Fig. 8. Confusion matrix for both training and testing set

4 Discussion and Conclusions

The study was to analyze the performance of an online tropical wood recognition system using YOLOv3 algorithm. 10 different tropical wood species have been selected and each of the species has 6 different wood cube samples. Both training set and testing set give 100% accuracy rate for each tropical wood species. The output probability accuracy rate generated by YOLOv3 for each tropical wood species is more than 90%. The average output probability accuracy rate of the prototype is approximately 95.63%. For future works, more wood species will be included to evaluate the performance of the KenalKayu prototype.

Acknowledgment. The authors would like to thank Ministry of Higher Education (MoHE), Malaysia for funding this research paper through a Fundamental Research Grant Scheme (FRGS) (vote number 5F265).

References

1. Bernama, Malaysia exports RM22.5b wood-based products in 2019, Malay Mail, Money, 7th September 2020
2. Khalid, M., Lee, E.L.Y., Yusof, R., Nadaraj, M.: Design of an intelligent wood species recognition. Int. J. Simul. Syst. Sci. Technol. IJSSST, **9**(3), 9–19, ISSN 1473-8031 (2008)
3. Rosli, N.R., Khalid, M., Yusof, R.: Wood species recognition based on Gabor filter image processing technique. In: 2008 Student Conference on Research and Development (SCORED 2008), pp.194-1–194-4 (2008)
4. Yusof, R., Rosli, N.R., Khalid, M.: Tropical wood species recognition based on Gabor filter. In: 2nd International Congress on Image and Signal Processing, vol. 1–9, pp. 2705–2709, ISBN 978-1-4244-4129-7. IEEE (2009)
5. Yusof, R., Rosli, N.R., Khalid, M.: Using Gabor filters as image multiplier for tropical wood species recognition system. In: 12th International Conference on Computer Modelling and Simulation (UKSIM), pp. 289–294, Article no. 5481196, ISBN 978-0-7695-4016-0. IEEE (2010).
6. Yusof, R., Rosli, N.R.: Tropical wood species recognition system based on Gabor filter as image multiplier. In: Proceeding 2013 International Conference on Signal-Image Technology and Internet-Based Systems (SITIS 2013), vol. 2013, pp. 737–743, ISBN 978-147993211-5. IEEE (2013)
7. Halim, W.A.H.W.A., Yusof, R., Rosli, N.R.: Integration of conveyor system with tropical wood species recognition system. In: National Conference on Wood Based Technology, Engineering and Innovation 2015 (NCWEI 2015), 17–18 November 2015 (2015)
8. Khairuddin, U., Yusof, R., Rosli, N.R.: A comparative study of k-NN and MLP-NN classifiers using GA-kNN based feature selection method for wood recognition system. In: 17th International Conference on Electrical, Control and Automation Engineering (ICECAE 2015), WASET, vol. 2, no. 4, pp. 2695–2699, eISSN 1307-6892 (2015)
9. Abdul Hamid, L.B., Rosli, N.R., Mohd Khairuddin, A.S., Mokhtar, N., Yusof, R.: Denoising module for wood texture images. Wood Sci. Technol. **52**(6), 1539–1554 (2018)
10. Yusof, R., Rosli, N.R., Bakar, F.N.A.: Non-wood anatomy image detector for tropical wood recognition system. In: 4th Kuala Lumpur International Agriculture, Forestry & Plantation Conference 2016 (KLIAP 2016), pp. 162–168, ISBN 978-967-2072-01-0 (2016)
11. Zamri, M.A.P., Cordova, F., Khairuddin, A.S.M., Mokhtar, N., Yusof, R.: Tree species classification based on image analysis using improved-basic gray level aura matrix. Comput. Electron. Agric. **124**, 227–233, ISSN 0168-1699 (2016)
12. Yusof, R., Ahmad. A., Ismail, M.N., Rosli, N.R.: Clustering the imbalance data sets using modified kohonen self-organizing map (KSOM). In: IEEE Technically Sponsores Computing Conference 2017, 18th–20th July 2017 (2017)
13. Yusof, R., Khairuddin, U., Ahmad. A., Khairuddin, A.S.M.: A study of feature extraction and classifier methods for tropical wood recognition system. In: Proceedings of TENCON 2018 - 2018 IEEE Region 10 Conference, Jeju, Korea, 28–31 October 2018 (2018)
14. Rosli, N.R., Khairuddin, U., Yusof, R., Ghapar, H.A., Khairuddin, A.S.M., Ahmad, A.: Online system for automatic tropical wood recognition. J. Elektrika **18**(3–2), 1–6 (2019)
15. Rajagopal, H., Khairuddin, A.S.M., Mokhtar, N., Ahmad, A., Yusof, R.: Application of image quality assessment module to motion–blurred wood images for wood species identification system. J. Int. Acad. Wood Sci. Wood Sci. Technol. **53**, 967–981 (2019)
16. Hafemann, L.G., Oliveira, L.S., Cavalin, P.: Forest species recognition using deep convolutional neural networks, pattern recognition (ICPR). In: 2014 22nd International Conference on at Stockholm, vol. 22 (2014)

17. Kwon, O., Lee, H.G., Lee, M.R., Jang, S., Yang, S.Y., Park, S.Y., Choi, I.G., Yeo, H.: Automatic wood species identification of korean softwood based on convolutional neural networks. J. Korean Wood Sci. Technol. **45**(6), 797–808 (2017). pISSN 1017-0715 eISSN 2233-7180
18. Lopes, D.J.V., Burgreen, G.W., Entsminger, E.D.: North American Hardwoods Identification Using Machine-Learning, 11 February 2020. www.mdpi.com/journal/forests. Accessed 5 Mar 2020, Published 7 Mar 2020
19. Menon, P.K.B.: Structure and identification of Malayan woods. Revised by Ani Sulaiman & Lim Seng Choon, Malayan Forest Records No. 25, Forest Research Institute Malaysia, Malaysia, p. 123 (1993)
20. Wheeler, F.A., Baas, P.: Wood identification – a review. IAWA J. **19**(3), 241–264 (1998)
21. Fuentealba, C., Simon, C., et al.: Wood products identification by internal characteristics reading. In: Proceedings of the IEEE International Conference on Industrial Technology, Hammamet (2004)
22. Baas, P., Wheeler, E., et al.: Dicotyledonous wood anatomy and the APG system of angiosperm classification. Bot. J. Linn. Soc. **134**(1–2), 3–17 (2000)
23. Tou, J.Y., Lau, P.Y., Tay, Y.H.: Computer vision-based wood recognition system. In: International Workshop on Advanced Image Technology (IWAIT), Jan 2007, Bangkok (2007)
24. Tou, J.Y., Tay, Y.H., Lau, P.Y.: A comparative study for texture classification techniques on wood species recognition problem. In: Fifth International Conference on Natural Computation (ICNC 2009) (2009)
25. Tou, J.Y., Tay, Y.H., Lau, P.Y.: Rotational invariant wood species recognition through wood species verification. In: Nguyen, N.T., Nguyen, H.P., Grzech, A. (Ed.) Proceedings 1st Asian Conference on Intelligent Information and Database Systems (ACIIDS 2009), Dong Hoi, pp. 115–120 (2009)
26. Tou, J.Y., Tay, Y.H., Lau, P.Y.: A comparative study for texture classification techniques on wood species recognition problem. In: Proceedings Fifth International Conference on Natural Computation (ICNC 2009) (2009)
27. Redmon, J., Divvala, S., Girshick, R., Farhadi, A.: You only look once: unified, real-time object detection. In: 2016 IEEE Conference on Computer Vision and Pattern Recognition (CVPR), pp. 779–788 (2016)
28. Mao, Q.C., Sun, H.M., Liu, Y.B., Jia, R.S.: Mini-YOLOv3: real-time object detector for embedded applications. IEEE Access **7**(99), 133529–133528 (2019)
29. Redmon, J., Farhadi, A.: Yolo9000: better, faster, stronger. In: 2017 IEEE Conference on Computer Vision and Pattern Recognition, CVPR, pp. 6517–6525. IEEE (2017)
30. Redmon, J., Farhadi, A.: YOLOv3: an incremental improvement. Comput. Sci. (2018)

Scalp Massage Therapy According to Symptoms Based on Vietnamese Traditional Medicine

Nguyen Dao Xuan Hai and Nguyen Truong Thinh(✉)

Ho Chi Minh City University of Technology and Education, Ho Chi Minh City, Vietnam
ndxuanhai@gmail.com, thinhnt@hcmute.edu.vn

Abstract. A massive growth in the modernize population, because of disabled people, patients after surgery those whom cannot have ability to move and even wash themselves. Tied on their own bed, has created many problems including the need for massage service and convenient healthcare to help patients body relax and shampoo themselves which reduce medical care staff, otherwise, most of them feel inferior when someone touches their head when they are sick or having dermatological problem. In this paper, on-going projects related to essential technology for reducing work force and overcrowding in Vietnamese hospital are discussed. We propose robotic autonomous scalp massage and shampoo combine with traditional Vietnamese's acupuncture massage to aid in resolving problems that people suffering from bed sick are experiencing, reports on design, service personalization and deployment of an assistive robot to support elderly people, disabled and patients with scalp care. Levenberg-Marquardt algorithm in artificial neural network are applied to train from collected acupuncture points to give the best trajectories following patient's symptoms, analyzed for diagnosing and evaluating hairy scalp issues for the neural network input variables associated with body values such as height, weight, age or race which plays an important role in the composition of the multi-layer perceptron. The experimented and analyzed from multi-modal data collection of the first ever longitudinal field trials from patients throughout each experiment, indoor environments demonstrate by people with unmovable needs care. From the experiments, it indicates the ability of combining the mechanical system with 2-DOFs, human scalp massages by water jets with trajectories predicted by machine learning based on physical factors to meet the desired position on human scalp, providing enhanced resilience to being treated patients.

Keywords: Scalp massage robot · Trajectory · Levenberg-Marquardt algorithm (LM) · Artificial neural network · Symptom · Acupuncture points · Acupressure

1 Introduction

The massage therapy helps to unclog meridians has been presented in the world with the evolution of human being as well the pressure points especially in Eastern countries may be mentioned as China, Singapore, Japan and Vietnam. Traditional massage method easily encounter as Acupuncture, Acupressure and Reflexology are now common that

J. A. Mat Jizat et al. (Eds.): iCITES 2020, AISC 1350, pp. 233–244, 2021.
https://doi.org/10.1007/978-3-030-70917-4_23

everybody recognition their positive aspects especially for patients, elderly, people with accidents or must undergo physical therapy. Obviously, when people get illness, the body's resistance to the environment as well as to all kinds of bacteria will decrease. Therefore, it is necessary to keep the patient's hygiene very well to contribute to the prevention of the disease, avoid potentially dangerous complications and shorten the treatment time. Especially, surgery patients who have to lie on bed, limited sedentary, facing with obstruction of the meridians also have problems with scalp hygiene and sweat gland secretions, for this reason can turn leads to the risk of infection, making the patient get worse.

The traditional currently method that chiefly uses the human resource and commonly hands massage or rub patient's hair in scalp area. Acupressure developed a firm finger by applying pressure to key healing points on the body while acupuncture uses needles, both therapy use the same pressure points and over 800 reflex points along energy lines run longitudinally the entire body [1] called meridians. Reflexology or acupressure are physical stimulation that directly affects the skin, nerves, blood vessels and receptors causing neurological, physical changes, translation, endocrine thereby improving the operating capacity of the nervous system, improve the nutrition process of the body [2]. Reflexology uses reflexes that are in an orderly arrangement resembling a shape of the human body on the feet, hands, outer ears, and also head area, blood stream will boost metabolism, bring oxygen to the cells, carry away waste products. Suzanna Maria, et al. are looked at the impact of acupressure on symptoms that frequently accompany fatigue in population: chronic pain, depression, anxiety and poor sleep. Compared with normal medicine, they observed that acupressure strengthened both of these symptoms. Survivors of breast cancer who indicated exhaustion were randomized to one of two forms of acupressure or normal treatment, including traditional methods of sleep control [3]. Hence, reflexology hybrid with acupressure is possible to take advantage of the external forces exerted on a particular area like human hairy scalp to eliminate slime, body waste when the patient takes antibiotics and impact on the long lying patients the main reason causing infectious disease. Moreover, the therapists is being put heavy burden that force them to work for hours and for many people. Last but not least, many people argued when the skin or finger nails touches on the patient's scalp not only can spread the Corona viruses and main cause of the contagious diseases from scratches or pores but also working in a long period can make therapists exhausted and the efficiency also decreases dramatically due to crowed patients. Knowing that need, with the development of modern science, machines are used to massage into society. It can be controlled by some buttons as well the remote control. In particularly, a combination of traditional massage methods and automatic cleaning system for the patient's hair and scalp in bed are now presenting in this paper and expected to perform healthcare missions. It is difficult to develop an intelligent robot from begin to end, which performs point and massage tasks autonomously on human scalp. Current massage technology makes necessary to rely on the human operator for providing overall task contact and interaction, and for handling special situations. The original dimension related to anthropology will be custom providing to system handle and static analysis. After that, taking advantage of force from water jet with high pressure will impact on acupuncture points also the area on the scalp. Trajectories of nozzle is a combination of a mechanical system and a processor, based on each patient's anthropology and symptoms will be interpolated by neural network to determine the best path to reach the target following multi symptoms and body size of the patient.

In the next section, the material and method to perform the mechanism of massage and washing human scalp robot. In Sect. 3, we will go into detail the mechanical design of massage robot and the fundamental kinetics problems. Cooperation between the robot and the human body will be seen in Sect. 4. Human points and massage therapy lines are displayed in Sect. 5. The experiments and debate are mentioned in Sect. 6. The conclusion and the references will be in the last part.

2 Methodology

For more than five thousand years, Acupressure, Acupuncture and Reflexology are known as an ancient art of healing. They are both "reflex" therapies in that they work with specific points or area on one part of the body to affect other parts of the body which benefit of "reflex" treatment has been described. However, acupuncture or acupressure points do not really coincide with the reflex points used in reflexology. In this paper treatment will focus on the three main meridians shown in Fig. 1.

A total of healthy at the age from 18 to 75 year-old Vietnamese (164 males and 136 females) were randomly selected to obtain dimension height and width of scalp dimension by the Mitutoyo electronic caliper model 551–204-10 has a length of 1000 mm (40 inches), the smallest division is 0.01 mm, the accuracy of $\pm$ 0.06 mm. This function is much higher than traditional images in 2D and it can be achieved quantitative and qualitative. However, within the limits, research focuses on studying only 58 acupuncture points, of which 24 are located on the right and left side of the vertical cross section of the body and 10 points are located along the line the body called the Governing Vessel (GV) shown as Table 1. From the newborn scans of skulls modern human, Ann Gibbons concludes that the form starts with elongated heads, the only time the human brain almost doubles in size in the first year of life known as maturity that the skull becomes globula [4] described in Fig. 2.

Table 1. Abbreviation of 34 acupoints on the one side head.

Meridians	The original order of the acupoints	The corresponding points in the artificial neural network respectively
Urinary Bladder (UB)	UB3 – UB10	11 – 18 and 11.0 – 18.0 for the opposite
Gall Bladder (GB)	GB4 – GB13 and GB15 – GB20	19 – 34 and 19.0 – 34.0 for the opposite
Governing Vessel (GV)	GV15 – GV24	1 – 10

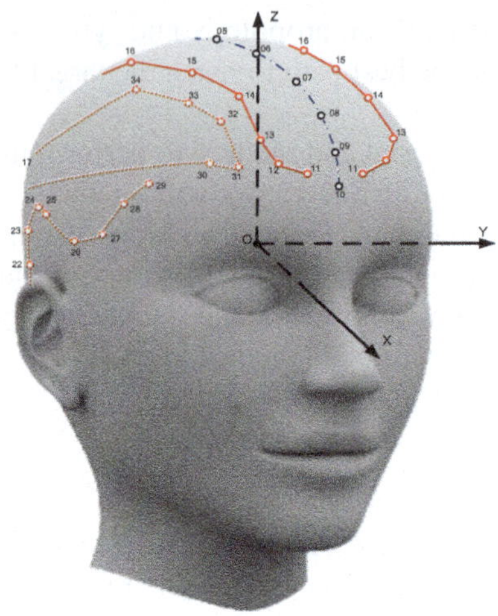

Fig. 1. Acu – point in 3 main Meridians namely Urinary Bladder (UB) – red continuous line, Gall Bladder (GB) – orange dot line, Governing Vessel (GV) – blue dash dot line and Cartesian coordinate is attached

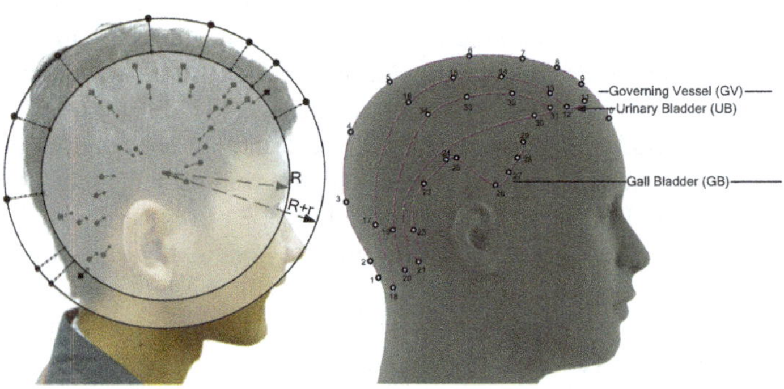

Fig. 2. Human head shape is shown to be round spherical

The robot system is capable of performing basic operations the same as a massage specialist by applied water force. The method for washing and massaging the scalp must define the working space of the actuator before describing the detailed structure of the system so that the appropriate configuration can be made. The washing and massage actuator must ensure that it operates on the whole head, including hair growth, without affecting other sensitive parts of the head, such as the ears, eyes and nose, we consider selected points as above. The requirement is to design an actuator that always has an axis towards the center of the skull, when we view the skull as a spherical shell analyzed in the next section with the actuator moves. The induced movement should move along the surface of the cranial cortex. The structure when designed should have a few degrees of freedom but still ensure the working features.

3 Massage Robot Mechanism

The paper proposes a robot system with 2 degree of rotate freedom and the actuator control releases force F linear to the surface of the scalp in the spherical coordinate system, shifting along a certain path to solve the problem of heath treatment and scalp hygiene. The combination of acupressure based-point and the regional reflexology impact force for searching algorithm applied for scalp massage by implementing a segmentation method related symptoms and neural networks on collected data.

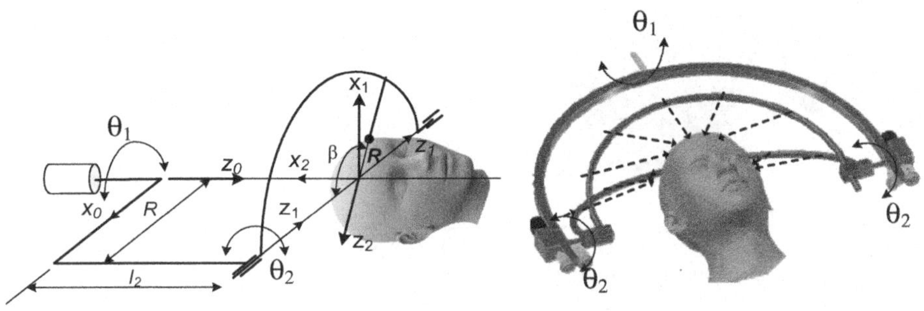

Fig. 3. Configurable robotic arm design performs shampooing and massage

Scalp massage involves dealing with sufficient pressure on the human scalp to relieves headache and shoulders pain symptom, stress and nervousness as well treat several health problems relating to blood pressure. Typical massage practices can be divided into 5 techniques known as pressing, kneading, rubbing, hammering and rolling, due to the human scalp has to many vulnerable nerves, therapists mainly use pressing and rubbing as 2 importance techniques for release meridian and supply oxygen for blood cells. With the specification requirements for cleaning the surface of the scalp of the patient, the affordable price, also working independently of the hospital bed, requires lightweight, compact. In running environments, clients lied on a massage table, sit in a massage chair, or lied on a mat on the base. Aquatic massage and shampoo are done directly on the human skin using hot-water jet therapy. The massage robot is serial structure that consist of separate controller with pump, connecting with the nozzles and linear actuator. Compared with parallel manipulated, potential advantages of this serial robot presents simply singularities and unlimited workspace. Therefore, the configuration of the robot ensure 2 degrees of freedom that is rotating around x and z axes and the water jet acts directly by y-axis in the direction towards the scalp surface as the Fig. 3. Robot arms are covered on the central of the head, these two moving heads must be independent of each other, one for mixture liquid from shampoo and water, the other arm will extrusion pure water connecting with high pressure pump. To ensure safety and avoid exceeding the permissible limit of robots, the mechanism reduces the optimal number of joints and easy to process from which the cost will also decrease. The movements of the massage nozzle must be easy to control and completely dependent on x and z axes movements. In order to make the robots protect the human scalp like the human arm, mathematical of dynamic issues need to be addressed. If the position of the nozzle depending on 2 other axes is determined, the angles of the robot arms need to be continuously adjusted to shift the nozzle to the exact position. This device circular section working spaces will travel around the sphere surface with stable speed and pre-calculated position like as Fig. 4.

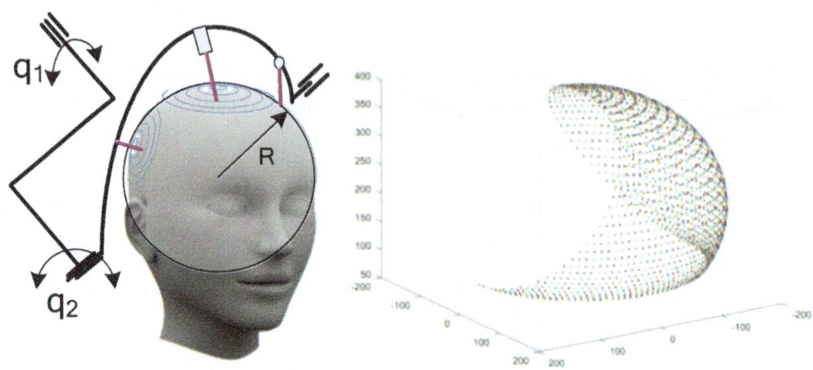

Fig. 4. System covering the human head as a sphere with radius R, the operating space of the serial 2 degree of freedom configuration

The coordinates of the nozzle point on the human head in global Cartesian frames are represented in the following Eqs. (1–3).

$$P_x = -r[\cos\theta_1 \cos\beta + \sin\theta_1 \cos\theta_2 \sin\beta] \tag{1}$$

$$P_y = r[-\sin\theta_1 \cos\beta + \cos\theta_1 \cos\theta_2 \sin\beta] \tag{2}$$

$$P_z = l_2 + r \sin\theta_2 \sin\beta \tag{3}$$

Robot arm in this case is a series system so the method to calculate the kinetic problem used in this topic is the Denavit-Hartebberg method, the results are shown in Table 2. With two angles respectively two joints connected directly to motors of joint 1 and joint 2. The water jet is directed directly to the center of the head and creates an angle θ_1, θ_2 with the axis Y_I (the motor axis at the second joint).

Table 2. Denavit – Hatenberg parameters

#	1	θ	d	α	Variable
1	0	π/2+θ1	12	π/2	θ1
2	0	−π/2+θ2	0	β	θ2

From here, the displacement matrix from the global reference namely axis 0 (the Cartesian coordinate of the human head) to the axis 2 system (Polar coordinate) is based on the Denavit-Hatenberg parameter table as Formula (4) follows:

$$
{}^{0}T_2 = \begin{bmatrix} -sin\theta_1 sin\theta_2 & cos\theta_1 sin\beta - cos\theta_2 cos\beta sin\theta_1 & cos\theta_1 cos\beta + cos\theta_2 sin\theta_1 sin\beta & 0 \\ \cos\theta_1 sin\theta_2 & sin\theta_1 sin\beta + cos\theta_1 cos\theta_2 cos\beta & cos\beta sin\theta_1 - cos\theta_1 cos\theta_2 sin\beta & 0 \\ -cos\theta_2 & cos\beta sin\theta_2 & -sin\theta_2 sin\beta & l2 \\ 0 & 0 & 0 & 1 \end{bmatrix} \tag{4}
$$

The scalp massage pathways decide the motion path for scalp washing and massage according to the defined pathway specifications. In the inverse kinematic problem, the parametric time Eq. (5–7) is normally replaced by the following:

$$
\sin\theta_1 = \frac{P_z - l_2}{r \sin\beta} \tag{5}
$$

$$
\cos\theta_2 = \sqrt{1 - \cos^2\theta_1} \tag{6}
$$

$$
\theta_2 = \arctan\left(\frac{\sin\theta_2}{\cos\theta_2}\right) \tag{7}
$$

Suppose we let the nozzle spray the surface of the head as a spiral equation (Spiral equation), this is a logarithm curve equation with the angle for the massage nozzle to impact on the head according to different orbits with time parameter, the parametric equation of the spiral curve on the head surface will have the following system of Eq. (8):

$$
x(t) = t.\cos(t);\ y(t) = t.\sin(t) \text{ and } z(t) = R.\cos(t) \tag{8}
$$

4 Neural Network Applied on Searching Path and Direction Associated with Symptoms

During the process of massage and scalp cleansing therapy for patients, robots executive structure that bring along the water nozzle was available applied a force of F on the scalp with orbits that could be spiral, zigzag or follow the arc curve of the head continuously from horizontal and vertical, motion for the trajectory requested shown in Fig. 5. Every acupoint has a specific function. In order to do that, we need to have knowledge about anthropometric and traditional medicine as well know the traditional unit, its name is "cun". According to traditional medicine, human body is divided into 38 horizontal and vertical sections. Everyone's height from head to toe is 75 cuns. The value of a cun is divided by 1/75 depending on the height of each person, combined with head size analysis to be able to infer the most reasonable going trajectories.

In this section, the structural design of the neural networks represented and combined their training processes. With the structure of the images extracted from the head region, the system will determine the head length and width along with the imaginary center created from the curvature radius of the head. Thus the network structure consists of 1 output which is trajectory from patient's symptoms and 2 hidden layers as Fig. 6. The

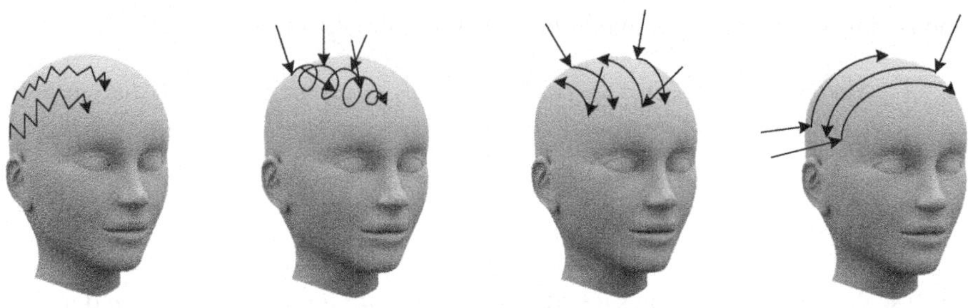

Fig. 5. Given trajectories for water jet nozzles generating hair wash and massage.

process of determining the hidden layer as well as the number of neurons of each hidden layer must undergo an experimental process to compare the error. For the first hidden layer, neural network will be taught to choose the acupressure points that correspond to the symptoms of the patient based on anatomy of human index in this neural are age, weight, head size, including: vertical, width of the head based on 3D point cloud [5] and also determine cun size from the index is the height of the body. The second layer will work as connect those acupuncture points according to not only specific symptom but also many different symptoms, so that the arrangement of known lines will be recombined again into a unified path, propagating across multiple layers, as gradients of shared weights are aggregated, making learning that is supervised on deep structures easier. Each network simultaneously predicts the required number of acupuncture points based on averages to reduce variance. The team analyzes three important factors in choosing a network structure that is the training process. Adding hidden layers increases the non-linearity from scratch input to output, and makes it possible to represent the associate between input and output accordingly. Secondly, for the neurons in the hidden layers, correcting the absolute value after the sigmoid activation function is effective can improve performance.

The repeated trajectory will be checked for accuracy by statistics from the symptom customer provided, which will produce a map of the acupuncture point with the initial collected results. On the other hand, it is depended on the patient's lying position compared to the manipulate nozzle head, when the patient is in active area of the robot arm, massage robot will be given an approximate position by the interpolation and projected onto the scalp surface with a radius equal to $R + r$, where r is distance of the nozzle heads to the scalp surface of the patient, which shown in the next section.

5 Result and Discussion

The scalp is cleaned and massaged by moving the unit in a defined trajectory based on acupuncture as well as scraping by manipulating two axes. The effect of the water jet on the human scalp is perpendicular to the scalp and matches the circular form of the human scalp. The patient will be asked to lie on their back with the nape of the neck supported by a stiff pillow so that the head stays in place like as Fig. 7.

Through experience in the proportions of parts of the body, the length of the human body can be determined by parameters of the height, length and width of the head.

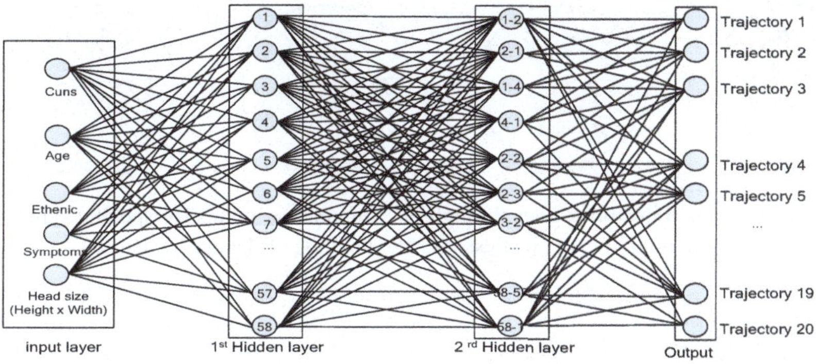

Fig. 6. Structure of Artificial Neural Network optimized by the LM algorithm for providing degree of location of acupuncture points to interpolate trajectories based on symptoms and size of "Cun"

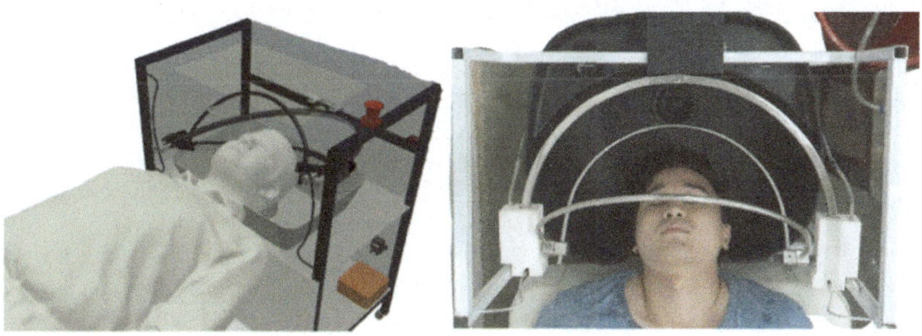

Fig. 7. Illustrate robotic mechanism in 3D modeling and reality product for experiment

Water jets have directly toward the center of the head. Work slowly with travel speed of 5 mm/s, water pressure from 7 to 10 N/cm^2 and depth of about 2 mm above human skin surface. Here, command input by neural network of human head was the position. There are several massage ways of robot trajectories synthesized from 58 points are being surveyed as Fig. 1 and Fig. 2, following 3 main meridians are Urinary Bladder (UB), Gall Bladder (GB), Governing Vessel (GV). For instant, trajectory for patients experiencing common headache symptoms namely as line 1 as Fig. 8: the robot head starts moving from the nape and goes spirally to Point 19(GB8) then stops there for 3–4 s, slowly curve on GB meridian forward to point 31 (GB 20), after that it reaches (GV15), follow arc curve of head and stop for a second at the (GV 14), then reverse at the same trajectory for 10 min of therapy. For line 2 (nausea combine with dizziness symptom): It also starts from nape then goes straight to Point 17(GB6) then stops there for a while at the (UB6), after that it arrives Point 4 and wait at the (UB08), until it reaches the nape and goes back on the contrary direction for process take 10 min long.

The 2 hidden layers used in this experience, with the number of neurons per hidden layer on 676 samples synthesized from 164 males, 136 females and performed normally training take more time than utilize with LM algorithm in backward propagation. Errors back-propagation have been checked sequentially and continuously by comparing from combine multi symptoms with different scalp size after each training session. Through

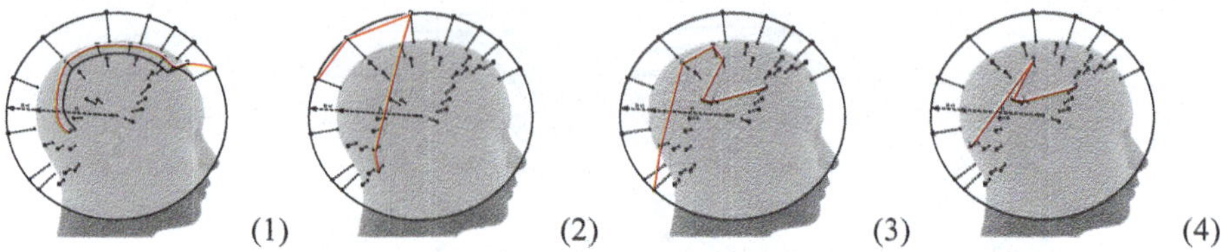

Fig. 8. The massage trajectories goes through the acupuncture points for specific symptoms (1) Headache, (2) Severe migraine, (3) restlessness and anxiety, (4) headache, nausea.

	1	2	3	4	5	6	7	8	9	10	11	12	13	14	15
Neurons in 1st layer	2	2	2	2	2	3	3	3	3	3	4	4	4	4	4
Neurons in 2nd layer	2	3	4	5	6	2	3	4	5	6	2	3	4	5	6
Error %	85.36	42.64	64.11	53.33	78.36	11.36	6.28	12.14	15.68	23.45	56.45	67.14	68.77	87.28	55.79

Fig. 9. Graph of the average error test optimized results of 2 hidden layers

observation, with the neural network demonstrated, error converge reached value at 3 when the number of neurons of 2 layers have been shown in Fig. 9.

Program interface is used programming languages Python and Java. Robot controller software have written to perform functions such as: Interpolate the size and head based on the AI application body proportions, communicate with the microcontroller to request performance as well as receive feedback signals. The program is named head_wash.py and has an interface as shown in Fig. 10 below. The software is intended for non-professionals but is still interactive to control robot by 2 functions, manual control and automatic control, in manual function, massage robot can be easily controlled by handle and resistance camera integrated inside the bath. This function allows patient to relax and not limit the time or technique of the exercise.

The robot uses a large enough force of water due to the difference in pressure in the tube and outside atmospheric pressure to perform the massage procedure. By examining the feelings of those who perform water massage therapy. Collected from the survey of experimented participants, the results are shown in Table 3 that the feeling of a person's pain threshold will have 3 ranges when did massage and washed on 150 people. About 80% people tickled lightly when water jet force below 5 N/cm^2, comfortable and relaxed feeling has been described by 60% when water force directly applied at above 5 N/cm^2 and not over 10 N/cm^2. Under 15N/cm^2, 63% felt that it was causing pain, while the rest

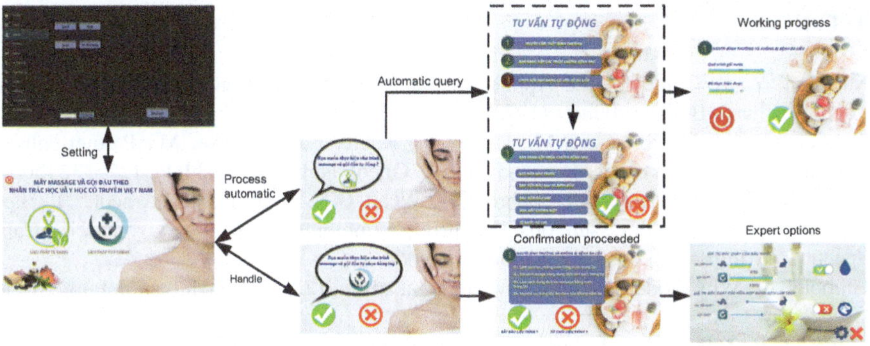

Fig. 10. Interactive screen of massage robot is built with 2 functions, automatic and manual.

Table 3. Evaluate pain threshold of water impact on human scalp associated with participants' pain sense

Water jet force (N/cm^2)	0–5	5–10	10–15
Sensing (feeling)	Slight pressure	Normal or slightly hurt	Pain

*The feeling is qualitative with no specific value [6], 1kg/cm2 ~ 9.8N/cm2.

felt normal. Thus, depending on the scalp and the thickness of the hair, the water force on the scalp will be weak or strong and the feeling of each person will be different. The sensation as well as the pain threshold is a qualitative parameter because each person will have a threshold different pain.

6 Conclusion and Future Work

Study has suggested a new approach for treating patients with water mixed with shampoo in order to preserve their grooming and fragile scalp and which used to take a lot of time to take care of a nurse or family member, thus lowering service prices easily allows patients not to be concerned with any procedure such as changing position or affect other people on hospital room. This treatment promises to bring better recovery for patients who lose mobility, patients with complications, patients who spend almost time recover after surgery, accidents, or the elderly in nursing homes. With few caregivers, the robot can be used for the home, it can adapted by its versatility, easy to move, simple, light and automated. The robot works smoothly with 2 degrees of freedom of rotation and can bring the actuator nozzle from points to points on the scalp area. With the 2 hidden layers, trajectory defined errors are still considerable, but for combining acupressure and reflexology Vietnamese traditional medicine as described in the introduction, the robot can help the patient not only for keep areas prone to infections but It is also possible to test the patient's ability to recover after using it.

References

1. Nakatani, Y.: A Guide for the Application of Ryodoraku Autonomous Nerve Regulatory Therapy. Chan's Books Products (1972)
2. Stephenson, N.L.N., Swanson, M., Dalton, J., Keefe, F.J., Engelke, M.: Partner-delivered reflexology: effects on cancer pain and anxiety. Oncol. Nurs. Forum **34**(1), 127–132 (2007)
3. Zick, S.M., Sen, A., Hassett, A.L., Schrepf, A., Wyatt, G.K., Murphy, S.L., Arnedt, J.T., Harris, R.E.: Impact of self-acupressure on co-occurring symptoms in cancer survivors. JNCI Cancer Spectr. **2**(4). https://doi.org/10.1093/jncics/pky064
4. Gibbons, A.: Why modern humans have round heads, p. 1229 (2018). https://doi.org/10.1126/science.aaw3660
5. Mejía, I.C.P., Zell, A.: Head measurements from 3D point clouds. In: 2016 Sixth International Conference on Image Processing Theory, Tools and Applications (IPTA). IEEE (2016)
6. Viitasalo, J.T., et al.: Warm underwater water-jet massage improves recovery from intense physical exercise. Eur. J. Appl. Physiol. **71**(5), 431–438 (1995)

Adsorption and Artificial Neural Network Modelling of Metolachlor Removal by MIL-53(Al) Metal-Organic Framework

Hamza Ahmad Isiyaka[1(✉)], Anita Ramli[1], Khairulazhar Jumbri[1,2], Nonni Soraya Sambudi[3], Zakariyya Uba Zango[1], and Bahruddin Saad[1]

[1] Department of Fundamental and Applied Sciences, Universiti Teknologi PETRONAS, 32610 Bandar Seri Iskandar, Perak Darul Ridzuan, Malaysia
hamza-18001996@utp.edu.my

[2] Department of Fundamental and Applied Sciences, HICoE-Centre for Biofuel and Biochemical Research, Institute of Self-Sustainable Building, Universiti Teknologi PETRONAS, 32610 Seri Iskandar, Perak Darul Ridzuan, Malaysia

[3] Chemical Engineering Department, Universiti Teknologi PETRONAS, 32610 Bandar Seri Iskandar, Perak Darul Ridzuan, Malaysia

Abstract. This study investigates the removal and prediction of metolachlor from aqueous medium using advanced prediction technique and porous adsorbent material (metal-organic framework (MOF)). The powdered X-ray diffraction (XRD), Field emission scanning electron microscopy (FESEM), and Brunauer, Emmett and Teller (BET) were used to characterize the synthesized MOF. High removal efficiency (97.75%) was recorded within short time (30 min) using only 10 mg of the MOF. The variation in the pH and concentration were also studied to understand the adsorption. The experimental data set for the training and validation were designed and predicted using the model of artificial neural network (ANN). The ANN model gave accurate prediction for training ($R^2 = 0.998$; $R^2adj = 0.991$ and RMSE = 0.311) and validation ($R^2 = 0.996$; $R^2adj = 0.991$ and RMSE = 0.441) at node five. The porous nature of the adsorbent (1104 $m^2\ g^{-1}$), fast equilibration time, % removal and accurate ANN prediction makes this MOF an effective material for remediating recalcitrant pollutants.

Keywords: Adsorption · Artificial neural network · Metal-organic framework · Metolachlor

1 Introduction

Recently, scientist have devoted attention and resources in the detection and treatment of new hazardous compounds. These emerging contaminants have threatened the quality of surface and ground water resources. Anthropogenic discharge of herbicides such as metolachlor (2-chloro-N-(2-ethyl-6-methyl-phenyl)-N-(1-methoxypropan-2-yl) acetamide) is widely applied as herbicide to kill weeds in variety of farmlands around the world. As at 2008 an estimate of 13.6 and 15.9 million kg of metolachlor were used in

J. A. Mat Jizat et al. (Eds.): iCITES 2020, AISC 1350, pp. 245–255, 2021.
https://doi.org/10.1007/978-3-030-70917-4_24

the United State of America (USA) to control weeds, ranking it the 4th most applied herbicide [1]. According to Otero et al. [2], over 50% of surface water and some percentage of groundwater (2%) in the USA are contaminated by metolachlor. It is among the five most detected compounds in European freshwaters. Metolachlor is a persistent organic pollutant with about 22–205 days half-life [3]. This make it potentially available in the environment through surface runoff, leaching and spray drift deposition that affected human health, plants and aquatic animals [4]. These pollutant can accumulate in the tissues of plants and later transferred to animals and birds through the food chain [5]. The United State Environmental Protection Agency (USEPA) have listed metolachlor as part of the Drinking water contaminant Candidate with possible carcinogenic and mutagenic health effects [3]. The maximum threshold of metolachlor that is permissible in drinking water is set as 0.01 mg L^{-1} by the World Health Organization (WHO) [2]. However, the most challenging anthropogenic induced contamination is the agricultural non-point source pollution that can emanate from unidentified and multiple sources. Several treatment techniques used for wastewater remediation do not completely remove these emerging contaminants in water [6]. This make it very difficult to track and control their distribution in the environment. It is therefore crucial to identify techniques that can be used for the effective removal of metolachlor in aqueous medium.

Over the years, methods have been developed and applied for the remediation of micro-pollutants in aqueous medium. Techniques such as advanced oxidation process, adsorption, bioremediation and membrane filtration were previously applied for the remediation of toxic contaminates in water [7]. From these methods mentioned, adsorption is one of the promising and practical technique due to its low cost, simple operations, high selectivity, environmental benign, convenient recycling and availability of alternative materials [8]. Different natural and synthetic porous materials such as Organosilica [9], Polymer [10] and Biochar [4] have recently been applied for the removal of metolachlor in aqueous medium. Yet the search for porous and stable materials becomes limitless.

Recently, metal-organic frameworks (MOFs) have captured the attention of scientist for application in the treatment of hazardous pharmaceuticals, dyes, heavy metals and pesticides [11]. MOFs are class of highly stable, crystalline and porous materials with ultra-high surface area and large pore [12]. MOF such as MIL-53(Al) are promising for the effective remediation of contaminants in water [13]. MIL-53(Al) was efficiently used by Gao et al. [14] for the treatment of antibiotics in aqueous medium. High removal efficiency and good stability was recorded after the third cycle. Guan et al. [15] applied MIL-53(Al) MOF for the adsorption of dyes in aqueous solution with fast equilibration time and high removal efficiency. Jia et al. [16] recorded fast removal of Triclosan using MIL-53(Al). This MOF was also applied for the treatment of Cu(II) by Li et al. [17].

This work applies porous adsorbent (MIL-53(Al)) for the efficient and effective remediation of metolachlor in aqueous medium. Optimization and modelling the adsorption of metolachlor using design for experiment and artificial neural network (ANN) provides an effective and accurate tools for predicting the impact of some critical parameters in the removal of pollutants. This method saves cost, time and helps to validate the experimental results gotten from the adsorption studies. The ANN is a robust and nonlinear model that learn the relationship and patterns in a dataset when trained [18].

2 Materials and Methods

Metolachlor (2-chloro-N- (2-ethyl-6-methyl-phenyl)-N-(1-methoxypropan-2-yl) acetamide) (98%) was sourced from Sigma-Aldrich (St. Louis, MO, USA). Aluminum nitrate nonahydrate ($Al(NO_3)_3{\cdot}9H_2O$, $\geq 98\%$), 1, 4-benzene dicarboxylic acid (H2BDC, 99%), ethanol (99.9%), dimethylformamide (DMF, 99%), methanol (99.5%), hydrochloric acid and sodium hydroxide were obtained from Avantis Laboratory (Perak, Malaysia).

2.1 Preparation of the MOF Material

The adsorbent was synthesized hydrothermally following a previous procedure [19]. 13 g of aluminum nitrate ($Al(NO_3)_3$) and 2.88 g of terephthalic acid (TPA) were dissolved into a 50 mL distilled water. The solution was mixed thoroughly by magnetic stirrer for 30 min and was transferred into an autoclave, caped and heated with the oven at 493K for 72 h. Then, the oven was allowed to cool down and the product was washed, filtered, several times with distilled water and DMF to get rid of the impurities and unreacted (1,4-benzenedicarboxylate) materials. The product obtained was dried at 353K for 2 h and was later activated for 8 h at 473K.

2.2 Characterization of the MOF Material

The ZEISS, ultra 55 was used to determine the morphology of the MOF using the Field emission scanning electron microscopy (FESEM). The crystallinity and structural properties of the synthesized MOF was ascertained by X-ray diffraction using Bruker D8 Advanced X-ray diffractometer. N_2 adsorption–desorption (Micrometrics ASAP 2020) was applied to ascertain the surface area of the MOF.

2.3 Experimental Procedures

Batch Adsorption Studies

Adsorption experiments were carried out by using 10 mg L^{-1} metolachlor solution in 50 mL volumetric flask. The adsorbent (10 mg) was prepared and put in the solution, and mixed at 150 rpm in a temperature regulated shaker (incubator ES 20/60, Biosan) at 35 °C and pH 6 for 30. 2 mL of the solution was sampled out in every 5 min, filtered with a 0.45 μm membrane filter. The absorbance was detected with the aid of UV-vis spectrophotometer (Shimadzu, Lamda 25) at 280 nm. All the adsorption data were recorded in triplicates from which the average values were calculated.

The adsorbed quantity of metolachlor solution (q_e), removal percentage (%R) and the quantity of solution adsorbed at a given time (q_t) were determined following the equation given below:

$$q_e = \frac{(C_o - C_e)V}{m} \tag{1}$$

$$Percentage\ removal(\%) = \frac{(C_o - C_t)}{C_o} \times 100 \quad (2)$$

$$q_t = \frac{(C_o - C_t)V}{m} \quad (3)$$

Where C_o, C_e, and C_t (mg/g) represent the initial concentration, equilibrium concentration and time respectively, V denotes the volume of metolachlor solution (L), and m is used to represent the quantity of adsorbent (g).

Artificial Neural Network Model (ANN)

ANN consist of intelligent algorithms that mimics the human brain by learning the data sets for modeling and predicting the relationship pattern of the experimental data when trained [20]. In this study, the back-propagation artificial neural network was effectively applied [21] to predict the removal efficiency of Metolachlor onto MIL-53(Al) MOF. The back-propagation artificial neural network was designed to comprise of several neurons that are arranged in layers. The number of hidden neurons were arrived at by a process of trial and error. 80% of the datasets were used for the training and 20% were used for the validation. The results were assessed using the diagnostic criteria of determination. The model was analyzed using the following equations:

$$R^2 = 1 - \frac{\sum (x_i - y_i)^2}{\sum y_i^2 - \frac{\sum y_i^2}{n}} \quad (4)$$

$$R^2adj = 1 - \left(1 - R^2\right)\left(\frac{n-1}{n-p}\right) \quad (5)$$

$$RMSE = \sqrt{\frac{1}{n} \sum \frac{n}{i=1} (x_i - y_i)^2} \quad (6)$$

Where x_i represents the observed data that was determined experimentally, y_i is the predicted data, n is the number of observations and p denotes the number of parameters.

3 Discussion of the Results

3.1 Adsorbent Characterization

Figure 1(a) and (b) describes the FESEM image of the adsorbent which correspond with that of previous studies. The EDX in Fig. 2 show the elemental composition of the adsorbent material that comprise of carbon, oxygen and aluminum. The surface area of the MOF is estimated to be around 1104 $m^2\ g^{-1}$. Figure 3 represents the XRD of the MOF with similar agreements with those reported in previous work [16] that correspond to a well formed crystalline structure.

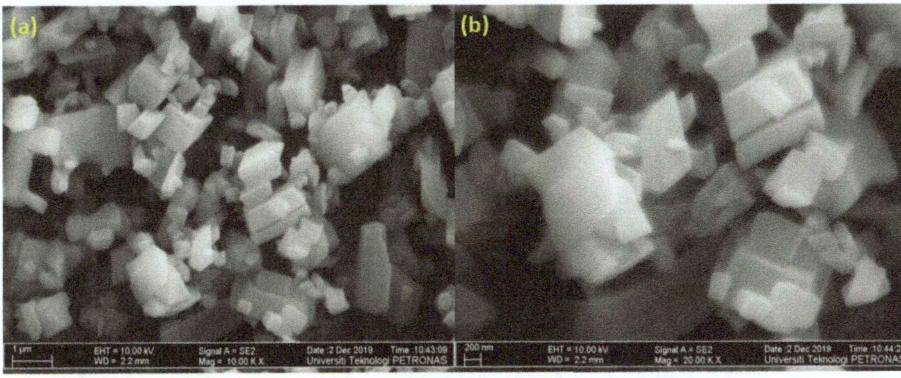

Fig. 1. (a) and (b) FESEM spectrum of the adsorbent

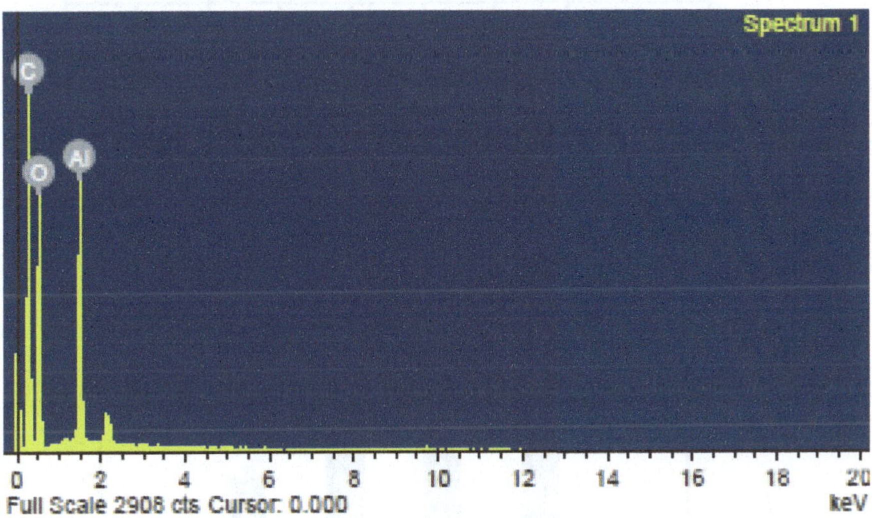

Fig. 2. EDX-Elemental composition of adsorbent material

3.2 Adsorption Studies

Impact of Solution pH

The pH of the solution is a key parameter for the adsorption process, since it can alter the electrostatic and non-electrostatic behavior of the analyte with the adsorbent material. The effect of pH was carried out at different pH values from 2–12, with other predefined experimental conditions (dosage, 10 mg; concentration of the analyte, 10 mg L^{-1} and temperature of 35 °C) in Fig. 4. A study by [14] show that the adsorbent MIL-53(Al) surface has a positive charge when the pH is low (2–6) and becomes negatively charged at high pH (> 6). Thus, in this work high removal of metolachlor was observed at acidic

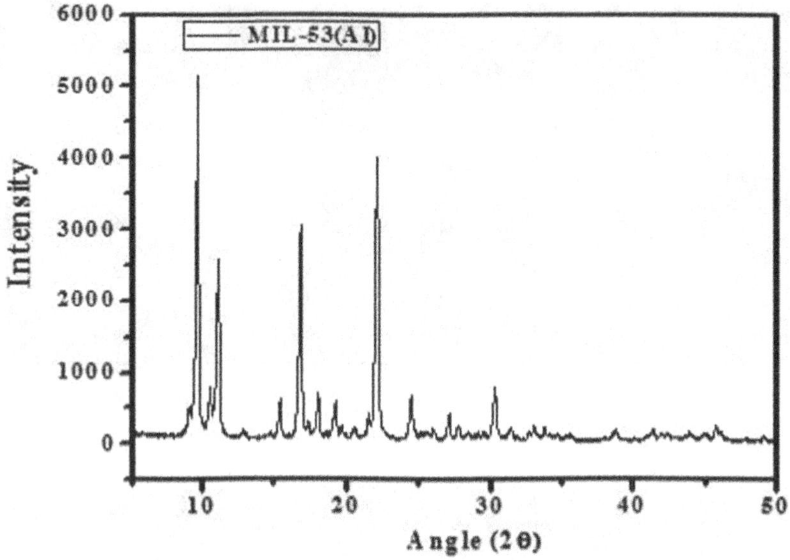

Fig. 3. XRD pattern of the adsorbent material

pH of 4–6 because of the attraction of the negatively charged pollutants molecules in the solution with the positive surface of the MOF adsorbent (electrostatic interaction). This process leads to the fast and efficient adsorption of the analyte. At alkaline pH (7–12),

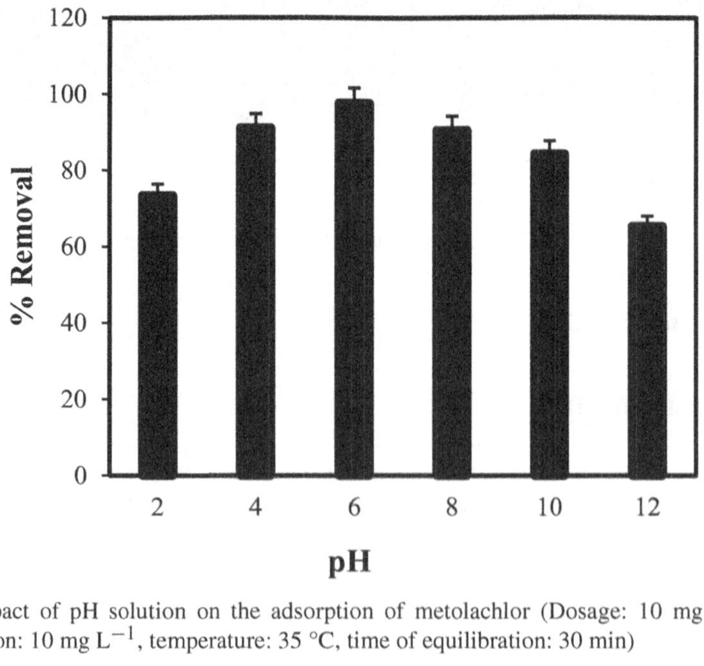

Fig. 4. Impact of pH solution on the adsorption of metolachlor (Dosage: 10 mg, rpm: 150, concentration: 10 mg L^{-1}, temperature: 35 °C, time of equilibration: 30 min)

the surface of the MOF has a charge that is negatively, leading to electrostatic repulsion with a gradual decrease in the adsorption capacity.

Initial Concentration Effect

The percentage removal of the adsorbent MOF in aqueous solution was also determined by varying the different concentration of metolachlor from 5–50 mg L^{-1} as shown in Fig. 5. High removal efficiency was noticed at low concentration from 5–10 mg L^{-1}. The percentage removal of metolachlor is 97.75% using 10 mg L^{-1}. When the concentration is increased (20–50 mg L^{-1}) the percentage removal begin to decrease because the pore of the adsorbent gets saturated.

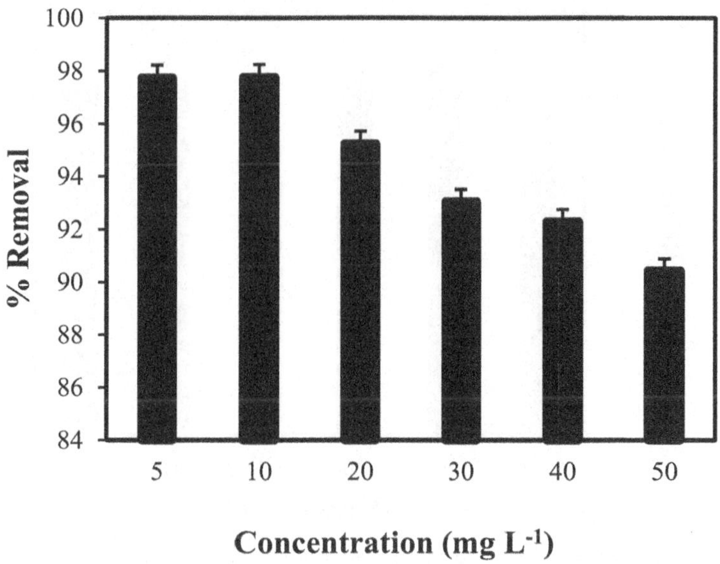

Fig. 5. Impact of analyte concentration for the removal of metolachlor (Dosage: 10 mg, temperature: 35 °C, equilibration time: 30 min, rpm: 150)

Effect of Dosage

The effectiveness of the adsorbent dose was evaluated by varying the loading from 10–50 mg in a solution of 50 mL using 10 mg L^{-1} of metolachlor. Based on the result in Fig. 6, high removal of the analyte (94.92 and 97.75%) was recorded with smaller dose of 5 and 10 mg, respectively due to the active sites and available vacant pores of the adsorbent. MIL-53(Al) is a porous adsorbent with large pour, giving it the capacity to adsorbed more molecules even at a small dose. Further increase in the dosage from 20–50 mg has limited effect on the adsorption efficiency. At 50 mg of the dose the removal efficiency is still high (96.40). Hence, 10 mg was retained as the dose for further experimental studies.

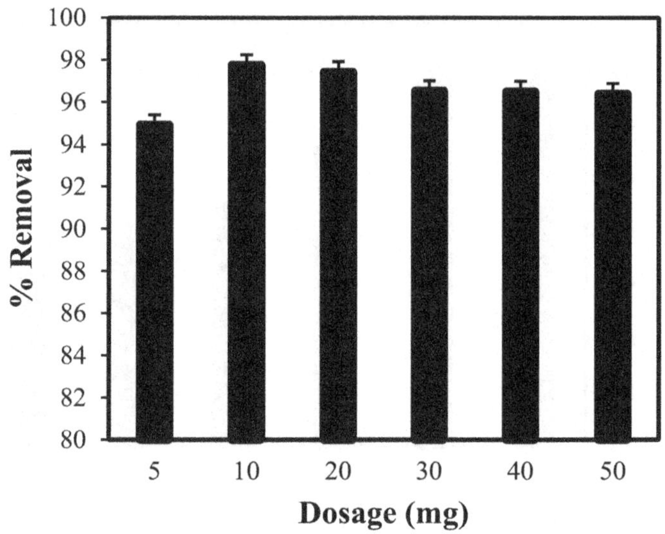

Fig. 6. Effect of adsorbent dose on the removal of metolachlor (Temperature: 35 °C, equilibration time: 30 min, Concentration: 10 mg L^{-1}, rpm: 150)

Prediction by Artificial Neural Network Model

The ANN model was applied to learn the pattern of the datasets and predict the adsorption capacity of metolachlor onto MIL-53(Al). Initial concentration, pH, adsorbent dosage, contact time and temperature were used to design the input layer of the ANN Architecture. The hidden layer is made up of four neurons and one output layer (% removal). The training process resulted in the formation of 5–4-1 topology with the best prediction model gotten at node five after series of trial and error. The resulted coefficient of determination is $R^2 = 0.998$, $R^2adj = 0.991$ and RMSE = 0.311. The detailed prediction values of the training and validation is described in Table 1. This result show that ANN can predict and validate the removal of metolachlor by MIL-53(Al) with high level of accuracy. Hence, the prediction removal efficiency of the ANN model is in good correlation with the experimental findings which implies the robustness of the ANN model. This is because the model when subjected to training can easily understand the multivariate relationship between the experimental and predicted variables.

Table 1. Optimization of the training and validation architecture for the ANN prediction

No	Neurons	Training			Validation		
		R^2	R^2adj	RMSE	R^2	R^2adj	RMSE
1	[3]	0.981	0.974	1.521	0.977	0.965	0.881
2	[4]	0.993	0.987	1.013	0.989	0.973	1.661
3	[5]	**0.998**	**0.991**	**0.311**	**0.996**	**0.991**	**0.441**
4	[6]	0.998	0.986	0.996	0.993	0.991	0.692
5	[7]	0.994	0.985	0.831	0.991	0.982	1.774
6	[8]	0.989	0.982	2.011	0.982	0.973	4.043
7	[9]	0.993	0.991	1.201	0.987	0.982	5.104
8	[10]	0.991	0.980	0.771	0.988	0.981	1.321

4 Conclusion

An in-depth study of the treatment and prediction of metolachlor by MIL-53(Al) MOF was carried out. The MOF adsorbent was synthesized by hydrothermal process, characterized and used for the removal of metolachlor in aqueous solution. The MOF characterization result show a relatively good surface area of 1104 m^2/g that represent the porous nature of the adsorbent. Different adsorption variables like effect of pH, dosage and initial concentration were studied. The adsorption mechanism is controlled by electrostatic interaction with high removal at pH 6. Fast equilibration time with high removal efficiency were recorded using small amount of the adsorbent material. The prediction technique was applied for the design and modeling of the removal of metolachlor. Five parameters that include initial concentration, pH, adsorbent dosage, contact time and temperature were used as input variables to predict the adsorption of metolachlor. The best prediction of the ANN model was arrived after several training at node five. The training, testing and validation results show an excellent prediction by the model of ANN. This model gave accurate prediction with very minimal error.

Acknowledgement. This work is funded by the Universiti Teknologi PETRONAS under the UIR-UTP International grant with cost centre 015MEO-038 and YUTP research grant with cost centre, 015LCO-211.

References

1. Grube, A., Donaldson, D., Kiely, T., Wu, L.: Pesticides industry sales and usage: 2006 and 2007 market estimates. U.S. Environ. Prot. Agency, pp. 1–41 (2011)

2. Otero, R., López, M.I., Esquivel, D., Fernández, J.M., Romero-Salguero, F.J.: Removal of S-metolachlor herbicide from aqueous solutions by meso and microporous organosilica materials. Microporous Mesoporous Mater. **278**, 35–43 (2019)
3. Liu, C., Chen, L., Ding, D., Cai, T.: Sulfate radical induced catalytic degradation of metolachlor: efficiency and mechanism. Chem. Eng. J. **368**, 606–617 (2019)
4. Wei, L., Huang, Y., Huang, L., Li, Y., Huang, Q., Xu, G., Muller, K., Wang, H., Ok, Y.S., Liu, Z.: The ratio of H/C is a useful parameter to predict adsorption of the herbicide metolachlor to biochars. Environ. Res. **184**, 109324 (2020)
5. Danner, Marie-Claire., Robertson, A., Behrends, V., Reiss, J.: Antibiotic pollution in surface fresh waters: occurrence and effects. Sci. Total Environ. **664**, 793–804 (2019)
6. Anastopoulos, I., Pashalidis, I., Orfanos, A., Manariotis, I., Tatarchuk, T., Sellaoui, L., Bonilla-Petriciolet, A., Mittal, A., Núñez-Delgado, A.: Removal of caffeine, nicotine and amoxicillin from (waste)waters by various adsorbents. a review. J. Environ. Manag. **261**, 110236 (2020)
7. Carmalin, S., Lima, E.: Removal of emerging contaminants from the environment by adsorption. Ecotoxicol. Environ. Saf. **150**, 1–17 (2018)
8. Singh, N.B., Nagpal, G., Agrawal, S., Rachna: Water purification by using adsorbents: a review. Environ. Technol. Innov. **11**, 187–240 (2018)
9. Otero, R., Esquivel, D., Ulibarri, M.A., Jiménez-Sanchidrián, C., Romero-Salguero, F.J., Fernández, J.M.: Adsorption of the herbicide S-Metolachlor on periodic mesoporous organosilicas. Chem. Eng. J. **228**, 205–213 (2013)
10. Fard, M., Vosoogh, A., Barkdoll, B., Aminzadeh, B.: Using polymer coated nanoparticles for adsorption of micropollutants from water. Colloids Surf. A: Physicochem. Eng. Aspects **531**, 189–197 (2017)
11. Dhaka, S., Kumar, R., Deep, A., Kurade, M.B., Ji, S., Jeon, B.: Metal – organic frameworks (MOFs) for the removal of emerging contaminants from aquatic environments. Coord. Chem. Rev. **380**, 330–352 (2019)
12. Rasheed, T., Hassan, A.A., Bilal, M., Hussain, T., Rizwan, K.: Metal-organic frameworks based adsorbents: A review from removal perspective of various environmental contaminants from wastewater. Chemosphere **259**, 127369 (2020)
13. Gao, Y., Liu, K., Kang, R., Xia, J., Yu, G., Deng, S.: A comparative study of rigid and flexible MOFs for the adsorption of pharmaceuticals: kinetics, isotherms and mechanisms. J. Hazard. Mater. **359**, 248–257 (2018)
14. Gao, Y., Kang, R., Xia, J., Yu, G., Deng, S.: Journal of Colloid and Interface Science Understanding the adsorption of sulfonamide antibiotics on MIL-53s : metal dependence of breathing effect and adsorptive performance in aqueous solution. J. Colloid Interface Sci. **535**, 159–168 (2019)
15. Guan, Y., Xia, M., Wang, X., Cao, W., Marchetti, A.: Inorganica Chimica Acta Water-based preparation of nano-sized NH 2 -MIL-53 (Al) frameworks for enhanced dye removal. Inorganica Chim. Acta **484**, 180–184 (2019)
16. Jia, Y., Chen, Y., Luo, J., Hu, Y.: Ecotoxicology and Environmental Safety Immobilization of laccase onto meso-MIL-53 (Al) via physical adsorption for the catalytic conversion of triclosan. Ecotoxicol. Environ. Saf. **184**, 109670 (2019)
17. Li, B., Zheng, J.Q., Guo, J.Z., Dai, C.Q.: A novel route to synthesize MOFs-derived mesoporous dawsonite and application in elimination of Cu(II) from wastewater. Chem. Eng. J. **383**, 123174 (2020)
18. Dolatabadi, M., Mehrabpour, M., Esfandyari, M., Alidadi, H.: Chemometrics and Intelligent Laboratory Systems modeling of simultaneous adsorption of dye and metal ion by sawdust from aqueous solution using of ANN and ANFIS. Chemom. Intell. Lab. Syst. **181**, 72–78 (2018)
19. Loiseau, T., Serre, C., Huguenard, C., Fink, G., Taulelle, F., Henry, M., Bataille, T., Ferey, G.: A rationale for the large breathing of the porous aluminum terephthalate, 1373–1382 (2004)

20. Isiyaka, H., Mustapha, A., Juahir, H., Phil-Eze, P.: Water quality modelling using artificial neural network and multivariate statistical techniques. Model. Earth Syst. Environ. **5**(2), 583–593 (2019)
21. Banerjee, P., Sau, S., Das, P., Mukhopadhayay, A.: Optimization and modelling of synthetic azo dye wastewater treatment using Graphene oxide nanoplatelets: characterization toxicity evaluation and optimization using artificial neural network. Ecotoxicol. Environ. Saf. **119**, 47–57 (2015)

A Review of Digital Watermarking Techniques, Characteristics and Attacks in Text Documents

Nur Alya Afikah Usop and Syifak Izhar Hisham(✉)

Faculty of Computer Science and Software Engineering, Universiti Malaysia Pahang, Gambang, Malaysia
alyausop@gmail.com, syifakizhar@ump.edu.my

Abstract. Security plays an important role in many sectors and industries. Nowadays, scams and illegal movements are spreading around the world. Copyright protection for PDF documentation is one of the focus in digital watermarking. Hence, a study was conducted on the related research of digital watermarking on PDF which is text documentation. This paper shows a review of watermarking techniques, characteristics, and possibility of attacks in text document format. This paper also discusses the comparisons between the existing scheme with different domains. The results can be seen from the dominant approaches in text digital watermarking which are the structural approach and hybrid approach. Based on the experimental phase, attacks are in a few formats such as insertion attack and removal attack. These are frequently used to test the robustness of digital watermarks that are embedded into an object.

Keywords: PDF document · Digital watermarking · Text watermarking technique · Text watermarking attack · Digital watermarking characteristic

1 Introduction

The Portable Document Format (PDF) is a file format which captures printed documents in an electronic form [1]. PDF files are frequently used in legal documents, such as tax papers, bank statements, and other forms that require users to fill in their data [2]. PDF has been an ISO standard since October 2005 [3]. This ISO standard represents an international consensus for best practice in conforming to a specification or standard. PDF is proven to meet the specifications for continuing and lasting recording. As an open standard, PDF does not depend on any specific platform or provider. The format is universally used and recognized in all industries.

However, there has been relatively little research on the authentication of text images or text watermarks. In multimedia files, authentication watermarking is a method to prove that a particular multimedia file is authentic and can be used legally. This is done by using the hidden copyright information since multimedia involves several digital media in many types of formats. Compared to digital images and videos, there is less research on textual documents [4, 5]. This paper surveys the characteristics of digital watermarking on text documents and attacks that may be harmful to PDF text documentation.

J. A. Mat Jizat et al. (Eds.): iCITES 2020, AISC 1350, pp. 256–271, 2021.
https://doi.org/10.1007/978-3-030-70917-4_25

2 Related Work

2.1 Techniques Used in Text Watermarking

Certain aspects such as grammar, messages, writing mode, volubility, and text values must be preserved during any transformation of the text. The algorithm of the text depends on text size, language, rules, grammar, conventions and writing styles (Jalil et al., 2009). The text watermarking techniques that have been proposed are image-based approach, syntactic approach, semantic approach, linguistic-based approach, structural-based approach, and hybrid approach (Brassil et al., 1995; Khadam et al., 2019).

Image-Based Approach. The image-based approach uses a text document image to embed the watermark. Some examples of a text document image are a line-shift algorithm, word-shift algorithm, feature coding algorithm, word classification, and inter-word space statistics, as mentioned by Jalil [6]. The image-based approach considers the image to embed the watermark into the cover image [7, 8]. From a watermark logo or image, it is then transformed into a text string pattern. The end product produces a watermarked data. This watermarked data is used as copyright protection and verification [8]. This technique may be safe from a formatting attack but has limited applicability since an Optical Character Recognition (OCR) attack [9] can destroy the hidden information.

According to the proposed scheme by Rizzo [10], a password is used as a technique to embed the watermark in the short text where the contents are preserved after the process of changing the text into an image. Blind watermarking is used as an invisibility view and preserving of the content's properties. Another scheme using an image-based technique was proposed by Tayan [11]. In this proposed technique, a hybrid technique is introduced based on zero watermarking where a transformation between string and image occurs before embedding it into the cover file. The deficiency of the scheme is that it needs large storage to store the Certified Authority (CA) keys. Brassil [12] proposed a line-shifting algorithm where it modifies the location of sentences in many directions depending on the embedded watermark. The position may be on the left or right, up or down.

Syntactic Approach. The composition of sentences in this concept generates the texts, while the composition of word generates the sentences where it could be nouns, verbs, and articles. Different syntactic structure depends on languages and sentence conventions [6]. This approach rearranges the sentences based on a certain order of the words.

Hassan [13] proposed a scheme of natural language watermarking algorithms by performing morphosyntactic alterations to the text. Then the text is transformed into a syntactic tree diagram by defining the hierarchy and dependencies before embedding the watermark. This process can be seen in Fig. 2.8 below. However, this technique is not preferable for English language due to the lack of spaces in watermark embedding [13].

Semantic Approach. The semantic approach focuses on text watermarking by applying a semantic text structure through modification of the text without changing the original meaning in order to apply the process of watermark embedding [6]. This method uses synonyms to be substituted into the sentences.

Atallah [14, 15] proposed a semantic watermarking scheme while the method of synonym substitution is continuous by replacing certain words without changing the original content of the text in order to embed the watermark [16]. Meanwhile, a method of noun-verb based approach was suggested by XingMing [17]. This approach exploits the sentence parsed with a grammar parser by using the semantic network. Figure 2.9 shows the parse tree for a noun-verb based transformation.

Linguistic-Based Approach. This concept shows the combination of the semantic and syntactic techniques where semantic is used to embed the watermark while the original meaning of the text does not change. In this technique, specific words are exchanged with synonym words for the process of data hiding [8].

In 2015, Liu [18] projected a mechanism which supported the convergent characteristic of Chinese text word string. Each selected information in the word is calculated, and through that selected information, the weight of each sentence is obtained. This proposed scheme shows a good performance on formatting attacks. A linguistic approach was proposed by Yingjie [19] where the scheme refers to the qualifications of composition writings. Representative words are used to generate a keyword, core verb set, and proportional feature of adjectives. In embedding watermarks, verbs, adjectives, nouns, and adverbs are applied. This technique is successful in proving low embedding capacity. Figure 1 shows the example of a linguistic approach.

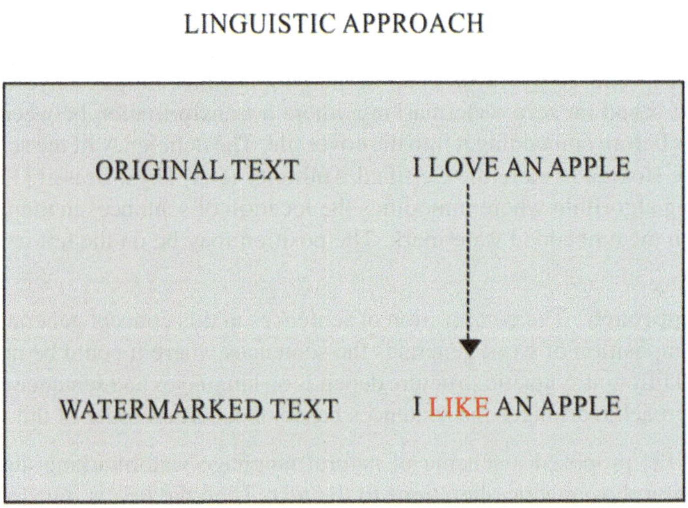

Fig. 1. The linguistic approach in digital watermarking.

Structural-Based Approach. In a structural-based approach, the necessary bits are unified with the text structures of the characteristics. The step for this approach in embedding a watermark into the original data is by utilizing the spaces between the lines and words. Unfortunately, the approach fails to help in handling the holder authentication's problems. Another obstacle is in managing the ownership authentication for every type of

PDF text document. The structural approach removes the spaces between paragraphs, lines, and words which will destroy the hidden data [8].

Based on the existing proposed system, Taha [21] suggested a new technique of language in Arabic which utilizes the Kashida extension character. This suggested technique also utilizes the extra small white space for watermarking. This technique does not show good robustness as it is not protected from formatting attacks. If the spaces are removed, then the hidden information will be lost. A new approach of zero watermarking was introduced by Ba-Alwi [22] which aimed at the probabilistic model. The spaces between the lines and words are used to embed the watermark. This approach showed a better performance on robustness where it helps in the reordering of attacks. Figure 2 below shows the example of a structural approach.

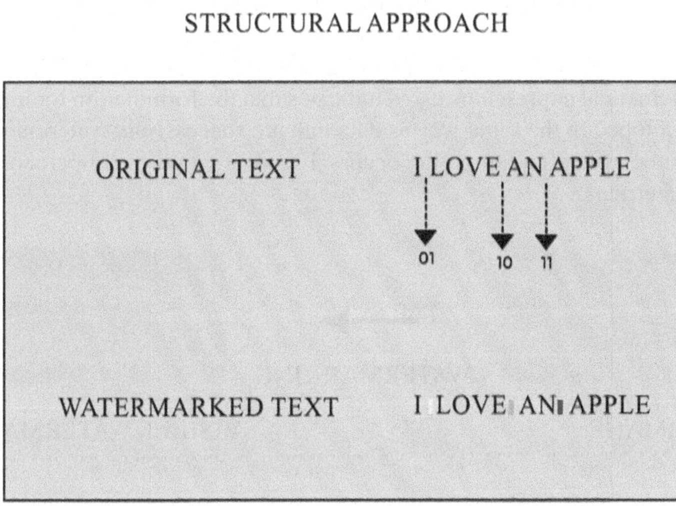

Fig. 2. Pattern of structural approach in digital watermarking.

Alotaibi and Elrefaei [23] proposed a pseudo-space technique focusing on Arabic verses. The connected letters that had been isolated by pseudo-space are used in the watermarks. Imperceptibility and robustness are proven on tampering attacks and formatting attacks; however, they do not have the ability to carry across retyping attacks. A new method was proposed by Hamdan and Hamarsheh [24] by applying an Omega network structure in information hiding which covers the text messages in texts. Meanwhile, Zhang [25] introduced a fragile watermark system for information unity security in IoT.

2.2 Characteristics of Digital Watermarking

Digital watermarking has few characteristics which is crucial in representing the watermarks. (Jalil and Mirza, 2010; Bandyopadhyay and Paul, 2010). These characteristics can be classified into few categories as mentioned below.

Imperceptibility. Imperceptibility has the capability of making watermarks unnoticed by the viewers' naked eyes. This characteristic can be obtained based on a huge level of affinity between watermarked data and primary data [17]. The perceptibility of a watermarked figure can be measured by the quality and fidelity itself. Fidelity is the measurement of before and after the embedding process is done. Fidelity is high if the watermarked figure has a high sameness to the original image [26].

Watermarked data is commonly imperceptible. Visibility can be measured based on whether the embedded digital watermark can be seen or not [17]. Visible watermarking is defined by an embedded watermark that is obvious and easily detected by the naked eye. Based on the existing scheme, this technique is commonly used in figure domains and recording domains in order to forbid illegitimate distribution. The signals shown by a watermark clearly states that the document is owned, and any tampering is not allowed. Invisible watermarking is needed since the text complacent is wielded freely and duplicated. An invisible embedded watermark cannot be detected by the naked eye, but the algorithm can check if a document has applied a watermark [17].

The mathematical expressions listed indicates that the formulation for imperceptibility can be developed in the same way as data that are successfully watermarked approximately the same even after tampering occurs. Figure 3 shows the imperceptibility of an embedded watermark.

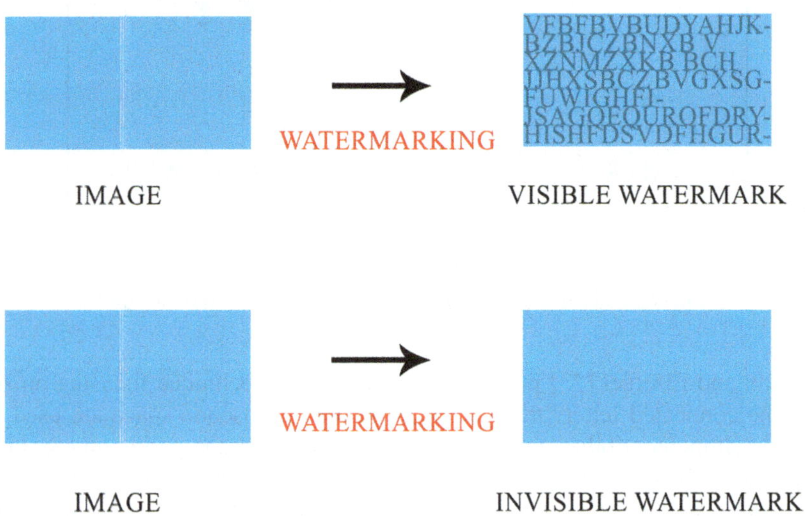

Fig. 3. Example of watermark invisibility after being embedded in the cover image.

Security. Security is a characteristic that avoids any unauthorized interruption or handling of the watermark [27]. The digital text can only be modified by the owner and it is secured in preventing any unauthorized removal of hidden data [9]. Any attacker that makes an attempt on the system without knowing the secret key is unable to adjust the hidden data [28]. The blind algorithm approach is useful and suitable for this characteristic where security is ensured, and the original cover data are not needed in order to validate the copyright text owner. Any manipulation in digital media content can be prevented by raising the effectiveness of the system and also limiting access to authorized users. Hence, the key needs to be well hidden since the attacker could manipulate the system by securing cryptographic security. In order to enhance the system of security, the cryptographic application and digital signature can be used for which the data will be hashed using a hash function.

In order to evaluate the security in a system, the attacker is assumed to know the approach that is used for watermark embedding. However, to ensure the security of the cryptography protocol, the key must remain hidden under it. Analyzation will be accessed by the attacker and the watermark will be destroyed. The attempt on content changing will not be successful if the attackers are unable to manipulate the watermarked data. In this situation, the watermark is considered as well secured. Security protocol must be great and provided with identified policies in order to improve the text watermarking effectualness. The usage of the hash function and cryptographic algorithm must be considered in the building of watermarking instrumentation [17].

Robustness. Robustness is a characteristic that refers to the capability of the watermark to withstand any modifications of image processing such as image cropping, transformation, or compression operations [29]. Robustness is a system that can withstand any attack that tries to retrieve or modify the invisible information [17]. In order to survive from attacks, the watermark should be firmly integrated into the covered image.

The watermark should be retrieved from the verification operations. The most crucial property of robustness is a watermark which cannot be exterminated without distortion or alteration of a cover file if there is a third party that can detect the watermark [17, 26]. A system of robustness could be classified into three classes. The first is robust, the second is fragile, and the third is semi-fragile. These classes refer to the stages of modification done in the hidden data [30, 31]. The robust technique is defined as securing and hiding data from illegitimate access or modifications. On the other hand, the fragile technique refers to a lack of robustness since it cannot withstand any attacks on the system. Fragile watermarking has an advantage in the authentication function since the watermark has been embedded; thus it can be considered as an unaltered image if it is correctly detected in the covered image [26]. The semi-fragile technique is half robust and half fragile because invisible data can only withstand certain attacks. Hence, good or bad text watermarking can be measured by the robustness characteristics through assessing the number of attacks that it can withstand [17]. An example of robustness can be seen in Fig. 4.

Capacity. Capacity means the peak quantity of embedded information invisible in the cover image over a specific period of time. Capacity is normally calculated in bit per second [17]. An increase in capacity is based on the increase of invisible data bits inside the cover image. A sufficient amount of information should be available for watermark embedding in the cover image. The quantity of content is based on function; for example, the information needed in the reversible watermarking scheme is higher compared to the watermarking system that requires an image hash value to be embedded for authentication goals [26].

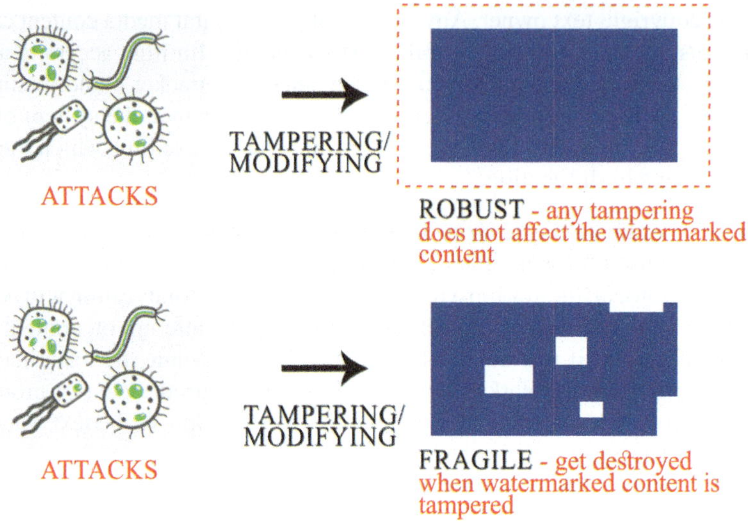

Fig. 4. Illustration of object after being tampered with malicious attacks.

High capacity refers to utilization of the entire system while the watermark is embedding. This situation may have positive effects in the beginning, but it might affect the watermark transparency in the data [17]. Hence, a great capacity structure should be large while the watermark visibility is not affected.

Effectively. Effectiveness is the possibility of detection of a message in a watermarked image and it ideally needs this probability to be 1 [29]. The levels of effectiveness can be measured by the various method of watermarking in scope of digital text security [4].

Image Fidelity. Watermarking is a process of alteration where the message is added into an original image, hence it automatically would influence the image element. To keep this reduction of the image's element to the lowest, there should be less possibility of difference in the image's fidelity [29]. Distortions of signals preserved fidelity only when perceptually significant areas continue to be undamaged, whereas the perceptually insignificant regions may cause sudden changes by small fidelity's effects [34].

Payload Size. The function of a payload size is to hold a message [29]. The size of the message header is not included and is usually essential as lots of schemes need a comparatively integrated large payload in the cover image. Capacity is the data payload which achieved the highest repetition within an image, and also message datas which contained in data payload [34].

False Positive. A false-positive occurs when the digital works number is defined to have a watermark embedded while in reality the embedding process did not occur. This type of property should have a low engagement in watermarking systems [29]. The wrong detection of watermark may be the reasons of false positive [34] (Table 1).

Table 1. Comparisons between existing schemes with different domain.

Author	Technique	Characteristics	Attacks	Domain
Iqbal et al. (2019)	Structural approach (Cryptographic, Visual Basic)	99% Imperceptible, secure, 99% robust, high capacity	Formatting, insertion,, replacement, removal	Digital text in word document
Alotai bi et al. (2016)	Structural approach (2 invisible blind watermarking method, Method 1 for Arabic text, Method 2 for any language)	Secure, robust	Insertion, removal. Tampering ratio 84%	Arabic text
Zhang et al. (2011)	Hybrid approach (Combination of TrueType font display properties with a few functions of Office Word - Embedding algorithm and extracting algorithm	Secure, robust	Formatting, Insertion, removal, recompose	Chinese Character
Jaseen a et al. (2011)	Hybrid approach (Combined image and text watermark, Zero text watermarking algorithm)	Secure, high capacity	Insertion, removal, reordering	Text document
Mirza et al. (2010)	Structural approach. Robust and fragile (Zero watermarking algorithm)	Secure, robust	Insertion, removal	Plain text
Jalil et al. (2010)	Linguistic-based approach. (Zero watermarking algorithm)	Fragile	Insertion, removal, reordering	Plain text

2.3 Attacks

Digital Watermark attacks are malicious and intentional attacks. There is the possibility for any individual to perform more than one unauthorized action, even if the performance is not completely done. In certain watermarking systems, some individuals may easily detect the presence of the mark, but it may be difficult to decode the message (Zhou et al., 2009).

Removal Attacks. Activities that are carried out by removal attacks are analysis of watermark data, watermark or host data estimation, data presenter, watermark separation, and also elimination of watermark. This type of attack works on removing all of the information in the original data without any watermark security damaged [32, 33]. Removal attacks are designed to extract the data contents from the watermarked substance. In reality, removal attacks are accomplished where normally the watermark is an additional noise indicator in the presenter indicator [34].

A removal attack is aimed to withdraw the incitation of the watermark from the watermarked cover figure without breaking the precaution of the watermark formulation [27]. This attack does not need to know the application of the technique of how the watermark is embedded. Encryption, noising, connivance, heighten, and bar graph leveling are categorized as removal attacks. A removal attack may damage the watermark information even if it fails to completely remove it. Figure 5 shows a situation where a removal attack occurs in texts.

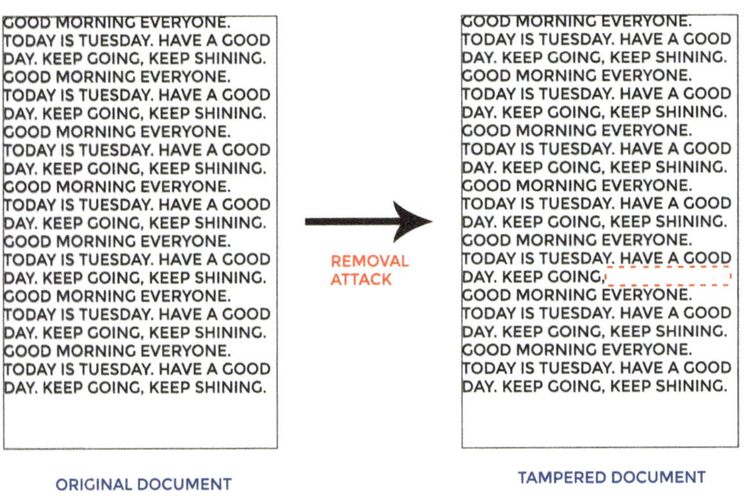

Fig. 5. Example of a removal attack.

Protocol Attacks. Protocol attacks are aimed at the whole concept of watermarking usage. There is a kind of protocol attack which refers to the idea of the invertible watermark, where the original watermark will be deducted by an attacker from the watermarked data and he/she demands to be the proprietor of the watermarked data [35].

Cryptographic Attacks. A cryptographic attack is a process where it solves a ciphertext without the recognition of cryptographic keys [35]. The information known by the cryptanalyst needs to be considered first to determine the hierarchy of the cryptographic attacks [3]. The cryptographer's renowned information indicates the sets of plaintexts or ciphertexts, and the cryptographic key which needs to be uncovered. Cryptanalysts may dominate the encipher key and discover the decipherment key in the case of an asymmetric cryptosystem [35].

The purpose of the cryptographic attack is to interrupt the safety degree inside the watermarking scheme. The integrated watermark is expelled or a deluging watermark is inserted once the security degree is broken. One type of proficiency which can attempt the watermark security is the Brute-force search. This attempt is to search for significant secret information using some known possible degree [26]. Figure 6 shows the illustration of a cryptographic attack.

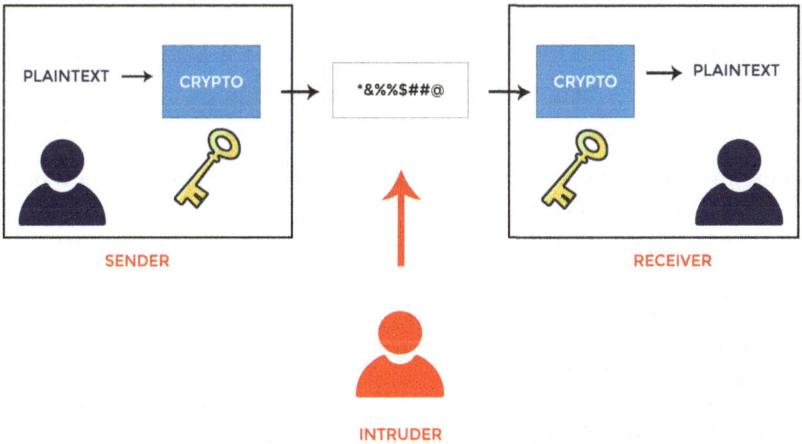

Fig. 6. Illustration of cryptographic attack.

Insertion Attacks. Another content changing attack is the insertion attack which functions to insert additional content into the original text document. Figure 3 shows the scenario of an insertion attack. IDS (security alert to administration) can accept a packet that an end-system rejects. An IDS that does this makes the mistake of believing that the end-system has accepted and processed the packet when it actually has not. An attacker can exploit this condition by sending packets to an end-system that it will reject, but that the IDS will think are valid. In doing this, the attacker is "inserting" data into the IDS. This situation can be seen in Fig. 7.

Combine Attacks. Combine attack is a combination of insertion attack and removal attack which will be deleting content while at the same time inserting additional details into the text documentation.

Replacement Attacks. A replacement attack is one of the new geometrical attacks that operates in content changing without decimating the text structure. It focuses on

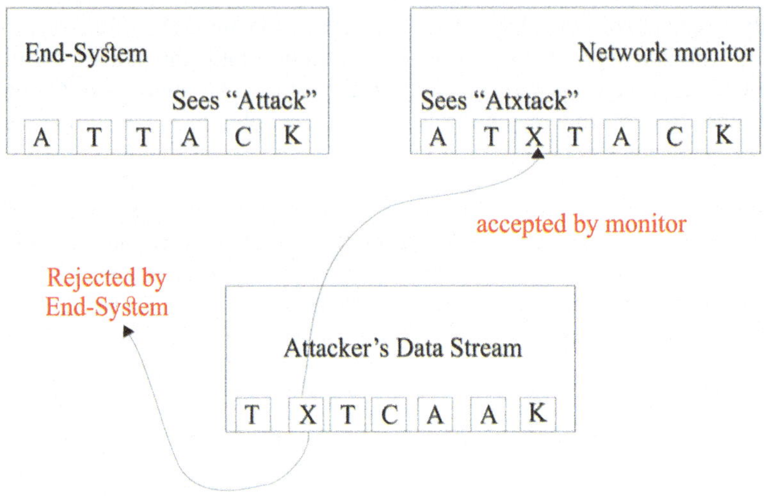

Fig. 7. Insertion attack in texts [36].

maintaining the localization of words in the text [37]. An example of a replacement attack can be seen in Fig. 8 below.

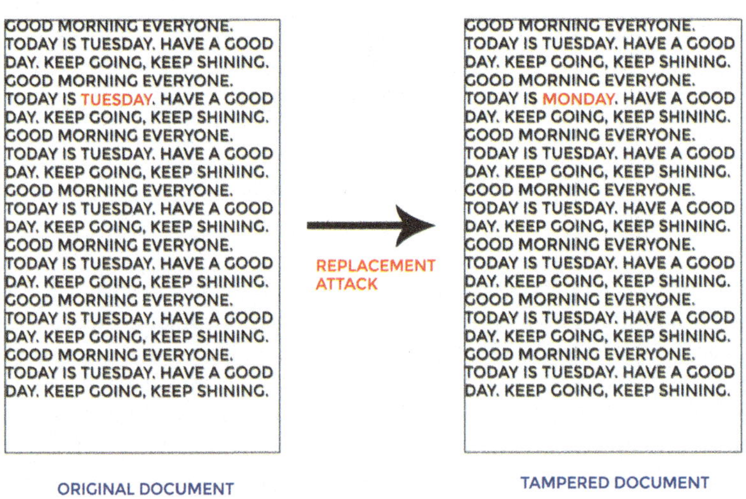

Fig. 8. Tampered document after being interrupted with a replacement attack.

Format Attacks. Format attack can be in the change of font, size, or paragraph background-colour. For example, the change of text size from the original content is used for manipulation.

3 Challenges of Digital Watermarking in Text Document Format

There is limited research on existing schemes for text such as inter-word, paragraph spacing, and word-shift [8]. With reference to the table comparison for a few of the existing proposed schemes, the focused domain were language characters such as Chinese, Arabic, and Indian text. Text on word document was also proposed in a few schemes but research in the PDF domain is too limited. Below are elaborations on the existing proposed schemes which focus on text documentations.

Based on a paper written by Iqbal [8] entitled 'A Robust Digital Watermarking Algorithm for Text Document Copyright Protection based on Feature Coding', the authors proposed some unique properties for watermarking to improve the security of text documentation legal right. After the process of encryption, information of the watermark is embedded in the unique properties. To hide the information without content interruption, the projected scheme applies the specific properties of an MS-Word document. The technique mentioned shows imperceptibility, robustness towards formatting attacks, achieves high embedding capacity, and is able to have copyright protection for delicate documents. The results show that the proposed technique is robust in various formatting attacks such as insertion, replacement and deletion attacks while the watermark can still be extracted. Moreover, the proposed technique manages to improve the embedding capacity of the watermark with the ability of imperceptibility.

Aloitaibi [23] proposed two invisible blind watermarking methods for Arabic text. The method starts by inserting the pseudo-space before and after the normal word space which refers to the dotting feature in Arabic text. Secondly, the method proposed inserts the pseudo-space and other three small or zero-width spaces to boost the capacity. Presence can be observed by bit "1" and absence is indicated by bit "0". This second proposed method can be used for any language. The first proposed method has the highest capacity and imperceptibility while the second method has the largest capacity, but slightly lower imperceptibility than the first method. They are also robust against electronic attacks such as copying, pasting, and formatting attacks.

Zhang [38] proposed a radical watermarking formula supported on new-defined property which is TrueType for Microsoft Word documents. The projected method shows a combination of demonstrate attributes of the TrueType font library with some Office Word operational tools. Referring to the stated explanation, the algorithm is able to locate the tampered location while high in imperceptibility. It is also robust to frequent attacks including the exchange of format, contraction, and typesetting attack. In order to improve the robustness, the suggested way is to embed the watermark repeatedly or embed the watermark in each character. However, the algorithm should be optimized where the watermark can be processed before embedding and the semantic approach is used to identify the positions of the embedded data. The processed version of TrueType shows an unsatisfactory performance from the results.

Jaseena [40] proposed a hidden watermarking formula supported by the presence of non-vowel ASCII characters. This method uses the combination of figure and text watermark for text document covering. The watermark is reasonably integrated in the text. Then, the watermark key is created. Attackers would face some difficulties when guessing the key due to the high key. Hence, the possibility of a brute force attack is decreased. However, the high key creates difficulties for the certificate authority (CA) to hold up the key, including the key transaction between the text holder and CA. Based on the algorithm, it shows that the watermark key has a high length. An advantage of high length is that it will be difficult for intruders to attack the watermark key. Hence, the possibility of a brute force attack will be lower. On the other hand, certificate authority (CA) will have difficulties to maintain the key and transferring between the owner and CA itself.

In 2010, Mirza [39] proposed a novel zero-watermarking algorithm for plain text document that incorporates the structural components in the text along with the image watermark. Before an image is extracted to prove ownership, the algorithm needs to embed the image of the watermark reasonably in the text. As a result, after the experiment is analysed on 12 samples of text, the technique proves to be sturdy, warranted, and able to withstand irregular tampering attacks such as insertion, deletion, and re-ordering attacks. A different key would be produced for two different documents even when using a similar watermark image since the owner of the original copyright provides input along with the watermark in the embedding process.

Jalil [6] proposed a novel zero-watermarking formula for substantiation of plain text. Based on the text content, the algorithm generates a watermark, and then it is extracted using an extraction algorithm in exchange for authenticity in a text document. A zero-watermarking algorithm does not change the characters of the original data but utilizes the characters of the original data to construct the original watermark information. The results show that tampering detection is successful after evaluation on 10 dispersed forms when the tampering volume is low.

Based on comparisons between the existing schemes, it shows that the PDF format has very limited research compared to text document. The method used in dominant researches of text document are most likely focusing on security and robustness where it prevents the attempt of intruders from reformatting the original text documents. Referring to the comparisons, we can conclude that Alotaibi et al. (2016), Mirza et al. (2010), and Jalil et al. (2010) have used good techniques since they managed to cover all types of languages and most text attacks can be covered. According to a previous review, PDF document is preferable with the fragile technique because it has authentication characteristics compared to robustness. Mirza et al. (2010) has a good technique of robust and fragile where it combines within images and text in order to embed the watermark into the data, while Alotaibi et al. (2016) have a good focused domain which is MS Word document type which may be compatible with the features of PDF document format. These two papers have good techniques and focused domains, hence we would suggest a new technique of watermarking scheme which can be improved better in the specific area of PDF format.

4 Conclusion

Digital watermarking is mostly focused on plain texts, language characters, and images. Previous research has rarely focused on textual contents in PDF format. Since PDF can be considered as a popular format to be applied in this era, robust watermarking schemes are needed. Thus, this paper focused more on the textual in PDF format only. Unlike images in whatever format they are, having a textual document in PDF, the area to embed is limited, while the background colour which is usually white makes the embedded data tricky to be hidden too. However, this paper also discusses the general techniques and characteristics of digital watermarking as the comparison to identify the advantages and disadvantages of the techniques used in text document watermarking. The comparison proves that the structural and hybrid approaches are often used in text digital watermarking. The method most used in the existing schemes has the characteristics of robustness which are able to protect watermarked data from attacks. However, it may cause some disadvantages on embedded watermark data. High robustness may distort watermarked data. Thus, the fragile watermarking method may be compatible with many categories in different domains including the PDF domain. Based on the study, the list of attacks that are possibly found in PDF digital watermarking and the attack frequently used in the existing schemes are format attacks which always interrupt the original data. These two types of attacks should be able to be handled in digital watermarking. A better technique of PDF watermarking would be suggested in future.

Acknowledgement. This research works is supported by a RACER grant entitled 'A New Embedding Algorithm using Hilbert-Peano Pattern in Enhancing Authentication Systems for Textual Documents' (RDU192623) and PGRS200369 supported by Universiti Malaysia Pahang.

References

1. Computer Hope. https://www.computerhope.com/jargon/p/pdf.htm. Accessed 2018
2. Yang, H., Kot, A.C.: Text document authentication by integrating intercharacter and word spaces watermarking. In: 2004 IEEE International Conference on Multimedia and Expo (ICME) (2004)
3. Mir, N.: Copyright for web content using invisible text watermarking. Comput. Human Behav. **30**, 648–665 (2014)
4. Kamaruddin, N.S., Kamsin, A., Por, L.Y., Rahman, H.: A review of text watermarking: theory, methods, and applications. IEEE Access **6**, 8011–8028 (2018)
5. Kim, Y.W., Oh, I.S.: Watermarking text document images using edge direction histograms. Pattern Recognit. Lett. **25**, 1243–1251
6. Jalil, Z., Mirza, A.M., Sabir, M.: Content based zero-watermarking algorithm for authentication of text documents. In: (IJCSIS) International Journal of Computer Science and Information Security, vol. 7, no. 2 (2014)
7. Hao, Y., Chuang, Q.F.L., Rong, D.: A survey of digital watermarking. J. Comput. Res. Dev. **7**, 1093–1099 (2005)
8. Khadam, U., Iqbal, M.M., Azam, M.A., Khalid, S., Chilamkurti, N., Rho S.: Digital watermarking technique for text document protection using data mining analysis. Special Edition on Data Mining for Internet of Things (2019)
9. Kaur, B., Sharma, S.: Digital watermarking and security techniques: a review. Int. J. Comput. Sci. Technol. **8**(2), 44–47 (2017)
10. Rizzo, S.G., Bertini, F., Motensi, D.: Content-preserving text watermarking through unicode homoglyph substitution. In: Proceedings of the 20th International Database Engineering & Applications Symposium, pp. 97–104 (2016)
11. Tayan, O., Kabir, M.N., Alginahi, Y.M.: A hybrid digital-signature and zero-watermarking approach for authentication and protection of sensitive electronic documents. Sci. World J. (2014).

12. Brassil, J.T., Low, S., Maxemchuk, N.F., Gorman, L.O.: Electronic marking and identification techniques to discourage document copying. IEEE J. Sel. Areas Commun. **13**(8), 1495–1504 (1995)
13. Meral, H.M., et al.: Natural language watermarking via morphosyntactic alterations. Comput. Speech Lang. **23**, 107–125 (2009)
14. Atallah, M., McDonough, C., Nirenburg, S., Raskin, V.: Natural language processing for information assurance and security: an overview and implementations. In: Proceedings 9th ACM/SIGSAC New Security Paradigms Workshop, Cork, Ireland, pp. 51–65 (2000)
15. Attallah, M., Raskin, V., Hempelmann, C.F., Karahan, M., Sion, R., Topkara, U., Triezenberg, K.E.: Natural language watermarking and tamperproofing. In: Fifth Information Hiding Workshop. LNCS, Noordwijkerhout, The Netherlands, vol. 2578. Springer (2002)
16. Topkara, U., Topkara, M., Atallah M.J.: The hiding virtues of ambiguity: quantifiably resilient watermarking of natural language text through synonym substitutions. In: Proceedings of ACM Multimedia and Security Conference, Geneva (2006)
17. Kamarudin, N.S., Kamsin, A.: Text watermarking using recursive decent semantic net parsers. In: Lecture Notes in Computer Science. LNCS, vol. 3612, pp. 958–961. Springer (2005)
18. Liu, Y., Zhu, Y., Xin, G.: A zero-watermarking algorithm based on merging features of sentences for Chinese text. J. Chin. Inst. Eng. **38**(3), 391–398 (2015)
19. YingJie, M., Huiran, L., Tong, S., Xiaoyu, T.: A Zero-watermarking scheme for prose writings. In: Proceedings of International Conference on Cyber-Enabled Distributed Computing and Knowledge Discovery (CyberC), pp. 276–282 (2017)
20. Brassil, J.T., Low, S., Maxemchuk, N.F.: Copyright protection for the electronic distribution of text documents. In: Proceedings of IEEE, vol. 87, no. 7, pp. 1181–1196 (1999)
21. Taha, A., Hammad, A.S., Selim, M.M.: A high capacity algorithm for information hiding in Arabic text. J. King Saud. Univ. Comput. Inf. Sci. **32**, 659–665 (2018)
22. Ba-Alwi, F.M., Ghilan, M.M., Al-Wesabi, F.N.: Content authentication of English text via internet using zero watermarking technique and Markov model. Int. J. Appl. Inf. Syst. **7**(1), 25–36 (2014)
23. Alotaibi, R.A., Elrefaei, L.A.: Improved capacity arabic text watermarking methods based on open word space. J. King Saud Univ. Comput. Inf. Sci. **30**(2), 236–248 (2018)
24. Hamdan, A.M., Hamarsheh, A.: AH4S: an algorithm of text in text steganography using the structure of omega network. Secur. Commun. Netw. **9**(18), 6004–6016 (2017)
25. Zhang, G., Kou, L., Zhang, L. Liu, C., Da, Q., Sun, J.: A new digital watermarking method for data integrity protection in the perception layer of IoT. In: Security and Communication Networks (2017)
26. Hisham, S.I., Muhammad, A.N., Zain, J.M., Badshah,G., Arshad, N.W.: Digital watermarking for recovering attack areas of medical images using spiral numbering. In: 2013 International Conference on Electronics, Computer and Computation (ICECCO), pp. 285–288. IEEE (2018)
27. Tyagi, S., Singh, H. V., Agarwal, R., Gangwar, S. K.: Digital watermarking techniques for security applications. In Proceedings of International Conference on Emerging Trends in Electrical Electronics & Sustainable Energy Systems (ICETEESES), pp. 379–382 (2016)
28. Cayre, F., Fontaine, C., Furon, T.: Watermarking security: theory and practice. IEEE Trans. Signal Process. **53**(10), 3976–3987 (2005)
29. Durvey, M., Satyarthi, D.: A review paper on digital watermarking. Int. J. Emerg. Trends Technol. Comput. Sci. (IJETTCS), **3**(4), ISSN 2278-6856 (2014)
30. Zhou, X., Wang, S., Zhao, W., Peng, R.: A semi-fragile watermarking scheme for content authentication of chinese text documents. In: Proceedings of 2nd IEEE International Conference on Computer Science and Information Technology, pp. 439–443 (2009)

31. Sharma, G., Coumou, D.: Watermark synchronization: Perspective and a new paradigm. In: Proceedings of 40th Annual Conference on Information Sciences and Systems, pp. 1182–1187 (2006)
32. Varshney, Y.: Attacks on digital watermarks: classification, implications, benchmarks. Int. J. Emerg. Technol. **8**(1), 229–235 (2017)
33. Dittmann, J., Wohlmacher, P., Nahrstedt, K.: Multimedia and security using cryptographic and watermarking algorithms, pp 54–65. IEEE (2001)
34. Singh, P., Chadha, R.S.: Review to digital watermarking and a novel approach to position the watermark in the digital image. Int. J. Eng. Adv. Technol. **2**(4) (2013)
35. Cangea, O., Moise, G.: A new approach of the cryptographic attacks. In: DICTAP 2011, Part I. CCIS, vol. 166, pp. 521–534 (2011)
36. Ptacek, T.H., Newsham, T.N.: Insertion, evasion, and denial of service: eluding network intrusion detection. In: Secure Networks Inc., Calgary, Alberta, (1998)
37. Bashardoost, M., Rahim, M.S.M., Saba, T., Rehman, A.: Replacement attack: a new zero text watermarking attack. 3D Research, **8**(1), 8 (2017)
38. Zhang, S., Yao, Z., Han, K.J., Meng, X., Liu, C.: New digital text watermarking algorithm based on new-defined characters. In: International Symposium on Computer, Consumer and Control (2014)
39. Jalil, Z., Mirza, A.M., Iqbal, T.: A zero-watermarking algorithm for text documents based on structural components (2010)
40. Jaseena, K.U., John, A.: An invisible zero watermarking algorithm using combined image and text for protecting text document (2011)
41. Jalil, Z., Mirza, A.M.: A review of digital watermarking techniques for text documents (2009)

Auditory Evoked Potential (AEP) Based Brain-Computer Interface (BCI) Technology: A Short Review

Md Nahidul Islam(✉), Norizam Sulaiman, Bifta Sama Bari, Mamunur Rashid, and Mahfuzah Mustafa

Faculty of Electrical and Electronics Engineering Technology, Universiti Malaysia Pahang, 26600 Pekan, Pahang, Malaysia
nahidul76.edu@gmail.com, {norizam,mahfuzah}@ump.edu.my, biftasama.eee@gmail.com, mamun110218@gmail.com

Abstract. Auditory evoked potential (AEP) is a sort of electroencephalographic (EEG) signal elicited by an acoustic stimulus from the brain scalp. An intelligent auditory perception level system helps the auditory system to analyze and assess its functional integrity. Initially, the AEP signals are utilized in accurate and fast hearing threshold level detection in early stage. However, this emerging psychological modality expands it applications like other EEG control signals. Recently, AEP responses are initiated into the brain-computer technology (BCI) technology. This paper presents a short review on AEP based BCI technology. First, a concise overview of AEP signal and its analysis procedure are discussed. Then the existing AEP based studies are reviewed and their summary in terms of data description, feature extraction, classification methods and performance are tabulated. Finally, the issues of the recent AEP signal analysis have been addressed and potential ways are suggested to alleviate existing problems. Finally, the issues to the existing AEP based BCI technology are presented, and the possible solutions are also recommended.

Keywords: Auditory Evoked Potential (AEP) · Electroencephalogram (EEG) · Brain-Computer Interface (BCI) · Machine learning

1 Introduction

Paralysis entails loss in one or more areas of the body of voluntary muscle activity due to nervous system damage. The nervous system consists of two parts: Central Nervous System (CNS) which includes the brain and spinal cord and the peripheral nervous system (PNS) which includes nerves from outside the CNS [1]. Because paralysis can happen with any muscle or group of muscles, it can affect many body functions. Some of the issues that may arise alongside paralysis include: Problems with blood flow, breathing, and heart rate, behavior and mood changes, Changes to muscles, joints, and bones [1]. Patients with locked-in syndrome (LIS) or other types of paralysis have trouble in communicating. The restoration of movement of chronic stroke or other brain injury

J. A. Mat Jizat et al. (Eds.): iCITES 2020, AISC 1350, pp. 272–284, 2021.
https://doi.org/10.1007/978-3-030-70917-4_26

patients is still a therapeutic challenge and there are no substantial changes to therapies available. Brain-computer interface (BCI) technology could be an effective alternative to mitigate these problems. BCI is an interaction between the brain and the device that enables brain signals to direct an external activity [2]. It is a computer-based system that acquires, analyzes and translates brain signals into commands transmitted to an output device to perform the desired action [3]. In order to capture the brain activity; different modalities including EEG, magnetoencephalography (MEG), functional magnetic resonance imaging (fMRI), Electrocorticography (ECoG) as well as Functional near-infrared spectroscopy (fNIRS) are employed. EEG is the commonly used non-invasive technique for BCIs to evoke various control signals, including SCP, steady-state visually evoked potential (SSVEP), motor imagery (MI), event-related potential (ERP), P300 and AEP. It is discernible that a large number of original papers and reviews have been reported in BCI over the last 15 years. A variety of noteworthy BCI reviews cover particular applications, such as wheelchair control [4], BCI mobile robot [5], emotion recognition using EEG [6], biometrics [7], as well as virtual reality and gaming [8]. The control signal and classifications of the electroencephalograms (EEG) have been addressed briefly [9]. However, it is noticeable that the details on AEP signals and its processing methods have not yet been taken into account when analyzing state-of-the art BCI solutions for both hardware and software. To authors best knowledge, there is no any review article related to the AEP based BCI system has been published. To fill the existing gaps, the present review reveals recent research in brain–computer interfaces (BCIs) with Auditory evoked potential (AEP) as promising solutions.

The purpose of this article is to review AEP-based BCI studies regarding data acquisition system, features extraction, classification algorithms and the evaluation metrics used. In addition, a succinct description of the AEP based BCI method is addressed to the readers. It is worth noting that significant issues relating to BCI systems based on AEPs are mentioned and potential solutions are also suggested for mitigating these issues. Additionally, relevant research gaps are also addressed in this paper for further investigation. This paper is organized as follows: Sect. 2 introduces a brief explanation on AEP signal used for EEG based BCI application. Recent advancements in BCI system using AEP signals are presented in Sect. 3. Section 4 states current issues with possible solutions in BCI system and followed by a conclusion in Sect. 5.

2 AEP Signals in BCI

The evoked potential is derived from the EEG and is a reaction to an electrophysiological stimulus. It can be divided into 2 categories, namely visual evoked potential and auditory evoked potential (AEP). The AEP signal is an electrical brain signal that is produced when a time-locked auditory stimulus is given. AEP consists of positive or negative spikes reproducibly repeated following the stimulus. Post-stimulus latency and peak amplitudes characterize the AEP response [10, 11]. In contrast to recorded EEG signals, the recorded AEP signals have a low signal-to-noise ratio. The AEP signal must be described on average during several single trials in order to enhance the signal-to-noise ration [12]. Shanks et al. [13] reported that speech intelligibility in unassisted environments decreases for listeners with relatively moderate hearing loss, but increases for listeners with moderate to extreme hearing loss. Over 5% of the world's population, comprising 360 million people (328 million adults and 32 million children), has hearing losses, according to the World Health Organization [14]. The amplitude of AEPs in contrast with the EEG signals is substantially lower [15, 16]. AEPs are being used to diagnose and locate the pathologies affecting brainstem pathways. It has been extensively used to assess auditory functioning and are used in child auditory screening as an effective diagnostic too [17]. BCIs based on AEP are an emergent topic in the BCI community in recent years.

The main components of a BCI system are: brain activity measurement, preprocessing, feature extraction, classification and translation into a command. Figure 1 represents a characteristic block diagram which shows the various stages of EEG signal processing for BCI. The brain activities of the targeted users are recorded via different types of EEG sensors in the brain activity acquisition process. The raw AEP includes a number of artifacts and in the pre-processing step, they are discarded. Feature extraction attempts to identify the signals by a few appropriate values called "features," which are also analyzed at this stage for the selection of essential features. The extracted features are then categorized in the classification process by various machine learning and deep learning algorithms. Lastly, classified findings are converted into device commands for developing BCI application in real life.

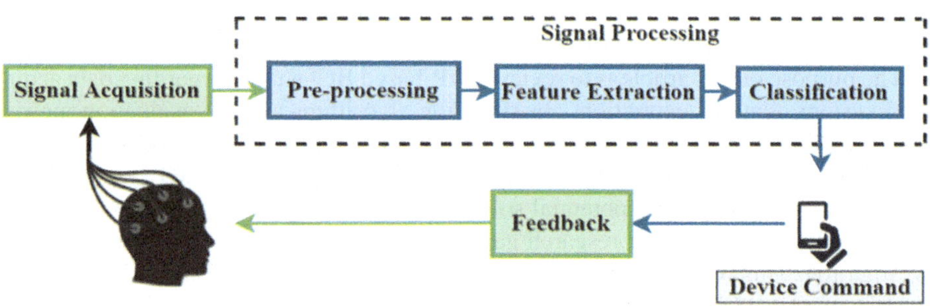

Fig. 1. Architecture of complete BCI technology.

A number of pre-processing, feature extraction, feature selection and classification algorithms are implemented to analyze the AEP signals. Figure 2 illustrate the summary of all employed algorithms and tools for AEP data analysis.

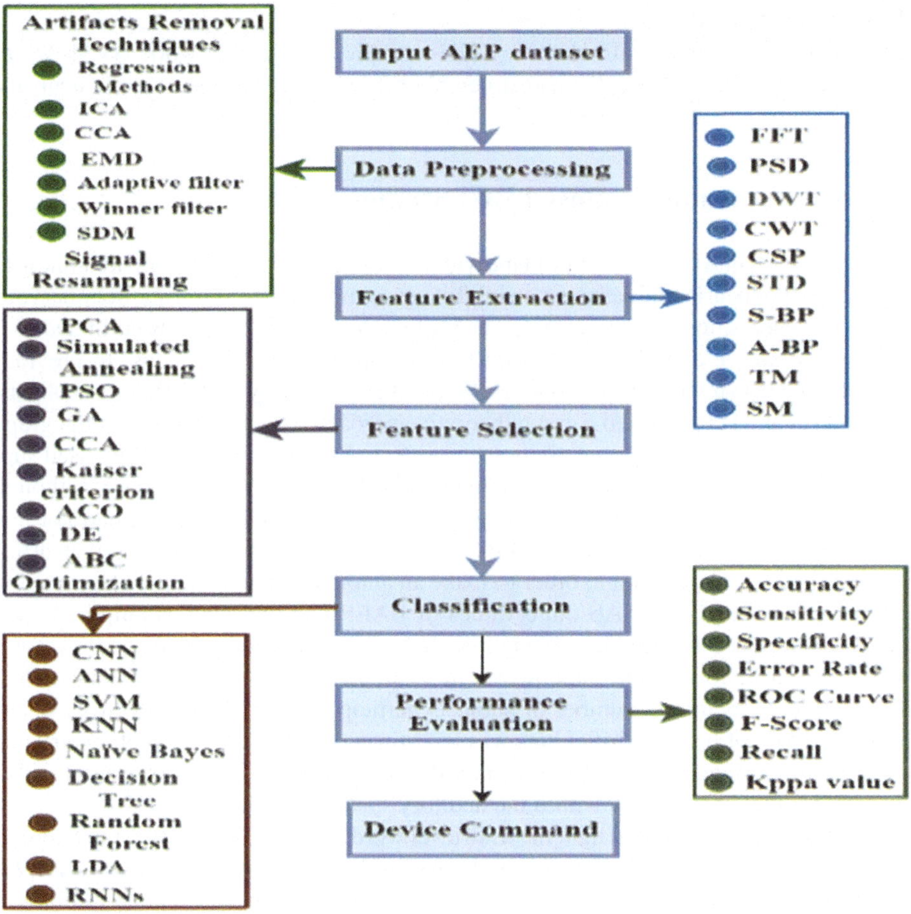

Fig. 2. Summary of AEP signal analysis procedures.

Associated acronyms:
Artifact removal technique:
ICA—Independent component analysis, CCA—Canonical correlation analysis, EMD—Empirical mode decomposition, SDM—Sparse decomposition methods.

Feature Extraction methods:
FFT—Fast Fourier transform, PSD—Power spectral density, DWT—Discrete wavelet transform, CWT—Continuous wavelet transform, CSP—Common spatial patterns, STD—Standard deviation.

Feature Selection methods:
PCA—Principal component analysis, PSO—Particle swarm optimization, CCA—Canonical correlation analysis, GA—Genetic algorithm, ACO—Ant colony optimization, DE—Differential entropy, ABC—Artificial bee colony.

Classification methods:
CNN—Convolutional neural network, ANN—Artificial neural network, SVM—Support vector machine, KNN—K-nearest neighbor, LDA—Linear discriminate analysis, RNN—Recurrent neural network.

3 Review of Auditory Based BCI Studies

The response of an AEP represents an individual's auditory ability. The updated findings on the impact of hearing impairment on inhibitory circuits of the central auditory nervous system has been summarized in brief. The past studies of AEP signal used in auditory nervous system are addressed here with their prospective pros and cons. In previous research, BCI researchers proposed several sorts of effective way for AEP signal analysis.

Sultan et al. [18] proposed a graph theoretic approach where global and nodal graph features was used in brain connectivity patterns from EEG hearing loss. They used support vector machine (SVM) for classification and achieved 85.71% accuracy. Although this system used a large number of subjects (32 subjects), their accuracy had not met the requirement of real-life AEP based BCI system. Edgar et al. [19] proposed an artificial neural network approach in order to make an automatic detection and classification of hearing loss conditions. Absolute values of BAEP features such as amplitude and latencies were used as a feature extraction method. The accuracy is above 90% for a variety of hearing loss condition, however they did not clearly mention the data collection procedures including number of subjects participated, number of electrodes. The age-related loss of hearing difficulties was investigated with the MR morphometry and diffusion tensor imagery (DTI) approach by Profant et al. [20]. This study reported that HL does not play major roles within the auditory system in older people [21].

A recent study [22] found that, the trial-level word recognition benefit from cingulo-opercular activity was equivalent for both hearing loss groups. Chaosheng el al. [23] proposed a novel identification system composed of a 3-level wavelet entropy, followed by MLP trained by TS-PSO for detecting hearing loss. They used wavelet entropy (WE) as a feature. Although a large number of subjects were used in their study, the achieved accuracy (86.17%) was not so high. Marco et al. [24] proposed a pattern-based classification (SPC) method in order to classify the auditory brainstem responses through symbolic pattern discovery and the overall accuracy was 99.4%. Ibrahim et al. [25] utilized multiple classifier where SVM classification algorithm achieves the highest classification accuracy compared to other classification algorithms. However, the number of subjects involved in data collection was very small. Sanjay et al. [26] proposed a method for auditory threshold perception in human beings. In the proposed study absolute threshold test (ATT) was used for feature extraction method. They used multiple classification method among all the classification method, the best accuracy was 93.94% with SVM classifier. Hubert et al. [27] used wavelet packet transform (WP) feature for detection of cochlear hearing loss and SVM classifier was used for classification these signals. They did a great job in the proposed study, but the overall accuracy was not well enough for detection the signal of hearing loss condition. Shanks et al. [13] reported that in unaided conditions, speech intelligibility lessens by enhancing speech levels for listeners with relatively small hearing loss whereas for listeners having severe to profound hearing loss, it increases. Thorpe et al. [28] used fast Fourier transform (FFT) and discrete Fourier transform as feature for classification of Speech. In this study, K-nearest neighbor was used in order to classify hearing impaired and achieved maximum 96% accuracy. Paulraj et al. [29] determined EEG based hearing threshold with the help of artificial neural network. In this study, Spatio-temporal domain features of three distinct bands were extracted from the recorded AEP signals. They achieved 95.6% accuracy with feed forward neural network classification method, however only ten subjects were used for data acquisition. Peiyuna et al. [30] used SVM model to analysis the signal of hearing-impaired patients using tongue movement. They used displacement, duration, maximum velocity, minimum velocity, mean velocity, standard deviation of velocity as the features. In this study, both the number of subjects used, and accuracy was low. Gautam et al. [31] proposed a real-time speech enhancement for hearing impaired listeners using smartphone with convolutional neural network classification method. In this study, log-power spectra (LPS) and Log melfilterbank energy features was used. The smartphone and the proposed algorithm offer an affordable portable platform that can be used by people with hearing loss. Zhang et al. [32] proposed a method to classify the auditory brainstem response and in this purpose, they combined wavelet analysis and Bayesian networks. Discrete wavelet transforms (DWT) was used as a feature extraction method and 78.80 testing accuracy was achieved with this feature. In this study, the accuracy is low compared to other studies. Moreover, they used only eight subjects in data collection which is not enough. Pei-Zhen et al. [33] proposed a twofold study in order to analyze the brain network for auditory disease. In this study, FBA (feature from brain area) was used to classify this signal with SVM classifier. Their accuracy is not so good. Rami et al. [34] used 2D photographs in order to identify the ear abnormality and in this purpose, CNN was used for classification these images. In this study the testing accuracy was 94.1% and they use a large number of subjects. Lirong et al. [35] analyzed sMRI and fMRI imaging data and invariant feature transform (SIFT) algorithm was employed to detect and describe the local features. The SVM classifier was used provide accurate disease markers for hearing impairment. Here, both the number of subjects and accuracy is low.

Thus, the existing AEP based BCI studies with its analysis procedure are summarized in Table 1.

Table 1. Summary of existing AEP based BCI studies.

Reference	Year	Data			Feature extraction	Classification	Performance
		No of subject	No. of electrode	No. of class			
[18]	2018	32	32	2	Global and nodal graph	SVM	Classification Accuracy = 85.71%
[19]	2019	–	–	2	Absolute values of BAEP	Multilayer perceptron (MLP)	Classification Accuracy = 90%
[23]	2018	180	–	3	Wavelet entropy (WE)	TS-PSO	The accuracy of the model = 86.17%
[25]	2019	20	8	2	ColmogEnt, SampleEnt, ApproxEnt	KNN, LDA, SVM, TREE, NB	Classification Accuracy = 100% (SVM)
[27]	2004	200	–	3	Wavelet packet transform (WP)	SVM	Detection rate for test data: NH group = 68.1% HF group = 74.7% PT group = 56.4%
[28]	2018	630	–	3	FFT and FT	K-Nearest Neighbour (K-NN)	Classification accuracy = 84% to 96%
[29]	2012	10	19	2	Spatio-temporal domain	Feedforward neural network	Classification accuracy = 95.6%
[30]	2018	22	–	2	Maximum, minimum, mean and standard deviation, displacement, duration	SVM and radial basis function network	Classification accuracy = 87.12% via (displacement, duration) using Random forest

(*continued*)

Table 1. (*continued*)

Reference	Year	Data			Feature extraction	Classification	Performance
		No of subject	No. of electrode	No. of class			
[31]	2019	–	–	–	Log-power spectra and Log Mel filter bank energy	CNN SE model on a smartphone processor	A platform with smartphone + proposed SE algorithm for hearing loss
[32]	2006	8	6	2	Discrete wavelet transform (DWT)	Bayesian network classification	Accuracy for Training data = 80.82% and testing data = 78.80%
[33]	2019	146	128	3	FFT	SVM	Regular control = 0.787%, Sudden deafness = 0.787% and Tinnitus = 0.787%
[34]	2019	671	–	2	Raw AEP data	Convolutional neural networks (CNNs)	Accuracy for Training data = 100% and testing data = 94.1%
[35]	2013	39	–	2	Invariant Feature Transform (SIFT)	SVM	AUC score of 0.90. Accuracy = 0.72 (sMRI), 0.74 (fMRI) and 0.87 (sMRI + fMRI)

4 Issues with Possible Solutions

4.1 Lack of Standard Auditory Stimulation Process

AEP based BCI data acquisition comprises of three phases: EEG headset, auditory stimulus, participant or subject. Several issues are address with the EEG headset which are required to be resolved. First and foremost, most EEG headsets need gel and liquid on electrodes that are incredibly inconvenient to the users. This problem can be overcome by dry electrodes. Some dry electrodes have already been available on the market however there are very few electrodes in these headsets and the technique is still not standardized. Thousands of devices are already produced for BCI applications by different companies. There are different electrodes (1–256) in different headsets, and these are not consistent with one another. Therefore, the number of electrodes for the AEP signal should be standardized. While the intended user of AEP-based BCI technology is hearing loss patients, data from healthy users with excellent expertise has been used in nearly all experiments. In addition, in some studies, the number of subjects is also quite inadequate. Therefore, experimental trials should be carried out for patients with hearing loss who currently need AEP dependent BCI technology and enhanced the number of experimental participants.

The BCI community should prescribe a threshold number of subjects including patients with healthy and hearing loss for the validation of the data as well as the system. It is obvious from the previous experiments that there is no standardized way to evoke stimulation of AEP responses. The authors have used many approaches to evoke AEP response stimulation. For example, loudspeakers or headphones for the generation of auditory stimulation in some research. Some researchers record the AEP signal when subjects listen various loud sound, pronunciation of words, music and story. Furthermore, the data collection environment is not explicitly recommended. It should be transparent and systematic on all these issues. If all BCI experiments based on AEP have the same norm, then it is very easy to compare and pick the best one.

4.2 Performance is Highly User Dependent

The signal quality of AEP is highly user dependent because the auditory signal elicited from the temporal area of the brain. Due to the different hearing ability of different subjects, the SNR of AEP signals are different. Moreover, non-stationary nature of AEP signal, it is quite impossible to collect the same pattern of AEP from the all individuals. Some other reasons may also affect the AEP signal like if the subject is fall into critical situation, depression or mentally not fit to generate a good signal. When the performance is highly subject dependent, it will create some difficulties for the AEP based BCI technology. If the system is not user independent, all the targeted users need to participate in the long training process. To overcome this issue, subject independent algorithm could be an excellent alternative which is already implemented in other BCI system.

4.3 Lack of Optimum Data Analysis Techniques

Since the majority of removal algorithms are efficient, there are different restrictions in current methods when used in a certain artefact-prone AEP response. Current algorithms for artefact removal concentrate on specific artefacts including eye blinking, EOG, ECG and EMG. For example, the removal algorithm for the artefact for eye blinking cannot be effective in minimizing an EOG artefact. Therefore, one of the potential goals of successful artefact mitigation is to create a single artefact removal algorithm that can minimize the AEP signal for all forms of artefacts. Furthermore, with the latest pattern of artefact removal, it can be inferred that machine learning and conventional methods for efficient automated artefact removal would be merged.

With regard to the techniques of feature extraction, a variety of features are extracted in order to obtain significant information from the AEP signal. The efficiency of FFT and its modified algorithms has been demonstrated reasonably encouraging when used in the AEP signal in previous studies. However, the ideal extraction technology for AEP signal may be defined too early.

SVM is the most robust classifier to characterize the high-dimensionality feature vector that has been noted from the previous studies. ANN will sometimes accommodate the high dimension feature set by evaluating the sequence of vectors. Low dimensional feature vector should be chosen to reduce computational complexity. However, in the case of high dimensional feature sets, feature reduction or selection algorithms may be considered. In order to classify AEP data, several approaches to deep learning have been implemented. The deep network architecture depends entirely on the layout of the network and the formalization of input. The existing studies indicate that deep learning approaches based on CNN outperform other methods of deep learning. Furthermore, when time series values or spectral images are used as inputs, CNN provided maximum accuracy.

4.4 Insufficient Study

Considering current scenery of AEP based study, it can be said that there are many scopes for AEP based BCIs. Very few studies have been conducted related to the AEP responses. Moreover, most of the conducted studies don't have depth analysis through which researchers may take decision. As the use of this technology has grown from medical to non-medical applications, extensive research related to the AEP response should be carried out.

5 Conclusion

Auditory evoked potential is one of the most promising developments for BCI technology, in particular for patients with impaired vision or for patients in LIS. This paper summarizes a variety of research that concentrate on the present state-of-the-art BCI auditory and explores its realistic implementation prospects. AEPs are required for the detection and assessment of moderate to extreme hearing loss. For the AEP response test, there is no anticipated form of special attention or cooperation by patients. A detailed evaluation is carried out by this article on the EEG-based AEP signal in BCI, specifically for investigating the methodological advantages and inconvenient and the necessary contributions needed for these fields of research. The data collection processes, control signals, feature extractions, selection and classification methods as well as performance evaluation metrics are analyzed for each BCI application. Lastly, it has discussed the possible complications of EEG-based AEP signal analyses in BCI system and proposed promising alternative solutions. The advancement of emerging AEP signal analysis can support auditory loss and paralysis patients. Since several outstanding developments has been addressed in BCI research, a number of issues still need to be resolved. A general standard for BCI should be declared to the BCI community, which is mandatory to follow by the researchers. Therefore, AEP is found to be an effective BCI tool for offering alternative means of contact with the outside world for LIS patients. This paper can provide valuable guidelines for helping the new researchers to implement a new as well as novel auditory system for physically impaired people.

Acknowledgment. This research is supported by Universiti Malaysia Pahang, Malaysia with the research grant RDU180396.

References

1. Paralysis: Definition and Patient Education. https://www.healthline.com/health/paralysis. Accessed 02 Oct 2020
2. Rashid, M., Sulaiman, N., Mustafa, M., Khatun, S., Bari, B.S., Hasan, M.J.: Recent trends and open challenges in EEG based brain-computer interface systems. In: Lecture Notes in Electrical Engineering, pp. 367–378. Springer (2020). https://doi.org/10.1007/978-981-15-2317-5_31
3. Wang, X.-Y., Jin, J., Zhang, Y., Wang, B.: Brain control: human-computer integration control based on brain-computer interface approach. Acta Autom. Sin. **39**, 208–221 (2013). https://doi.org/10.1016/s1874-1029(13)60023-3
4. Al-qaysi, Z.T., Zaidan, B.B., Zaidan, A.A., Suzani, M.S.: A review of disability EEG based wheelchair control system: coherent taxonomy, open challenges and recommendations (2018). https://doi.org/10.1016/j.cmpb.2018.06.012
5. Bi, L., Fan, X.A., Liu, Y.: EEG-based brain-controlled mobile robots: a survey. IEEE Trans. Hum.-Mach. Syst. **43**, 161–176 (2013). https://doi.org/10.1109/TSMCC.2012.2219046
6. Al-Nafjan, A., Hosny, M., Al-Ohali, Y., Al-Wabil, A.: Review and classification of emotion recognition based on EEG brain-computer interface system research: a systematic review. Appl. Sci. **7**, 1239 (2017). https://doi.org/10.3390/app7121239
7. Alariki, A.A., Ibrahimi, A.W., Wardak, M., Wall, J.: A review study of brian activity-based biometric authentication (2018). https://thescipub.com/abstract/jcssp.2018.173.181. https://doi.org/10.3844/jcssp.2018.173.181
8. Cattan, G., Mendoza, C., Andreev, A., Congedo, M.: Recommendations for integrating a P300-based brain computer interface in virtual reality environments for gaming. Computers **7** (2018). https://doi.org/10.3390/comput
9. Ramadan, R.A., Vasilakos, A.V.: Brain computer interface: control signals review. Neurocomputing **223**, 26–44 (2017). https://doi.org/10.1016/j.neucom.2016.10.024
10. Plourde, G.: Auditory evoked potentials (2006). https://pubmed.ncbi.nlm.nih.gov/16634420/. https://doi.org/10.1016/j.bpa.2005.07.012
11. Jewett, D.L., Williston, J.S.: Auditory-evoked far fields averaged from the scalp of humans. Brain **94**, 681–696 (1971). https://doi.org/10.1093/brain/94.4.681
12. Handbook of Auditory Evoked Responses - James W. Hall, III - Google Books. https://books.google.com.my/books/about/Handbook_of_Auditory_Evoked_Responses.html?id=lf56AAAACAAJ&redir_esc=y. Accessed 02 Oct 2020
13. Shanks, J.E., Wilson, R.H., Larson, V., Williams, D., Henderson, W.D.: Speech recognition performance of patients with sensorineural hearing loss under unaided and aided conditions using linear and compression hearing aids. Ear Hear. **23**, 280–290 (2002). https://doi.org/10.1097/00003446-200208000-00003

14. Deafness and Hearing Loss | Center for Parent Information and Resources. https://www.parentcenterhub.org/hearingloss/. Accessed 02 Oct 2020
15. Picton, T.W., John, M.S., Purcell, D.W., Plourde, G.: Human auditory steady-state responses: the effects of recording technique and state of arousal. Anesth. Analg. **97**, 1396–1402 (2003). https://doi.org/10.1213/01.ANE.0000082994.22466.DD
16. Hurley, A.: Human auditory evoked potentials (2012). https://doi.org/10.1097/AUD.0b013e3182498db9
17. Boston, J.R.: Spectra of auditory brainstem responses and spontaneous EEG. IEEE Trans. Biomed. Eng. **BME-28**(4), 334–341 (1981). https://doi.org/10.1109/TBME.1981.324801
18. Mahmud, M.S., Yeasin, M., Shen, D., Arnott, S.R., Alain, C., Bidelman, G.M.: What brain connectivity patterns from EEG tell us about hearing loss: a graph theoretic approach. In: 10th International Conference on Electrical and Computer Engineering, ICECE 2018, pp. 205–208. Institute of Electrical and Electronics Engineers Inc. (2019). https://doi.org/10.1109/ICECE.2018.8636698
19. Mosqueda Cárdenas, E., de la Rosa Gutiérrez, J.P., Aguilar Lobo, L.M., Ochoa Ruiz, G.: Automatic detection and classification of hearing loss conditions using an artificial neural network approach. In: Lecture Notes in Computer Science, pp. 227–237. Springer (2019). https://doi.org/10.1007/978-3-030-21077-9_21.
20. Profant, O., Škoch, A., Balogová, Z., Tintěra, J., Hlinka, J., Syka, J.: Diffusion tensor imaging and MR morphometry of the central auditory pathway and auditory cortex in aging. Neuroscience **260**, 87–97 (2014). https://doi.org/10.1016/j.neuroscience.2013.12.010
21. Study of functional connectivity in patients with sensorineural hearing loss by using resting-state fMRI. https://www.ncbi.nlm.nih.gov/pmc/articles/PMC4358486/. Accessed 02 Oct 2020
22. Vaden, K.I., Kuchinsky, S.E., Ahlstrom, J.B., Teubner-Rhodes, S.E., Dubno, J.R., Eckert, M.A.: Cingulo-opercular function during word recognition in noise for older adults with hearing loss. In: Experimental Aging Research, pp. 86–106. Routledge (2016). https://doi.org/10.1080/0361073X.2016.1108784
23. Tang, C., Lee, E.: Hearing loss identification via wavelet entropy and combination of Tabu search and particle swarm optimization. In: International Conference on Digital Signal Processing, DSP. Institute of Electrical and Electronics Engineers Inc. (2019). https://doi.org/10.1109/ICDSP.2018.8631839
24. Molina, M.E., Perez, A., Valente, J.P.: Classification of auditory brainstem responses through symbolic pattern discovery. Artif. Intell. Med. **70**, 12–30 (2016). https://doi.org/10.1016/j.artmed.2016.05.001
25. Ibrahim, I.A., Ting, H.N., Moghavvemi, M.: Formulation of a novel classification indices for classification of human hearing abilities according to cortical auditory event potential signals. Arab. J. Sci. Eng. **44**, 7133–7147 (2019). https://doi.org/10.1007/s13369-019-03835-5
26. Sanjay, H.S., Hiremath, B.V., Prithvi, B.S., Dinesh, P.A.: Machine learning based assessment of auditory threshold perception in human beings. SN Appl. Sci. **2**, 1–0 (2020). https://doi.org/10.1007/s42452-019-1929-7
27. Dietl, H., Weiss, S.: Detection of cochlear hearing loss applying wavelet packets and support vector machines. In: Conference Record - Asilomar Conference on Signals, Systems and Computers, pp. 1575–1579 (2004). https://doi.org/10.1109/acssc.2004.1399421
28. Thorpe, B., Dussard, T.: Classification of speech using MATLAB and K-nearest neighbour model: aid to the hearing impaired. In: Conference Proceedings - IEEE SOUTHEASTCON. Institute of Electrical and Electronics Engineers Inc. (2018). https://doi.org/10.1109/SECON.2018.8479223

29. Paulraj, M.P., Yaccob, S.B., Adom, A.H.B., Subramaniam, K., Hema, C.R.: EEG based hearing threshold determination using artifical neural networks. In: 2012 IEEE Conference on Sustainable Utilization and Development in Engineering and Technology, pp. 268–270 (2012). https://doi.org/10.1109/STUDENT.2012.6408417
30. Xue, P., Bai, J., Wang, Q., Zhang, X., Feng, P.: Analysis and classification of the nasal finals in hearing-impaired patients using tongue movement features. Speech Commun. **104**, 57–65 (2018). https://doi.org/10.1016/j.specom.2018.09.008
31. Bhat, G.S., Shankar, N., Reddy, C.K.A., Panahi, I.M.S.: A real-time convolutional neural network based speech enhancement for hearing impaired listeners using smartphone. IEEE Access **7**, 78421–78433 (2019). https://doi.org/10.1109/ACCESS.2019.2922370
32. Zhang, R., McAllister, G., Scotney, B., McClean, S., Houston, G.: Combining wavelet analysis and Bayesian networks for the classification of auditory brainstem response. IEEE Trans. Inf. Technol. Biomed. **10**, 458–467 (2006). https://doi.org/10.1109/TITB.2005.863865
33. Li, P.Z., Huang, L., Wang, C.D., Li, C., Lai, J.H.: Brain network analysis for auditory disease: a twofold study. Neurocomputing **347**, 230–239 (2019). https://doi.org/10.1016/j.neucom.2019.04.013
34. Hallac, R.R., Lee, J., Pressler, M., Seaward, J.R., Kane, A.A.: Identifying ear abnormality from 2D photographs using convolutional neural networks. Sci. Rep. **9**, 1–6 (2019). https://doi.org/10.1038/s41598-019-54779-7
35. Tan, L., Chen, Y., Maloney, T.C., Caré, M.M., Holland, S.K., Lu, L.J.: Combined analysis of sMRI and fMRI imaging data provides accurate disease markers for hearing impairment. NeuroImage Clin. **3**, 416–428 (2013). https://doi.org/10.1016/j.nicl.2013.09.008

Rotated TOR-5P Laplacian Iteration Path Navigation for Obstacle Avoidance in Stationary Indoor Simulation

A'qilah Ahmad Dahalan[1(✉)] and Azali Saudi[2]

[1] Centre for Defence Foundation Studies, National Defence University of Malaysia, Kuala Lumpur, Malaysia
a.qilah@upnm.edu.my

[2] Faculty of Computing and Informatics, Universiti Malaysia Sabah, Kota Kinabalu, Malaysia

Abstract. Lately, autonomous navigation has undergone constant research to develop superior path searching capabilities for self-directed navigation. This study attempt to enhance the path navigation competence of a mobile robot by using the computational technique, namely the Half-Sweep Two-Parameter Over Relaxation 5-Point Laplacian (HSTOR-5P) method, to solve pathfinding problems iteratively. The harmonic functions, which are the solution of Laplace's equation, are the sources for generating the potential function of the mobile robot's configuration space. As a result, the experiment shows that a smooth and optimal path in a given space is capable to create from any departure to the target point. Furthermore, it is proved that the HSTOR-5P approach was more effective than its predecessors in solving mobile robot pathfinding problems concerning iterations count and time taken.

Keywords: Mobile robot path finding · Optimal path · Collision free · Half-sweep technique

1 Introduction

These days, the robotic field is winning lots of places in our routine lives and in different fields of modern industrial and cyber-physical automation. This needs implanting intelligence inside the robots to ensure the optimal solution in the execution of any task. Such intelligence is mainly related to path planning and navigation. The robustness of this intelligence is one of the most challenging issues in robotics applications. In the interest of forming a fully self-directed mobile robot, it should be capable, efficiently and precisely, of constructing a path from departure to destination without bumping into any obstacles between them. Proficient algorithms for sorting out this kind of complexity have significant utilizations in the fields of computer animation, robotics manufacturing, architectural design, home maintenance, or security, defence, and surveillance [1].

This paper intends to simulate point-robot steering in the configuration space using numerical potential functions in accordance with the theory of thermal conduction [2]. Such thermal conduction paradigm causes a setting that has no local minima and its

J. A. Mat Jizat et al. (Eds.): iCITES 2020, AISC 1350, pp. 285–295, 2021.
https://doi.org/10.1007/978-3-030-70917-4_27

advantageous to the robot steering control. Throughout this analysis, the problem of thermal transmission is representation of Laplace's equation. Laplace's equation solutions are best recognized as harmonic functions, thus specifying the temperature values for the path generation model in the configuration space. Miscellaneous techniques have been employed for the acquisition of harmonic functions, yet mostly the general approach is the numerical technique, owing to the obtainability of rapid processing mechanisms and their ingenuity and competence in resolving the issue. A number of experiments have been performed in this study to examine the efficacy of accelerated iterative techniques used in producing paths of mobile robots for multi-size environments.

1.1 Path Planning Structure

Planning the route or path navigation in a closed environment engaged by a robot is able to be accomplished in a diversity of ways. Some of the approaches used to determine the path include reactive planning, landmark-based navigation, and countless path-planning algorithms. In a predefined environment, a path-planning algorithm can undeniably obtain a series of coordinates to be followed by a robot. Usually, as a way of evaluating different algorithms, a grid of a predefined size is made indicating where "passable" is on the region. It is fair to believe, for research, that all boundaries of the grid can be met by the robot.

Within the framework of this experiment, the concept of robot vehicle movement is simulated using a dot stirring in a predetermined environment. The problem of finding the direction of the robot is designated as a problem of steady-state heat transfer. The target is to be perceived in the analog of thermal conduction as a heat-drawing sink. The boundaries and barriers of the region are known as sources of heat and are set at constant temperatures. Through the thermal conduction manner, the temperature dispersal evolves, and the thermal flow lines streaming into the sink fill the configuration region. This situation can be viewed as a means of transmission between the target, obstacles/barriers, and the robot. The temperature dispersal in the environment subsequently is exploited as a director for the mobile robot to travel upon the beginning to the point of destination by following the thermal flux of high-temperature bases into the low-temperature spot in the region. The temperature dispersal of the environment is figured by engaging the harmonic function to model the configuration space arrangement.

Harmonic Functions. Arithmetically, in domain $\Omega \subset R^n$, a harmonic function is a function that fulfills Laplace's equation. For production of the robot path, the boundary of Ω contains of the internal and external border walls, all obstacles/barriers in the configuration region, initial points and target point.

$$\begin{aligned} \nabla^2\phi &= \sum_{i=1}^{n} \frac{\partial^2\phi}{\partial x_i^2} = 0 \\ x_i &= i\text{-th Cartesian coordinates} \\ n &= \text{dimension} \end{aligned} \tag{1}$$

The Laplace equation can be efficiently solved using a numerical technique where Jacobi and Gauss-Seidel (GS) are standard solutions to the problem. Since the harmonic functions comply with the min-max principle, Laplace's Eq. (1) is employed to restraint the formation of functions used, which prevents the formation of a false local minimum within the boundary field. Harmonic functions are convenient in robotic navigation since they contribute a thorough path-planning algorithm and the routes extracted out of it are typically smooth [3]. Laplace's equation was resolved numerically under traditional methods [4] i.e. Successive Over-Relaxation (SOR), GS, and Jacobi. Whereas Eq. (1) has been solved in this paper by means of an accelerated iterative scheme, Two-parameter Over-Relaxation (TOR), for rapid computation.

The potential field is measured in a global manner in the global approach to pathfinding problems. Harmonic solutions for the Laplace equation are utilized to map the trail lines for a robot to travel between any departure position up till a given target location without hitting any obstacles during movement. Obstacles/barriers are reflected as current sources. The target with the lowest allocated potential value is regarded to be the sink. It is similar to exploit Dirichlet boundary conditions. Afterwards, a sequence of points with lower potential values moving to the point with the lowermost potential value (i.e. target point) is discovered by executing the standard Gradient Descent Search (GDS).

In summary, this study seeks to follow the resemblance of temperature for the potential and the thermal stream for the trail line in illustrating the solution of Laplace's Eq. (1). The test is implemented on a two-dimensional (2D) domain with various types of obstacles/barriers, internal and external walls (see Fig. 3). The HSTOR iterative scheme is employed to address Eq. (1) and to gain temperature values at every nodal point. In addition, the Full-Sweep SOR (FSSOR), Full-Sweep Accelerated Over Relaxation (FSAOR), Full-Sweep TOR (FSTOR), Half-Sweep SOR (HSSOR), and Half-Sweep Accelerated Over Relaxation (HSAOR) were measured for efficacy assessment determinations.

2 Materials and Techniques

Let the equation of 2D Laplace's as set out in Eq. (1) be seen as

$$\frac{\partial^2 U}{\partial x^2} + \frac{\partial^2 U}{\partial y^2} = 0. \tag{2}$$

The approximation of Eq. (2) shall be simplified over the five-point second-order finite difference scheme as commonly referred as

$$U_{i-1,j} + U_{i+1,j} + U_{i,j-1} + U_{i,j+1} - 4U_{i,j} = 0. \tag{3}$$

For Laplace's Eq. (2), this iterative practice plainly involves swapping each node value repeatedly with the median of its four neighbors. These node values are kept fixed, reflecting the internal and external boundaries, obstacles/barriers, and the target point. Meanwhile, Eq. (3) is essentially a full-sweep (FS-) iteration, where the computation would consider all nodes in grid points. The technique of the half-sweep (HS-) scheme initially led by Abdullah [5] by means of the Explicit Decoupled Group (EDG) approach for addressing 2D Poisson equations. In addition, this scheme is useful for unraveling partial differential equations [6–8]. The improved version of this approach to solving the diffusion equation is also investigated [9].

2.1 Formulation of the Proposed Method

Rather than calculating entire points in the region, merely half of the points for the HS-method will be measured, thereby decreasing the overall computational time. Figure 1 demonstrates the use of the HS-technique (b) which only evaluates the inner points (black dot) in contrast with the FS-technique (a) that iterates all-inclusive points in the configuration area. In the field of robotics, the HS-practice was first implemented in [10]. Mainly, the iteration of the HS-technique is built on the five-points rotated of the finite difference approximation equation for resolving Eq. (1). The key feature of such an iterative manner is to lessen the complexity of computations by seeing just partially of the over-all node points. The mesh point for FS- and HS-iteration approaches is presented in Fig. 2.

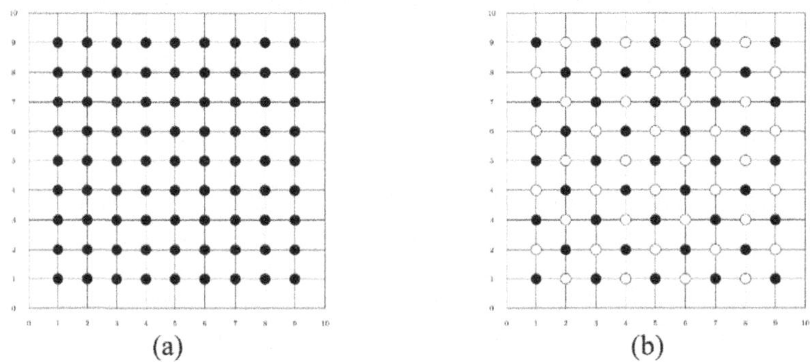

Fig. 1. The computational mesh of (a) FS- and (b) HS-approach, respectively.

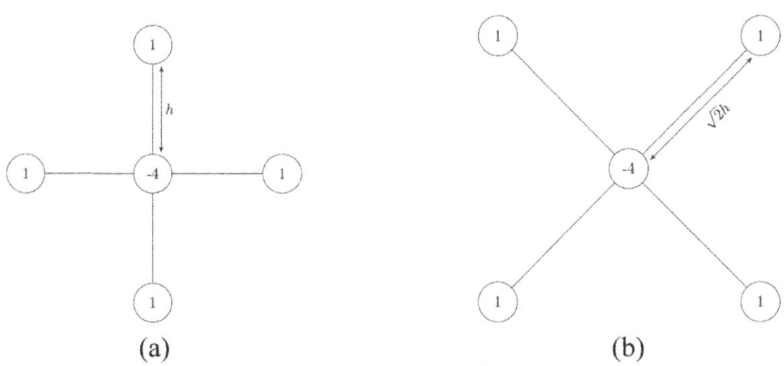

Fig. 2. The computational molecules of (a) FS- and (b) HS-approach, respectively.

As a matter of fact, the HS-approximation equation is in accordance with the cross-orientation operator that able to be achieved on turning the $i - j$ plane axis clockwise by 45°. This will lead to the formula of five-point approximation rotated as

$$U_{i-1,j-1} + U_{i+1,j-1} + U_{i-1,j+1} + U_{i+1,j+1} - 4U_{i,j} = 0. \tag{4a}$$

This scheme can be remodeled and set as

$$U_{i,j}^{(k+1)} = \frac{1}{4}\left[U_{i-1,j-1}^{(k+1)} + U_{i+1,j-1}^{(k+1)} + U_{i-1,j+1}^{(k)} + U_{i+1,j+1}^{(k)}\right]. \tag{4b}$$

When SOR is implanted into the Eq. (4a and 4b) by inserting an optimal parameter [11], the operation of the HS-iteration be presented as

$$u_{i,j}^{(k+1)} = \frac{\omega}{4}\left(u_{i-1,j-1}^{(k+1)} + u_{i+1,j-1}^{(k+1)} + u_{i-1,j+1}^{(k)} + u_{i+1,j+1}^{(k)}\right) + (1-\omega)u_{i,j}^{(k)}. \tag{5}$$

In order to derive the AOR method for this calculation as described in [12], the nodal points of $u_{i-1,j-1}^{(k+1)}$ and $u_{i+1,j-1}^{(k+1)}$ are interchanged to $u_{i-1,j-1}^{(k)}$ and $u_{i+1,j-1}^{(k)}$ respectively, while the terms $\frac{r\left(u_{i-1,j-1}^{(k+1)} - u_{i-1,j-1}^{(k)}\right)}{4}$ and $\frac{r\left(u_{i+1,j-1}^{(k+1)} - u_{i+1,j-1}^{(k)}\right)}{4}$ are appended. By dividing the optimal parameter ω from the Eq. (5) into two new optimal parameters, r and ω, the HS-scheme of the AOR iterative technique is given as

$$\begin{aligned} u_{i,j}^{(k+1)} &= \frac{r}{4}\left(u_{i-1,j-1}^{(k+1)} - u_{i-1,j-1}^{(k)} + u_{i+1,j-1}^{(k+1)} - u_{i+1,j-1}^{(k)}\right) \\ &+ \frac{\omega}{4}\left(u_{i-1,j-1}^{(k)} + u_{i+1,j-1}^{(k)} + u_{i-1,j+1}^{(k)} + u_{i+1,j+1}^{(k)}\right) + (1-\omega)u_{i,j}^{(k)} \end{aligned}. \tag{6}$$

The Half-Sweep Two-Parameter Over-Relaxation Iterative Scheme. The TOR iterative method, which is generalization of the AOR method, is implemented to enhance the convergence speed. The HSTOR scheme contains three different optimum relaxation parameters defined as r, r' and ω then written as

$$\begin{aligned} U_{i,j}^{(k+1)} &= \frac{r}{4}U_{i+1,j-1}^{(k+1)} + \frac{r'}{4}U_{i-1,j-2}^{(k+1)} + \frac{\omega}{4}\left(U_{i-1,j+1}^{(k)} + U_{i+1,j+1}^{(k)}\right) \\ &+ \left(\frac{\omega - r}{4}\right)U_{i+1,j-1}^{(k)} + \left(\frac{\omega - r'}{4}\right)U_{i-1,j-1}^{(k)} + (1-\omega)U_{i,j}^{(k)} \end{aligned}. \tag{7}$$

The ambiguous optimum values of all three r, r' and ω did not limit the minimum iteration counts. Ali and Pin [10] and Hadjidimos [13] have stated that the values r and r' are commonly selected to be as near as the value of ω for the corresponding SOR. All four possible acceleration constant in this study is defined in the range of [1, 2).

3 Experiments and Results

In this analysis, the environmental configuration involves four different sizes: 300 × 300, 600 × 600, 900 × 900 and 1200 × 1200, with four different situations (see Fig. 3). The target point was set at fixed and lowermost temperature values, although no particular temperature values are applied to all three starting points. Various obstacles with different shapes have been installed in the environment. Dirichlet boundary condition was applied in the initial setting, where the obstacles and walls were mounted at high-temperature values. All the other points were set to zero temperature value, except for the target point at the lowest temperature values.

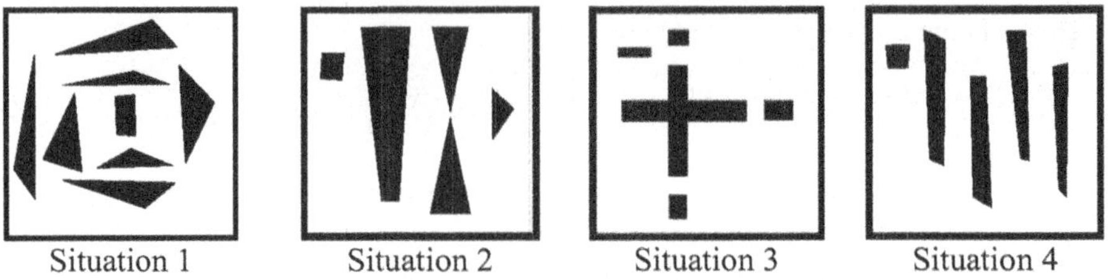

Fig. 3. Different situations in configuration space to navigate [14].

The computing process was executed with AMD A10-7400P Radeon R6, 10 Compute Cores 4C+6G operating at a speed of 2.50 GHz with 8 GB of RAM. The iteration process for numerically measuring temperature values at all points continues as long the condition of the stoppage is met. If the temperature values do not indicate any further changes, the loop will be ended where the variance of calculation values was really small, i.e. 1.0^{-15}. This high accuracy was essential in the solutions to prevent the development of a flat area, also called saddle points, from failing to generate paths.

Tables 1 and 2 express the iteration counts and the time taken, respectively, required for each numerical approaches in the experiment to analyze whole temperature values in the region. Obviously, the HSTOR iterative method has been confirmed to be really fast (in terms of iteration counts and time taken) when equated with the other proposed approaches.

For the purpose of simplifying, Fig. 4 displayed the line chart of iteration counts and time taken for Situation 4. The other 3 situations presented the same trend where it shows that HSTOR iterative approach illustrated the best results relative to the others in producing and finishing the path, for both iterations count and time taken. From the results table and the line chart, it can be inferred that almost and more than 50% reduction has been made by using HS-technique compared with the standard procedure.

3.1 Discussion

As soon as the temperature values have been gained, the path was created by implementing the steepest descent search upon the starting to the target spot. The progression of path generation was really quick, in which the algorithm simply picked the lowest temperature value of its adjacent points from the current point. This activity continues until the target point has been reached. Figure 5 illustrates that the paths were positively generated in an obstacle setting relying on the temperature dispersal outline gained by the numerical computation. Every starting point (square/green point) was successfully completed within specified target point (round/red point) and escaped all different obstacles that were set in the environment. The simulations only consider the known static two-dimensional indoor environments through Robot 2D Simulator [15].

Table 1. Outcome of the measured techniques for iteration counts.

	Methods	N × N			
		300	600	900	1200
Situation 1	FSSOR	1728	8117	17831	31346
	FSAOR	1591	7529	16594	28984
	FSTOR	1656	7815	17199	27895
	HSSOR	837	4108	9086	15892
	HSAOR	759	3803	8420	14768
	HSTOR	797	3949	8721	14234
Situation 2	FSSOR	2228	8776	19254	33558
	FSAOR	2006	7973	17538	30573
	FSTOR	1893	7553	16642	29008
	HSSOR	1071	4438	9813	17149
	HSAOR	944	4023	8924	15614
	HSTOR	877	3811	8461	14813
Situation 3	FSSOR	3624	14644	33004	57484
	FSAOR	3236	13165	29680	51738
	FSTOR	2843	11685	26393	46021
	HSSOR	1780	7445	16856	29418
	HSAOR	1568	6681	15149	26456
	HSTOR	1349	5909	13463	23523
Situation 4	FSSOR	2507	9868	21654	37762
	FSAOR	2288	9025	19840	34601
	FSTOR	2067	8217	18052	31519
	HSSOR	1212	5000	11036	19288
	HSAOR	1097	4555	10098	17670
	HSTOR	967	4141	9180	16085

Table 2. Outcome of the measured techniques for time taken (in second).

	Methods	N × N			
		300	600	900	1200
Situation 1	FSSOR	8.13	227.95	1134.25	3728.92
	FSAOR	8.61	230.17	1148.87	3692.74
	FSTOR	7.60	233.91	1188.08	3565.09
	HSSOR	2.39	81.24	404.15	1375.27
	HSAOR	1.72	73.76	369.91	1247.65
	HSTOR	2.55	84.84	413.84	1335.52
Situation 2	FSSOR	10.69	251.72	1270.23	4077.22
	FSAOR	10.27	248.24	1226.66	3976.33
	FSTOR	9.39	233.83	1194.50	3732.02
	HSSOR	2.95	86.77	445.70	1423.27
	HSAOR	2.75	76.79	403.25	1263.63
	HSTOR	2.70	82.42	401.42	1326.65
Situation 3	FSSOR	16.22	427.27	2190.45	7432.68
	FSAOR	18.66	418.45	2073.25	7254.02
	FSTOR	15.20	369.55	1927.30	6300.13
	HSSOR	5.16	154.79	783.72	2634.52
	HSAOR	4.80	137.18	721.94	2300.84
	HSTOR	4.30	135.81	661.90	2262.25
Situation 4	FSSOR	11.02	281.85	1441.47	4853.57
	FSAOR	12.52	281.78	1423.54	4743.21
	FSTOR	10.91	255.82	1292.23	4269.42
	HSSOR	3.58	102.16	510.22	1686.65
	HSAOR	3.08	92.44	471.17	1511.93
	HSTOR	2.99	93.87	458.45	1527.54

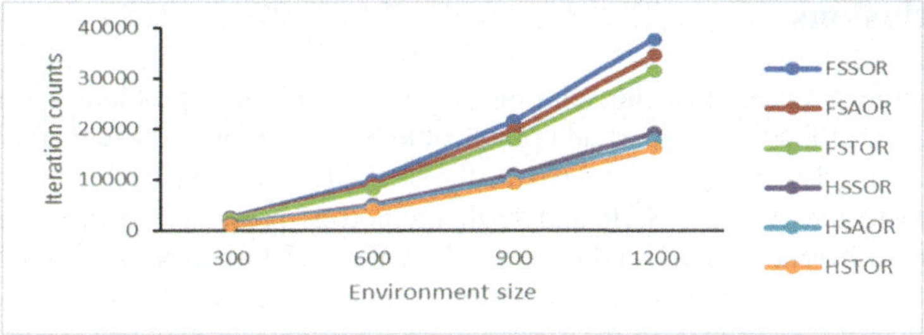

(a)

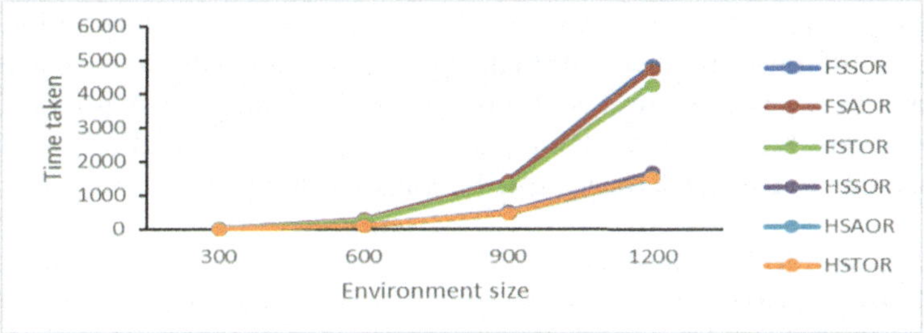

(b)

Fig. 4. The (a) iteration counts and (b) execution time in varying environment sizes.

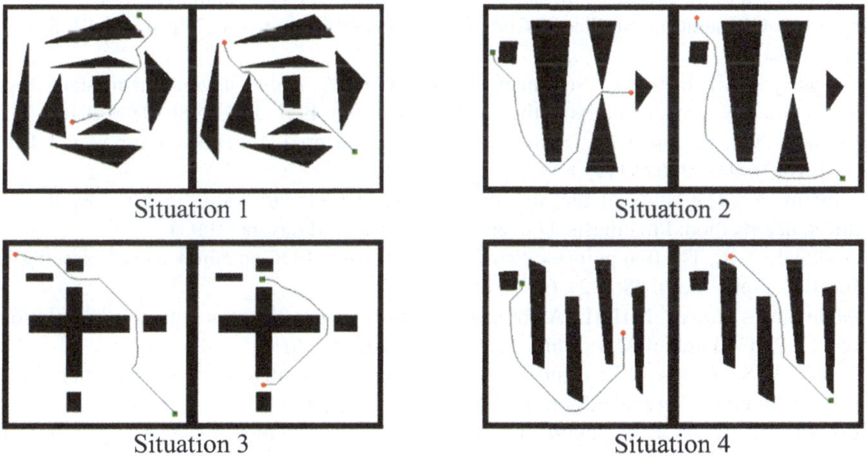

Fig. 5. The generated paths from several different start (square/green point) and target (round/red point) positions for various environment.

4 Conclusions

This experiment reveals that the solution of robot pathfinding problem via numerical approaches is, in fact, very smart and practicable by reason of the recent advanced and newly discovered techniques, along with the accessibility of fast machines today. In comparison to the standard SOR and AOR method as presented in the results table, the TOR iterative method verified to be faster with regard to iteration counts and time taken. An increment in the amount of obstacles/barriers does not negatively influence the results, as the computation process is simply getting faster as the occupied zones by obstacles are disregarded/overlooked during the calculation. The advantage of the developed algorithm is that, regardless of the obstacles in its shape and position/location, the robot can still travel from the start to the target point, not only safely, but also on the shortest path. In essence, the results of this study can be categorized as a family of FS- and HS-iterations. Further research on quarter-sweep [16, 17] iterations will boost the convergence rate of the standard suggested iterative techniques.

Acknowledgement. This research is fully supported by FRGS grant, FRGS/1/2018/ICT02/UPNM/03/1. The authors fully acknowledge Ministry of Higher Education (MOHE) and National Defence University of Malaysia for the approved fund which makes this important research viable and effective. The researchers declare that there is no conflict of interest regarding the publication of this study.

References

1. Koubaa, A., Bennaceur, H., Chaari, I., Trigui, S., Ammar, A., Sriti, M.F., Alajlan, M., Cheikhrouhou, O., Javed, Y.: Introduction to mobile robot path planning. In: Robot Path Planning and Cooperation. Studies in Computational Intelligence, vol. 772, pp. 3–12. Springer, Cham (2018)
2. Dahalan, A.A., Saudi, A., Sulaiman, J., Din, W.R.W.: Robot navigation in static indoor environment via accelerated iterative method. Adv. Sci. Lett. **24**(2), 986–989 (2018)
3. Connolly, C.I., Gruppen, R.: On the applications of harmonic functions to robotics. J. Robot. Syst. **10**(7), 931–946 (1993)
4. Ibrahim, A.: The study of the iterative solution of boundary value problem by the finite difference method. PhD thesis, Universiti Kebangsaan Malaysia (1993)
5. Abdullah, A.R.: The four point explicit decoupled group (EDG) method: a fast Poisson solver. Int. J. Comput. Math. **38**(1–2), 61–70 (1990)
6. Abdullah, A.R., Ali, N.H.M.: A comparative study of parallel strategies for the solution of elliptic PDEs. Parallel Algorithms Appl. **10**, 93–103 (1996)
7. Yousif, W.S., Evans, D.J.: Explicit de-coupled group iterative methods and their parallel implementations. Parallel Algorithms Appl. **7**(1–2), 53–71 (1995)
8. Dahalan, A.A., Saudi, A., Sulaiman, J., Din, W.R.W.: Autonomous navigation in static indoor environment via rotated Laplacian operator. In: AIP Conference Proceedings, vol. 1974, p. 0200352018 (2018)
9. Sulaiman, J., Hassan, M.K., Othman, M.: The half-sweep iterative alternating decomposition explicit (HSIADE) method for diffusion equation. In: Zhang, J., He, J.H., Fu, Y. (eds.) Computational and Information Science. LNCS, vol. 3314, pp. 57–63. Springer, Heidelberg (2004)

10. Saudi, A., Sulaiman, J.: Half-Sweep Gauss-Seidel (HSGS) iterative method for robot path planning. In: Proceeding of the 3rd International Conference on Informatics and Technology (2009)
11. Young, D.M.: Iterative methods for solving partial difference equations of elliptic type. Trans. Am. Math. Soc. **76**, 92–111 (1954)
12. Ali, N.H.M., Pin, F.K.: Modified explicit group AOR methods in the solution of elliptic equations. Appl. Math. Sci. **6**(50), 2465–2480 (2012)
13. Hadjidimos, A.: Accelerated overrelaxation method. Math. Comput. **32**(141), 149–157 (1978)
14. Shiang, F.C.: Collision-free path planning. Retrospective theses and dissertations, Iowa State University (1997)
15. Saudi, A.: Robot path planning using family of SOR iterative methods with Laplacian behaviour-based control. Ph.D. thesis, Universiti Malaysia Sabah (2015)
16. Othman, M., Abdullah, A.R.: An efficient four points modified explicit group Poisson solver. Int. J. Comput. Math. **76**, 203–217 (2000)
17. Dahalan, A.A., Saudi, A., Sulaiman, J.: Numerical technique for autonomous robot path planning based on QSAOR iterative method using indoor environment. J. Eng. Appl. Sci. **13**(20), 8414–8418 (2018)

Healthy Diet Food Decision Using Rough-Chi-Squared Goodness

Riswan Efendi[1(✉)], Dadang S. S. Sahid[2], Emansa H. Putra[2], Mustafa M. Deris[3], Nurul G. Annisa[1], Karina[1], and Indah M. Sari[1]

[1] Faculty of Science and Technology, UIN Sultan Syarif Kasim Riau, Pekanbaru 28293, Indonesia
riswan@uthm.edu.my
[2] Politeknik Caltex Riau, Pekanbaru 28265, Indonesia
[3] Faculty of Computer Science and Information Technology, UTHM, 86400 Batu Pahat, Malaysia

Abstract. Rough sets approximations have been implemented to handle categorical attributes from various domain problems. However, few studies investigate the combination between dependency degree of rough sets and chi-square test in decision support of the categorical data analysis. In this paper, we are interested to decide the healthy status based on healthy diet foods based on rough sets and chi-square test. To improve the decision accuracy, the data reduction of rough sets is also considered. The gathering data was taken from 20 adult persons using question list form. Results showed that dependency degree and chi-square test after data reduction is better than before. Therefore, the final healthy status (Good) is determined by combination food type, level and dependency degree = {(Food A; High; 50%), (Food; Less; 50%), (Food C; Normal; 44.4%) and (Food F; Less; 27.8%)}. In this case, the accuracy on decision making is really influenced by inconsistent information from the conditional attributes and objects in the data set, thus both can be solved by data reduction strategy. Additionally, the dependency degree is very helpful to support chi-square test for investigating relationship between conditional and decision attributes, especially limited or small data sizes.

Keywords: Rough sets · Dependency degree · Chi-square test · Data reduction · Healthy diet food · Decision making

1 Introduction

The rough set approach is frequently utilized with other methods in tuning on attributes of data set. The power of analysis using rough set are un-nosy and conspicuous to enhance data in non-numerical type. This condition is suitable to apply to real-life effectively [1]. Nowadays, many complex and uncertain problems can be solved using the rough set. Areas and studies that frequently use this method include medical diagnosis, pattern recognition, classification, intelligent control systems, and prediction purposes [1]. Implementation of the rough set approach in medical field, it can be used to help

J. A. Mat Jizat et al. (Eds.): iCITES 2020, AISC 1350, pp. 296–308, 2021.
https://doi.org/10.1007/978-3-030-70917-4_28

inexperienced doctors. This application mainly helps doctors in making diagnoses with a decision support system from clinical symptoms of disease [1–5]. For medical purpose, a decision support system can be used to check whether a patient requiring further investigation, consultation, or may be discharged.

Some decision support model have the potential to provide accuracy recommendation as good as medical experts [1]. There are existing rough set applications in medical diagnostic procedure to detect diseases [4, 5], such as, dengue [2], diabetes mellitus [2, 3], chikungunya [2], and other. Anyway, research related to the process and stages to determine the fit rules for application in medical diagnosis is still an interesting topic. This is because the ultimate goal of this application is to find and improve a more accurate level of accuracy.

Rough set method is useful in analysing database attributes and controlling process. In application research, Rough set is widely used for fault diagnosis, image processing, massive data processing, etc. [18]. Moreover, some integration studies between rough sets and statistical approaches are also implemented to fault diagnosis and attribute selection such as, medical diagnosis [1–5, 11–13], purchase decision [14], financial wellbeing [15], e-learning [16, 17].

The aim of this paper is to present the contribution data reduction strategy of Rough Set Theory (RST) for improving the dependency degree and chi-square approaches in decision support system. Furthermore, application both approaches is implemented to investigate the relationship between food groups and healthy status. Additionally, developing rules are also provided to predict healthy diet decision. This paper is organized as follows. In Sect. 2, the fundamental theories are presented. In Sects. 3 and 4, the proposed decision rules and implementation are discussed. The summary is concluded in Sect. 5.

2 Theoretical Background and Methodology

2.1 Rough Sets Theory (RST)

The methodology of RST is related to classification and analysis of inconsistent, imprecise, knowledge, incomplete information and it is considered one of the first non-statistical approaches in data analysis [6, 7]. Information system can also be name as information. Within the table there are objects and attributes. By using this system, knowledge representation in rough sets is done which can show the relationship between object and attribute value in tabular form as presented in Fig. 1.

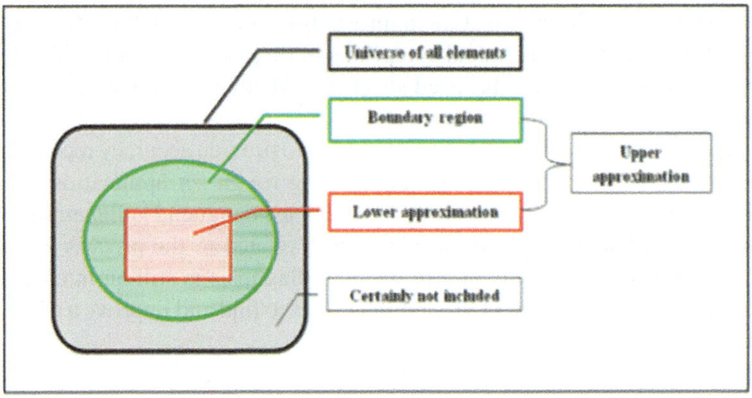

Fig. 1. RST and its information system

Indiscernibility relation is the main concept in rough set theory that may identifies a group of data. Indiscernibility relation is considered as a relation between multiple objects where all the values are identical in relation to a subset of considered attributes. The idea behind rough sets is to construct approximations of sets. The area of approximation consists of ordered pair U and R, where U is the universe that define as a finite and non-empty set of elements. R represents as indiscernibility relation which is an equivalence relation for U [1].

- Lower Approximation (B_*): The lower approximation (B_*) is the collection of objects which able to be classified surely as the members of the set A, by using the attribute of set B [1, 6, 7]
- Upper Approximation (B^*): The upper approximation (B^*) is the collection of objects that may possibly to be classified as the members of set A [1, 6, 7]
- Boundary Region (BR): The boundary region is the difference between the upper approximation and lower approximation. BR of X is written as [1, 6, 7]:

$$\mathrm{BND}(X) = B * (X) - B * (X). \tag{1}$$

- Decision Tables and Decision algorithms: The decision table consists of two types attributes which is condition attribute and decision attribute. Every row of the table determines a decision rule which specifies the decisions that should be taken when conditions are point out by condition attributes. The decision rules are always shown as implications in the form of **"If... Then..."** [1].
- Dependency of Attributes: Intuitively, the set of attribute D depends entirely on the attribute set C, denoted by, if the attribute value of C uniquely determines the attribute value of D. In other words, D depends entirely on C, if there is a functional dependence between the values of C and D. Hence, if and only if rule is true in A for every $x \in U$. D can depend partly on C. Formally such dependence can be defined in the following way. It is said that D depends on C to degree $k(0 \le k \le 1)$ and denoted as [8, 9]:

$$l = \gamma(D, C) = \frac{\sum_{x \in U/D} \left| \underline{C}(X) \right|}{|U|}; C, D \subseteq A \wedge C \cap D = \emptyset. \tag{2}$$

Note that:

If $l = 1$, D (Decision attribute) depends totally on C (Conditional attribute).

If $l < 1$, D (Decision attribute) depends partially (in a degree l) on C (Conditional attribute).

- Reduction of Attributes: The process of reducing an information system as the set of attributes of the reduced information system is independent and none of the attribute can be eliminated further. Some of the information from the system might be lose if the elimination continues [1].

2.2 Chi-Square Test

Chi-square test is one type of non-parametric comparative test which is carried out on two variables, where the data scale of the two variables is nominal. Chi-square test is one of the most widely performed non-parametric tests. The chi-square test is used to test the variance homogeneity of several populations. There are several other problems that can be solved by taking the benefits of the chi-square test, including: testing the proportion for multinomial data, the mean similarity of the Poisson data, independent between two factors in contingency, conformity. Between the observed data with the distribution model from which the data was thought to be taken, and the distribution model based on the observed data. The chi-square test can be formulated as follows [10]:

$$x^2 = \sum_{i=1}^{k} \frac{(O_i - E_i)^2}{E_i}. \quad (3)$$

Based on Eq. (3), x^2 is denoted as chi-square test, O_i is the i-th observation value and E_i is the i–th expected value. The computational steps of chi-square test are:

Step 1: State the hypotheses.

H_0: There is no significant relationship between the two variables.
H_a: There is a significant relationship between the two variables.

Step 2: Calculate chi-square (x^2) using Eq. (3).
Step 3: Determine the level of significance.
Step 4: Make a decision based on criteria:

If Sig. > 0.05 then H_0 is accepted.

Step 5: Write a conclusion.

3 Association Healthy Diet Food and Healthy Status

In this case study, the daily food consumed by a group of people is recorded. Table 1 shows the quantity based on group of food consumed by our experimental subject within a month. The columns of the table represent the attributes, which is conditional attribute and decision attribute. The rows represent our experimental subject. The entries in the table are the attribute values. From Table 1, where P are all of the objects or patients of the system, given set $P = \{p1, p2, p3, p4, p5, p6, p7, p8, p9, p10, p11, p12, p13, p14, p15, p16, p17, p18, p19, p20\}$, the set conditional attributes is represented by $F = \{$Food A, Food B, Food C, Food D, Food E, Food F$\}$ for Vegetable Group, Fruit Group, Carbohydrate Group, Dairy Group, Protein Group, Fat Group and the set G represented the decision attribute, where $H = \{$Healthy Status$\}$. For example, the patient p5 is characterized by the value set (Food A, Less), (Food B, Normal), (Food C, High), (Food D, Normal), (Food E, High), (Food F, Less) and (Healthy Status, Bad), which gives information about subject p5.

Table 1. Patient profile and food consuming

Patient code	Food A	Food B	Food C	Food D	Food E	Food F	Healthy status
p1	L	H	N	N	H	L	N
p2	H	L	H	N	N	L	G
p3	L	L	L	H	N	H	B
…	…	…	…	…	…	…	…
p20	L	N	N	L	H	H	B

From Table 1, the conditional attributes can be organized into the relations with their nominal values. In the Table 2, nominal values of attributes is shown as:

Table 2. Healthy diet attribute and its nominal values

Attributes	Healthy diet attributes	Nominal values of attribute
Conditional attributes	Food A, B, C, D, E, F	(L: Less: 1), (N: Normal: 2), (H: High: 3)
Decision attribute	Healthy status	(B: Bad: 1), (N: Normal: 2), (G: Good: 3)

From Tables 1 and 2, the indiscernibility (IND) elementary sets can be grouped for each attribute in Table 3. Each attribute is divided into 3 categories.

Table 3. Indiscernibility elementary sets for food group and healthy status

Conditional and decision attributes	Indiscernibility elementary sets
Food A	INDA ({Food A}) = {(Less), (Normal), (High) = {{p1, p3, p5, p8, p12, p13, p14, p20}, {p7, p10, p11, p16, p17, p18}, {p2, p4, p6, p9, p15, p19}}
…	…
Food F	INDA ({Food F}) = {(Less), (Normal), (High) = {{p1, p2, p5, p6, p11, p16, p19}, {p4, p9, p13, p15, p19}, {p3, p7, p8, p10, p12, p14, p18, p20}}
Healthy status	INDA ({Healthy status}) = {(Bad), (Normal), (Good) = {{p3, …, p20}, {p1, …, p18}, {p2, …, p19}}

Table 3 shows the registration object or element for each attribute based on set and sub-sets by following criteria given in Table 2.

3.1 Association Based on Chi-Square Test

Based on Sub-sect. 2.2, the association between attributes before data reduction using Chi-square test can be presented in Table 4.

Table 4. Chi-square test before data reduction

Attributes/variables	Value	df	Significance ($p < 0.05$)
Food A	6.000[a]	4	0.199 (Not significant)
Food B	4.578[a]	4	0.333 (Not significant)
Food C	4.095[a]	4	0.393 (Not significant)
Food D	5.333[a]	4	0.255 (Not significant)
Food E	5.333[a]	4	0.255 (Not significant)
Food F	18.032[a]	4	0.001* (Significant)

Table 4 shows the Food F (Fat Group) is very significant attribute if compared with other food groups. This indicates there is a strong association between Food F and Healthy status of patients. In this case, there is no elimination object and attribute, respectively. Moreover, the further association is also measured after data reduction as presented in Table 5.

Table 5. Chi-square test after data reduction

Attributes/variables	Value	df	Significance ($p < 0.05$)
Food A	4.082[a]	4	0.395 (Not significant)
Food B	5.569[a]	4	0.234 (Not significant)
Food C	5.247[a]	4	0.263 (Not significant)
Food F	13.454[a]	4	0.009* (Significant)

From Table 5, the most significant variable is also Food F if compared with Food A, B and C. In this Table, Food D, E attributes and also (p7, p18) patients already removed using data reduction of rough sets. Therefore, we are really confident to claim that Food F has a strong relationship with healthy status based on both tables given. Theoretically, Fat food groups are an essential part of our diet and important for good health. There are different types of fats, with some fats being healthier than others. To help make sure you stay healthy, it is important to eat unsaturated fats in small amounts as part of a balanced diet. When eaten in large amounts, all fats, including healthy fats, can contribute to weight gain. Fat is higher in energy (kilojoules) than any other nutrient and so eating less fat overall is likely to help with weight loss [20].

3.2 Association Based on Dependency Degree

By using Sub-Sect. 2.1 and data reduction results, the dependency degree between conditional and decision attributes can be shown in Table 6.

Table 6. Dependency degree between food type and healthy status

Conditional attribute		Decision attribute	Intersection between conditional and decision attributes	Dependency degree
Food type	Category	Healthy status		
Food A	Less	Bad	$\{p3, p5, p11, p13, p18\}$	0.5 (50%)
	Normal	Normal	$\{p16\}$	
	High	High	$\{p2, p6, p17\}$	
Food B	Less	Bad	$\{p3, p8, p9\}$	0.5 (50%)
	Normal	Normal	$\{p4, p12, p14\}$	
	High	High	$\{p6, p10, p15\}$	
Food C	Less	Bad	$\{p3, p8, p9\}$	0.44 (44.4%)
	Normal	Normal	$\{p1\}$	
	High	High	$\{p2, p6, p7, p10\}$	
Food F	Less	Bad	$\{p5\}$	0.278 (27.8%)
	Normal	Normal	$\{p4, p12, p16\}$	
	High	High	$\{p7\}$	

Table 6 shows the healthiest status (level) depends partially 0.5 (50%) to the Food A (vegetable group) and B (fruit group). Otherwise, (Food C, Food F) contribute (44.4%, 27.8%) to the healthy status. These percentages indicate that the patient's healthy level is very much influenced by food composition such as, vegetables, fruits, carbohydrates and fats. This result is also supported previous work that the adequate intake of carbohydrate and fats have a relationship to the nutritional status of employees [19].

3.3 Evaluation of Associations

In this section, both associations have been obtained from two different approaches (Sub-Sects. 3.1 and 3.2). The intersection is Food F attribute between Chi-square and dependency degree. However, the healthy status is also partially influenced by Food A, B and C, respectively. In this case, we consider dependency degree of rough sets to measure the relationship among attribute, while chi-square test is not fully appropriate to approximate this association because small data size issue.

4 Generating Rules for Healthy Diet Food Decision

4.1 Generated Rules Before Data Reduction

Based on Sub-Sect. 2.1, the object boundary region (BR) is p7, p9, p15 and p18. Hence, the patients p7, p9, p15 and p18 cannot be uniquely classified with available knowledge. Because they possess the same characteristics, but with differing conclusions differ in the decision attribute. After removing these patients, then equivalent information can be arranged in Table 7.

Table 7. The equivalent information after removing of the inconsistent information

Patient code	Food A	Food B	Food C	Food D	Food E	Food F	Healthy status
p3	L	L	L	H	N	H	B
p5	L	N	H	N	H	L	B
p10	N	L	L	L	H	H	B
p11	N	H	H	N	L	L	G
p16	N	H	N	N	N	L	G

Based on Table 7, the necessary decision rules that help to the determine healthy status can be generated by reduction of information shown as above. The rules are presented in Table 8:

Table 8. Proposed rule without attribute reduction

Rule No.	Proposed rule without attribute reduction
Rule 1	**If** patient consumed **Food E with level H (High)** and **Food F with level also H (High)**, **then** his/her **Healthy status is B (Bad)**
Rule 2	**If** patient consumed **Food E with level H (High), Food F with level N (Normal) or Food E with level N (Normal)**, and **Food F with level H (High)**, then **his/her Healthy status is B (Bad)**
Rule 3	**If** patient consumed **Food A with level H (High) and Food B with level H (High)**, then **his/her Healthy status is G (Good)**
Rule 4	**If** patient consumed **Food A with level H (High), Food B with level N (Normal)** or **Food A with level N (Normal), Food B with level H (High)**, then **his/her Healthy status is G (Good)**
Rule 5	**If** patient consumed **Food C with level H (High), Food D with level H (High)**, then **his/her Healthy status is N (Normal)**
Rule 6	**If** do not match with rules **R1, R2, R3, R4**, or match with multiple rules in contradiction result **except R5**, then **Healthy status is N (Normal)**

From Table 8, there are six generated rules for healthy diet food decision. In this table, we assumed that each criterion has two proposed rules such as, R3 and R4 can be used to decide the healthy criteria "Good".

4.2 Generated Rules After Data Reduction

Food E and F are removed from data set, while p7 and p18 are also eliminated from data set. In this section, we are interested to investigate the effect of removing conditional attributes and elements in rule development. The generated rules are presented in Table 9.

Table 9. Proposed rules with removing attributes and elements

Rule No	Proposed rule with attribute reduction and elements
Rule 1	**IF** Food A = L AND Food B = H AND Food C = N AND Food F = L **THEN** Healthy status → N
Rule 2	**IF** Food A = H AND Food B = L AND Food C = H AND Food F = L **THEN** Healthy status → G
Rule 3	**IF** Food A = L AND Food B = L AND Food C = L AND Food F = H **THEN** Healthy status → B
Rule 4	**IF** Food A = H AND Food B = N AND Food C = L AND Food F = N **THEN** Healthy status → N
…	…
Rule 15	**IF** Food A = N AND Food B = H AND Food C = N AND Food F = L **THEN** Healthy status → G
Rule 16	**IF** Food A = N AND Food B = L AND Food C = H AND Food F = N **THEN** Healthy status → N

4.3 Evaluation of Accuracy Rules Before-After Data Reduction

Generated rules from Sect. 4.1 and 4.2 are implemented to approximate the healthy status of patients. In Table 10, the predicted healthy status without removing the inconsistent attribute and with removing the inconsistent objects.

In this research, we also provide a special tool namely, RWR (Rough Wan Regression) application, this application is very useful to analyze categorical data based on rough-regression approach. Through this application, the inconsistent objects (p7, p18) and attributes (Food D and Food E) are removed from data set. The comparison between actual and predicted healthy status is illustrated in Fig. 2.

Figure 2 shows the accuracy of prediction on healthy decision is 88.88%. This percentage indicates that proposed rules could predict the better decision if compared with generated rules in Table 10. It caused by removing the inconsistent information from conditional attributes and also elements. While, the attribute reduction is not considered in Table 10.

Table 10. Actual vs predicted decision without removing inconsistent attributes

Patient code	Actual healthy status	Predicted healthy status
p1	Normal	Normal
p2	Good	Normal
p3	Bad	Bad
p4	Normal	Normal
p5	Bad	**Normal**
...	...	...
p19	Good	**Normal**
p20	Bad	Bad
	Accuracy	65%

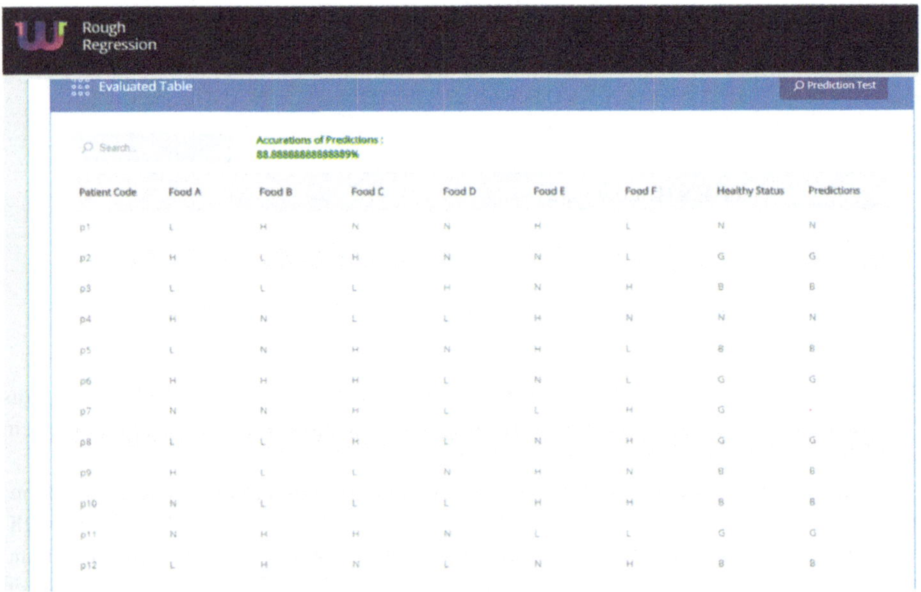

Fig. 2. Actual vs predicted healthy status based on RWR application

5 Conclusion

In this paper, we considered two different approaches in healthy diet decision. Firstly, Chi-square approach is the most frequently used to investigate the relationship between two categorical attributes/variables or more. Based on this approach, only Food F (Fat group) attribute has a strong association to the healthy status. This attribute always appears before and after data reduction. While, two conditional attributes are reduced using rough sets, namely Food D and E. Based on dependency degree, the rest of attributes partially contribute to the healthy decision included Food F. Therefore, there is a similarity food attribute between dependency degree and chi-square test which related to healthy status.

Through this paper, we adopted rough sets approximations, namely dependency degree to chi-square test. In the beginning stage, the removing the inconsistent information from elements or conditional attributes is considered. This type of information disturbed the decision attribute significantly. Statistically, chi-square approach is very significant to check the association between categorical variables, however it could not be used to measure the value of association. Thus, the dependency degree of rough sets is considered to measure the association above. In applications, both approaches support each other to provide the better decision. Our recommendation, the data size should be increased in terms of attribute and object numbers to show the performance both approaches for decision support system.

Acknowledgements. This work was supported by the Grant of Fundamental Research Scheme from Ministry of Research and Technology – National Research and Innovation Agency, Republic of Indonesia 2020.

References

1. Rissino, S., Torres, G.L.: Rough set theory-fundamental concepts, principals, data extraction and application. In: Ponce, J., Karahoca, A. (eds.) Data Mining and Knowledge Discovery in Real Life Applications, p. 438. I-Tech, Vienna, Austria (2009)
2. Thivigar, M.L., Richard, C., Paul, N.R.: Int. J. Inf. Sci. **2**, 33–36 (2012)
3. Ali, R., Hussain, J., Siddiqi, M.H., Hussain, M., Lee, S.: Sensors **15**, 15921–15951 (2015)
4. Tsumoto, S.: Stud. Comput. Intell. **708**, 139–156 (2017)
5. Tripathy, B.: Application of rough set based models in medical diagnosis. In: Handbook of research on Computational Intelligence Applications and Bioinformatics, pp. 108–118 (2016)
6. Pawlak, Z.: Theoretical Aspect of Reasoning about Data, 1st edn. Springer, The Netherlands (1991)
7. Pawlak, Z.: Int. J. Comput. Sci. **11**, 341–356 (1982)
8. Klosgen, W., Zytkow, J. (eds.): Handbook of Data Mining and Knowledge Discovery, pp. 328–333. Oxford University Press, Oxford (2002)
9. Herawan, T., Deris, M.M., Abawajy, J.H.: Knowl.-Based Syst. **23**, 220–231 (2010)
10. Daniel, W.W.: Applied Non-Parametric Statistic, 2nd edn. Georgia State University, Atlanta (1990)
11. Efendi, R., Deris, M.M.: Recent Advances on Soft Computing and Data Mining, pp. 243–251 (2018)
12. Efendi, R., Samsudin, N.A., Deris, M.M., Ting Y.G.: J. Phys. Conf. Ser. (2018)
13. Efendi, R., Samsudin, N.A., Deris, M.M.: ACM Proceeding on High Compilation, Computing and Communications, pp. 35–39 (2018)
14. Rasyidah, Nawi, N.M., Efendi, R.: Proceeding 7th ICCCE-IEEE Xplore, Kuala Lumpur, Malaysia, pp. 264–267 (2018)
15. Efendi, R., Mu'at, S., Arisandi, N., Samsudin, N.A.: Proceedings of the 2019 8th International Conference on Software and Computer Applications-ACM Proceedings, pp. 87–90 (2019)
16. Sahid, D.S.S., Efendi, R., Putra, E.H.: IOP Conf. Ser.: Mater. Sci. Eng. **732**, 1–7 (2020)

17. Sahid, D.S.S., Efendi, R., Putra, E.H., Wahyudi, M.: Int. J. Adv. Comput. Sci. Appl. **11**, 292–298 (2020)
18. Zhang, Q., Xie, Q., Wang, G., Trans, C.A.A.I.: Intell. Tech. **1**, 323–333 (2016)
19. Aritonang, E., Siregar, E.I.S., Nasution, E.: Int. J. Adv. Sci. Eng. Inf. Tech. **6**, 104–106 (2016)
20. https://www.eatforhealth.gov.au/food-essentials/fat-salt-sugars-and-alcohol/fat

Effect of Moisture Content on Crack Formation During Reflow Soldering of Ball Grid Array (BGA) Component

Syed Mohamad Mardzukey Syed Mohamed Zain[1], Fakhrozi Che Ani[1], Mohamad Riduwan Ramli[1], Azman Jalar[2,3](✉), and Maria Abu Bakar[2]

[1] Western Digital®, SanDisk Storage Malaysia Sdn. Bhd., Plot 301A, Persiaran Cassia Selatan 1, Taman Perindustrian Batu Kawan, MK13, Batu Kawan, 14100 Seberang Perai Selatan, Penang, Malaysia

[2] Institute of Microengineering and Nanoelectronics, Universiti Kebangsaan Malaysia, 43600 Bangi, Selangor, Malaysia
azmn@ukm.edu.my

[3] Department of Applied Physics, Faculty of Science and Technology, Universiti Kebangsaan Malaysia, 43600 Bangi, Selangor, Malaysia

Abstract. Semiconductor packaging assembly process involves many variables such as temperature, pressure, humidity, chemical interaction together with various type of materials. Such processes are prone to the occurrence of defects. It is essential to control process parameters in order to minimize product defects. This study is to investigate the occurrence of crack on the BGA component prior to the reflow soldering process with different moisture content. Different humidity test exposure periods have been used to induce moisture content. The moisture content of the BGA component increased up to 0.185% for 168 h. It is found that the crack observed along the epoxy mold compound after reflow soldering process of BGA component that exposed to humidity test.

Keywords: Crack · Epoxy mold compound · Humidity test · Reflow soldering · BGA component

1 Introduction

One of the established technologies in semiconductor packaging and assembly is ball grid array (BGA) technology. This normally involved a BGA component is assemble into a printed circuit board (PCB). Figure 1 shows a schematic of ball grid array component on printed circuit board. Solder ball is used to establish the joining between the BGA component and PCB via reflow soldering process. Among the popular lead-free solder ball have been used in electronic packaging are from Sn-based alloy such as Sn-Ag-Cu (SAC), Sn-Cu and Sn-Ag [1–3]. Typical solder ball diameter ranged from 0.15 mm up to 0.75 mm. Reflow soldering process is one of soldering process used in surface mount and technology (SMT) to established joining for various type of component and PCB [4, 5].

J. A. Mat Jizat et al. (Eds.): iCITES 2020, AISC 1350, pp. 309–314, 2021.
https://doi.org/10.1007/978-3-030-70917-4_29

Typical reflow soldering process involves from ramp to soak, preheat, ramp to peak, reflow and cooling process (Fig. 2) [6]. Solder joint provides electrical and mechanical connection for structural integrity and functionality of the components. Consequently, the reliability of the components or electronic packaging is always given attention. There are a lot of factors affecting the reliability of the electronic packaging. Among of the popular test normally being conducted to investigate the performance of overall packaging such as thermal cycle test, pressure cooker test, high temperature storage, to name a few [7, 8]. One of the typical reliability issues is crack either occur inside of the component or on the solder joint area. This paper investigates the effect of humidity test of the BGA component prior to reflow process. This involves effect of temperature and humidity simultaneously.

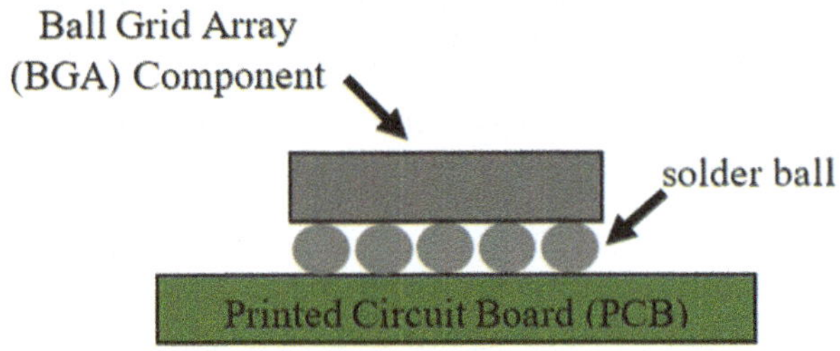

Fig. 1. Schematic of ball grid array component on printed circuit board

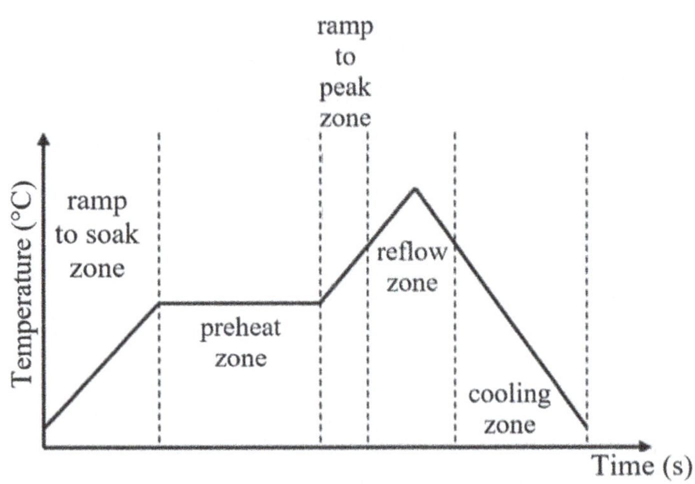

Fig. 2. Schematic of typical reflow soldering process

2 Materials and Method

An active BGA component with dimension of 17.0 mm × 17.0 mm × 0.8 mm were used in this study. This BGA component consist of silicon die, epoxy mold compound

(EMC) as encapsulant material, copper leadframe and solder interconnection. EMC is contributed about 90% of the total material for this BGA component. EMC consists of epoxy materials (30%) and filler materials (70%).

This BGA component were placed into a humidity chamber with temperature of 30 °C and 60% RH for 24 h, 48 h, 72 h, 96 h, 120h, 144 h and 168 h. BGA component without humidity test was used as controlled sample. After completion of humidity test on each period, the BGA component were taken out from the humidity chamber. The weight of the BGA component were measured before and after the humidity test. Next, all the BGA components (tested samples and controlled sample) were attached on the printed circuit board (PCB) using reflow oven (BTU Pyramax150 12Z) via reflow soldering process with maximum temperature of 250 °C.

Metallographic technique was adopted to prepare the cross-section of the sample. The sample were cut cold mounting process. The samples were mounted using epoxy resin and hardener. Next, the samples ground using silicon carbide (SiC) abrasive paper with grit size of 180, 240, 600, 800 and 1200. After completion of grinding process, the samples were polished using polishing cloth and Alumina colloid with particles size of 1 μm followed prior by 0.3 μm. Microstructure of the cross-section were observed using Field Emission Scanning Electron Microscope (FESEM).

3 Results and Discussion

Figure 3 shows the moisture absorption of the BGA component after humidity test at 30 °C with 60% RH from 24 h up to 168 h. It is clearly seen that the humidity test has an influence on the moisture content inside BGA component. The moisture content increase linearly up to 0.185% within the testing range time i.e. up to 168 h. In this study, the BGA component was exposed to 60% RH, this mean that the chamber possessed 60% water vapor as indicate by relative humidity. It is believed that moisture from the chamber were absorbed into the BGA component as shown by weight gain of the component. It is not clear the absorption route from the chamber into the BGA component, as the component fully encapsulated by the epoxy mold compound. In a typical semiconductor package, even the package is fully encapsulated, it has been reported that very tiny gap between materials and/or encapsulant interface exist. To some extent, delamination has also been reported [9].

This paper focused on the effect of the moisture level of the BGA component prior reflow soldering process. Reflow process is necessary to establish metallurgical interconnection between component and PCB. Therefore, BGA component with 'induced moisture' via humidity test has been introduced as shown in Fig. 3. This will make feasibility study on the effect of moisture content in the BGA component and its effect due to reflow soldering. Then it became point of interest to compare effect of reflow soldering between BGA component with and without humidity test. Figure 4 shows a FESEM image of the BGA component after subjected to humidity test at 30 °C with 60% RH for 168 h. It is clearly seen that crack were appear on the epoxy mold compound area.

Figure 5 show the comparison between controlled BGA sample (without humidity test) and sample which experienced humidity test for 168 h after reflow soldering. Figure 5(a) shows no occurrence of crack for controlled sample while (b) showing crack

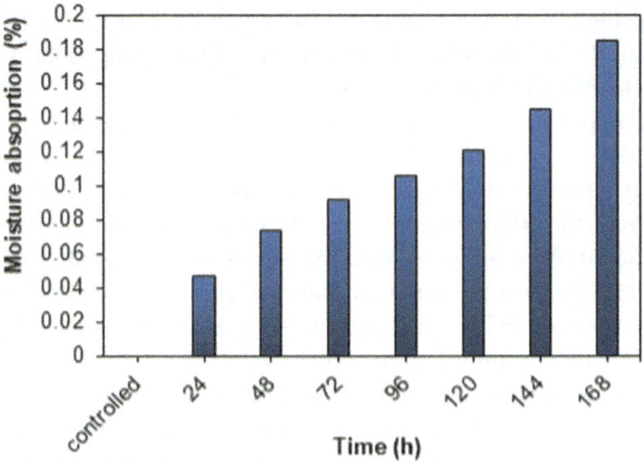

Fig. 3. Moisture absorption of the BGA component after humidity test at 30 °C with 60% RH from 24 h up to 168 h

as a line throughout the EMC and interface between EMC and passive component for sample after subjected to humidity test for 168 h. This indicates that the moisture plays significant role in crack promotion and propagation [10]. Detail studies need to be carried out to investigate how moisture content that trapped in the BGA packaging structure. Initial observation on the crack morphology suggests that moisture trapped induced internal stress during reflow soldering process. The reflow profile involve variation of temperature from low temperature and varied until exceeding melting temperature of solder ball. On top of that, various time with different temperature zone may also contribute to the activity of trapper moisture, leading to the stress concentration. From crack morphology, it is appearing that probably the interface between die and EMC is the initial point of moisture trapped activity before crack propagation. Therefore, a crack line is observed throughout the EMC.

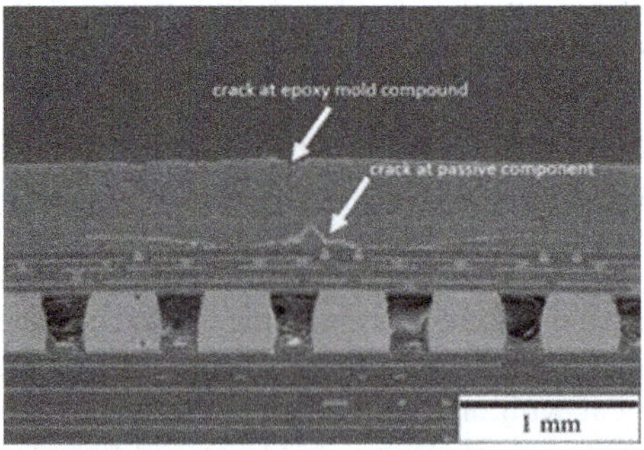

Fig. 4. FESEM image of the crack on the BGA component after subjected to humidity test at 30 °C with 60% RH for 168 h

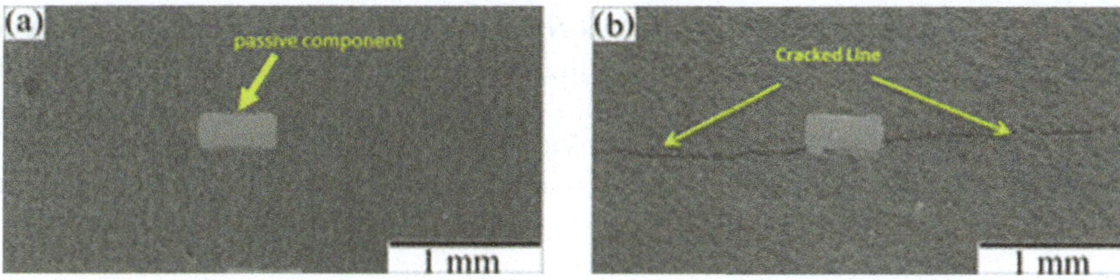

Fig. 5. FESEM images of (a) controlled sample (b) BGA component after subjected to humidity test at 30 °C with 60% RH for 168 h and underwent reflow soldering process

4 Conclusions

The humidity test carried out has been used in the BGA package to induce moisture content. In order to compare the structural integrity of the BGA component due to moisture content, reflow soldering was performed. The moisture content sample (after the 168 h humidity test) reveals signs of crack propagation in the EMC. Origin of crack is believed to be associated with moisture trapped that induced stress concentration.

Acknowledgement. The authors would like to acknowledge the financial support provided by Western Digital via SanDisk Storage Malaysia Sdn. Bhd through research grant (RR-2020-004) and collaboration activities with Universiti Kebangsaan Malaysia.

References

1. Ani, F.C., Jalar, A., Ismail, R., et al.: Reflow optimization process: thermal stress using numerical analysis and intermetallic spallation in backwards compatibility solder joints. Arab. J. Sci. Eng. **40**, 1669–1679 (2015)
2. Che Ani, F., Jalar, A., Ismail, R., et al.: Backward compatibility solder joint formation at high peak reflow temperature for aerospace applications. Arab. J. Sci. Eng. **41**, 1813–1823 (2016)
3. Li, Y., Sun, M., Ren, S., Ling, H., Hang, T., Hu, A.-M., Li, M.: The influence of non-uniform copper oxide layer on tin whisker growth and tin whisker growth behavior in SnAg microbumps with small diameter. Mater. Lett. **258**, 126773 (2020)
4. Ismail, N., Jalar, A., Abu Bakar, M., Ismail, R., Ibrahim, S.: Effect of flux functional group for solder paste formulation towards soldering quality of SAC305/CNT/Cu. Solder. Surf. Mount Technol. **32**(3), 157–164 (2020)
5. Che Ani, F., Jalar, A., Saad, A.A., Khor, C.Y., Abas, M.A., Bachok, Z., Othman, N.K.: Characterization of SAC – x NiO nano-reinforced lead-free solder joint in an ultra-fine package assembly. Solder. Surf. Mount Technol. **31**(2), 109–124 (2019)
6. Jalar, A., Bakar, M.A., Ismail, R.: Temperature dependence of elastic-plastic properties of fine-pitch SAC 0307 solder joint using nanoindentation approach. Metall. Mater. Trans. A **51**, 1221–1228 (2020)
7. Xu, L., Liu, S., Wang, M., Zhou, S.: Crack initiation and propagation mechanism of Al2O3-DBC substrate during thermal cycling test. Eng. Fail. Anal. **116**, 104720 (2020)
8. Ho, C.-Y., Lin, C.-W., Lee, Y.-Y., Cheng, S.-C.: Interfacial evolution and mechanical properties of Au–Sn solder jointed Cu heat sink during high temperature storage test. Mater. Lett. **275**, 128103 (2020)

9. Kwatra, A., Samet, D., Rambhatla, V.N.N.T., Sitaraman, S.K.: Effect of temperature and humidity conditioning on copper leadframe/mold compound interfacial delamination. Microelectron. Reliab. **111**, 113647 (2020)
10. Jansen, K.M.B., Zhang, M.F., Ernst, L.J., Vu, D.-K., Weiss, L.: Effect of temperature and humidity on moisture diffusion in an epoxy moulding compound material. Microelectron. Reliab. **107**, 113596 (2020)

Objective Tool for Chili Grading Using Convolutional Neural Network and Color Analysis

Yap Soon Hing, Wong Yee Wan, and Hermawan Nugroho(✉)

Department of Electrical and Electronic Engineering, University of Nottingham Malaysia, Jalan Broga, 43500 Semenyih, Selangor, Malaysia
hermawan.nugroho@nottingham.edu.my

Abstract. Currently chili is classified manually by farmers into Grade A or B. This manual sorting process is prone to errors. Farmers usually sell Grade A, and set aside Grade B. In this paper, the aim is to develop an objective tool which can perform chili grading and determine the level of ripeness of the chili. Chili's image is first captured by the camera and then is classified into either grade A or B. Next, the chili, categorized as A, is further processed to calculate its ripeness level. A Convolutional Neural Networks (CNNs)-based method is employed in grading the chili. A combination of LAB color transformation and K-Means Clustering is developed to determine the ripeness level of the chili. Data used in this project are provided by a chili farm located in Nilai, Malaysia. Results from the experiments, show that the proposed method can accurately classify the chili with 86% accuracy and determine its ripeness objectively.

Keywords: Convolutional Neural Network (CNN) · RGB · CIELAB

1 Introduction

Chili, considering its taxonomy, is from the genus Capsicum which are members of the nightshade family (Solanaceae). In Malaysia [1], chili or chili pepper are the main ingredients of several dishes. It is usually used in cooking to increase the spiciness level of the dish and is a crucial supply of vitamins [2]. The hottest chili is the Bhutanese Chili which has over 1 million Scoville units.

Before releasing the chili into market, chili is classified manually by farmers into grade A and grade B. Figure 1 shows how currently chili is graded by farmers in Malaysia. The illustration shows how farmers measure the length of the chili. To differentiate between grade A and B, farmers consider the length, the shape and the ripeness level of the chili. For example, grade A is usually defined as chili which length is more than 4 inches, the shape is straight, and it is almost ripe. If these conditions are not fulfilled, the chili is classified as grade B. Since the process is highly subjective, it is prone to errors. In addition, the process is repetitive and could be time-consuming. Manipulation of chili grade may also occur as grade A and B have different price. The manipulation

J. A. Mat Jizat et al. (Eds.): iCITES 2020, AISC 1350, pp. 315–324, 2021.
https://doi.org/10.1007/978-3-030-70917-4_30

can affect the quality of chili in the market and inflict a financial loss to both farmers and customers [3].

To improve the quality of chili in the market, we propose the use of an artificial intelligence (AI) approach which can objectively determine and classify chili. An Android-based application can then be developed for farmers to take real-time images using mobile phone equipped with the AI program. This will enable farmers and collection centers better chili grading accuracy and efficiency.

Fig. 1. (a) Current manual chili grading process by farmer; (b) Conventional way where the length of chili is measured

2 Related Work

Indrabyu et al. propose a system for detecting and counting the ripe chilies on trees. In their proposed system, the images are taken in gardens using a digital camera of 18.0 Mega Pixels resolutions. Data are collected from a chili garden in Soppeng, Indonesia. There are 180 images included in the study. The algorithm steps are preprocessing, segmentation, detection, and ripe chili counting process. They perform RGB to HSV color transformation, thresholding, image morphology, and to count the chili, blob analysis method is employed. It is reported that the proposed system can achieve 89.7% accuracy [4].

Padariya et al. develop a convolutional neural networks (CNN)-based method to classify dry chilies. The classification is developed based on the color image analysis of the dry chilies. It is reported that their methods can achieve precision between 65 and 75%, for separate class tests [5].

Fitri Nur Janati et al. Build an image capturing box for chili analysis. The box is developed to ensure that the analysis process is objective. It is reported that the developed box can achieved an average score of 4.6 from the calculated RGB values (Each RGB range is scored from 1 to 5 depending on the ideal value for each colour) indicating a promising result [6].

Md Khurram Monir Rabby et al. conduct research using image processing procedure to identify two different classes and types of fruits. In their study, the proposed method recognizes fruit by extracting color and shape feature. A modified Canny Edge Detection (MCED) algorithm is used as a fruit recognition method. There are two fruits involved in the study, namely apple and orange. The proposed method achieves high accuracy and it is reported that it is computationally easy to implement within limited training dataset [7].

Research on counting the number of fruits on a plant is conducted by Y. Song, et al. They propose a method to recognize and count fruits from images in cluttered greenhouses. There are a total of 28,000 color images of over 1000 experimental plants involved in the study. A two-step method is proposed to locate and count pepper fruits. The first step is to find fruits in a single image using a bag-of-words model, and the second step is to aggregate the estimated features from multiple images using a novel statistical approach. This proposed method achieves 74.2% accuracy without any linear adjustment for a large dataset [8].

We found that most of the methods are not suitable for our objectives. A different approach is required for not only to classify the chilli into different grades, but also to determine its ripeness. Therefore, in our work, different methods are applied.

3 Proposed Method

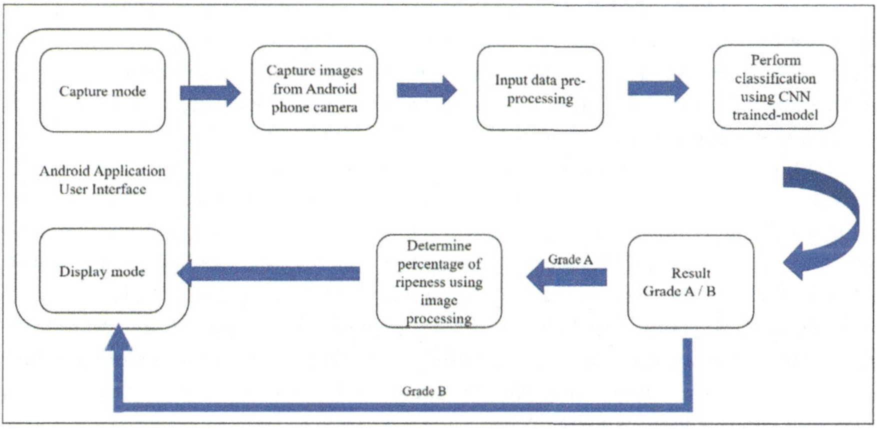

Fig. 2. Proposed method

Figure 2 shows the block diagram of the method. The proposed method consists of a chili detection module, a chili classification module and ripeness determination module. Grade B will not be considered in ripeness determination module. Firstly, in the Android-based application, the method will capture the chili images from the camera and the system will then detect the chili. Next, it will classify the chili using CNN-based model. The classification result will then send to display mode showing the result, if it is classified as grade B. If it is classified as grade A, the system will calculate the percentage of ripeness and send the result to display mode.

3.1 Chili Classification

In this report, we will focus on the classification and the ripeness determination modules. To detect the chili, the transfer learning method of TensorFlow Object Detection API is used [9]. Using this API, the training model is fed with the prepared dataset. Each image in the dataset is labeled, with a total of 450 chili categories and attributes. These will ensure that the system is able to recognize the chili in different angles.

The output from chili detection module will next be classified by the chili classification module. In the development of the chili classification module, two models are examined and compared in classifying the chili. The first model used is based on the MobileNet trained model [10] and the second model is developed based on CNN layers design inspired by VGG architecture [11]. The MobileNet model is considered since the system is planned to be run on mobile application. The pre-trained MobileNet model can be extracted from TensorFlow Hub. As most modules in the TensorFlow Hub, the pre-trained MobileNet model can be accessed to its internal layers, indicating that it can be used based on different cases [8]. Using the pre-trained model, we train the model using the training images of grade A and B. The size of the input images is fixed to 224-by-224 pixels.

The second model is shown in Fig. 3. Figure 3 shows the proposed CNN layer design as the chili classification module. In the design, convolution 2D (conv2D) has four parameters, namely x, filters, kernel_size and activation, and maxpooling 2D (max_pool2d) has two parameters, which are x and kernel_size. For the Conv2D, the parameters refer to the number of filters of which the convolutional layer will gain knowledge of. It indicates that layer closer to the input data will learn less with less filters required whereas a deeper layer will require more to learn.

To find out the perfect number of convolution blocks to be used, we examine different settings of the layers. For example, 2 convolution blocks will be used first to see the training accuracy and after that 3,4,5,6 convolution blocks will be examined and compared. The typical of value for kernel_size is 1, 3 and 5. In our design, kernel_size of 5 is selected because the the size of input is found to be greater than 128-by-128 pixels. This will enable the convolution layer to learn using larger spatial filters and reduce its size. As part of the experiment, we increase the number of epochs to examine whether the training accuracy is improved. We also increase the size of the layer to observe and discover the perfect layer for the design.

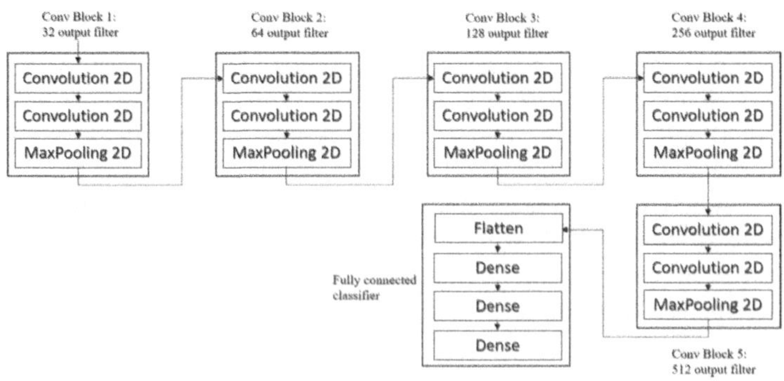

Fig. 3. Proposed CNN layer design inspired by VGG architecture

3.2 Determining the Percentage of the Ripeness of the Chili

The ripeness level of chili can be defined as "how red the chili is" [12]. Before the chili is fully grown (ripe), it is usually green in color. The formula is given as below,

$$Ripeness\ of\ chili = \frac{\text{Area of red color in pixel}}{\text{Total pixel of chili}} \times 100\% \tag{1}$$

Colors are generally represented in 3-dimensional color spaces. In image processing, we use few models to define the color, such as RGB color model and HSV color model. Human eye can see the color due to its receptors which are combinations of the primary colors of red, green, and blue (RGB). In RGB color mode, pixels of each color layer normally have intensity values ranges from 0 to 255. Taking into consideration its combination, there will be many types of red colors such as maroon, crimson, firebrick, and many others, which have different RGB values. Therefore, in the study, it is important to find the ranges of red color considered as ripe for chili.

International Commission on Illumination (CIE) defines a color space known as CIELAB. L color space represents Lightness, A color space represents the spectrums of green to red and B represents the spectrums of blue to yellow. In CIELAB, colors can defined in the "ab" plane whereby a specific color is defined by a triplet (L, a, b) [13]. In the LAB color code, the pure color can be viewed by the pairs (a, b) where L value indicates how bright that color is if it is seen by the human eye. + a represents the red value of the image while -a can represent green value of the image. + b represents yellow value and -b can represent blue value. Using the CIELAB color code, we can find the range of the color of chili areas which are considered as not ripe and ripe areas.

After color conversion to CIELAB, K-means clustering algorithm is employed to group and cluster pixels of the image into a few groups [9]. The aim of this algorithm is to separate and group *n* data points into *k* clusters. A cluster will be allocated to each of the *n* data points based on its nearest mean, while the center of the cluster is defined as its means/average. The "*k*" represents the number of groups or clusters. In our system, the color of an image mostly is red, and green with some background color.

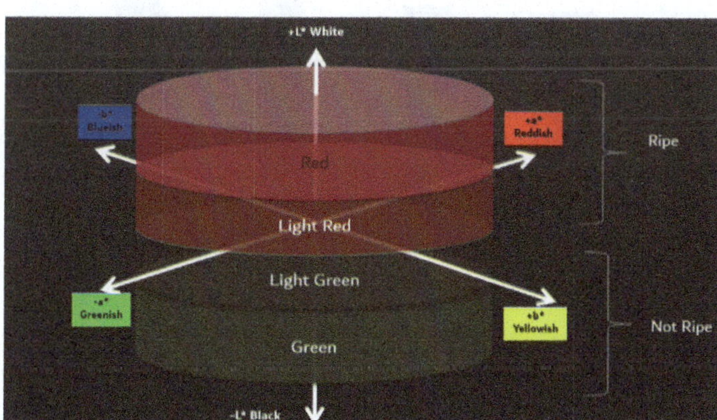

Fig. 4. The ripe and not ripe color in CIELAB color model

From experiment, the most suitable value of k to be used is 5. Figure 4 shows the color ranges and its definition based on K-means clustering.

4 Result and Discussion

For this development study, data are collected from a chili farm located in Nilai, Malaysia. To standardize the data, chili images are captured from 20 cm height with a white color background. The images used for training are 80 images of grade A and 94 images of grade B. The test set consists of 16 images. The TensorFlow Object Detection API used to train for the chili detection module i has an accuracy of 95%.

For the chili classification module, the first model (MobileNet) has only 41.7% accuracy. So this method is not considered for the study. The second model based on the proposed CNN layer design are then examined with different settings [14, 15]. Figure 5 shows the performance of different convolution layers applied into the CNN design. As shown in Fig. 5, there are no difference in accuracy can be seen between 2 layers to 4 layers with 60 epochs. With 5 layers, we observe an increase in accuracy. Adding more layers however does not always improve the system's accuracy since the idea of adding layer is to get more feature information. Nevertheless, there is a limit on how many layers we can add as it may cause overfitting. Overfitting can lead to error such as higher rates of false positive. Therefore, it is important to find the balance between number of layers used and its accuracy [16]. From Fig. 5, we can conclude that in our case, a set of 5 convolution layers and 1 fully connected layer is the best possible solution as it can achieve an accuracy as high as 93% from the training set. When we run the algorithm on the test data, the accuracy is dropped to 86%. Figure 6 shows examples of classification results.

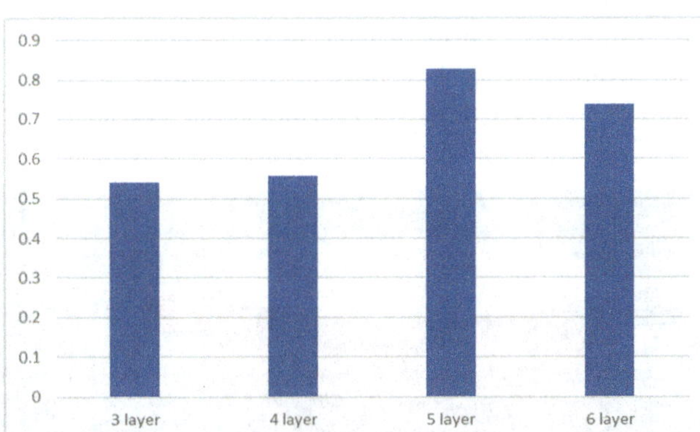

Fig. 5. Accuracy of different layers at 60 epochs

For determining the percentage of the ripeness of the chili, we first sample areas of not ripe and ripe chili randomly. To standardize the process, the images are processed with white balancing based on Python PIL and Numpy built-in functions [17]. From Fig. 7, we find that the chili areas considered as ripe have color ranges of + 20 to + 65,

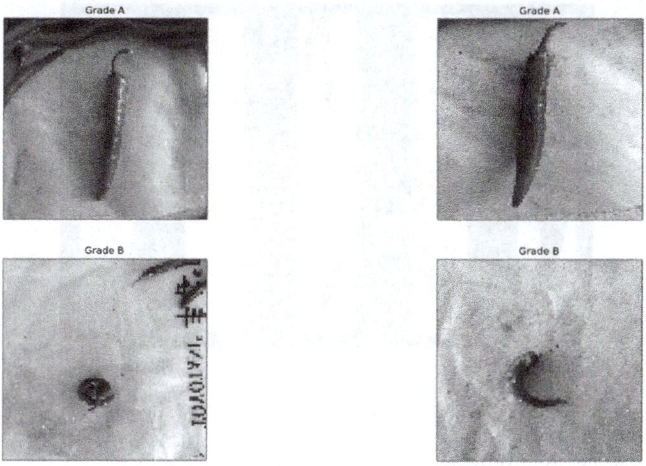

Fig. 6. Classification/grading result

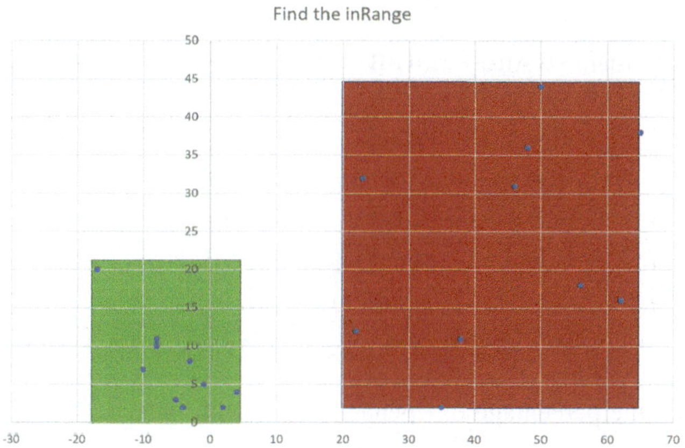

Fig. 7. Color ranges of ripe (red) and not ripe (green) chili in Lab color code

while the not ripe areas have color in the ranges of −17 to 4. Average values of samples areas are shown as blue dots in the Figure. For the calculation, any values outside these ranges will be ignored. Next, we cluster the image into regions/areas with different colors. These regions represent the pixels value with its respective color. Figure 8 shows example of chili are clusterd into 5 color regions.

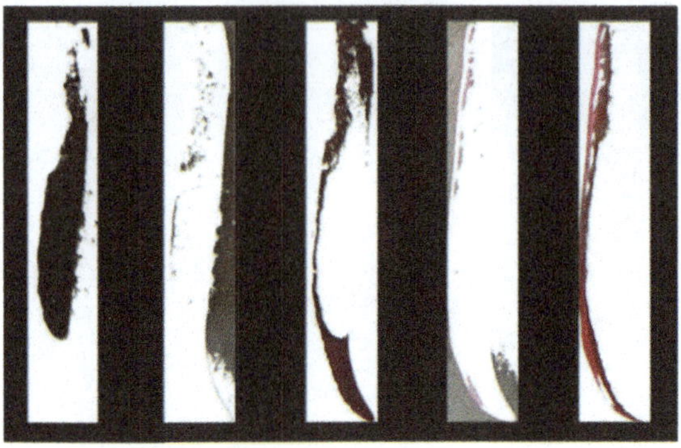

Fig. 8. Chili image clustering

The cluster areas calculation of Fig. 1 are shown in Table 1.

Table 1. Clustering result

Cluster	%Area (pixels)	R-G-B	Lab	Category
1	10.3	18-11-6	3-2-3	Green
2	9.1	71-59-50	26-5-8	–
3	19.07	57-17-16	12-20-11	Red
4	8.9	158-145-146	61-5-1	–
5	15.15	157-68-79	41-39-13	Red

It shows that the percentage of ripeness of this chili is 54.64%. The percentage is calculated using Eq. 1.

$$Ripeness\ of\ chili = \frac{Area\ of\ red\ color\ in\ pixel}{Total\ pixel\ of\ chili} \times 100\% \tag{2}$$

$$= \frac{19.07 + 15.15}{10.3 + 9.1 + 19.07 + 15.15 + 8.9} \times 100\% = 54.64\%$$

Table 2 shows the ripeness calculation of test data (from 16 test images, we have 10 grade A chilis and 6 grade B chilis).

As shown in Table 2, we can find differences of ripeness level determination estimated by farmers and by our proposed system. This is following our expectation since the judging process of the farmer is subjective while ours can objectively determine the ratio of ripe areas of the chili. Further analysis on the accuracy of the ripeness level calculation will be explored in the future work.

Table 2. Comparison result between estimation by farmer and by program

No	Ripeness estimated by farmers (%)	Ripeness estimated by Program (%)
1	Between 90%–100%	100%
2	Around 80%	76.13%
3	Around 80%	88.34%
4	Around 80%	73.22%
5	Around 70%	54.64%
6	Around 70%	31%
7	Around 70%	45.52%
8	Between 90%–100%	95.67%
9	Around 80%	83.49%
10	Around 70%	64.86%

5 Conclusion

Chili is an important ingredients in Malaysia dishes. Usually in chili farm, chili is classified into two grades, grade A and grade B. The classification process is conducted manually by the farmer. Hence it is prone to error as the decision and the judgement are based on the farmer's experience which is subjective. In this project, we propose an objective tool based on AI which can objectively classify the chili and determine the percentage of the ripeness of the chili.

Convolutional Neural Networks (CNNs) based method is developed to classify the chili. Results show that the method is able to achieve 86% accuracy. A color analysis based on CIELAB color space and k-means clustering are also developed to determine and segregate chili surface areas into ripe and not-ripe regions/areas. Results from the experiments show that the proposed system is able to objectively determine the percentage of the ripeness of the chili. We can conclude that our proposed approach is promising to be used as an objective tool in assessing chili grades.

References

1. Noor, N.M., Zakaria, L.: Identification and characterization of Colletotrichum spp. associated with chili anthracnose in peninsular Malaysia. Eur. J. Plant Pathol. (2018). https://doi.org/10.1007/s10658-018-1431-x
2. Chomchalow, N.: Spice Production in Asia - An Overview * Spices in Asia (1996, unpublished)
3. Mariyono, J.: Marketing channels of chili in central java: players, levels, and segmentations. J. Econ. Bus. Account. Ventura **22**, 167–176 (2019). https://doi.org/10.14414/jebav.v22i2.1688
4. Indrabayu, Mar'atuttahirah, Areni, I.S.: Automatic counting of chili ripeness on computer vision for industri 4.0. In: Proceedings - 2019 IEEE International Conference on Industry 4.0, Artificial Intelligence, and Communications Technology, IAICT 2019 (2019). https://doi.org/10.1109/ICIAICT.2019.8784858

5. Padariya, N., Patel, N.: Quality grading classification of dry chilies using neural network. Lecture Notes on Data Engineering and Communications Technologies (2021). https://doi.org/10.1007/978-981-15-4474-3_20
6. Jannati, F.N., Khuriyati, N., Falah, M.F.: Designing an image capturing box for red chilies (Capsicum annuum L.) by using smartphone camera. In: Proceedings - 2019 5th International Conference on Science and Technology, ICST 2019 (2019). https://doi.org/10.1109/ICST47872.2019.9166243
7. Rabby, M.K.M., Chowdhury, B., Kim, J.H.: A modified canny edge detection algorithm for fruit detection classification. In: ICECE 2018 - 10th International Conference on Electrical and Computer Engineering (2019). https://doi.org/10.1109/ICECE.2018.8636811.
8. Song, Y., Glasbey, C.A., Horgan, G.W., Polder, G., Dieleman, J.A., van der Heijden, G.W.A.M.: Automatic fruit recognition and counting from multiple images. Biosyst. Eng. (2014). https://doi.org/10.1016/j.biosystemseng.2013.12.008
9. Kanungo, T., Mount, D.M., Netanyahu, N.S., Piatko, C.D., Silverman, R., Wu, A.Y.: An efficient k-means clustering algorithms: analysis and implementation. IEEE Trans. Pattern Anal. Mach. Intell. (2002). https://doi.org/10.1109/TPAMI.2002.1017616
10. Howard, A.G., Zhu, M., Chen, B., Kalenichenko, D., Wang, W., Weyand, T., Andreetto, M., Adam, H.: MobileNets: efficient convolutional neural networks for mobile vision applications (2017)
11. Ren, S., He, K., Girshick, R., Sun, J.: Faster R-CNN: towards real-time object detection with region proposal networks. IEEE Trans. Pattern Anal. Mach. Intell. **39**, 1137–1149 (2017). https://doi.org/10.1109/TPAMI.2016.2577031
12. Spence, C.: Crossmodal contributions to the perception of piquancy/spiciness. J. Sens. Stud. (2019). https://doi.org/10.1111/joss.12476
13. Luo, M.R.: CIELAB. In: Encyclopedia of Color Science and Technology (2015). https://doi.org/10.1007/978-3-642-27851-8_11-1
14. Habibi Aghdam, H., Jahani Heravi, E.: Guide to convolutional. Neural Netw. (2017). https://doi.org/10.1007/978-3-319-57550-6
15. Krizhevsky, A., Sutskever, I., Hinton, G.E.: ImageNet classification with deep convolutional neural networks. Commun. ACM. (2017). https://doi.org/10.1145/3065386
16. Morgado, P., Vasconcelos, N.: NetTailor: tuning the architecture, not just the weights. In: Proceedings of the IEEE Computer Society Conference on Computer Vision and Pattern Recognition (2019). https://doi.org/10.1109/CVPR.2019.00316
17. Olimid, R.: Python implementation of visual secret sharing schemes. J. Inf. Syst. Oper. Manag. **5**, 538–550 (2011)

Person Identification System for UAV

Bonnie Lu Sing Chen, Dik Son Cheah, Kok Wei Chan, and Hermawan Nugroho(✉)

Department of Electrical and Electronic Engineering, University of Nottingham Malaysia, Jalan Broga, 43500 Semenyih, Selangor, Malaysia
bonnie.12.lsc@gmail.com, cheahdikson@gmail.com, kokwei217@gmail.com, Hermawan.Nugroho@nottingham.edu.my

Abstract. The usage of Unmanned Aerial Vehicle (UAV) is preferred in certain situations, such as when sending in human can be dangerous or if we want to deliver important payloads to a target person in a crowded area. In this paper, we demonstrate and present a development of an Unmanned Aerial Vehicle (UAV) system that is capable of an autonomous flight and identifying a target (person) from a group of people. The autonomous flight system consists of an offline path planning module and a real time object detection module. Results show that the path planning allows the UAV to plan its path with latitude and longitude accuracy of 0.072 m and 0.087 m respectively using the A* algorithm. Experiments also show that the object detection module allows depth estimation with an 83.5% accuracy using a single camera. For identifying target, the UAV employs facial recognition algorithm based on FaceNet which extract embedding features as representative of the face. The system can recognize a person up to a horizontal distance of 3.8 m and vertical distance of 3.0 m with a minimum FPS of 6.4 when the number of faces in the frame is 10. Results show that the develop UAV platform is potential to be used for identifying a person in a crowd.

Keywords: Path planning · Object detection · Depth estimation · Facial recognition · Unmanned Aerial Vehicle (UAV)

1 Introduction

Generally, Unmanned Aerial System (UAS) comprises of three different systems, namely Unmanned Aerial Vehicle (UAV), Ground Control System (GCS) and the Command and Communication module. The UAV can be defined as an aircraft that does not have human operator inside the aircraft and is operated remotely or autonomously.

The autonomous UAV is able to fly autonomously using various sensors and camera as input and pre-programmed flight plans. Depending on its design and flight plans, it will determine the range of payloads that the UAV can be carried. Beside military flight reconnaissance, researchers find that UAV can be used for surveillance, search and rescue operation and product delivery [1].

The usage of UAV can be helpful, especially when sending in human can be dangerous or if we want to deliver fast to a target person in a crowded area. These functions, however, are not optimised if the USV is controlled by an operator remotely. The remote

J. A. Mat Jizat et al. (Eds.): iCITES 2020, AISC 1350, pp. 325–335, 2021.
https://doi.org/10.1007/978-3-030-70917-4_31

control-based UAVs are susceptible to latency problems which can trigger more complex situation. For example, loss of instructions received from the operator may cause lagging which can comprise the whole operation and in worst cases, the loss of the UAV. Moreover, to find a target within a crowd through video footage of the UAV can be very demanding for the operators/pilots of the UAV [2].

In this paper, the aim is to develop an autonomous system on the UAV which can identify a target person in a crowd. Besides for search operation, such UAV-based system is also applicable to monitoring and reconnaissance in finding a target.

This paper is organized as follows Sect. 2 presents briefly explains our method of developing a fully autonomous UAV. Section 3 presents our path planning approach for the UAV to determine a collision free path from a layout of the area as well as avoiding stationary objects that are in its path during its flight using object detection and depth estimation. Section 4 presents our approach to facial detection and recognition to identify the person within a group. The results of the previous sections are presented in Sect. 5 and the conclusions is presented in the last section.

2 Framework and Method

Figure 1 shows the details of components used in the Unmanned Aerial Vehicle (UAV) and the ground unit. The UAV consists of a quadcopter kit, brushless motors connected to propellers, a power system (batere), a PixHawk flight controller, a telemetry and GPS module. For image processing module, a webcam connected with Raspberry Pi 4 (added by Intel Neural Compute Stick 2) is used. A laptop is used as the ground station.

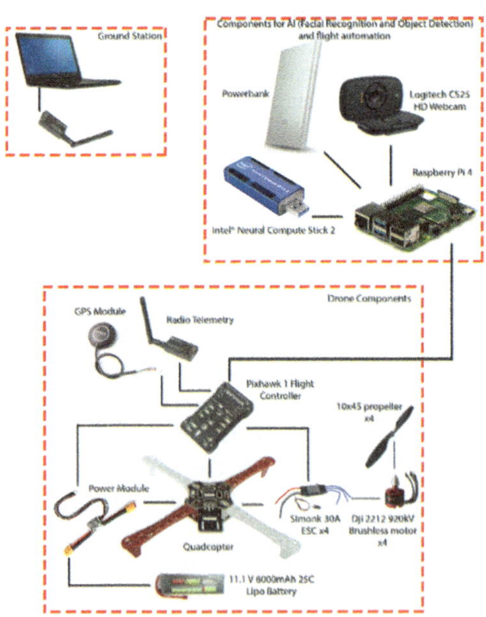

Fig. 1. Hardware used for the Autonomous UAV

The developed UAV has the capability of computing its own collision free path to a required location, obstacle avoidance while flying through depth estimation using computer vision and identification of a person within a group. It consists of three modules, namely path planning module, depth estimation module, and facial recognition module. The flowchart is shown in Fig. 2 below.

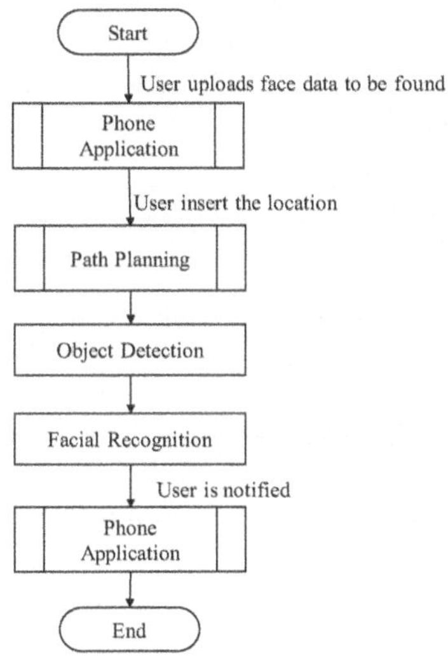

Fig. 2. The framework of the UAV

The path planning module is responsible for computing a collision free path and converting the path into usable input for the UAV. This path planning algorithm uses a 2-Dimensional map of the area that has been classified into either the free space or obstacle space, in which the obstacle space are classified using the color of the buildings and trees in the area. To compute a collision free path, a grid-based search algorithm based on the A* algorithm is used on the map that returns a list of X and Y coordinates representing the path. Since the X and Y represent the pixel coordinates, it is transformed into the GPS coordinates using the nautical miles model. The GPS coordinates are then used by the UAV to fly the computed path at a constant altitude.

The object detection and depth estimation module are responsible for avoiding obstacles during the flight that are missed by the path planning module (i.e. a small group of trees with incorrect color codes in the map). For the object detection system, the system adopts the convolutional neural network of YOLOv3 for real time detection. Using the depth estimation through mono vision image processing and input from the flight controller, such as the inertial measurement unit, we re-orientate UAV away from the obstacle when the obstacle is within a minimum distance and move in that direction for a certain period to pass the obstacle before returning to the original path.

The facial detection and recognition module are then executed when the UAV reaches the end point. The aim of the module is to identify a target (person) within a group of people. This module uses a face database to recognize a target face in which the target's face is captured manually from the ground station or by upload it on a mobile phone application. The module uses an openVINO pre-trained face detection model developed using SqueezeNet light architecture to detect faces and its bounding boxes. The model will predict five facial landmarks such as the eyes, nose and the corners of the lips which are then used to align the faces in a desired position based on these landmarks coordinate. The facial recognition based on the FaceNet approach is then applied to identify the target's face.

A phone application is developed to send the face of the target to the UAV via Google Firebase and once the target is identified, the UAV will notify user via through the phone application.

3 Path Planning

In the the path planning, we implement the A* algorithm to the matrix form of the image which has the size of M $\times$ N. Thus, to differentiate between pixels in the map that is either obstacle or free, any colored pixel is converted to 1 to represent the obstacle space and 0 to represent the free space. This algorithm will then compare all possible paths and ensures that the shortest possible path between the starting and end point is computed [3]. The algorithm will then generate a list of coordinates based on the pixel position of the map based on the outcome.

In the development, we find the list coordinates cannot be used directly. These coordinates are needed to be converted into coordinates that represents the real world, i.e. GPS coordinates which can be used by the UAV. Three parameters are required to perform this conversion which are the vertical distance per pixel, horizontal distance per pixel and a reference coordinate which is the top left corner of the map (Eq. 1)

$$\textit{Vertical Distance } (m) = \textit{Latitude Difference} \times 60 \times 1851 \tag{1}$$

$$\begin{aligned} \textit{Latitude Difference} &= \textit{Latitude Reference} - \textit{Unknown Latitude} \\ \textit{Horizontal Distance } (m) &= \textit{Longitude Reference} \times 60 \times 1852 \\ \textit{Longitude Difference} &= \textit{Unknown Longitude} - \textit{Longitude Reference} \end{aligned}$$

Using the definition of nautical miles, we can determine the subsequent unknown GPS coordinates required to make the vertical or horizontal distance in the real world using one reference GPS coordinates only (Fig. 3).

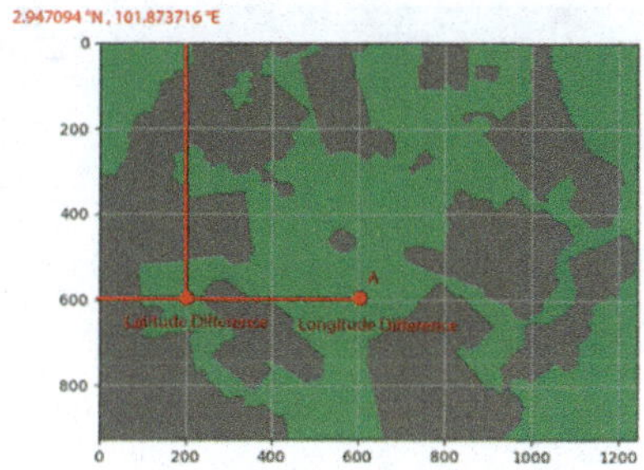

Fig. 3. Calculating latitude and longitude of point A

3.1 Depth Estimation

When the UAV is flying, it is essential to determine the stationary objects that are in the flight path of the UAV. Therefore, we used a method proposed by Rahman et al. to estimate the distance between UAV and object when it is pre-calibrated using the object depth and lateral distance [4] (Fig. 4).

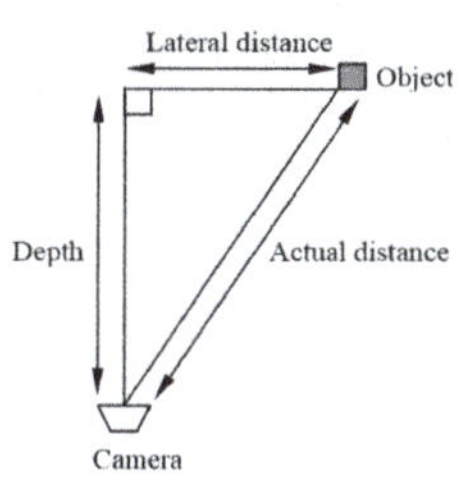

$$\textit{Actual distance} = \sqrt{\textit{object depth}^2 + \textit{lateral distance}^2}$$

Fig. 4. Object depth estimation

The relationship between the pixel height of the boundary box at a different depth determines the object depth. As the object is getting further from the camera, the size of the object will be relatively smaller, i.e. the pixel height of the boundary box decreases. Therefore, this relationship determines the object depth between camera and object. The boundary box (bounding box) is estimated using YOLOv3 [5]. The You Only Look Once (YOLO) algorithm is mainly used for object detection. The algorithm expresses object detection as a single regression problem, using image pixel to predict bounding box and class probabilities. It divides the input image into grid of cells with S-by-S size. The prediction of bounding boxes and confidences scores of objects are computed from each grid cells (Fig. 5).

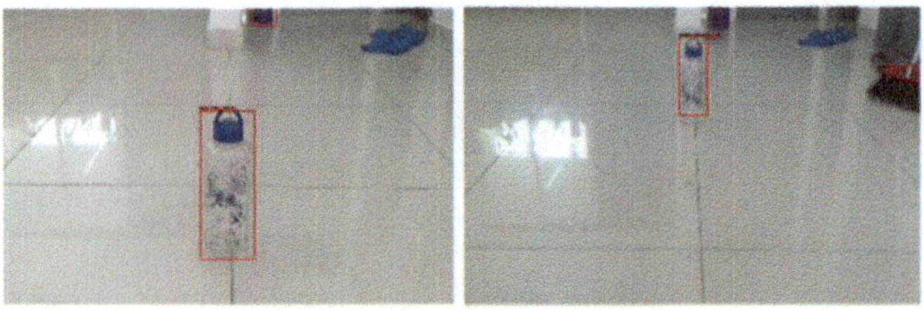

Fig. 5. Size of Bounding Box when object is near camera (left) - Far from camera (right)

The lateral distance is determined by using the relationship between the lateral pixel of the boundary box and lateral distance, which is the distance from the object to the center line of the camera. When the object is moved sideways from the center line of the camera, the ratio of the lateral pixel and lateral distance is found to be constant. Thus, the lateral distance can be ignored (Fig. 6).

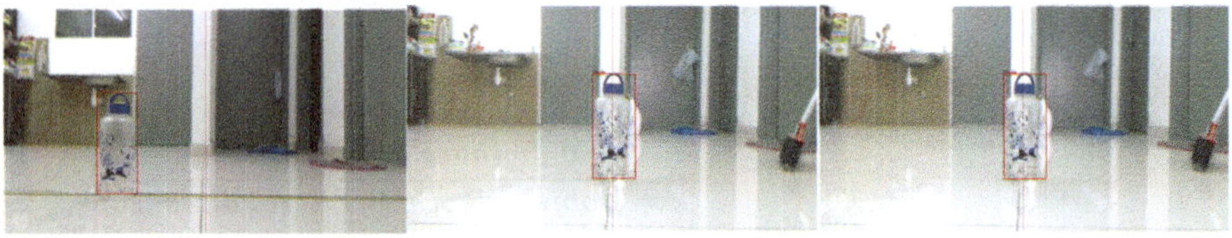

Fig 6. Lateral distance at different positions

4 Facial Detection and Recognition

4.1 Face Detection

As in previous section, to detect face and predict their bounding boxes, an OpenVINO pre-trained face detection model is used. For each face detected, the model will produce the coordinated bounding box and its confident level. If the confidence level exceeds a certain probability threshold (which is set to 60%), a bounding box will be drawn over the detected face (Fig. 7).

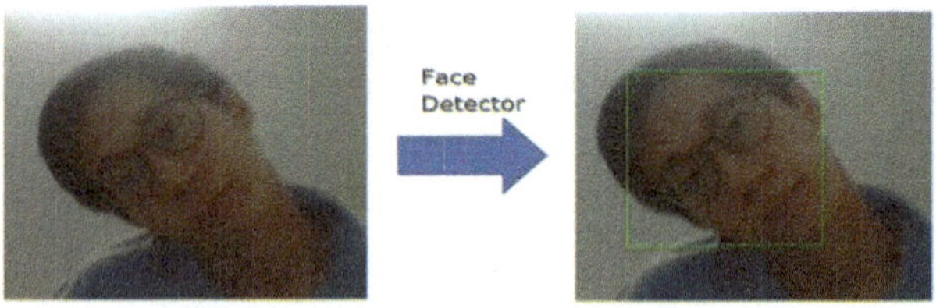

Fig. 7. Output of the face detector

4.2 Landmark Detector

In a pipeline process, the output from the face detector model is used as the input for another pre-trained model to predict five facial landmarks: the eyes, nose and the corners of the lip. For each successful detection, we will obtain 10 floating point values for the five landmarks coordinate in the form of (x0, y0, …, x5, y5) [6]. These landmarks coordinates are then applied as a reference to align the face in desired position.

4.3 Face Embedding

The FaceNet approach is used to represents a face [7]. The neural network model will then be loaded onto the NCS using OpenVINO's MYRAID plugin. The model generates a 256-dimensional embedding which will be adjacent in cosine distance for similar faces and far for different faces. It is reported that the model works best if the input face is aligned and frontally oriented [8].

4.4 Face Recognition

After obtaining the face embeddings, the module will compare it with target's face stored in the system. Then, a matrix consists of cosine distances between detected face and each identity can be determined. Equation 2 shows the matrix calculation.

$$\textit{Minimum distance} = [\textit{number of detected faces}] \times [\textit{number of identities}] \tag{2}$$

$$M = \begin{bmatrix} X_0 \\ \vdots \\ X_q \end{bmatrix} \times \begin{bmatrix} Y_0 \\ \vdots \\ Y_q \end{bmatrix}$$

Let X be the detected faces, q be the number of detected faces, Y be the identity in database and q be the number of identities. After obtaining the matrix, we use Hungarian algorithm to determine the best matching identification of the detected faces [9]. The identity of the detected face is defined as the matrix which yields the minimum total distance. And if the distance between the detected face and its assigned identity is less than a specified threshold, then the algorithm will consider it as a successful matching.

5 Experimental Analysis

5.1 Path Planning

The accuracy of the conversion to GPS coordinates is measured by comparing the expected distance calculated from the distance per pixel to the actual distance between the two GPS coordinates and is increased every 50 m to see its effect on the accuracy over a larger distance. The standard deviation of the vertical and horizontal errors are 0.072 m and 0.087 m respectively (Table 1) (Fig. 8).

Table 1. Distance accuracy

Vertical distance			Horizontal distance		
Expected (m)	Actual (m)	Error (m)	Expected (m)	Actual (m)	Error (m)
50.07	50.09	0.02	50.13	50.08	−0.05
100.15	100.16	0.01	99.97	99.95	−0.02
149.93	150.03	0.10	150.10	150.18	0.08
200.00	200.21	0.21	199.94	199.91	−0.03
250.08	250.15	0.07	250.08	249.90	0.18

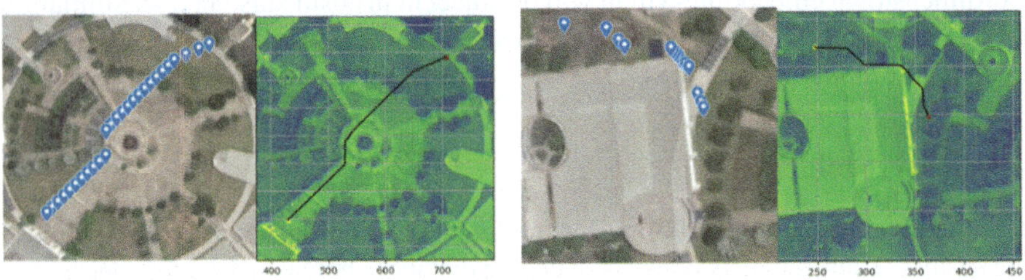

Fig. 8. Actual Path taken by the UAV (Left) and Simulated Path taken by the UAV (Right)

5.2 Depth Estimation

The estimated distance is compared against the actual distance of the object from the camera starting from 60 cm and is repeated with an interval of 30 cm to determine its accuracy with an increase in distance (Fig. 9).

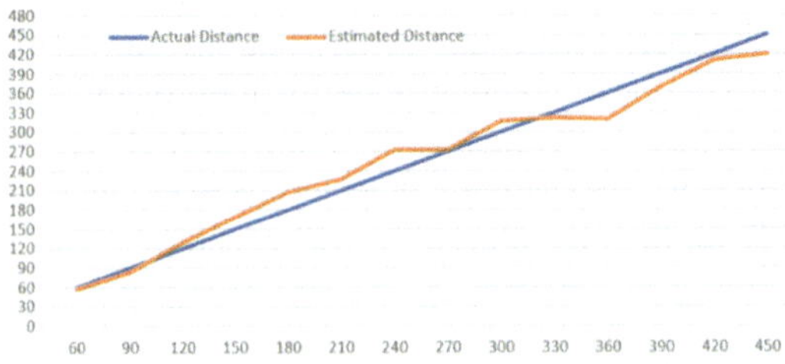

Fig. 9. Actual distance and estimated distance comparison

By calculating the average of the percentage error at 30 cm interval, the method has accuracy of 83.5% which is promising in comparison with most of the advanced vision distance estimated methods.

5.3 Face Recognition

For the face recognition module, we created four face galleries that contain a pre-set number of images for each unique identity. The images in the galleries are as follows (Tables 2 and 3) (Fig. 10):

Gallery A: 1 front view image
Gallery B: Gallery A with 2 additional side view images
Gallery C: Gallery B with 2 additional top view images
Gallery D: Gallery C with 5 additional images from different angles

Table 2. Number of images for each unique identity of the four galleries

Gallery	Number of images for each identity	Number of identities	Total number of images
Gallery A	1	3	3
Gallery B	3	3	9
Gallery C	5	3	15
Gallery D	10	3	30

Table 3. Facial recognition results

Gallery	Confidence level of successful facial recognition		
	Front view	Side view	Top view
Gallery A	87.84%	Recognition Failed	Recognition Failed
Gallery B	87.16%	71.83%	Recognition Failed
Gallery C	88.30%	71.63%	76.49%
Gallery D	87.38%	76.75%	80.07%

From the experiment, it is observed that the minimum number of sample images for successful facial recognition from all angles is 5. The confidence level, however, will be at the minimum (with just 5 samples). By increasing the number of samples to 10, the confidence level of side view recognition increases from 71.63% to 76.75% and the confidence level of top view recognition increases from 76.49% to 80.07%.

We run the algorithm on laptop to serve as a benchmark. The laptop should have an Intel's processor which is supported by the OpenVINO SDK (Table 4).

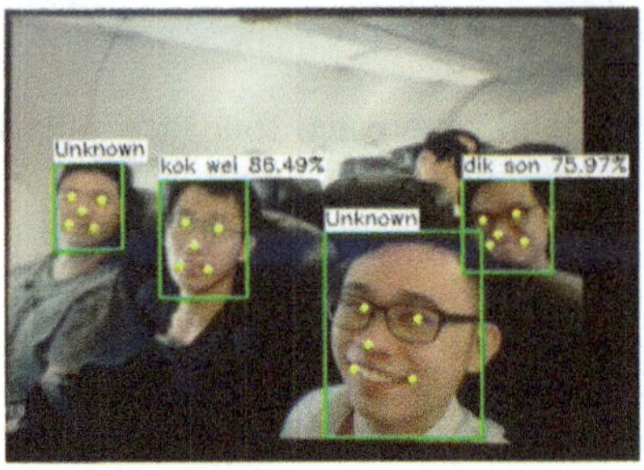

Fig. 10. Example of test images

Table 4. FPS of facial recognition system with different system and configuration

Number of faces in frame	Frame rate per second (FPS)			
	Laptop, I5-5200U		Raspberry Pi 4, 4GB Ram	
	CPU	1 NCS	2 NCS(OpenVINO)	2 NCS (queue setting)
1	28.6	17.7	17.6	17.7
2	22.7	14.9	15.1	14.6
3	15.2	10.8	10.7	13.7
4	13.5	10.0	10.0	11.7
5	11.1	7.9	9.5	10.1
6	10.0	7.5	9.3	8.8
7	8.8	6.2	8.2	8.4
8	7.7	5.7	7.6	7.7
9	7.3	5.0	7.0	7.2
10	6.5	4.8	6.7	6.4

6 Conclusion and Future Work

In this paper, we propose and develop an application of UAV which is capable of an autonomous flight and identifying a person from a group of people. The UAV employs an edge computing approach with the face recognition module run onboards. To enable such approach, the algorithm is developed on a Rasperry Pi and Intel NCS 2. Besides facial recognition module, the UAV is also equipped with depth estimation (object detection) and path planning modules for the autonomous flight. Experiments are conducted on University of Nottingham Malaysia and shows promising results.

During the experiments, we find that the Raspberry Pi often goes into undervolted state when the face recognition module is running. This might be due the use of mobile power bank as power supply. For future work, a HAT (Hardware Attached on Top) can be employed to improve the stability of power source without sacrificing mobility. If the power issue solved, the Raspberry Pi can operate at its normal clock speed. This will improve the FPS rate which is beneficial for the facial recognition module.

Acknowledgement. Authors would like to thank Prof. Haider Abbas Almurib for the support of the UAV unit. Authors would also like to thank Intel Malaysia for the Intel NCS 2.

References

1. Hildmann, H., Kovacs, E.: Review: using unmanned aerial vehicles (UAVs) as mobile sensing platforms (MSPs) for disaster response, civil security and public safety. Drones **3**, 59 (2019). https://doi.org/10.3390/drones3030059
2. Drone Efficacy Study Evaluating the Impact of Drones for Locating Lost Persons in Search and Rescue Events (2018)
3. Braun, J., Brito, T., Lima, J., Costa, P., Costa, P., Nakano, A.: A comparison of A* and RRT* algorithms with dynamic and real time constraint scenarios for mobile robots. In: SIMULTECH 2019 - Proceedings of the 9th International Conference on Simulation and Modeling Methodologies, Technologies and Applications, pp. 398–405. SciTePress (2019). https://doi.org/10.5220/0008118803980405
4. Rahman, A., Salam, A., Islam, M., Sarker, P.: An image based approach to compute object distance. Int. J. Comput. Intell. Syst. **1**, 304–312 (2008). https://doi.org/10.1080/18756891.2008.9727627
5. Redmon, J., Farhadi, A.: YOLO9000: better, faster, stronger. In: Proceedings of the 30th IEEE Conference on Computer Vision Pattern Recognition, CVPR 2017, 2017-January, pp. 6517–6525 (2017). https://doi.org/10.1109/CVPR.2017.690
6. landmarks-regression-retail-0009 - OpenVINO Toolkit
7. Schenkel, T., Ringhage, O., Márki, A., Bae, J.: A Comparative Study of Facial Recognition Techniques With focus on low computational power. Bachelor Degree Project in Information Technology Basic level 30 ECTS Spring Term 2019
8. Face-reidentification-retail-0095 - OpenVINO™ Toolkit
9. Kuhn, H.W.: The Hungarian method for the assignment problem. In: 50 Years of Integer Programming 1958–2008: From the Early Years to the State-of-the-Art (2010). https://doi.org/10.1007/978-3-540-68279-0_2

Artificial Neural Network Modelling for Slow Pyrolysis Process of Biochar from Banana Peels and Its Effect on O/C Ratio

Neoh Jia Hsiang, Anurita Selvarajoo(✉), and Senthil Kumar Arumugasamy(✉)

Department of Chemical and Environmental Engineering, Faculty of Science and Engineering, University of Nottingham Malaysia Campus, Jalan Broga, 43500 Semenyih, Selangor, Malaysia
{Anurita.Selvarajoo,Senthil.Arumugasamy}@nottingham.edu.my

Abstract. Banana peels are one of the most underutilized agriculture wastes, having a great potential to be converted into various alternative products in many industries. In agriculture, biochar derived from banana peels can be used as soil amendment and soil conditioner. In this study, slow pyrolysis was chosen among the several thermochemical techniques to produce biochar, due to its ability to achieve low oxygen to carbon (O/C) ratio. Artificial neural network (ANN) modelling of the slow pyrolysis process was implemented based on the experimental data gathered. Feedforward artificial neural network (FANN), a type of ANN, was used to create the model based on a single hidden layer network using MATLAB™ software. A combination of up to 25 hidden neurons in the hidden layer as well as 10 different training algorithms were tested using iterative training method. Levenberg-Marquardt training algorithm with 24 hidden nodes showed the best results, with the highest determination coefficient (R^2) of 0.99858, low output MSE of 0.04467 and lowest validation MSE of 0.00001.

Keywords: Biochar · Slow pyrolysis · Banana peels · Feedforward artificial neural network

1 Introduction

In Malaysia, agricultural waste presents itself as an issue because most of it are not being utilized after their fruits or crops have been removed for further processing. As reported by Zafar (2019), agricultural crops that can be found in peninsular Malaysia such as rubber, oil palm, rice, cocoa and coconut, from which only 27% of the total waste residue is repurposed [1]. Additionally, the remaining waste is disposed either by burning or ending in landfill. Besides that, huge areas of land in Malaysia are presently being used for plantation, with 3.87 million hectares for oil palm, 147 thousand hectares for coir, 34 thousand hectares for banana, and 15 thousand hectares for pineapple [2]. These plantation areas alone account for approximately 12% of the total land in Malaysia, which makes any wastage a substantial issue. Out of these crop plantations, bananas have risen as one of the crops being largely consumed and cultivated not only in Malaysia, but worldwide, with a gross production exceeding 110 million tons in the year 2018,

J. A. Mat Jizat et al. (Eds.): iCITES 2020, AISC 1350, pp. 336–350, 2021.
https://doi.org/10.1007/978-3-030-70917-4_32

according to statistics by the Food and Agriculture Organization of the United Nations [3]. In Malaysia, there are several small and medium enterprises (SME) involved in banana production, such as banana crisps manufacturers, fried banana stalls, as well as banana jam manufacturers. These SMEs generate huge amounts of waste especially in the case of banana peels [2]. Instead of being disposed as municipal waste and ending up in landfills, banana peels have great potential as an alternative product.

Presently, extensive research has been conducted on banana peels in various sectors to manufacture value-added products from it. According to Nasution et al. (2012), banana peels can be used as a substitute for whole meal flour in the making of whole meal bread, which had apparent increase in its total dietary fiber [4]. In addition, banana peels have also found to be a suitable substrate in manufacturing ethanol, which is heavily utilized as a renewable fuel and solvent in industries [5]. Moreover, by using banana peels instead of conventional crops such as wheat and corn, the cost of producing renewable fuel can be reduced. Apart from that, industrial wastewater containing high concentrations of heavy metals such as chromium, can be treated with banana peels [6]. In a study, banana peels were found to be more effective in removing chromium (VI) metals from an aqueous solution, as compared to other agricultural biowastes, due to its high sorption properties [7]. Lastly, in terms of agriculture, banana peels have the potential to enhance soil fertility as well as provide long-term carbon storage, when converted into biochar [8]. Besides that, biochar can also be used for soil amendment, due to it being a comparatively more efficient fertilizer since it also reduces the nitrogen loss in soil [9]. Therefore, with biochar's growing potential in conditioning soil, the conversion of banana peel into banana peel-based biochar (BPB) and its application is further studied in this report.

Owing to its high porosity and natural nutrient availability, biochar is effective as a growth promoter for plants. When applied to soil, biochar is reportedly able to raise the pH of soil, improve the capacity for holding moisture and reduce the occurrence nutrients leaching from the soil to surface and ground water [10, 11]. However, there are certain requirements that dictate the quality and production of biochar. Both the European Biochar Certificate (EBC) and International Biochar Initiative (IBI) suggested that biochar produced should have an oxygen to carbon (O/C) ratio of <0.4, as a safety and quality assurance precaution. This requirement also ensures that forests are not cut down for the sole purpose of producing biochar as well as preventing excessive amounts of pollutant emitted during biochar production [9]. Apart from that, biochar is dissimilar to other carbonaceous compounds like charcoal and char. This is because according to EBC, biochar is obtained through the pyrolysis of biomass, which is acquired by sustainable means, under controlled operating conditions and its application does not involve rapid mineralization into carbon dioxide [12, 13].

The conventional modeling of pyrolysis process done using lumped kinetic model [14], biomass thermal decomposition kinetic model [15] or numerical model in terms of heat and mass transfer [16]. However, these models face similar problems such as difficulty in modelling more complex systems with multiple input parameters, hence requiring an alternate method. Using the experimental data, neural networks are capable of simulating and analyzing imprecise and complex data sets and can also generate trends and patterns from the data provided [17]. A widely used neural network, artificial neural networks (ANN), is a prognostic method which has the capability to obtain solutions, whereby other means of statistical methods are not possible. ANN is analogous to the human brain, from which the basic element in the structure is the neuron, also known as perceptron, which transfers information through electrical signals and is represented by numerical figures in the case of ANN [18]. In ANN, a neuron functions as the basic building block for constructing the entire neural network, whereby outputs are obtained after the inputs have been processed by hierarchical sets of neurons [19], as shown in Fig. 1. ANN's main advantage is its ability to model complex and non-linear relationships from the input and output data provided, by undergoing multiple layers of learning [20].

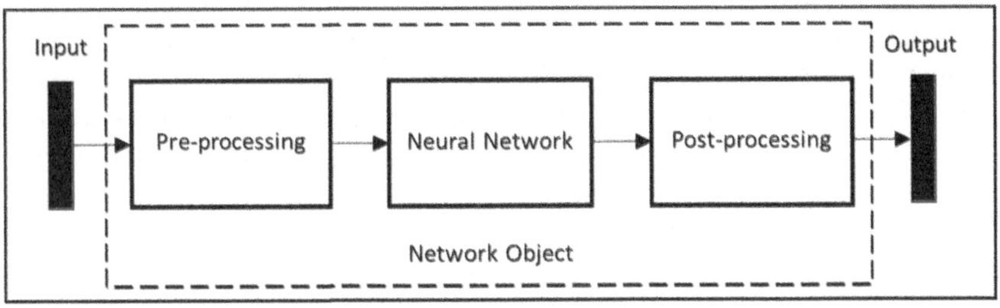

Fig. 1. Simplified block diagram of neural network processing [29]

In ANN, feedforward artificial neural network (FANN), also known as multilayer perceptrons (MLP), is among the commonly used neural network types, which generally consists of input, hidden and output layers [20]. In MLP, each of the artificial neurons is connect to a layer of nodes in the hidden layer, with no connections formed between the neurons of the same layer [18]. The connection between the hidden layer and input layer is by a link, another basic component in FANN, which contains a weight and bias [17]. Next, the intermediate output of the hidden layer then passes through the final layer (output layer) to obtain a network output, which represents the best solution for the neural network modelling. The feedforward neural network can be visualized as shown in Fig. 2.

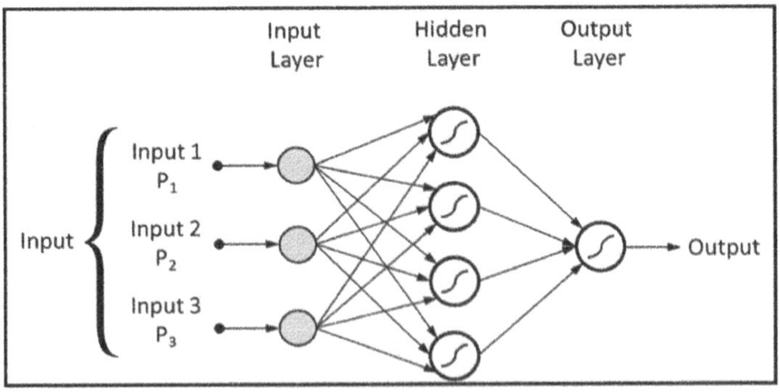

Fig. 2. Feedforward ANN with input, output and one hidden layer [21]

Additionally, previous works on biomass pyrolysis have proven that slow pyrolysis of banana peel biomass using ANN modelling, can be done. Aydinli et al. successfully modelled the pyrolysis of various plant-based wastes with product yield as the output, using ANN modelling [22]. Other ANN models were also developed to analyze the yield output for: biochar from various biomass [23], activated carbon from pine nut shell [24], char from pine sawdust [25] and biochar from banana peels [26], char from rambutan peels [27] and biochar from palm fibre [28].

Based on literature findings, previous studies mainly determined the effect of pyrolysis parameters on biochar yield, whereas its effect on O/C ratio of biochar were not identified. Studies on the modelling of slow pyrolysis process for converting banana peels into biochar have previously been conducted. However, using feedforward artificial neural network (FANN) modelling, the consideration of various training algorithms has not been done to effectively determine the most optimal modelling method. Hence, this report aims to investigate the effect of heating temperature, heating rate and residence time on the O/C ratio. Furthermore, by using various training algorithms in FANN modelling, this report also seeks to determine the most accurate and reliable training algorithm for modelling the slow pyrolysis process of turning banana peel biomass into biochar.

2 Materials and Method

2.1 Sample Preparation and Pyrolysis Experiment

Approximately 6 kg of banana peels were obtained from a local fried banana stall near Taman Tasik Semenyih, Semenyih, Selangor. The banana peels were repeatedly rinsed with distilled water to remove impurities and the remaining fruit of the banana. Next, the washed banana peels which have a moisture content of about 60–90%, were torn to thin strips and dried in an oven (Memmert Universal Oven UF260) at 80 °C for 24 h. This procedure produced a dry biomass of 10–15 wt% moisture, which were then grinded using a rotor beater mill (Retsch SM100) to a particle size of $\leq$5 mm and further sieved to 1–2 mm.

Using the dry biomass of 1–2 mm obtained, slow pyrolysis was conducted in a horizontal tube furnace (Carbolite CTF 12 & TZF 12 Wire Wound) to produce biochar as shown in Fig. 3. The furnace's tube component is made up of stainless-steel, having a length of 120 cm, internal diameter of 5 cm and external diameter of 6 cm. At the end of the tube furnace, the outlet was connected to a stoppered 2 L conical flask containing distilled water, using a rubber tube, which acts as a water bath to cool condensable gaseous product during the experiment. Besides that, the first conical flask was connected to another 2 L conical flask filled with silica gel. The second conical flask functions to scrub the gaseous product before being released to the atmosphere. By using RSM approach with CCD, a total of 20 experimental runs was required to be executed, each with varying heating temperature of 63.64–736.36 °C, heating rate of 1.59–18.41 °C/min and residence time of 19.09–220.91 min. For each run, 20 g of banana peel biomass was weighed using an electronic mass balance and loaded onto two thin stainless-steel metal boats, each with dimensions of 15 cm × 4 cm × 2 cm. Next, the metal boats containing biomass, were inserted into the center of the tube furnace and enclosed. When enclosing the tube furnace entrance, vacuum grease was applied onto the flanges and shut tight with screws and nuts, to prevent leakage and avoid other gases from entering the tube furnace. Then, the tube furnace was preset to the intended heating temperature, heating rate and residence time of each run, and initiated by activating the switch.

In characterizing biomass and biochar samples, a scanning electron microscope (FEI Quanta 400F FESEM) coupled with an energy dispersive X-ray (Oxford-Instruments INCA 400 with X-Max Detector), also known as FESEM/EDX, was used. From FESEM data, the oxygen and carbon content will be used to find the O:C ratio.

Fig. 3. Experimental setup for pyrolysis experiment [29]

2.2 Neural Network Training Method

In feedforward artificial neural network (FANN) modelling, the workflow of the modelling process generally included seven steps: (i) data collection, (ii) creating the network, (iii) network configuration, (iv) initializing weights and biases, (v) network training, (vi) network validation, and (vii) using the network [30]. The workflow of the modelling was summarized as shown in Fig. 4.

Data Collection. Firstly, the input and output data for modelling slow pyrolysis were extracted from the operating parameters and results of the experimental work conducted. In this experiment, the input data were the slow pyrolysis parameters, which are heating temperature, heating rate and residence time, while the output data was the corresponding O/C ratio of each experimental run. Therefore, a total of 20 data points was incorporated into this FANN model. Prior to inputting the data, data pre-processing was implemented to ensure the model's accuracy, by reducing outliers and avoiding any influence in data dissimilarity (Nawi et al., 2013) [20]. Since the input and output data have large differences in magnitudes, ranging from 63.64–736.36 °C for heating temperature, 1.59–18.41 °C/min for heating rate, 19.09–220.91 min for residence time and 0.2044–0.7144 for O/C ratio, hence data standardization was required. There are several functions that normalize data such as 'mapminmax', 'mapstd', 'processpca', 'fixunknowns' and 'removeconstantrows' [30] Among these functions, min-max normalization method was used to scale the values of data to range from 0–1.

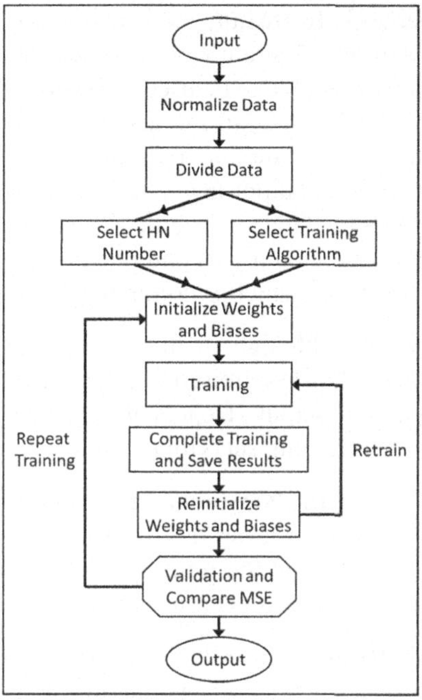

Fig. 4. FNN modelling workflow

Then, the displayed data which has been normalized, as shown in Table 1, were further divided into three subsets (i.e. training, validation and testing). The training set computes the gradient of the network and updates the weights and biases. Next, the validation set identifies the error from the training process, while the testing set functions to compare different models of the neural network [30]. In this experiment, the normalized data were divided into the subsets randomly, with 70% allocation to training, and 15% each to validation and testing.

Table 1. Input and output parameters for slow pyrolysis modelling

Input/Output parameter	Values	Normalized values	Units
Heating temperature	63.64–736.36	0–1	°C
Heating rate	1.59–18.41	0–1	°C/min
Residence time	19.09–220.91	0–1	mins
O/C ratio	0.2044–0.7144	0–1	–

Creating and Configuring the Network. After the normalized data have been distributed randomly into their respective subsets, the next step in modelling, is to create and configure the neural network. In creating the network, one hidden layer was enough to provide good results according to literature because according to Hornik et al. (1989), single-hidden-layered neural network can be used to approximate continuous functions to achieve high accuracy [31]. By default, the activation function in hidden layers is tangent-sigmoid, and linear transfer function, purelin for the output layer. The network involves one hidden layer and one output layer. The input parameters first entered the hidden layer, where it was weighted with a specific IW and a bias vector was simultaneously added. The sum of the weighted inputs and biases then passed through a transfer function, tangent-sigmoid function, which is commonly used for multilayer networks.

Initialize Weights and Biases. Although in the configuration step, the weights and biases were already initialized, however retraining the network was required to obtain a more accurate and optimum output. Hence, the weights and biases need to be reinitialized, which repeats the training set with the new weights and biases.

Training, Validating and Using the Network. Prior to begin the training, early stopping was implemented to avoid overfitting the feedforward neural network. This was also done to speed up and standardize the training process, since the network is to be trained with different number of hidden neurons and various training algorithms.

Methods to Improve ANN Results. In order to improve the results obtained, Beale et al. suggested two additional approaches, which are increasing the number of hidden neurons in the hidden layer and substituting other training functions. For the first approach, the number of hidden nodes was manipulated from 1–25 because the more neurons present in the hidden layer, the more flexibility was given for the network to optimize the input parameters. In the next approach, different training functions yield different training results due to their various training properties. During the training of multilayer networks, Back Propagation Artificial Neural Network (BP-ANN) is commonly used [32]. MATLAB© software's Neural Network Toolbox contains several Back Propagation or Jacobian-based training functions, as shown in Table 2.

For small networks having fewer weights, both Levenberg-Marquardt (LM) and BFGS Quasi-Newton (BFG) are fast training algorithms which provide good results. LM is also commonly used in neural network modelling due to its fast convergence and flexibility but has a drawback of requiring large storage for its training process. Bayesian Regularization (BR) is also chosen because of its ability to continue validation until its targeted optimal performance value is reached [33]. Besides that, for larger networks, Resilient Backpropagation (RP) is suitable because its memory requirement is small and can develop results faster than certain gradient descent algorithms [30]. Apart from these training functions, other algorithms as shown in Table 2 were used in this experiment to ultimately compare and determine the most suitable model, with the consideration of hidden neuron number in the hidden layer.

2.3 Neural Network Performance Evaluation

Once the network model has been trained, its performance and reliability were determined by three criteria, the determination coefficient (R^2), mean squared error (MSE) of the output data as well as the validation performance.

Table 2. Training functions and algorithms (Beale et al. 2016)

Training function	Training algorithm
trainlm	Levenberg-Marquardt (LM)
trainbr	Bayesian Regularization (BR)
trainbfg	BFGS Quasi-Newton (BFG)
trainrp	Resilient Backpropagation (RP)
traincgf	Fletcher-Powell Conjugate Gradient (CGF)
traincgp	Polak-Ribiére Conjugate Gradient (CGP)
trainoss	One Step Secant (OSS)
traingdx	Variable Learning Rate Gradient Descent (GDX)
traingd	Gradient Descent (GD)
traingdm	Gradient Descent with Momentum (GDM)

Determination Coefficient (R^2). The first method of evaluating the network was by creating a regression plot, which indicates the relationship between the actual outputs of the experiment and the predicted target using the neural network model. In common practice, the training set is not perfect, hence the relationship between the outputs and targets will also not be perfect. As an indication of the relationship as well as the accuracy of the predicted model, the magnitude of determination coefficient (R^2) was used. In this experiment, the R value, which denotes the fitting accuracy between the outputs and targets, were obtained for the training set, validation set, testing set and output data, by plotting regression lines for each set of data. From the magnitude of the R value obtained for the output data regression line, R^2 was obtained through Eq. (1), by taking the ratio of the sum of squared output difference to the sum of squared network output, then obtaining its deviation from 1, which is the perfect scenario of outputs [19].

$$R^2 = 1 - \left\{ \frac{\sum_i (t_i - a_i)^2}{\sum_i (a_i)^2} \right\} \tag{1}$$

where,

$\sum_i (t_i - a_i)^2$ is the sum of the squared difference between target and network outputs
$\sum_i (a_i)^2$ is the sum of the squared network outputs

Mean Squared Error (MSE). The next evaluation was based on the mean squared error (MSE) of the target output and network output. As the default performance function in feedforward neural networks, MSE, defined as the average squared difference between the target and network outputs, like R^2, indicates the degree of error between the actual and predicted output values, as defined in Eq. (2).

$$MSE = \frac{1}{N}\sum_{i=1}^{N} (e_i)^2 = \frac{1}{N}\sum_{i=1}^{N} (t_i - a_i)^2 \tag{2}$$

where,
N is the total number of output data in the model
e_i is the difference between the target output and the network output
t_i is the target output
a_i is the network output

Validation Performance. Another method of evaluating the network's performance was by observing the validation performance. By default, the validation performance is quantified using MSE as shown in Fig. 4, and was obtained by plotting the performance of the modelled network with the '*plotperform*' function shown in Fig. 4. This function indicated the number of epochs (iterations) at which the minimal validation performance was achieved.

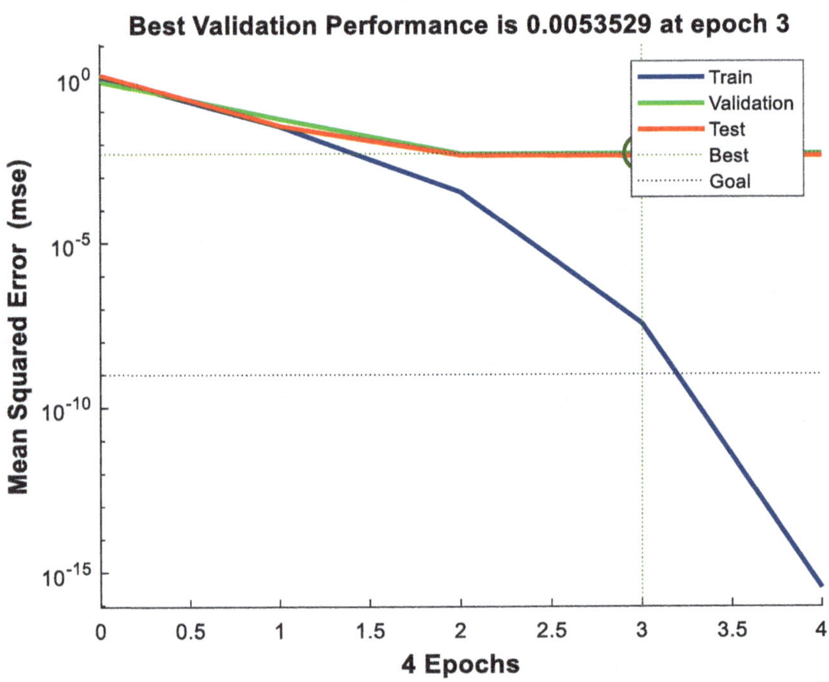

Fig. 5. Validation MSE for Levenberg-Marquardt with 24 hidden nodes

Figure 5 showed that the minimum validation performance was reached at 3 epochs, when Levenberg-Marquardt training algorithm with 24 hidden nodes was used for the feedforward neural network. Based on this figure, the training was successful because the pattern between the validation and test curves were quite similar, with almost equal gradient between the epochs number 1 to 3. Moreover, according to Beale et al. (2016), a significant increase in the test curve as compared to its validation curve, would indicate the occurrence of overfitting (Beale et al., 2016) [28]. However, a major increase was not observed because both curves showed downward trends, as shown in Fig. 4. In addition, the best validation performance's MSE value, which indicates the average squared error between the actual output and target output, was also used determine the most suitable FNN model, whereby a lower value is preferred.

3 Results and Discussion

Artificial neural network (ANN) modelling was implemented in this study to investigate the effect of number of hidden nodes as well as different training functions. For each training function, 1 to 25 hidden nodes (HN) were iteratively trained, whereby five to ten models were trained through trial and error method, in order to obtain the most optimum result for each number of hidden nodes. Using determination coefficient (R^2), output mean squared error (MSE) and validation MSE as performance criteria, the results were obtained for Levenberg-Marquardt training algorithm and tabulated as shown in Table 3.

Based on Table 3, all the hidden nodes provided satisfactory determination coefficient (R^2) of above 0.8, which proved that Levenberg-Marquardt training algorithm was suitable in modelling the slow pyrolysis of banana peel biomass. The high R^2 values of each hidden node also represented good correlation between the experimental output and the simulated ANN output. In order to determine the optimal number of hidden node for Levenberg-Marquardt, each hidden node's performance was further evaluated using graphical method as shown in Fig. 6. Referring to Fig. 6, the highest R^2 value of 0.99858 was obtained at HN24, with low output and lowest validation MSE value of 0.04467 and 0.00001, respectively. Although its output MSE value was not the lowest, however its individual R test value for training, validation and test data were the highest. Hence, based on its comparatively high R^2 value, which is close to 1, as well as the lowest validation MSE, HN24 was selected as the best model for Levenberg-Marquardt technique.

Based on Fig. 6, an optimal ANN model was achieved using 25 hidden nodes for Levenberg Marquardt training algorithm. Besides that, no significant change was observed for R^2 and output MSE between number of hidden nodes 9 to 25. Therefore, it was deduced that 25 hidden nodes were sufficient to develop a functional neural network to model the slow pyrolysis of banana peels to biochar, for any training algorithm. Using the same ANN modelling method as Levenberg-Marquardt, the results for the other training algorithms, as mentioned in Table 2, were obtained as shown in Table 4. Similarly, the performance of each training function was plotted in Fig. 7 to determine the most appropriate and suitable training function.

According to Table 4 and Fig. 7, Variable Learning Rate Gradient Descent (GDX) training algorithm with 4 hidden nodes, exhibited the lowest R^2 value of 0.75353, but with the lowest output MSE of 0.03584. On the other hand, Levenberg-Marquardt (LM) with 24 hidden nodes, showed the highest R^2 value of 0.99858, lowest validation MSE value of 0.00001 and comparatively moderate output MSE value of 0.04467. Moreover, its R value representing the deviation of regression line to expected model output line, for training, validation and testing data were the highest among the training algorithms. This can be clearly seen from Fig. 8(a), (b) and (c), whereby the individual R values were either 1 or approximately 1, indicating a very high correlation between the experimental

Table 3. ANN modelling result using Levenberg-Marquardt training algorithm

No. of HN	R				R^2	Output MSE	Validation MSE
	Training	Validation	Test	Output			
1	0.91003	0.99982	0.95297	0.92324	0.85237	0.10820	0.00009
2	0.91977	1.00000	0.92206	0.90309	0.81557	0.05670	0.01208
3	0.99991	0.92177	1.00000	0.94341	0.89002	0.03833	0.05142
4	0.99801	0.98557	1.00000	0.96207	0.92558	0.04956	0.00525
5	0.98929	0.90876	0.99977	0.90818	0.82479	0.04191	0.05596
6	1.00000	1.00000	0.99904	0.97908	0.95860	0.02844	0.00174
7	0.92899	0.99101	0.98106	0.92051	0.84734	0.07998	0.04716
8	1.00000	1.00000	1.00000	0.96069	0.92293	0.03519	0.03996
9	0.92341	1.00000	0.92874	0.90862	0.82559	0.04030	0.02722
10	1.00000	0.90461	0.95804	0.99139	0.98285	0.03025	0.00730
11	0.98377	1.00000	1.00000	0.96014	0.92187	0.03551	0.00276
12	0.99431	1.00000	0.99716	0.92505	0.85572	0.04340	0.01375
13	1.00000	0.99869	0.96547	0.99159	0.98325	0.03117	0.00303
14	0.97778	0.98392	0.98988	0.96904	0.93904	0.03762	0.00471
15	1.00000	0.95657	0.96187	0.91629	0.83959	0.04670	0.08306
16	1.00000	0.94641	1.00000	0.96591	0.93298	0.04225	0.02854
17	0.94069	0.99916	0.99986	0.94232	0.88797	0.05002	0.00176
18	1.00000	1.00000	0.91867	0.91991	0.84623	0.05132	0.08386
19	0.96588	1.00000	1.00000	0.92216	0.85038	0.05273	0.06454
20	1.00000	1.00000	1.00000	0.97315	0.94702	0.04674	0.00160
21	0.99911	1.00000	0.98482	0.90925	0.82674	0.05806	0.10026
22	1.00000	0.99740	0.99929	0.96246	0.92633	0.04865	0.00602
23	0.95872	1.00000	0.98772	0.96443	0.93013	0.04886	0.00028
24	1.00000	1.00000	1.00000	0.99929	0.99858	0.04467	0.00001
25	1.00000	1.00000	0.91494	0.95999	0.92158	0.05288	0.01593

data and training data. Furthermore, the regression line in Figure (d) suggested that the experimental data output also had a high correlation to the ANN model developed using Levenberg-Marquardt training algorithm with 24 hidden nodes.

Overall, this research is on modelling of slow pyrolysis process for converting banana peels into biochar. In this study the major objective was to determine the effect of pyrolysis parameters on O/C ratio of biochar. Feedforward artificial neural network (FANN) modelling, was done with the data obtained from the slow pyrolysis considering various training algorithms. Trial and error procedure for various activation functions and several hidden neurons were conducted. The performance criterions MSE, R, R^2

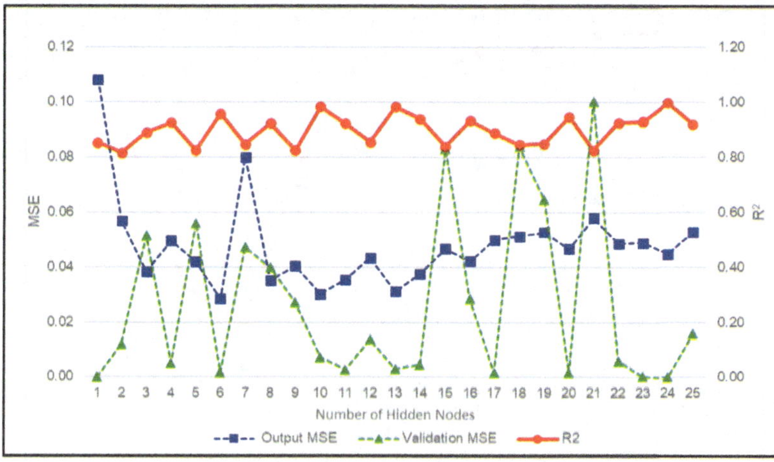

Fig. 6. Graph of Levenberg-Marquardt training algorithm result

Table 4. Summary of ANN results for various training algorithm

Training function	No. of HN	R				R^2	Output MSE	Validation MSE
		Training	Validation	Test	Output			
trainlm	24	1.00000	1.00000	1.00000	0.99929	0.99858	0.04467	0.00001
trainbr	16	0.99891	–	1.00000	0.99836	0.99672	0.04281	0.00035
trainbfg	16	1.00000	1.00000	0.78353	0.99205	0.98416	0.05402	0.01054
trainrp	22	0.99431	0.96864	0.74130	0.99129	0.98266	0.05036	0.00025
traincgf	18	0.98057	0.78548	0.98167	0.98010	0.96060	0.06972	0.00104
traincgp	12	0.99116	1.00000	0.78353	0.98956	0.97923	0.05404	0.00024
trainoss	16	0.95740	0.98298	1.00000	0.94365	0.89048	0.04622	0.01518
traingdx	4	0.89029	1.00000	0.82921	0.86806	0.75353	0.03584	0.00566
traingd	8	0.86697	1.00000	0.98764	0.88742	0.78751	0.04343	0.07166
traingdm	15	0.99281	0.73819	1.00000	0.95703	0.91591	0.04207	0.00732

were used to compare the results obtained from various training algorithms. Out of all the training algorithms tried, the best results were obtained for Levenberg Marquardt training algorithms which provided the lowest MSE value (closer to 0) and the highest R and R^2 (closer to 1). The implications of the ANN modeling is that it has been successful in predicting the experimental response (O:C) that were obtained from the slow pyrolysis experiments. The modeling is able to evaluate or predict the interactions between a pyrolysis parameter and its impact on the final biochar properties.

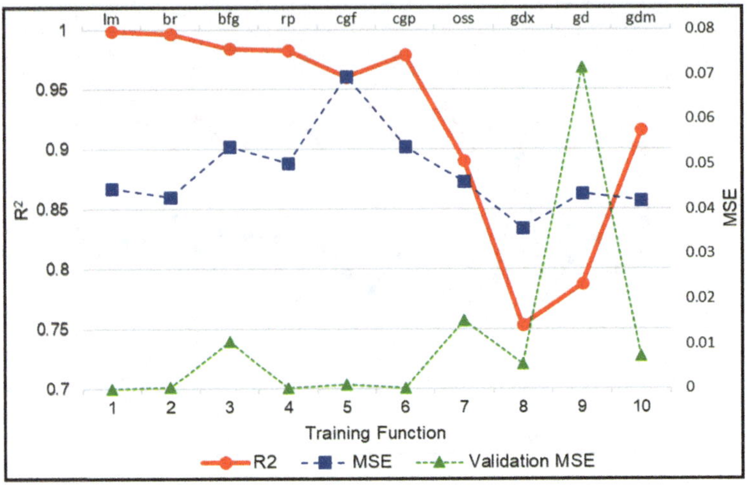

Fig. 7. Performance criteria of various training functions

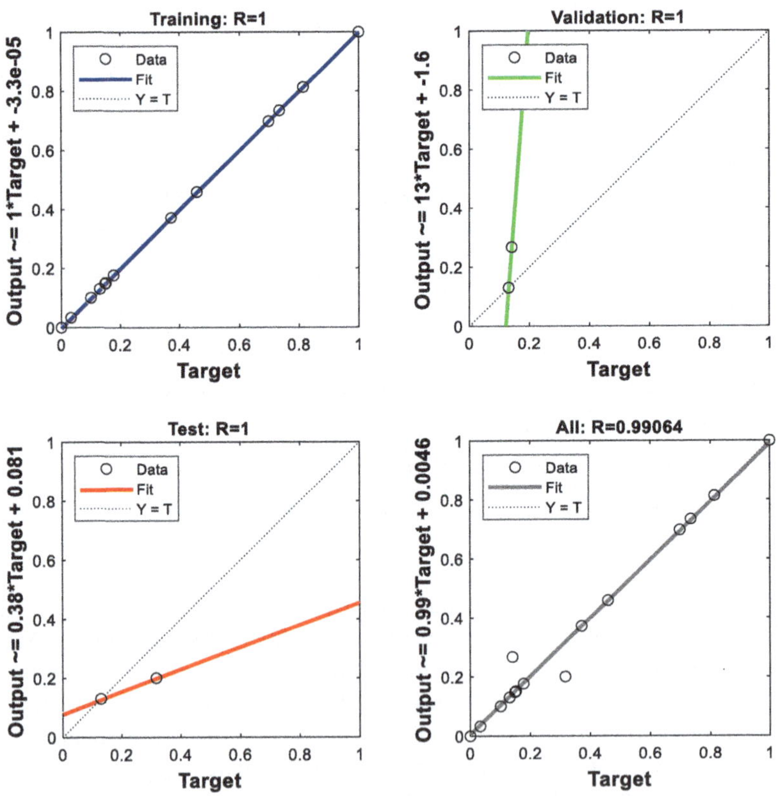

Fig. 8. Regression line for training, validation, testing and output data for Levenberg-Marquardt at HN24

4 Conclusion

In this study, banana peels were used to investigate the effect of pyrolysis parameters such as heating temperature, heating rate and residence time, on the yield and oxygen to carbon (O/C) ratio of biochar. Artificial neural network (ANN) using Levenberg-Marquardt training algorithm with 24 hidden nodes successfully trained and modelled the slow pyrolysis of banana peels, resulting in the highest determination coefficient (R^2) of 0.99858, low output MSE of 0.04467 and lowest validation MSE of 0.00001. Therefore, it was concluded that ANN is compatible in the modelling of pyrolysis works to predict yield and other physicochemical characteristics of biochar. Thus, ANN model can be used to optimally predict the O:C for selected operating condition (temperature, heating rate and residence time).

References

1. Agricultural Biomass in Malaysia (2019). https://www.bioenergyconsult.com/agricultural-biomass-in-malaysia/
2. Yusuf, N., Rosly, E., Mohamed, M., Bakar, B., Yusoff, M., Sulaiman, M., Ahmad, M.: Mater. Sci. Forum **840**, 394–398 (2016)
3. Fao.org. FAOSTAT. https://www.fao.org/faostat/en/#data/QC. Accessed 7 Oct 2020
4. Nasution, Z., Lim, R., Wan Hafiz, W.: Banana peel flour: an alternative ingredient for whole-meal bread. In: Universiti Malaysia Terengganu 11th International Annual Symposium on Sustainability Science and Management (2012)
5. Padam, B., Tin, H., Chye, F., Abdullah, M.: J. Food Sci. Technol. **51**(12), 3527–3545 (2012)
6. Memon, J., Memon, S., Bhanger, M., El-Turki, A., Hallam, K., Allen, G.: Colloids Surf. B: Biointerfaces **70**(2), 232–237 (2009). https://doi.org/10.1016/j.colsurfb.2008.12.032
7. Park, D., Lim, S., Yun, Y., Park, J.: Biores. Technol. **99**(18), 8810–8818 (2008)
8. Montanarella, L., Lugato, E.: The application of biochar in the EU: challenges and opportunities. Agronomy **3**(2), 462–473 (2013)
9. Fryda, L., Visser, R.: Agriculture **5**(4), 1076–1115 (2015)
10. Laird, D.: Agron. J. **100**(1), 178–181 (2008)
11. Speratti, A., Johnson, M., Martins Sousa, H., Nunes Torres, G., Guimarães, C.E.: Agronomy **7**(3), 49 (2017)
12. Schmidt, H., Bucheli, T., Kammann, C., Glaser, B., Abiven, S., Leifeld, J.: Guidelines European Biochar Certificate for a Sustainable Production of Biochar. European Biochar Certificate (2012). https://www.europeanbiochar.org/en/download
13. Daful, A., Chandraratne, M.: Encycl. Renew. Sustain. Mater. **4**, 370–378 (2018)
14. Branca, C., Blasi, C.: Thermochim. Acta **614**, 68–75 (2015)
15. White, J., Catallo, W., Legendre, B.: J. Anal. Appl. Pyrol. **91**(1), 1–33 (2011)
16. Wardach-Święcicka, I., Kardaś, D.: Arch. Thermodyn. **34**(2), 53–71 (2013)
17. Chen, J., Do, Q., Hsieh, H.: Algorithms **8**(2), 292–308 (2015)
18. Stangierski, J., Weiss, D., Kaczmarek, A.: Eur. Food Res. Technol. **245**(11), 2539–2547 (2019)
19. Nalbant, M., Gokkaya, H., Toktaş, İ.: Modelling and Simulation in Engineering, pp. 1–14 (2007)
20. Nawi, N., Atomi, W., Rehman, M.: Procedia Technol. **11**, 32–39 (2013)
21. Kabir, G., Hasin, M.: Int. J. Fuzzy Syst. Appl. **3**(1), 1–24 (2013)
22. Aydinli, B., Caglar, A., Pekol, S., Karaci, A.: Energy Explor. Exploit. **35**(6), 698–712 (2017)
23. Merdun, H., Sezgin, I.: Int. J. Renew. Energy Rese.-IJRER **8**(2), 1178–1185 (2018)

24. Liao, M., Kelley, S., Yao, Y.: Biofuels, Bioprod. Biorefin. **13**(4), 1015–1027 (2019)
25. Sun, Y., Liu, L., Wang, Q., Yang, X., Tu, X.: J. Anal. Appl. Pyrol. **120**, 94–102 (2016)
26. Selvarajoo, A., Muhammad, D., Arumugasamy, S.K.: Model. Earth Syst. Environ. **6**(1), 115–128 (2019)
27. Selvanathan, M., Khoo, T.Y., Chang, H.C., Selvarajoo, A., Arumugasamy, S.K., Sethu, V.: Water, Air & Soil Pollution (2017)
28. Arumugasamy, S.K., Selvarajoo, A., Tariq, A.: Mater. Sci. Energy Technol. **3**, 868–878 (2020)
29. Selvarajoo, A., Oochit, D.: Effect of pyrolysis temperature on product yields of palm fibre and its biochar characteristics. Mater. Sci. Energy Technol. **3**, 575–583 (2020)
30. Beale, M., Hagan, M., Demuth, H.: Neural Network Toolbox™ User's Guide, pp. 2–29. The MathWorks, Inc. (2016)
31. Hornik, K., Stinchcombe, M., White, H.: Neural Netw. **2**(5), 359–366 (1989)
32. Thimm, G., Moerland, P., Fiesler, E.: Neural Comput. **8**(2), 451–460 (1996)
33. Mohd Najib, N., Sethu, V., Arumugasamy, S.K., Selvarajoo, A.: Environmental Processes (2020)

Effect of Potting Encapsulation on Crack Formation and Propagation in Electronic Package

Azman Jalar[1,2], Syed Mohamad Mardzukey Syed Mohamed Zain[1,3], Fakhrozi Che Ani[3], Mohamad Riduwan Ramli[3], and Maria Abu Bakar[1(✉)]

[1] Institute of Microengineering and Nanoelectronics, Universiti Kebangsaan Malaysia, 43600 Bangi, Selangor, Malaysia
maria@ukm.edu.my

[2] Department of Applied Physics, Faculty of Science and Technology, Universiti Kebangsaan Malaysia, 43600 Bangi, Selangor, Malaysia

[3] Western Digital®, SanDisk Storage Malaysia Sdn. Bhd., Plot 301A, Persiaran Cassia Selatan 1, Taman Perindustrian Batu Kawan, MK13, Batu Kawan, 14100 Seberang Perai Selatan, Penang, Malaysia

Abstract. Potting encapsulation inside plastic housing of electronic components has been investigated to study the encapsulation failure. Crack is one indicator of encapsulation failure. This type of defect causing the rejection of the product; thus, the crack occurrence must be assessed with respect to the potting encapsulation. Potting material was established inside the plastic housing of the surface-mount device (SMD) transformer. Cross-sectional observation of the soldered component was studied through optical and electron microscopy. Results reveal that the presence of cracks at the corner of the SMD transformer was detected. The further cross-sectional analysis suggested the presence of voids in the region around the cracks. Further crack formation and propagation was found from potting encapsulation to the plastic housing region. This behavior is observed in the corner of the SMD transformer.

Keywords: Crack · Potting process · Potting materials · Epoxy · Encapsulant failure

1 Introduction

Encapsulation technology plays a significant role in ensuring the functionality and structural integrity of the electronic product. For example, in encapsulation technology, the epoxy mold compound is used in semiconductor packaging to protect silicon, leadframe, and interconnection from the environment. On the other hand, some of the electronic products used epoxy materials as potting material. Various type of potting materials widely uses such as polybutadiene, polycarbonate (PC), polypropylene (PP), polyurethane (PUR), polyamide, polyvinyl chloride (PVC) and etc. [1]. Some reports also investigated about silicon-based compound for potting encapsulation [2].

J. A. Mat Jizat et al. (Eds.): iCITES 2020, AISC 1350, pp. 351–357, 2021.
https://doi.org/10.1007/978-3-030-70917-4_33

Potting technology is considered as end-stage of the electronic packaging process. The technology involves a process of filling a complete electronic assembly with a suitable epoxy material to form a protective encapsulant for electronics component from shock and vibration, exclusion of liquid, moisture, or corrosive agents [3, 4]. Furthermore, this kind of encapsulation also protects sensitive electronic components from impact, vibration, and loose interconnection. It's also ensured long term electronic component performance and functionality by providing continuous electrical supply.

The potting encapsulation is widely used for electronic components for a harsh environment, such as aerospace and automotive applications [5]. For example, in the automotive application, various electronics components may be exposed to dirt, liquids, moisture, vibration, and extreme temperatures. The encapsulation materials are very crucial in electronic devices. Significant attention has been given on this matter where as Phua et al. [6] reported on development of high temperature polymeric encapsulation material for extreme environment electronics packaging.

Potting encapsulation on the electronics component can significantly improve reliability and protect against premature aging and failure. It has been suggested that the failure of potting encapsulation can occur either during the package assembly process or in service/device operation [7–9]. In particular, during the process of potting encapsulation and assembly of microelectronic package on printed circuit board, the package normally undergoes high temperature of solder reflowing process. These steps may induce delamination of the potting encapsulant. On the other hand, the external forces and load also facilitate the delamination to occur. Therefore, combination factors of existing vapor, moisture, process temperature and internal/external load influence the delamination and may extended to failures such as cracks. This paper describes the incidence of a crack in a part of a surface-mount device (SMD) transformer. In this paper, crack formation and propagation will be further discussed.

2 Materials and Method

The surface mount device (SMD) transformer is used in this study, containing 12 pieces of lead, as shown in Fig. 1. Figure 2 below shows the process flow for potting encapsulations. The SMD transformers were then connected to a printed circuit board (PCB) after the potting encapsulation process using lead-free solder joint based Sn-Ag-Cu material. According to the Pb-free standard reflow profile, the SMD transformers were reflowed at 250 °C, as shown in Fig. 3 [10, 11]. The cracked samples were examined after the reflow soldering process using an optical microscope and the Field Emission Scanning Electron Microscope (FESEM). Cross-section of the samples were prepared prior characterization of FESEM using standard metallographic technique. The samples were cold-mounted using resin and hardener, grinding process using abrasive paper of silicon carbide (SiC) from 800 grit up to 2000 grit and final process of polishing using polishing cloth and diamond spray containing 0.25 μm diamond particles [12–14].

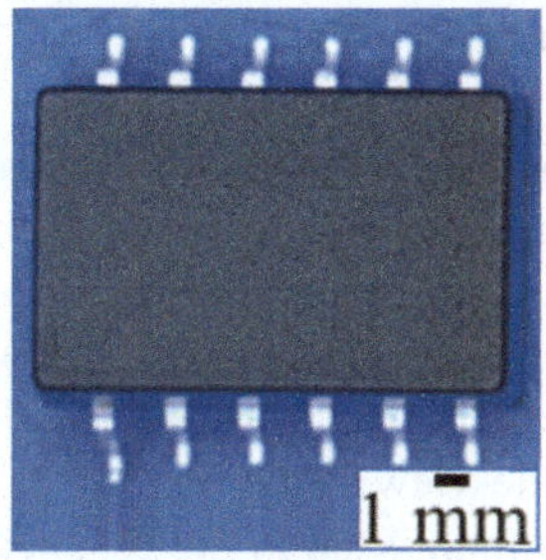

Fig. 1. Surface mount device (SMD) transformer

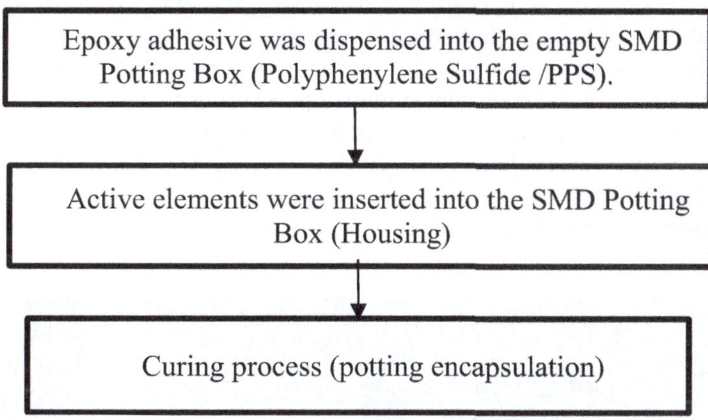

Fig. 2. Process flow for potting encapsulations

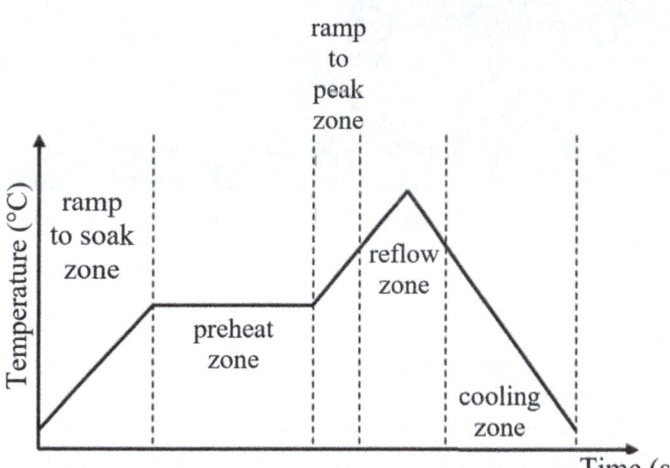

Fig. 3. Schematic of the typical Pb-free reflow soldering process

3 Results and Discussion

Figure 4 shows the top-view of the SMD transformer after the potting encapsulation process. It is observed that some of the crack lines are visible on the top-view of the SMD transformer along the lead area. It is clearly shown that cracks have been propagated on the SMD transformer's edge or corner side (Fig. 4(b)). Figure 5 shows the side-view of SMD transformer: (a) before cross-section and (b) after cross-section on side-view and (c) after cross-section on side-view vertically.

The void formation is a common issue in the IC encapsulation process in whatever types of encapsulation method. The presence of a void in the package can lead to the package cracking (popcorning). Hence, this kind of defect causing the rejection of the product. Several potential sources lead to the package's void formation during the potting encapsulation process, such as moisture absorbed, handling procedure, inadequate baking process/parameters, or exposed to vapor. Other points to consider from encapsulation process such as inadequate temperature of post curing process, unmatched potting encapsulant ratio and mismatch properties of potting encapsulant. Moreover, the curing quality of the potting encapsulant is affected by mixing process. Thus, air trap will occur during initial mixing process of potting materials together with viscosity of encapsulant and high condition of moisture content.

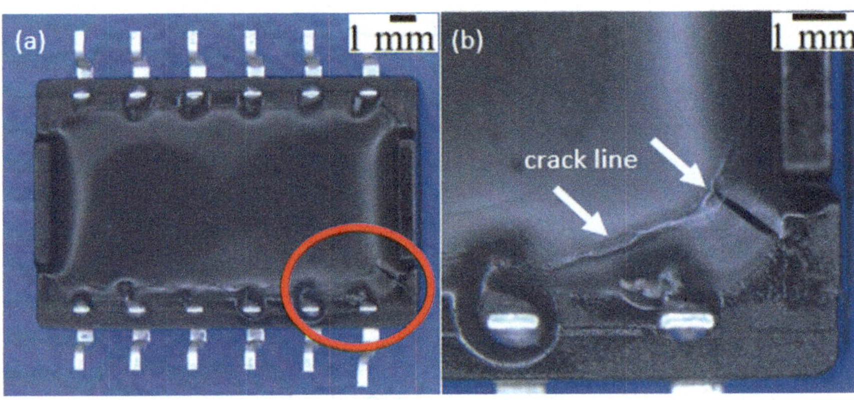

Fig. 4. SMD transformer after potting encapsulation process: (a) top-view and (b) enlarge the image of the red circle in (a)

During the microelectronic package's assembly process, the variation of temperatures involved from room temperature and up to more than melting point of solder, 220 °C. The glass transition temperature of typical epoxy mold compound approximately between 110 °C to 200 °C. This process such as surface mount technology requires high temperatures in order to complete the microelectronic package assembly process. During the reflow process at high temperature, the pre-existing moisture that readily available inside the potting encapsulant will evaporates into vapor state. This condition induced the physical and mechanical instability due to vapor pressure, thermal mismatch, surface-interface incompatible and hygroscopic swelling leading to delamination and de-adhesion at the interface of potting encapsulant. Furthermore, vapor pressure and internal stress induced by temperature profile of reflow soldering will also cause the occurrence of crack or also known as popcorning [15, 16]. The popcorning described the trapped of high-pressure vapor in the potting encapsulation escaped out of the package and induced crack formation and propagation [17].

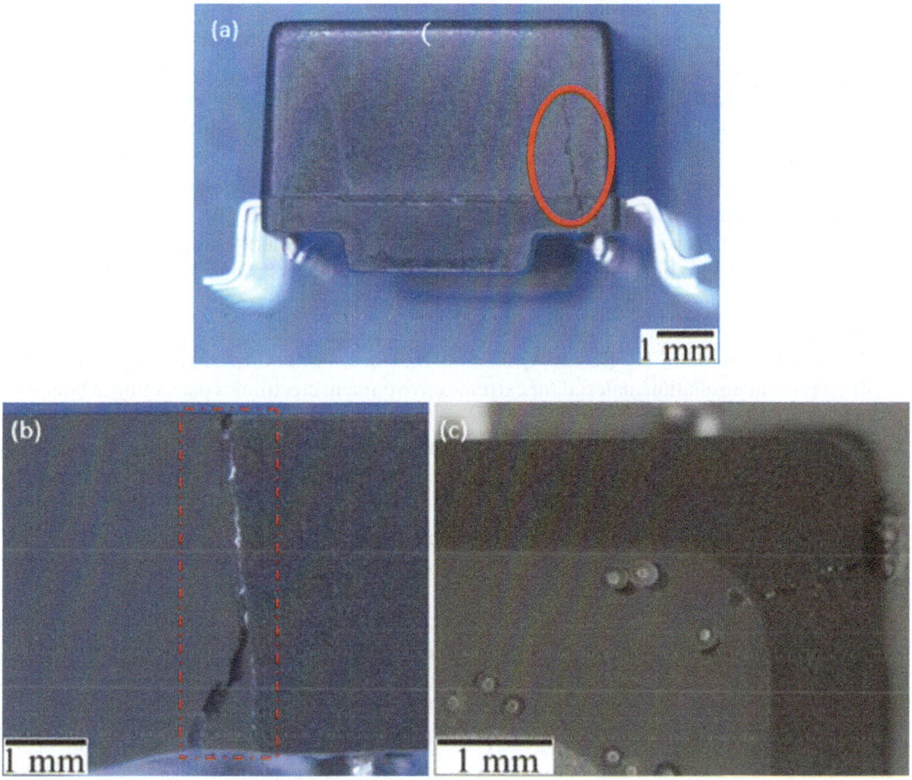

Fig. 5. Side-view of SMD transformer: (a) before cross-section and (b) after cross-section and (c) after cross-section on side-view vertically

4 Conclusions

The crack occurrence was observed at the corner of SMD transformer. The cross-sectional analysis suggested that voids appear in the surrounding area of the cracks. Potting materials factor may induce voids formation, and then the voids coalesce to make popcorning possible. This led to crack formation and propagation towards plastic housing from potting encapsulation. This behavior is observed at the corner of SMD transformer.

Acknowledgement. The authors would like to acknowledge the financial support provided by Western Digital via SanDisk Storage Malaysia Sdn. Bhd through research grant (RR-2020-004) and collaboration activities with Universiti Kebangsaan Malaysia.

References

1. Kögler, P.: Thermally conductive potting solutions for electric motors. Adhes Adhes Sealants **16**, 20–23 (2019)
2. Srivastava, T., Katari, N.K., Ravuri, B.R., et al.: Investigation of high-temperature stability and thermal endurance of silicone potting compound by thermo-gravimetric analysis. Silicon (2020)
3. Gladkikh, S.N., Basharina, E.N., Troitskaya, O.L., et al.: Development and tests of thermally stable potting and impregnating compounds. Polym. Sci. Ser. D **5**, 145–149 (2012)
4. Lall, P., Islam, S., Dornala, K., Suhling, J., Shinde, D.: Nano-underfills and potting compounds for fine-pitch electronics. In: Morris, J. (ed.) Nanopackaging. Springer, Cham (2018)
5. Sonderhoff Holding GmbH: Sealing, gluing, potting - multiple solutions for the automotive industry. Adhes. Adhes. Sealants **16**, 14–19 (2019)
6. Phua, E.J.R., Liu, M., Cho, B., Liu, Q., Amini, S., Hu, X., Gan, C.L.: Novel high temperature polymeric encapsulation material for extreme environment electronics packaging. Mater. Des. **141**, 202–209 (2018)
7. Ardebili, H., Zhang, J., Pecht, M.G.: Defect and failure analysis techniques for encapsulated microelectronics (Chap. 8). In: Ardebili, H., Zhang, J., Pecht, M.G. (eds.) Encapsulation Technologies for Electronic Applications. Materials and Processes for Electronic Applications, 2nd edn., pp. 317–373. William Andrew Publishing, Norwich (2019)
8. Zhang, H., Che, F., Lin, T., Zhao, W.: Reliability and failure analysis of encapsulated packages (Chap. 5). In: Zhang, H., Che, F., Lin, T., Zhao, W. (eds.) Modeling, Analysis, Design, and Tests for Electronics Packaging beyond Moore. Woodhead Publishing Series in Electronic and Optical Materials, pp. 245–292. Woodhead Publishing, Sawston (2020)
9. Pei, C., Fu, G., Wan, B., Zhong, L., Jia, M.: Failure analysis of plastic encapsulated circuits after autoclaving. Eng. Fail. Anal. **66**, 421–431 (2016)
10. Jalar, A., Bakar, M.A., Ismail, R.: Temperature dependence of elastic-plastic properties of fine-pitch SAC 0307 solder joint using nanoindentation approach. Metall. Mater. Trans. A **51**, 1221–1228 (2020)
11. Ismail, N., Jalar, A., Abu Bakar, M., Ismail, R., Ibrahim, S.: Effect of flux functional group for solder paste formulation towards soldering quality of SAC305/CNT/Cu. Solder. Surf. Mount Technol. **32**(3), 157–164 (2020)
12. Che Ani, F., Jalar, A., Saad, A.A., Khor, C.Y., Abas, M.A., Bachok, Z., Othman, N.K.: Characterization of SAC-x NiO nano-reinforced lead-free solder joint in an ultra-fine package assembly. Solder. Surf. Mount Technol. **31**(2), 109–124 (2019)
13. Wan Yusoff, W.Y., Ismail, N., Safee, N.S., Ismail, A., Jalar, A., Abu Bakar, M.: Correlation of microstructural evolution and hardness properties of 99.0Sn-0.3Ag-0.7Cu (SAC0307) lead-free solder under blast wave condition. Solder. Surf. Mount Technol. **31**(2), 102–108 (2019)
14. Ani, F.C., Jalar, A., Ismail, R., et al.: Reflow optimization process: thermal stress using numerical analysis and intermetallic spallation in backwards compatibility solder joints. Arab. J. Sci. Eng. **40**, 1669–1679 (2015)
15. Lau, J., Chen, R., Chang, C.: Real-time popcorn analysis of plastic ball grid array packages during solder reflow. In: Electronics Manufacturing Technology Symposium, pp. 455–463 (1998)

16. McCluskey, P., Munamarty, R., Pecht, M.: Popcorning in PBGA packages during IR reflow soldering. Microelectron. Int. **14**(1), 20–23 (1997)
17. Gannamani, R., Pecht, M.: An experimental study of popcorning in plastic encapsulated microcircuits. IEEE Trans. Compon. Packag. Manuf. Technol. Part A **19**(2), 194–201 (1996)

Novel Approach of Class Incremental Learning on Internet of Things (IoT) Framework

Swaraj Dube(✉), Yee Wan Wong, Jeen Ghee Khor, and Hermawan Nugroho

University of Nottingham Malaysia Campus, 43500 Semenyih, Selangor, Malaysia
{kecy3dsm,hermawan.nugroho}@nottingham.edu.my,
yeewan6823@gmail.com, jeenkhor@gmail.com

Abstract. Class incremental learning and Internet of Things (IoT) are related but have not yet been extensively integrated. With the exponential rise in the number of IoT devices, the total amount of data produced is massive, thus it is infeasible to send all the data directly to the cloud. Training deep learning models in an incremental manner can help saving the training time and reduce the computational load required to train the model. In this work, we develop and demonstrate an incremental class learning system using only features of images to reduce large memory footprints in comparison with using raw images. Our experiments show that the proposed method achieves similar accuracy within a shorter time frame. We also develop a distributed training method of convolutional neural networks (CNNs) between an IoT edge device and a cloud. The IoT edge device carries out feature extraction to reduce the transmission load to the cloud which also eases the training cost. The cloud carries out the classifier training.

Keywords: Class incremental learning · Internet of Things · Convolutional neural networks

1 Introduction

In the computer vision domain, deep learning has shown great amount of success and in some tasks even surpassing human accuracy level. A lot of this success has been obtained in the offline setting in which all the data is already present before training. The deep learning model is trained on this dataset and then it is deployed. In the real world however, not all data can be available beforehand. Even if it is available, to train a model with huge dataset will require a powerful hardware system and will consume a lot of time. Another problem with training a deep learning model offline is that once it is trained and deployed, the model will stop learning its parameters in the future. In practice however, deep learning models should be able to learn in a continuous environment when new data arrives.

Class incremental learning refers to a deep learning framework which can learn new tasks over time for e.g. classification. For example, class incremental learning can refer to model to classify cars which after some time, the same learning model is able to learn to classify bikes while retaining knowledge of cars. The biggest challenge with continuous

J. A. Mat Jizat et al. (Eds.): iCITES 2020, AISC 1350, pp. 358–367, 2021.
https://doi.org/10.1007/978-3-030-70917-4_34

learning is a phenomenon known as catastrophic forgetting [1] whereby representations of the old classes are lost when the model is fine-tuned on the new data. The most basic way to mitigate this problem is to use a combination of both the old and new data and use this combined dataset to train the model. This approach however needs high memory demand at every increment because at every increment, images will need to be saved. Researchers have used memory modules to store part of exemplars from the old task [2], and generative model to generate exemplars to not only avoid additional memory footprint but also avoid data privacy issues [3].

With majority of class incremental learning research are carried out on a centralized computer, there is a clear need for incremental learning to be approached from IoT perspective. Because in the real world, IoT devices that are collecting data, can be geographically spread out. In a real IoT environment, it is infeasible to make all the devices send raw data to the cloud because of bandwidth costs and the sheer amount of load the cloud would have to handle. Since the IoT edge device is the first device that receives new raw data, it will be much more feasible if a certain parts of the deep learning model are deployed on the IoT edge device in an effort to assist the cloud in the entire continuous learning process [4].

When a given deep learning model is trained in a distributed manner between 2 or more devices, it will come with certain challenges such as 1) deciding how many and which specific layers of the model to run on the edge and cloud; 2) How to carry out the training in a distributed manner i.e. in an setting whereby an IoT edge device and the cloud are present.

In this paper, we present an approach of class incremental on cloud-edge architecture via data sampling of new classes. Most methods are able to perform data sampling only for tasks that have been learnt by AI models whereas there is not much work available that demonstrates the effect of data sampling from completely new classes i.e., tasks for which the AI model has no knowledge about. We use image features as exemplars to perform class incremental learning and develop a distributed learning by partitioning CNNs between the IoT edge device and the cloud.

2 Related Work

In the incremental learning context, there are 3 main approaches: regularization, rehearsal, and dual memory system approaches. In the regularization approach, the loss function is designed to retain the representations of the old classes, i.e., by not changing the values of the important parameters. In the rehearsal approach, the focus is on using a mixture of both the old and new data in appropriate proportions. It is assumed that using the old data, the old knowledge is retained.

Incremental classifier and representation learning (iCaRL) [2] uses both rehearsal and regularization approaches. This approach uses exemplars of the old classes together with the new training data for learning. Since it is not feasible to keep storing exemplars, every time a task is learnt, the number of exemplars per class are decreased by selecting a limited number of exemplars that are closest to their respective class mean. The downside of this approach is that the more the number of classes that are observed, the lesser the number of exemplars will be available for each old task. They also store raw images as exemplars which is highly memory demanding. Another drawback is that their method involves end-to-end learning i.e., all the layers of the model are being modified throughout the entire training process. In such scenarios, for every mini batch of data, transmitted to the cloud, the gradients must also be transmitted back to the IoT edge device followed by backpropagation on the IoT edge device. The computational expense will grow in direct proportion to the number of layers on the IoT edge device which makes the approach unattractive for edge-cloud approach.

Incremental classifier learning with GAN (ICwGAN) [5] uses a similar approach to iCaRL [2]. They however use GAN to generate images instead of using raw images. Though this approach solves the issue of memory and privacy, the computational overhead of training and running a GAN especially on IoT edge devices will be very demanding. REMIND your neural network to prevent catastrophic [6] is another approach that tackles the problem of catastrophic forgetting using rehearsal however, they store quantized tensors for rehearsal which is a lot less memory demanding. As their model learns incrementally, at the end of each incremental training step, they pass these images through the convolutional layers. The output features are then quantized and stored as exemplars. To apply the approach to edge-cloud setting, a further investigation on data partition should be performed. Another method, Autoencoder-Based Incremental Class Learning without Retraining on Old Data [7], proposes to use an autoencoder as a classifier instead of the traditional SoftMax classification layer. The reason is that when using a SoftMax layer, neurons must be added to the layer whenever a new class is encountered. Using an autoencoder will avoid this issue. They first perform base training whereby the model is trained on a few initial classes and only after this stage, the incremental learning begins where Synaptic intelligence [8] or Memory aware synapses [9] loss is added to reduce the effect of forgetting. The autoencoder however could not be generic and should be designed according to the input/output requirements.

3 Methodology

Figure 1 shows our proposed system flowchart. Feature extractors of CNNs are run on the IoT edge device and the classifier is run on the cloud. The weights of the classifier are randomly initialized. The choice of such a partition is because in a deep CNN, majority of the parameters reside in the fully connected layers which implies a huge number of matrix multiplications. It means a high computational cost especially during backpropagation training, especially when dealing with high dimensional data such as images. The learning rate of the feature extractor is set to 0 indicating that the feature extractor parameters will not be modified any further. This also implies that once backpropagation takes place on the cloud, there is no longer need to transmit the gradients back the IoT edge device. At every incremental training batch, n number of classes are trained together at a time. Once new class data arrives, a certain percentage of the overall data on the IoT edge device is discarded randomly after which the selected data samples are forward propagated through the feature extractor. The output (feature data) is then converted to a string format and the string converted feature map output tensor is concatenated to the buffer that stores all the tensor converted strings. This buffer is then transmitted to the cloud via the TCP/IP protocol. The size of this buffer reduces in direct proportion to the number of samples discarded randomly which leads to reduction in the transmission cost.

The cloud on the other hand listens to the incoming stream from the IoT edge device. Once the cloud received the bytes from the IoT edge device, it converts the stream to tensors after which the classifier training takes place. At the end of the training, the size of the SoftMax classification layer increases with respect to the number of tasks to be learnt at a time for e.g., if 2 tasks are being learnt at a time then 2 neurons are added in the SoftMax classification layer. The message format is shown in Table 1.

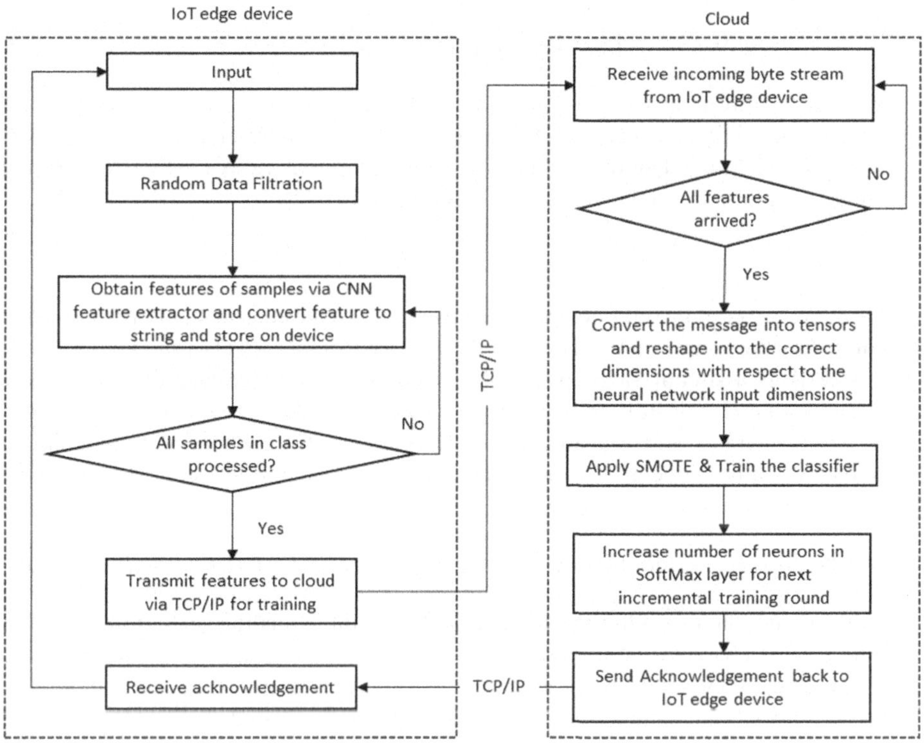

Fig. 1. Flowchart of our framework.

Table 1. Message formats of events to transmit to the cloud.

Event	Message format
Training	*<tensor converted string>t!*
Append tensors	*<tensor converted string>a!*
Process termination	*d!*

3.1 Communication Format Between the IoT Edge Device and the Cloud

Since multiple feature maps can be sent to the cloud, let FM_n represent the feature at index n in the feature map. And let D_n, W_n, H_n be the depth, width, and height respectively of feature n. Along with the feature maps, the associated labels must also be transmitted to the cloud. Since each feature map has 1 unique label, let l_n be the label of feature n, let L_{str} be the string of labels and let N will be the total number of feature maps. The label format can be written as shown in Eq. (1).

$$L_{str} = l_n, l_{n+1}, l_{n+2}, \ldots l_{n+N} \tag{1}$$

A single feature map will have a depth of 1 but the same width (W_n) and height (H_n) as the overall feature map. Let S_n be the size of the feature map n which in fact is $1{*}W_n{*}H_n$ and let $val_{n,i}$ denote the value of feature map n at index i. The string format of a single feature map can be expressed as shown in Eq. (2).

$$FM_n = val_{n,i}, val_{n,i+1}, val_{n,+2}, \ldots val_{n,S} \tag{2}$$

All values are comma separated when converted to string and each feature map string is separated by the character '*k*'. The overall message format (M) is as shown in Eq. (3). Since TCP/IP is a stream based protocol and not a message based protocol, there is no guarantee that all transmitted bytes will be received at the recipient device which is why a unique character must be attached to the message string whereby if the recipient device receives this character, it knows that all the bytes have been received. We use the character '*!*' as the unique character.

$$M = D_n, W_n, H_n, < label\ string >, FM_n, k, FM_{n+1}, k, \ldots FM_N, < action >! \tag{3}$$

On the cloud, the incoming stream is accepted, and the reading operation continues until the unique character is not detected. Once the unique character is detected, the action character is obtained and based on that the appropriate process is carried out. For example, if the action character is '*t*' then training will be carried out, and if the action character is '*d*', then the entire training process will stop on the cloud.

At the end of every action that is carried out on the cloud, an acknowledgement message ("*ok!*") is sent to the IoT edge device. And upon receiving this message, the IoT edge device proceeds with processing the next mini batch in the dataset i.e., converting the images and labels to string and this process continues until all the samples in the dataset are processed.

3.2 Incremental Learning on the Cloud

For illustration purposes, Fig. 2 shows how neurons are added to the final SoftMax classification layer in our incremental learning algorithm. Figure 2 shows the case where 2 classes are being learnt at a time however, any number of classes can be learnt at a time depending on the application at hand and the number of classes being learnt at a time is equal to the number of neurons being added to the final SoftMax classification layer.

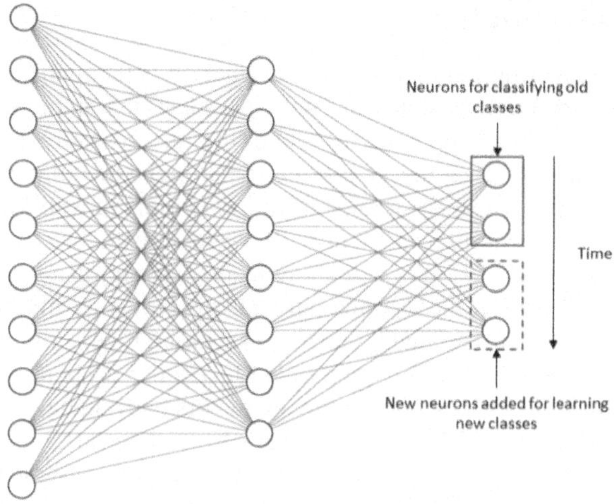

Fig. 2. Dynamically adding neurons to the final SoftMax classification layer for learning new classes.

We carry out the learning algorithm on the cloud due to the well-known computational costs of training fully connected layers especially on high dimensional data such as images or even the features of images. We use the rehearsal technique to carry out incremental learning i.e., using all of the training samples of the old class and combining them with the samples from the new class and then jointly carrying out the training. The mathematical representation of this form of data concatenation is shown in Eq. (4).

$$E_t = \varnothing(x_t) \cup \varnothing(x_{t-1}) \cup \varnothing(x_{t-2}) \cup \ldots \varnothing(x_0) \tag{4}$$

In Eq. (4), let x be all the samples of an incremental training batch and $\varnothing(x)$ be the output of the feature extractor and t be the current incremental iteration. E_t is defined as the exemplar set at iteration t and represents the new set of batches being fed into the classifier. The new batch of feature maps and their respective labels are randomly shuffled every epoch to maintain the classification accuracy. Next, the combined feature maps are passed through the fully connected layers. During the backpropagation process, only the fully connected layers are updated (we freeze the feature extractor layers). This is known as joint training and is one of the most common techniques to alleviate catastrophic forgetting.

4 Experiments

4.1 Datasets

We evaluate our proposed edge-cloud approach on the CIFAR-100 [10] dataset and, CUB-200 [11] dataset. CUB-200 [11] dataset contains 50000 RGB training images of size 32 × 32 × 3 whereby CUB-200 [11] has a total of 200 classes which has a different number of samples per class and different sized images.

4.2 Implementation Details

The final SoftMax layer initially has n neurons where n is the number of classes to be learnt at a time and the number of neurons grow per incremental training batch. For the communication between the edge and the cloud, the features were encoded to a utf-8 format and then transmitted via the TCP/IP protocol. Since the aim of this work is not representation learning, we use pre-trained feature extractors of well-known CNNs such as SqueeezeNet [12] and ShuffleNet V2 [13]. These feature extractors reside on the IoT edge device that carry out the job of extracting features from images, no learning takes place on the IoT edge device, the learning only takes place on the cloud where the fully connected layers reside. The Adam Optimizer [14] is used together with the categorical cross entropy loss function. The fully connected layers use dropout with a probability of 0.25. We use the SqueeezeNet [12] feature extractors to evaluate on CIFAR-100 [10] dataset and the ShuffleNet V2 [13] feature extractor to evaluate on CUB-200 [11] dataset. When evaluating on CIFAR-100 [10] dataset, the learning rate is set to 0.0001, the total number of epochs are set to 50 whereas when evaluating on CUB-200 [11], the total epochs are set to 25 and the learning rate is set to 0.0015. The PyTorch [15] library is used to program the approach and the experiments. Google cloud platform is used as the cloud whereby the hardware used is the Nvidia Tesla K80 GPU. A laptop with a quad core i7 processor, is chosen as the IoT edge device. SqueeezeNet [12] and ShuffleNet V2 [13] are chosen as the CNN models as they are lightweight models and more suitable for embedded devices.

5 Results and Discussions

Table 2 and Table 3 show the classification accuracy results when learning different number of classes at every incremental training round on the CIFAR-100 [10] dataset and using the feature extractor of the pre-trained SqueezeNet [12] model. For each class we choose 500, 400 and 300 training samples to be transmitted to the cloud.

Table 2. Learning 20 classes at a time using SqueezeNet [12] on CIFAR-100 [10] dataset.

Classification accuracy after learning every 20 classes (%)						
Number of samples	20	40	60	80	100	Total training time (s)
500	83.75	71.70	65.28	60.75	57.07	288.22
400	84.15	71.53	64.73	60.43	56.07	230.17
300	83.45	70.23	63.38	59.06	55.58	173.65

The purpose of this experiment is to observe the relationship between the classification accuracy of the model with respect to the number of samples discarded at the IoT edge device. Training the model with 500 samples per class is the baseline because that's the maximum number of samples in each class in the CIFAR-100 [10] dataset. From the results, we can observe that when 100 or even 200 samples are discarded at the IoT

Table 3. Learning 50 classes at a time using SqueezeNet [12] on CIFAR-100 [10] dataset.

Classification accuracy after learning every 50 classes (%)			
Number of samples	50	100	Total training time (s)
500	67.44	57.28	152.65
400	67.32	56.55	113.97
300	65.72	55.48	82.39

edge device, the classification accuracy at every incremental training round is within 2% as compared to the baseline. Discarding 100 and 200 samples at the IoT edge device is equivalent to 20% and 40% discarding rate in exchange of only a 2% difference in the model performance. It can also be observed that in some cases, even after data sampling, the classification accuracy at certain incremental training rounds is better than the baseline. The reason for this is the stochasticity of neural networks, every time incremental learning takes place, the weights associated with the newly assigned neurons are randomly initialized which causes slight variances in the learning thus slightly affecting the accuracies. Table 4 and Table 5 show the classification accuracy results when learning different number of classes at every incremental training round on the CUB-200 [11] dataset and using the feature extractor of the pre-trained ShuffleNet V2 [13] model. For each class we transmit all samples to the cloud and for data sampling we discard 5% of the training samples in each class.

Table 4. Learning 40 classes at a time using ShuffleNet V2 [13] on CUB-200 [11] dataset.

Classification accuracy after learning every 40 classes (%)						
Number of samples discarded per class (%)	40	80	120	160	200	Total training time (s)
0	71.99	59.94	49.26	41.98	37.15	182.21
5	72.09	58.62	47.59	40.46	35.92	174.95

Table 5. Learning 50 classes at a time using ShuffleNet V2 [13] on CUB-200 [11] dataset.

Classification accuracy after learning every 50 classes (%)					
Number of samples discarded per class (%)	50	100	150	200	Total training time (s)
0	68.97	56.60	44.52	38.43	156.32
5	68.90	55.27	44.03	37.51	146.71

The CUB-200 [11] dataset contains 200 classes which is double the amount of the total classes available in the CIFAR-100 [10] dataset however the number of samples in each class in the CUB-200 [11] dataset is a lot lesser which is why it is important to test whether data sampling is effective for datasets with small number of samples per class and from the results, it can be seen that even in this case, the classification accuracy at every incremental training round is within 2% of the baseline. It all the experiments, it can be seen that the training time on the cloud reduces with respect to data sampling at the IoT edge device.

6 Conclusion

In this work we present a class-incremental learning method on IoT framework where the approach is applied on edge-cloud architecture. The proposed model uses image features as exemplars for training instead of raw images. The computing and learning process is partitioned between an IoT edge device and cloud where the IoT edge device's role is to extract features from raw images. To evaluate the performance of the proposed approach, we run the approach on CIFAR-100 [10] and CUB-200 [11] datasets using feature extractors of SqueezeNet [12] and ShuffleNet V2 [13] respectively. Experiments show the significance of the proposed approach which is that it is possible to discard certain number of samples of each class on the IoT edge device to reduce the transmission cost between the IoT edge device and the cloud and still maintain the performance of the incremental learning model. Along with this it can also be seen from the experiments conducted that data sampling also reduces the training time on the cloud.

7 Future Work

An algorithm is required that can dynamically discard samples per category irrespective of the dataset size instead of discarding just a fixed number of samples. A loss function must also be designed that can place importance on the new arriving classes on the cloud in order to get rid of oversampling as oversampling can be computationally expensive especially if the imbalance between classes is very high.

References

1. French, R.M.: Catastrophic forgetting in connectionist networks. Trends Cogn. Sci. **3**, 128–135 (1999). https://doi.org/10.1016/S1364-6613(99)01294-2
2. Rebuffi, S.A., Kolesnikov, A., Sperl, G., Lampert, C.H.: iCaRL: incremental classifier and representation learning. In: Proceedings - 30th IEEE Conference on Computer Vision and Pattern Recognition. CVPR 2017. January 2017, pp. 5533–5542 (2017). https://doi.org/10.1109/CVPR.2017.587
3. Kemker, R., Kanan, C.: FearNet: brain-inspired model for incremental learning. In: 6th International Conference for Learning Representations. ICLR 2018 - Conference Track Proceedings, pp. 1–164 (2018)
4. Chen, J., Ran, X.: Deep learning with edge computing: a review. Proc. IEEE (2019). https://doi.org/10.1109/JPROC.2019.2921977
5. Wu, Y., Chen, Y., Wang, L., Ye, Y., Liu, Z., Guo, Y., Zhang, Z., Fu, Y.: Incremental classifier learning with generative adversarial networks (2018). https://arxiv.org/abs/1802.00853

6. Hayes, T.L., Kafle, K., Shrestha, R., Acharya, M., Kanan, C.: REMIND your neural network to prevent catastrophic forgetting (2019). https://arxiv.org/abs/1910.02509
7. Choi, E., Lee, K., Choi, K.: Autoencoder-based incremental class learning without retraining on old data (2019). https://arxiv.org/abs/1907.07872
8. Zenke, F., Poole, B., Ganguli, S.: Continual learning through synaptic intelligence. In: 34th International Conference on Machine Learning. ICML 2017, no. 8, pp. 6072–6082 (2017)
9. Aljundi, R., Babiloni, F., Elhoseiny, M., Rohrbach, M., Tuytelaars, T.: Memory aware synapses: learning what (not) to forget. Lecture Notes in Computer Science (Including Subseries Lecture Notes in Artificial Intelligence and Lecture Notes in Bioinformatics) LNCS, vol. 11207, pp. 144–161 (2018). https://doi.org/10.1007/978-3-030-01219-9_9
10. Krizhevsky, A.: Learning multiple layers of features from tiny images (2009). https://www.cs.toronto.edu/~kriz/cifar.html
11. Welinder, P., Branson, S., Mita, T., Wah, C., Schroff, F.: Caltech-UCSD Birds 200, Caltech-UCSD Technical report 200, pp. 1–15 (2010). CNS-TR-2010-001. http://www.vision.caltech.edu/visipedia/CUB-200.html
12. Iandola, F.N., Han, S., Moskewicz, M.W., Ashraf, K., Dally, W.J., Keutzer, K.: SqueezeNet: AlexNet-level accuracy with 50x fewer parameters and <0.5 MB model size, pp. 1–13 (2016). https://arxiv.org/abs/1602.07360
13. Ma, N., Zhang, X., Zheng, H.T., Sun, J.: ShuffleNet V2: practical guidelines for efficient cnn architecture design. Lecture Notes in Computer Science (Including Subseries Lecture Notes in Artificial Intelligence and Lecture Notes in Bioinformatics) LNCS, vol. 11218, pp. 122–138 (2018). https://doi.org/10.1007/978-3-030-01264-9_8
14. Kingma, D.P., Ba, J.L.: Adam: a method for stochastic optimization. In: 3rd International Conference on Learning Representations. ICLR 2015 - Conference Track Proceedings, pp. 1–15 (2015)
15. Paszke, A., Gross, S., Massa, F., Lerer, A., Bradbury, J., Chanan, G., Killeen, T., Lin, Z., Gimelshein, N., Antiga, L., Desmaison, A., Köpf, A., Yang, E., DeVito, Z., Raison, M., Tejani, A., Chilamkurthy, S., Steiner, B., Fang, L., Bai, J., Chintala, S.: PyTorch: an imperative style, high-performance deep learning library (2019). https://arxiv.org/abs/1912.01703

The Development of Monitoring Germination Through IoT Automated System

Suhaimi Puteh, Nurul Fadhilah Mohamed Rodzali, Nur Ameerah Hakimi, Nik Nurin Qistina Saiful Johar, Amirul Asyraf Abdul Manan, Nur Fatin Farisha Abdullah, and Mohd Azraai Mohd Razman(✉)

Universiti Malaysia Pahang (UMP), Pekan, Pahang, Malaysia
mohdazraai@ump.edu.my

Abstract. Irrigation is defined as the science of the artificial water application to the land to fulfil the crops' water requirements. The proposed project is to develop a Germination Monitoring Automated System (GiMAmS) to solve and improve precision plant growth and irrigation system. This system's primary purpose is to monitor the plant's growth rate by collecting and tracking the plants' information data such as water availability, soil condition, and temperature. This device utilizes both smart irrigation, monitoring sensors, and Internet of Things (IoT) systems. Arduino MEGA will be the central controller for this system. It will assimilate several sensors that comprise temperature and humidity sensors, light sensors, PH meters, moisture sensors, and pressure sensors. All the data collected from the sensors will then be stored in a cloud-based system. One of the essential factors is moisture content in the soil. The system can be triggered if the content is less than 70.0%. Thus, the system will be watering the plant until it can recover the moisture more than 80.0%. According to the data, the plants can achieve 90.1% of soil moisture content after being watered. Furthermore, variance and standard deviation are 0.05167 and 0.2273, respectively. In conclusion, this project had successfully monitored and obtained the data in real-time and provided better information for management decisions.

Keywords: Irrigation system · IoT · Agriculture · Monitoring · Real-time data

1 Introduction

Agriculture is one of the most crucial roles in the economy of Malaysia. Thus, it is essential to refine the method of farming industry operation. Based on the Department of Statistics Malaysia, agriculture had contributed 7.1% to the Gross Domestic Product (GDP) in 2019, which is equal to RM 101.5 billion [1]. According to the World Water Assessment Programme, agriculture is one of the highest users for water consumption [2]. The World Water Development Report (WWDR) predicted that the global water consumption for agriculture to increase by 19% or to reach 8,515 km^3 per year by 2025 [3]. Irrigation is the method used for agriculture to control the amount of water supplies to plants [4]. The irrigation system plays a vital role in agriculture. It utilizes to assist in

J. A. Mat Jizat et al. (Eds.): iCITES 2020, AISC 1350, pp. 368–377, 2021.
https://doi.org/10.1007/978-3-030-70917-4_35

the growing of crops [5], maintenance of landscapes, and revegetation of disturbed soils in dry areas and during periods of inadequate rainfall [6, 7]. Such an automated system has been developed not just for agriculture but also for aquaculture [8–10]. Hence, it is noteworthy that automated with a monitoring system is crucial in understanding the needs of the crops in real-time basis. Green Mustard or known as "Sawi" is the most widely grown leafy plant in Malaysia and it is a suitable vegetable grown throughout the year regardless of the season. The mustard takes about 27 days to 32 days to harvest in Malaysia compared to other countries that need about 40 days to gather. In Malaysia, the green mustard cultivation area is reported up to 8,310 hectares in 2011, in which it generates 120,160 mt with a production value of RM 350 million [11].

The developed system is to solve and improve precision plant growth and irrigation system as it shown numerously by other researchers [12]. The main focus would be for green mustard plantations on a small scale for farmers or agriculturists to nurture and observe the plant. The system can monitor the plant's growth rate by collecting and tracking the plants' information data such as water availability, soil condition, and temperature. This information is vital as it would provide the prediction or forecasting of the crop's needs to achieve optimum growth rate. Besides, this system also can enhance the farmer's ability to work more efficiently and increase production. The device includes WiFi connectivity, allowing the collected data to be stored and charted in cloud storage. It enables the users to monitor each plant's conditions on their smartphones through the designed mobile app. Not only that, but the app will also alert the user regarding the process and failures faced by the system such as connectivity failure and low water volume in the water tank. This device utilizes both smart irrigation, monitoring sensors and IoT systems.

2 Methodology

This chapter will provide the method to complete this project, where the outline of the project development is demarcated into hardware and software segment. The monitoring operation and also the application of IoT system will be provided.

2.1 Development Process

Three main parts need to be focused on developing this monitoring system: irrigation, mechanical and electrical. In mechanical parts, the general material that was used is mild steel and aluminium sheet. These two materials are used to build the frame structure of the project. Next, the piping system Polyvinyl chloride (PVC) pipe was used to flow the water and fertilizer from the tank to all the pots. 12 V DC water pump is the necessary pump to transport water. 12 V Solenoid valves used to allow automatic control in the system. It is connected to through a 5 V relay to the microcontroller. An external 12 V DC supplied power will be used to switch up the relay in the Arduino-based system (Fig. 1).

The electrical circuit's construction is divided into two parts: a master controller and control box. The master controller read all the data from sensors and activated the relay switches that trigger the actuators. Besides that, it also decides the time when the plants

Fig. 1. Illustration of the product of the project

will be fertilized. In the system, the time had been set that the plants will be fertilized twice a day. On the other hand, the control box allows the users to visually see the system's process with the help of indicators. It also gives the user access to control the operation manually with the use of switches. Furthermore, an emergency button within the system will cut off the power supply of the valves and pump for emergencies.

Irrigation is the artificial application of water to land for agricultural production [13]. Adequate irrigation will influence the entire growth process from seedbed preparation, germination, root growth, nutrient utilization, plant growth and regrowth, yield and quantity. Studies show that a water reduction of up to 90% is realized using the drip irrigation system [14]. Hence, the irrigation method that was applied within the system is the dripper irrigation method [15].

As shown in Fig. 2, the construction of the piping system can be seen clearly. Valve 1(V1) is used to control the water flow to Plant 1 and 2 while Valve 2 (V2) controls the water flow to Plant 3 and 4. Similarly, Valve 3 (V3) manages the flow of water, and Valve 4(V4) is the flow of the fertilizer to the plant. Next, this irrigation system also can be used manually, where the user will be fully involved. The user can turn on the water pump as they are required to water or fertilize the plant as a manual control box also provided with the product.

On the next page, Fig. 3 is the flowchart for the irrigation system for GiMAmS when the automatic system is involved. The condition may vary when the plant needs water or fertilizer. So, the first condition is where the usage of water is required for the plant. When the Arduino detects the signal "Water plant", it shows that the soil is too dry, and

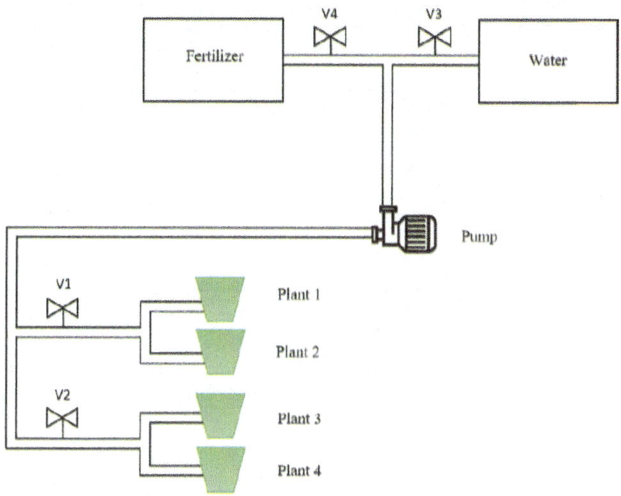

Fig. 2. Piping structure of the system

the plants need water. The system detects the signal from a particular plant's sensor, the valve leading to that plant section will open, and then the pump will on. Then, the water will flow to that plant. After the plants get the right amount of water, the pump will shut off, and all the valves from that section will be closed. The main valve that controls the water flows in and out is V3.

Next, the condition when fertilizer is needed for each day. Rather than water, diluted fertilizer is more important for the plant as the diluted fertilizer is a combination of 1 L of water and 20 ml of AB type of concentrated fertilizer. V1, V2 and V4 will open to flows the fertilizer to the plants. The suitable amount of fertilizer for each plant is 250 ml per session. Hence, the pumps will off once fertilizer flows out approximately 1 L per session. There will be two sessions every day, which are in the morning and the evening.

2.2 System Overview

Arduino IDE is open-source programming software for code writing for integrated Arduino circuit boards such as Arduino MEGA [16]. It is used to create the program for GiMAmS that enables the circuit board to read all the data from sensors, and the relay switches activation for triggering the actuators. There will be four types of sensors used within the system: soil moisture sensor, humidity sensor, light sensor, and ultrasonic sensor. The data collected from the system will be stored in the cloud as the microcontroller will be integrated with the ESP-01 WIFI module to allow the system to assess the internet. 12 V Solenoid valves are used to enable automatic control in the system. It is connected through 5 V relay to the microcontroller. An external 12DC supplied power will be used to power up the relay in the Arduino-based system (Fig. 4).

An android app was created to observe the plant's condition and sensor values remotely. The interface made using MIT App Inventor. From this app, users can monitor

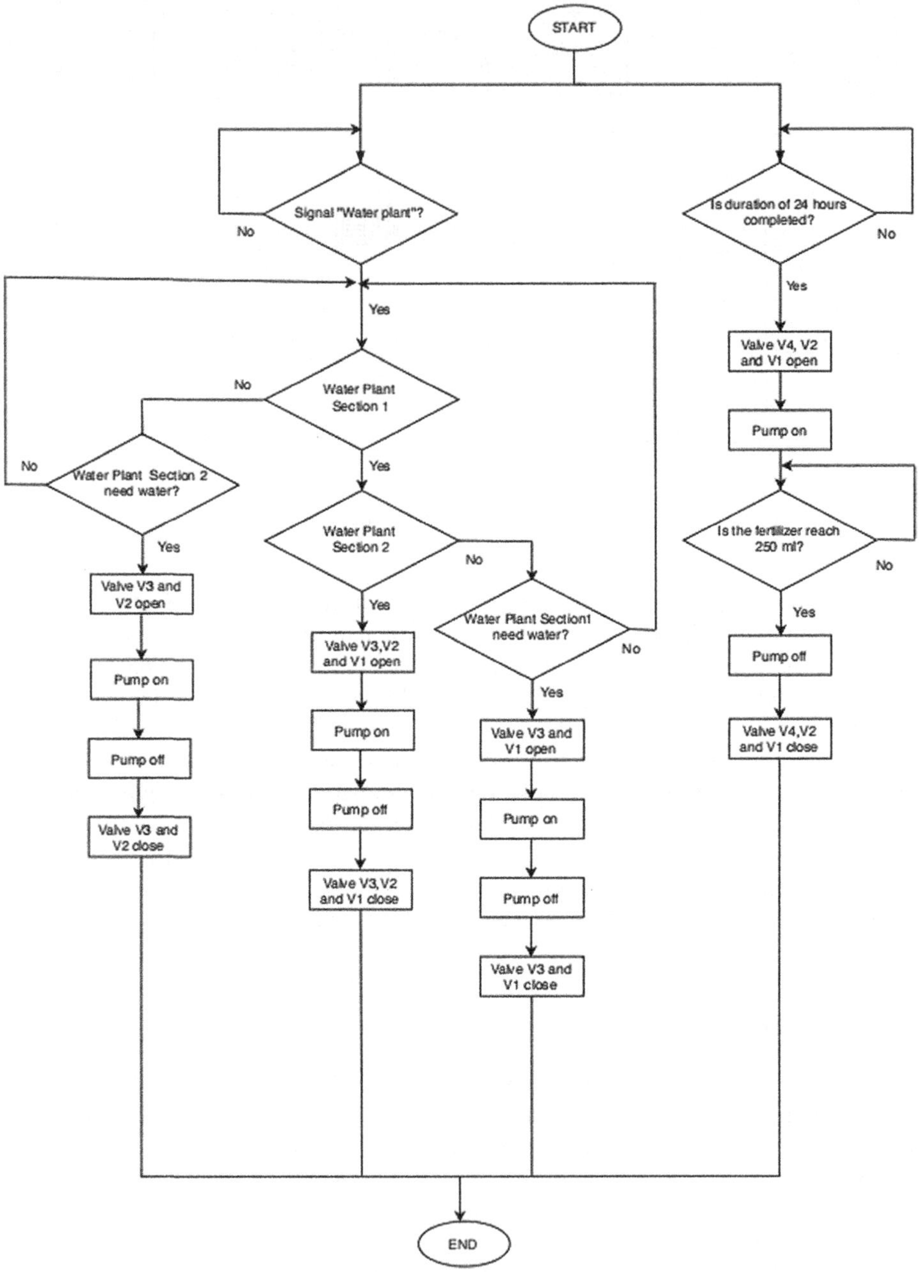

Fig. 3. Flowchart of irrigation system

the plants' status and be notified about the process that the system currently operates. The status of the actuators should be sent to the cloud. So, the idea will be to capture those data from sensors. The cloud used is called ThingSpeak, which was used as the

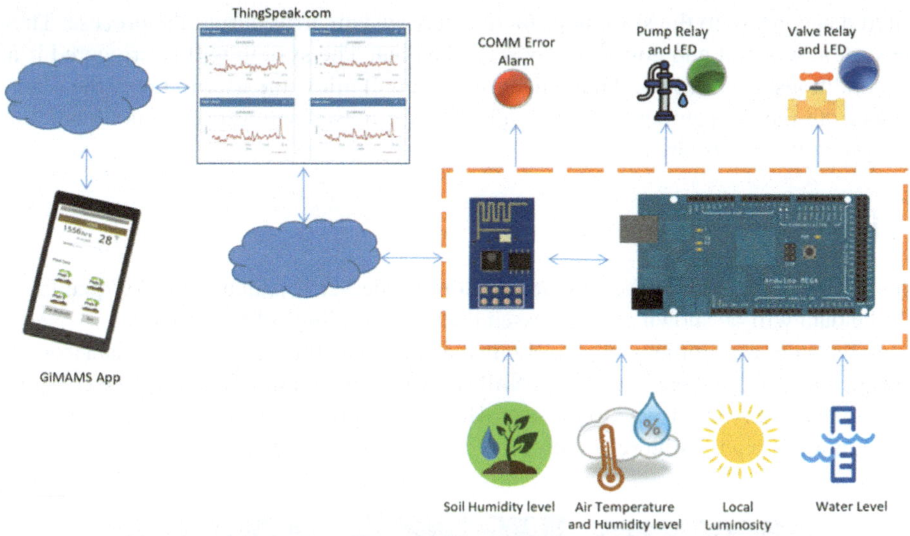

Fig. 4. Illustration of the system architecture

data storage collected from the sensors. Based on the data, the interface will update the user based on two statements, when the soil humidity is too low and when valve 4 is open.

For this research, the Green Mustard was used as the tested subject. The age for the plant is approximately 40 days. The soil moisture sensors data were collected from a sample of four pots of Green Mustard. Before collecting data, each pot was assigned a number from 1 to 4. Thus, data from soil moisture content from each pot are according to the number of the pots. The water pump data will be collected when it is switched to 'ON' or 'OFF'. For each pot, the water should be 250 ml per session. As the system runs simultaneously, the tanks should have at least 2.5 L water and fertilizer respectively. The time for the water to flow in from the tank and complete the 2.0 L each session is around 57 s. This is dues to the delay of 2 s to open and close the pump and valves. The humidity and temperature sensors were installed on the master box. The readings from these sensors will also be sent to the cloud when the system starts. A set of data from 10 readings of the sensor had been taken approximately 35 to 40 min, as shown in Table 1.

In short from all of the above, this is the procedure of the system. The system will start as soon as the user pushes the button start 'ON'. Then, the system indicator also will be 'ON'. Before the water or fertilizer flow to the plants, the system will read the sensors to send the cloud signals. Then, according to the data, the system will be activated. The first condition is the value of the Light Dependent Resistor (LDR). When the value is more than 40, the system will proceed to read the temperature data. If not, the valve and water pump will remain closed. If the temperature requirement were fulfilled, the system would detect the percentage of soil moisture content. If the moisture content is less than 70.0%, the valve and the water pump will be open. After the water or fertilizer flows to the plants, the valve and pump will be closed. In the meantime, the sensors will send the data to the cloud regarding the plant's condition whether the plants were watered according to the system, or there is any system error during the process. Then, the system will run and continuously send the data. The system will be triggered if all the conditions are fulfilled. An application used will allow the user to know the plant's status from the data storage in the cloud. Besides, a user will also get a notification of the system if there is any problem.

3 Result and Discussion

This chapter will present the research findings, whether it is qualitative or qualitative. All the data will be shown and discussed the major findings of the research.

IoT system was implemented to obtain real-time data for the plants. The data from the sensors that store in the cloud storage will show the information regarding the condition of the plant in a simpler way to the user. Hence, an application was created for the user to monitor the plants. Figure 3 shows the interface of the application (Fig. 5).

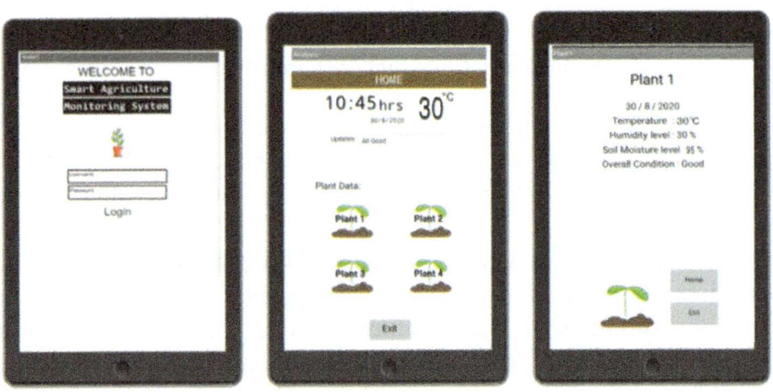

Fig. 5. The interface of the application when the user wants to view plant status.

The data that was stored in the cloud where the real-time data of the plants. For this project, the ThingSpeak channel includes eight fields of data: temperature, humidity, four areas for each moisture sensor, water pump status, water level, and light level. The data also can be seen in the table by download as an Excel file or further analysis of the data can be done on the recorded information using MATLAB.

Table 1 shows the simplified data that had been obtained from the system. The data from the sensors were sent to the cloud for several seconds. Hence, there will be a bulk of data in a brief amount of time. Nevertheless, the data succeed transfer and stored in the cloud via WiFi. Table 1 suggests there is only once a time the water pump status was equal to one, which signaled that the pump was on. This is because the time to fertilize the plant is already set (twice per day). Thus, the moisture level in each of the soil increasing after the plant had been fertilized. Otherwise, the system will only have triggered when the condition set for the temperature and soil moisture content were fulfilled. The average relative humidity for the whole time is 94.95%.

Figure 6 shows the graph of the soil moisture against time. As shown, there are four different data on soil moisture. This is because each of the soil moisture sensors was

Table 1. The data of the soil moisture sensors every 5 min from 16:25 until 17:10

Created_at	Pump status	Air Temperature (°C)	Relative Humidity (RH %)	Water Level (%)	Soil Moisture 1	Soil Moisture 2	Soil Moisture 3	Soil Moisture 4
16:25:15	0	30	95	50	89	89	89	88.5
16:30:30	0	31	95	50	89	90	89	89
16:35:10	0	30	95	50	89.5	91	90	89
16:40:06	1	30	95	25	90	90	90	91
16:45:38	0	29	95	25	90	90	91	89.5
16:50:12	0	30	94.5	25	90	90	90	89.5
16:55:09	0	30	95	25	91	91	90	90
17:00:20	0	30	95	25	90	91	90	90
17:05:12	0	30	95	25	91	91	91	91
17:10:16	0	30	95	25	91	91	91	91

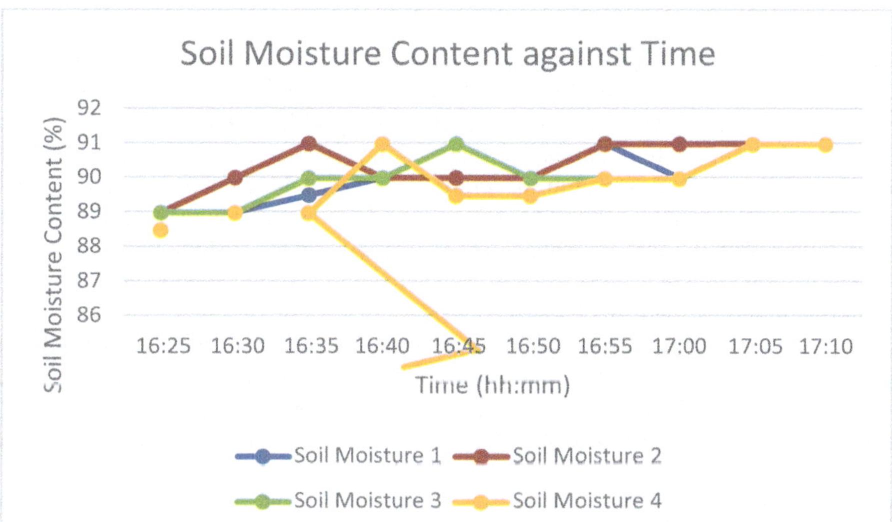

Fig. 6. The graph based on the data of the sensors every 5 min.

placed in each of the pots. In this project, four pots of Green Mustard were used. The highest level is when there is 91.0% moisture in the soil based on the data obtained. The time taken to the peak is different for each soil moisture. The time taken for Soil Moisture 1 to the peak is 15 min, while the reading for Soil Moisture 2 is quite unstable as each suddenly reached the peak before being fertilized. But, it got to the peak after 15 min. Soil Moisture 3 took about five minutes but the reading still not stable because it was dropping again. After about ten minutes, it starts to stable. Lastly, the Soil Moisture 4 straightly got to peak during the fertilization time, and after five minutes, the reading dropping. It takes about 20 min to reach the peak.

The difference between the minimum and maximum soil moisture content is almost 3.0% from the data given. The plants will need to be watered if the soil moisture content is below 70.0% or 80.0% accordingly. From the data out of four soil moisture content, the maximum value is 91.0% while the minimum is 88.5%. The mean from four sets of data is 90.1%. The variance and the standard deviation are 0.05167 and 0.2273, respectively. A smaller value standard deviation indicated that the data would not spread out over an extensive range of values, and it tends to be very close to the mean. Thus, the smaller the value, the better the results. Therefore, after it receives water or fertilizer, the soil moisture content is enough for the plant to grow as the moisture content will be almost 90.1%.

4 Conclusion

In conclusion, data from the soil moisture sensors are precious for this system since the system can only be triggered if the moisture content is below 70.0% due to the requirement that had been set up for the system. Moreover, soil water or fertilizer content is a crucial factor for the plants to grow. According to the result, moisture content can be recovered almost to 91.0% whenever the moisture level drops below 70.0%. Hence, there is no problem for the plant to get enough water and fertilizer.

The monitoring system can ensure that the crops and soil receive optimum health and productivity. Next, this system successfully increased the accuracy and efficiency of agricultural application by setting the watering cycle according to the soil moisture. The data obtained from the sensors managed to trigger and run the system. Furthermore, the crop production efficiency and crop performance optimization were obtained by closely monitoring plants' growth since the seed stage.

Nevertheless, there is still space for improvements. In terms of electrical and electronic design, maybe the system can allow the auto-detection of plants and activate it only when the plants are present for further studies. Another improvement that can be made is by enabling users to monitor the plant's physical growth by installing cameras.

Acknowledgement. The authors would like to acknowledge Universiti Malaysia Pahang for funding this study under the Research Grant (RDU200332).

References

1. DOSM: Department of Statistics Malaysia Press Release, pp. 5–9. Department of Statistics Malaysia (2020).
2. Demri, S.: 3-SAT = SAT for a class of normal modal logics (1995)
3. UNESCO: Managing Water under Uncertainty and Risk, Paris (2012). 150 p.
4. Campos, I., Balbontín, C., González-Piqueras, J., González-Dugo, M.P., Neale, C.M.U., Calera, A.: Combining a water balance model with evapotranspiration measurements to estimate total available soil water in irrigated and rainfed vineyards. Agric. Water Manag. **165**, 141–152 (2016)
5. Al-Ghobari, H.M., Mohammad, F.S.: Intelligent irrigation performance: evaluation and quantifying its ability for conserving water in arid region. Appl. Water Sci. **1**, 73–83 (2011)

6. Top Article from 2019 - Irrigation Technology in Agriculture: How New Technologies Overcome Challenges—AgriTechTomorrow
7. Hinnell, A.C., Lazarovitch, N., Furman, A., Poulton, M., Warrick, A.W.: Neuro-Drip: estimation of subsurface wetting patterns for drip irrigation using neural networks. Irrig. Sci. **28**, 535–544 (2010)
8. Taha, Z., Razman, M.A.M., Adnan, F.A., Abdul Ghani, A.S., Abdul Majeed, A.P.P., Musa, R.M., Sallehudin, M.F., Mukai, Y.: The identification of hunger behaviour of Lates calcarifer through the integration of image processing technique and support vector machine. In: IOP Conference Series: Materials Science and Engineering (2018)
9. Ogunlela, A.O., Adebayo, A.A.: Development and performance evaluation of an automatic fish feeder. J. Aquac. Res. Dev. **07**, 1–4 (2016)
10. Razman, M.A.M., Susto, G.A., Cenedese, A., Abdul Majeed, A.P.P., Musa, R.M., Abdul Ghani, A.S., Adnan, F.A., Ismail, K.M., Taha, Z., Mukai, Y.: Hunger classification of Lates calcarifer by means of an automated feeder and image processing. Comput. Electron. Agric. **163**, 104883 (2019)
11. Malaysia Mustard Greens Market Guide - Tridge
12. Jha, K., Doshi, A., Patel, P., Shah, M.: A comprehensive review on automation in agriculture using artificial intelligence. Artif. Intell. Agric. **2**, 1–12 (2019)
13. Water—Farm management—Agriculture Victoria
14. Irrigation Systems Malaysia—Wastech Group
15. Abrisqueta, I., Conejero, W., Valdés-Vela, M., Vera, J., Ortuño, M.F., Ruiz-Sánchez, M.C.: Stem water potential estimation of drip-irrigated early-maturing peach trees under mediterranean conditions. Comput. Electron. Agric. **114**, 7–13 (2015)
16. Software—Arduino

The Diagnosis of COVID-19 Through X-Ray Images via Transfer Learning Pipeline

Amiir Haamzah Mohamed Ismail[1], Muhammad Amirul Abdullah[1], Ismail Mohd Khairuddin[1], Wan Hasbullah Mohd Isa[1], Mohd Azraai Mohd Razman[1], Jessnor Arif Mat Jizat[1], and Anwar P. P. Abdul Majeed[1,2](✉)

[1] Innovative Manufacturing, Mechatronics and Sports Laboratory, Faculty of Manufacturing and Mechatronics Engineering Technology, Universiti Malaysia Pahang (UMP), 26600 Pekan, Pahang Darul Makmur, Malaysia
amajeed@ump.edu.my

[2] Centre for Software Development and Integrated Computing, Universiti Malaysia Pahang (UMP), 26600 Pekan, Pahang Darul Makmur, Malaysia

Abstract. Transfer Learning (TL) opens new possibilities of detection of disease through radiography as compared to conventional machine learning as well as deep learning methods. The extraction of features through pre-trained Convolutional Neural Networks (CNN) and the tuning of the fully connected layers of the CNN model is the core for the development of a transfer learning pipeline. The present study investigates the diagnosis of COVID-19 through X-ray images by means of three TL models, namely Inception V3, VGG-16, and the VGG-19 for feature extraction along with heuristically fine-tuned fully connected layers. It was demonstrated through this preliminary work that both the VGG-16 and VGG-19 tuned pipelines could achieve a train and test classification accuracies of 99.8% and 94%, respectively.

Keywords: VGG-16 · VGG-19 · Inception v3 · Transfer learning · X-ray Images · COVID-19

1 Introduction

Humanity has been exposed to tough times confronting multiple pandemics and epidemics, such as the 1918 flu, the Asian flu, severe acute respiratory syndrome (SARS), and Ebola. However, with the rapid advancement of technology, humanity was caught off-guard with a new strain of virus, i.e., the severe acute coronavirus 2 respiratory syndrome (SARS-CoV-2), also known as COVID-19. The virus is thought to have originated from zoonotic roots in China's Hubei Province in early December 2019 [1]. The outbreak of COVID-19 is catastrophic, as the World Health Organization (WHO) announced it to be a pandemic on the 11th March 2020 [2]. There is an increase in suspected cases with 141,533 confirmed cases with 559 confirmed deaths in Malaysia [3]. As a result of the ever-growing cases of COVID-19, it appears that effective testing methods are required to curb the spread of this virus. Polymerase chain reaction (RT-PCR) kits designed to

J. A. Mat Jizat et al. (Eds.): iCITES 2020, AISC 1350, pp. 378–384, 2021.
https://doi.org/10.1007/978-3-030-70917-4_36

genetically detect SARS-CoV-2 are the most popular test kits for COVID-19. However, false positive and false negative detections have been recorded for this form of test kit. In order to improve the current test process, more COVID-19 diagnostics are emerging, such as CRISPR, LAMP, RT-LAMP, but such methods are somewhat time-consuming in delivering results and could possibly suffer from shortages. It is worth noting that in Hubei, patients were scanned using computed tomography (CT) as COVID-19 patients had pneumonia like infections [4].

A substantial amount of research has been carried out on the utilization of transfer learning models on detection of COVID-19 through X-ray images. The application of deep learning for rapid detection of COVID-19 in X-Rays was reported by Panwar et al. [5]. The authors built a VCG16 CNN-based model dubbed as the NCOVnet. Opensource datasets was used in the study in which it consists of 142 images of COVID-19 positive patients. Moreover, the authors augmented the dataset by including images from Kaggle's Chest X-Ray Images (pneumonia) and subsequently, three classes are demarcated, namely 'Natural', 'Viral Pneumonia' and 'Bacteria Pneumonia'. The 70:30 hold-out ratio was used. It was shown from the study that the proposed model was able to detect a positive COVID-19 patient in less than five seconds while achieving a true positive rate exceeding 90%.

Apostolopoulos and Mpesiana attempted to employ TL models for the automatic detection of COVID-19 from X-ray images [6]. Two datasets from public medical repositories were then used by the authors, 2896 in total for confirmed COVID-19, confirmed bacterial pneumonia, and normal status class. The authors utilize a combined VGG19 and MobileNet V2 pre-trained models for feature extraction. A 10-fold cross validation was used for the training results with 10 epochs. The proposed model has been shown to have a classification accuracy (CA), sensitivity and specificity of 96.78%. 98.66% and 96.46%, respectively.

Zhang et al. has proposed a workaround by using a two-step transfer learning model called COVID19XrayNet [7]. The authors split the dataset into 70%, 10% and 20% for the training, validation, and testing dataset from a collection of 5860 chest X-ray images. The model developed is based on the pre-trained ResNet34 optimized using Adam optimizer with a learning rate 0.0001. The data trained for the proposed model undergoes feature smoothing layer (FSL) and feature extraction layer (FEL). It was demonstrated by the authors that their proposed COVID19XrayNet could provide a CA of 91.08% and was shown to be better than that of the conventional ResNet34 model.

The present study aims at employing different TL models, namely VGG-16, VGG-19 as well as InceptionV3 along with its fully connected layers fine-tuned (heuristically) to diagnose patients that are affected by Covid-19 through X-Ray images. Different activation functions are varied on the final layer to investigate its effect towards the classification accuracy. It should be noted the identification of the suitable TL along with its tuned fully connected layers are non-trivial towards the development of a robust pipeline for of such a disease.

2 Methodology

2.1 Data Collection

The X-ray images for COVID-19 patients and non-COVID-19 patients were retrieved from the Extensive COVID-19 X-Ray and CT Chest Images Dataset, curated by Walid El-Shafai, and Fathi Abd El-Samie [8]. The aforementioned dataset consists of 17099 X-ray and CT images of patients with and without COVID-19 infection. For the purpose of this research, the directory for X-ray images was used, in which 4044 X-ray images are designated for Covid, whilst 5500 X-ray images are for Non-Covid. It is crucial for the transfer learning (TL) models to be trained with a balanced dataset to avoid any form of skewness. Therefore, the dataset is further reduced to 4000 X-ray images, which contains 2000 images for each class. It is worth noting that the images were converted into.PNG file format and resized to fit the TL models minimum size, i.e., 299×299 for the Inception v3 model and 224×224 for both the VGG models evaluated.

2.2 Feature Extraction: Transfer Learning Model

Transfer learning deals with how programs can easily adjust to new scenarios, new tasks, and new environments. It allows machine learning systems the ability to exploit auxiliary data and models when there is only a limited amount of data available in the target domain to help solve target issues [9]. Pre-trained Convolutional Neural Network (CNN) usually makes up the Transfer Learning models. These pre-trained models are well known for the robustness in a myriad of applications [10]. In the present study, three common TL models are employed viz. Inception v3, VGG-16, and VGG-19. These TL models are used exclusively for the purpose of feature extraction.

2.3 Classification: Fully Connected Layer (FCL)

The architecture of CNN contains several layers such as the input layer, the convolutional layers, the fully connected layers and the output layer. The feature learning is represented through the convolutional layers, in which in the present study, it is defined through the aforesaid TL models. The classification segment of the CNN model is governed by the fully connected layers as well as the output layer [11]. The extracted features are fed into the fully connected layers. A single hidden layer with a hundred neurons were used, with the output neurons being two according to the classes set. The activation function is varied between sigmoid (logistic), rectified linear unit (ReLU) as well as hyperbolic tangent (tanh). The regularization parameter was set to 0.0001 with the optimization algorithm used is Adam. A 70:30 hold-out ratio was employed on the dataset for training and testing, respectively. Hence, a total of nine (9) pipelines were developed through the pairing of the different TL models with different activation function hyperparameter. The pipelines are evaluated through the classification accuracy as well as the confusion matrix. The models were developed via Python-based Anaconda IDE, specifically Orange v3.

3 Results and Discussion

Figures 1(a) and (b) depicts the images of the two categories examined, i.e., Covid and Non-Covid, respectively. The results of the different pipelines evaluated is shown in Fig. 2 below. It is apparent that in general the VGG-based pipelines outperformed the Inception v3 pipeline with an exception for test classification accuracy (CA) of the VGG-16-logistic pipeline which is slightly lower than that of the Inception v3-ReLu pipeline by 0.09%. It is also evident that the best performing pipelines are the VGG-16-tanh and VGG-19-ReLu pipelines with both having a test CA of 94%. The worst performing pipeline is the Inceptionv3-logistic pipeline with a test CA of 92.50%. Nevertheless, it is worth noting that the results are rather comparable to those reported in the literature, bearing in mind that extensive hyperparameter optimisation was not extensively carried out.

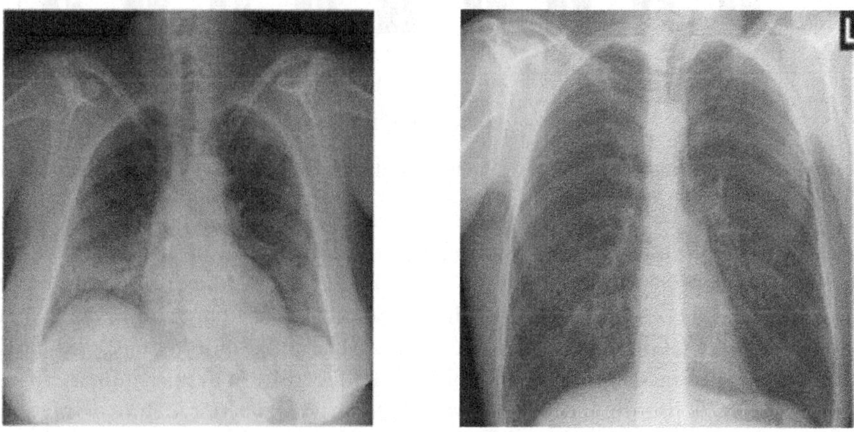

(a) (b)

Fig. 1. X-ray images of (a) Covid-19 (b) Non-Covid-19

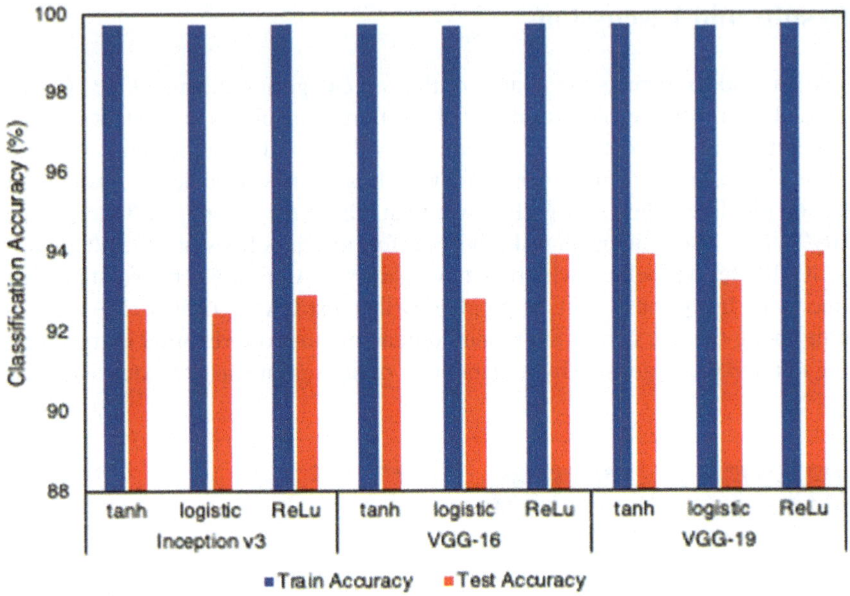

Fig. 2. Classification accuracy of evaluated fine-tuned fully connected layer pipelines.

Figures 3 and 4 show the confusion matrix of the test dataset of the best pipeline identified, namely VGG-16-tanh and the VGG-19-ReLu pipelines. It appears that the misclassification rate between the two is almost identical, but a more conservative model appears to be the VGG-19-ReLu pipeline as it misclassified more Non-Covid case as Covid as shown in Fig. 4. Moreover, it could be deduced from the preliminary investigation that the VGG pre-trained models are able to provide a better feature extraction of the X-ray images.

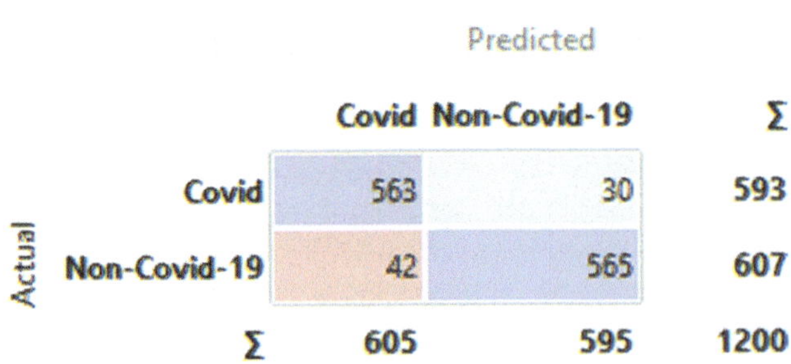

Fig. 3. Confusion Matrix of the VGG16-tanh pipeline.

Actual \ Predicted	Covid	Non-Covid-19	Σ
Covid	564	29	593
Non-Covid-19	43	564	607
Σ	607	593	1200

Fig. 4. Confusion Matrix of the VGG19-ReLu pipeline.

4 Conclusion

In this preliminary study, several TL pipelines for the classification of Covid-19 through X-ray images was investigated. It was demonstrated that the VGG-16-tanh pipeline and VGG-19-ReLu pipeline could both produce a reasonably accurate test CA of 94%. The findings are rather promising and future work will further be focused on optimizing other hyperparameters apart from evaluating other pre-trained CNN models for feature extraction. Moreover, this study does suggest that the employment of transfer learning models is a one step closer to the possibility of facilitating in the diagnosis of Covid-19 from X-ray images.

References

1. Quigley, M.C., Attanayake, J., King, A., Prideaux, F.: A multi-hazards earth science perspective on the COVID-19 pandemic: the potential for concurrent and cascading crises. Environ. Syst. Decis. **40**, 199–215 (2020). https://doi.org/10.1007/s10669-020-09772-1
2. World Health Organisation: WHO Coronavirus Disease (COVID-19) Dashboard—WHO Coronavirus Disease (COVID-19) Dashboard (2020)
3. Kementerian Kesihatan Malaysia: Terkini—COVID-19 MALAYSIA. In: Kementeri. Kesihat. Malaysia (2020)
4. Udugama, B., Kadhiresan, P., Kozlowski, H.N., Malekjahani, A., Osborne, M., Li, V.Y.C., Chen, H., Mubareka, S., Gubbay, J.B., Chan, W.C.W.: Diagnosing COVID-19: the disease and tools for detection. ACS Nano **14**, 3822–3835 (2020). https://doi.org/10.1021/acsnano.0c02624
5. Panwar, H., Gupta, P.K., Siddiqui, M.K., Morales-Menendez, R., Singh, V.: Application of deep learning for fast detection of COVID-19 in X-Rays using nCOVnet. Chaos Solitons Fractals **138**, 109944 (2020). https://doi.org/10.1016/j.chaos.2020.109944
6. Islam, M.Z., Islam, M.M., Asraf, A.: A combined deep CNN-LSTM network for the detection of novel coronavirus (COVID-19) using X-ray images. Inform. Med. Unlocked **20**, 100412 (2020). https://doi.org/10.1016/j.imu.2020.100412
7. Zhang, R., Guo, Z., Sun, Y., Lu, Q., Xu, Z., Yao, Z., Duan, M., Liu, S., Ren, Y., Huang, L., Zhou, F.: COVID19XrayNet: a two-step transfer learning model for the COVID-19 detecting problem based on a limited number of chest X-ray images. Interdiscip. Sci. Comput. Life Sci. **12**, 555–565 (2020). https://doi.org/10.1007/s12539-020-00393-5
8. El-Shafai, W., Abd El-Samie, F.: Extensive COVID-19 X-Ray and CT Chest Images Dataset, Mendeley Data, V3 (2020)

9. Yang, Q., Zhang, Y., Dai, W., Pan, S.J.: Transfer Learning. Cambridge University Press, Cambridge (2020)
10. Shapiee, M.N.A., Ibrahim, M.A.R., Mohd Razman, M.A., Abdullah, M.A., Musa, R.M., Abdul Majeed, A.P.P.: The classification of skateboarding tricks by means of the integration of transfer learning and machine learning models. In: Mohd Razman, M., Mat Jizat, J., Mat Yahya, N., Myung, H., Zainal Abidin, A., Abdul Karim, M. (eds.) Embracing Industry 4.0. Lecture Notes in Electrical Engineering, vol. 678, pp. 219–226. Springer, Singapore (2020)
11. Joshi, A.V.: Machine Learning and Artificial Intelligence an Introduction. Springer, Heidelberg (2020)

Development of Skill Performance Test for Talent Identification in Amateur Skateboarding Sport

Aina Munirah Ab Rasid[1], Noor Aishah Kamarudin[2], Muhammad Amirul Abdullah[3], Muhammad Ar Rahim Ibrahim[3], Muhammad Nur Aiman Bin Shapiee[3], Mohd Azraai Mohd Razman[3], Anwar P. P. Abdul Majeed[3], Mohamad Razali Abdullah[4], and Rabiu Muazu Musa[1(✉)]

[1] Centre for Fundamental and Continuing Education, Universiti Malaysia Terengganu, 21030 Kuala Nerus, Terengganu, Malaysia
rabiu.muazu@umt.edu.my
[2] Faculty of Health Science, Universiti Sultan Zainal Abidin, 21300 Kuala Nerus, Terengganu, Malaysia
[3] Innovative Manufacturing, Mechatronics and Sports Laboratory, Universiti Malaysia Pahang, 26600 Pekan, Pahang Darul Makmur, Malaysia
[4] East Coast Environmental Research Institute, Universiti Sultan Zainal Abidin, 21300 Kuala Nerus, Terengganu, Malaysia

Abstract. Talent identification is a vital process of mapping out athletes with the potentiality to excel in the future sporting career. To ensure success in this process, appropriate skill test in tandem to the capability of the athlete level is required to avoid dropout and demotivation of the athletes. A variety of skateboarding tricks are available for testing and identification of potential skateboarders, however, many of such skills are hard to deliver properly particularly for amateurs. The present investigation is aimed at identifying the suitable skateboarding tricks that could be used for mapping out talent in amateur skateboarding. The most common skateboarding tricks that consisted of Ollie, Kickflip, Shove-it, Nollie and Frontside 180 were identified while an experienced amateur skateboarder executed each skill five times. A customized ORY skateboard integrated with IMU sensors were used to stream the data in real-time during the performance of the tricks. The average datasets from the acceleration and angular velocity of the x, y and z-axis were obtained and a Principal Component Analysis (PCA) was used to study the dimension of the related data tricks as well as to identify the important trick manoeuvres that could be suitable for the level of the athlete's performance. The results revealed that the dataset contained two dimensions based on the Eigenvalue analysis of the PCA whilst Ollie and Nollie tricks were identified to be the most important tricks due to their higher factor loading (>0.80). It is therefore postulated that the Ollie and Nollie tricks could be used as a skill related test for the identification of talented amateur skateboarders. This may be invaluable to coaches and talent scouters in saving time, effort as well as manpower during the talent identification program in this sport.

Keywords: Skateboard · Talent identification · Trick manoeuvres · Skill test

J. A. Mat Jizat et al. (Eds.): iCITES 2020, AISC 1350, pp. 385–390, 2021.
https://doi.org/10.1007/978-3-030-70917-4_37

1 Introduction

Skateboarding is categorised under the umbrella of extreme sports. The athlete in this sport is required to execute tricks that consisted of Ollie, Nollie, Kickflip, Heelflip, Pop Shove-it and 360-flip by using the skateboard. The sport is highly competitive at both local, regional, and international level. For instance, the sport was first introduced in the 2018 Asian Games and in the year 2019, the sport was announced to debut in the now-suspended 2020 Summer Olympic Games which was slated to be held in Tokyo. Similarly, the sport is used locally as a means of leisure, entertainment as well as for transportation purposes [1, 2]. It is also important to note that the skateboarding industry is presently having an estimated value of USD 4.8 billion [3, 4]. These factors contributed to the rise in the popularity of this sport all over the world thus prompted the need for an identification of talent to provide an early sporting edge in this sport.

Talent identification has been defined as the technique of recognising individuals with the potential to succeed in a specific sport whilst talent development could be seen as the process of providing most suitable learning surrounding to discover this potential [5, 6]. It has been generally acknowledged that the TID programme could serve as a means of assisting individuals to realise their full potential [7–9]. However, to ensure success in this process, appropriate skill test in tandem to the capability of the athlete level is are required in order to avoid dropout as well as maintain the motivation of the athletes. In essence, once an appropriate means of TID program is identified, talented athletes could be scouted which could promote the judicial as well as the optimal deployment of the scarce available resources to permit the realisation of the needs and the development of the athletes.

A variety of skateboarding tricks are available for testing and identification of potential skateboarders, however, many of such skills are hard to deliver properly particularly for amateurs as well as less experienced skateboarders. Most of the tricks are meant for advanced and highly skilled skateboarders as such, the less experienced skateboarders may not be able to execute such tricks appropriately. Thus, the identification of the suitable tricks that could be used to identify talented amateur skateboarders is nontrivial in which the present investigation embarked upon.

2 Methodology

2.1 Instrumentation and Data Collection

The instrumentation as well as the design of the IMU device modelling were developed using CATIA. A Zortrax M200 Plus 3D printer was utilised to develop the casing of the device in a suitable form such that it does not interfere with the movement of the skateboard. All the materials as well as the microcontroller system utilised in the present study are documented in the earlier research [10]. An experienced amateur skateboarder (23 years, 170 cm in height, 54kg weight and ±5 years of skateboarding experience) was selected due to his ability to deliver all the skills perfectly and instructed to execute each of the skill (Ollie, Kickflip, Shove-it, Nollie and Frontside 180) five times whilst the data that consisted of the acceleration and angular velocity of the motions were collected in real-time. The axis of the rotation from the goofy stance direction was used to obtain the datasets from the signals retrieve from the IMU in a six degree of freedom (acceleration x,y, z as well as the angular velocity of the same reference points).

2.2 Principal Component Analysis Application in the Study

PCA is a mathematical method that is mostly used to recognize the sequence of the dataset from the group of variables observed [11]. PCA Provides insights into the essential variables that could reflect a specific dataset by observing the spatial and temporal heterogeneity of the whole dataset. The method of removing information from the PCA is carried out by eliminating the data consisting of least important component thereby subsequently, maintaining the most useful information from the data [12, 13]. The use of PCA is non-trivial in selecting the most vital information from a vast amount of dataset which is essential as the analysis could serve to save effort, expense and time, as the original information is usually maintained. In the present investigation, the PCA was employed to determine the suitable tricks that could be used to identify talented amateur skateboarders. The data obtained through the process outlined in the previous sections were used in achieving the objective of this study. Thus, a loading factor that is equal to or greater than 0.80 ($\geq$0.80) was considered to be important while a variable that is smaller than the threshold value set, is not considered to be necessary [14].

3 Results and Discussion

Table 1 tabulates the descriptive analysis of the tricks examined. The total number of the observed tricks (the frequency of the signals from the IMUs), the minimum, maximum observed frequencies, the mean as well as the standard deviation are displayed. It worth noting that before preceding the PCA analysis, the test of sample size adequacy via the KMO analysis was performed and demonstrated that the sample observed would be adequate for the PCA analysis as suggested by the preceding investigators [15].

Table 1. Descriptive statistics of the tricks investigated

Tricks	Obs	Min	Max	Mean ± SD	
FS180	64	−36.248	34.546	1.818	10.492
KICKFLIP	64	−16.663	117.887	5.901	21.727
NOLLIE	64	−137.891	11.067	−12.087	29.127
OLLIE	64	−34.077	17.301	0.205	7.824
POP SHUVIT	64	−80.576	60.550	−4.277	18.649

Figure 1 shows the scree plot of the Eigenvalues for the PCA analysis. It could be observed from the figure that the PCA demonstrated that a total of two dimensions are attributed to the tricks related performance dataset. These dimensions are identified and considered sufficient to explain the variability of the whole dataset due to their relatively higher Eigenvalues greater than 1. These identified dimensions were retained and subsequently used as inputs parameters for further analysis i.e., varimax rotation.

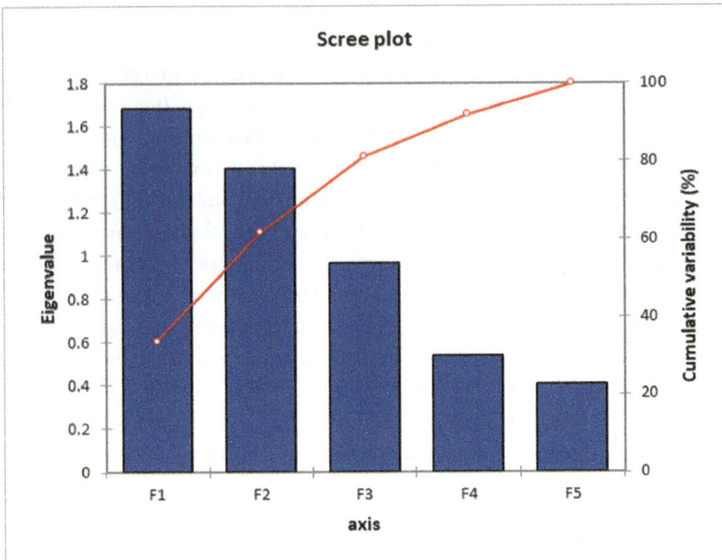

Fig. 1. The Eigenvalue of the primary PCA Analysis

Table 2. The PCA results after varimax rotation analysis

Tricks	D1	D2
FS180	0.535	– 0.362
KICKFLIP	0.679	– 0.090
NOLLIE	– 0.843*	– 0.126
OLLIE	0.181	0.886*
POP SHUVIT	– 0.358	0.770
Eigenvalue	1.842	1.308
Variability (%)	36.839	26.163
Cumulative %	36.839	63.002

*Important trick(s)

Table 2 indicates the PCA analysis after the varimax rotation. It could be seen from the table that in each of the identified 2 dimensions, some important tricks performance is revealed. These essential tricks are identified due to satisfying the predetermined factor loading threshold i.e., greater, or equal to the 0.80 that could be seen with a star (*) from the table. Likewise, it could be observed that a total of two tricks (Nollie and Ollie) out of the five initially examined are identified in all the two dimensions as most pronounced as well as vital in amateur skateboarding performance.

Since skateboarding related tricks could differ significantly in the difficulty to successfully execute them, it is therefore vital to establish the common tricks that could be used to identify amateur skateboarders with the potential to excel in the future. It has been established from the PCA analysis carried out that the Ollie as well as Nollie tricks could be suitable for the level of amateur skateboarders. The Ollie trick is a hop in which the front wheels are first elevated from the ground. This action is accomplished by snapping the tail from the rear of the foot whilst at the same time sliding one's front foot forward to attain some altitude. It is important to note that several technical abilities are put together to accomplish this trick such as the kickflip, heelflip as well as the 360-flip. Conversely, the Nollie trick is executed when the back wheels first leave the ground through nipping the nose of the board whilst the back-foot is gliding in the direction of the tail [16]. It could then be inferred that these two skills are an integral aspect of evaluating the performance ability of an amateur skateboarder.

4 Conclusion

The current investigation was primarily conducted to identify the skateboarding tricks that could be used to evaluate the performance ability of an amateur skateboarder. Several technical skills are required to showcase the prowess and the ability of a skateboarder's in this sport. Nonetheless, many of such skills are complex and beyond the level of an amateur skateboarder hence may not be suitable for early identification of talent in the sport. Therefore, investigating the suitable tricks that could be used to identify early talented athletes in this sport could not be overemphasised. The results from the present investigation revealed that the Ollie and Nollie tricks are the most important tricks in amateur skateboarding. It is therefore suggested that the Ollie and Nollie tricks could be administered as a skill performance test for the identification of talented amateur skateboarders. This may be invaluable to coaches and talent scouters in saving time, effort as well as manpower during a talent identification program in this sport.

Acknowledgement. The authors would like to acknowledge the Ministry of Education Malaysia for supporting this study under the Fundamental Research Grant Scheme (FRGS/1 /2020/TK0/UMT/03/2).

References

1. Groh, B.H., Fleckenstein, M., Kautz, T., Eskofier, B.M.: Classification and visualization of skateboard tricks using wearable sensors. Pervasive Mob. Comput. **40**, 42–55 (2017)
2. Shapiee, M.N.A., Ibrahim, M.A.R., Razman, M.A.M., Abdullah, M.A., Musa, R.M., Hassan, M.H.A., Majeed, A.P.P.A.: The classification of skateboarding trick manoeuvres through the integration of image processing. In: InECCE2019: Proceedings of the 5th International Conference on Electrical, Control & Computer Engineering, Kuantan, Pahang, Malaysia, 29 July 2019, p. 347. Springer (2020)

3. Corrêa, N.K., de Lima, J.C.M., Russomano, T., dos Santos, M.A.: Development of a skateboarding trick classifier using accelerometry and machine learning. Res. Biomed. Eng. **33**, 362–369 (2017)
4. Ibrahim, M.A.R., Shapiee, M.N.A., Abdullah, M.A., Razman, M.A.M., Musa, R.M., Majeed, A.P.P.A.: The classification of skateboarding trick manoeuvres: a frequency-domain evaluation. In: Embracing Industry 4.0, pp. 183–194. Springer (2020)
5. Vaeyens, R., Lenoir, M., Williams, A.M., Philippaerts, R.M.: Talent identification and development programmes in sport. Sports Med. **38**, 703–714 (2008)
6. Musa, R.M., Majeed, A.P.P.A., Taha, Z., Abdullah, M.R., Maliki, A.B.H.M., Kosni, N.A.: The application of Artificial Neural Network and k-Nearest Neighbour classification models in the scouting of high-performance archers from a selected fitness and motor skill performance parameters. Sci. Sports **34**, e241–e249 (2019)
7. Abbott, A., Collins, D.: Eliminating the dichotomy between theory and practice in talent identification and development: considering the role of psychology. J. Sports Sci. **22**, 395–408 (2004)
8. Musa, R.M., Abdul Majeed, A.P.P., Taha, Z., Chang, S.W., Nasir, A.F.A., Abdullah, M.R.: A machine learning approach of predicting high potential archers by means of physical fitness indicators. PLoS One **14** (2019). https://doi.org/10.1371/journal.pone.0209638
9. Charles, M.A.G., Abdullah, M.R., Musa, R.M., Kosni, N.A., Maliki, A.B.H.M.: The effectiveness of traditional games intervention program in the improvement of form one school-age children's motor skills related performance components. J. Phys. Educ. Sports **17**, 925–930 (2017)
10. Abdullah, M.A., Ibrahim, M.A.R., Shapiee, M.N.A., Majeed, A.P.P.A., Razman, M.A.M., Musa, R.M., Zakaria, M.A.: The classification of skateboarding tricks by means of support vector machine: an evaluation of significant time-domain features. In: Embracing Industry 4.0, pp. 125–132. Springer (2020)
11. Musa, R.M., Abdullah, M.R., Maliki, A.B.H.M., Kosni, N.A., Haque, M.: The application of principal components analysis to recognize essential physical fitness components among youth development archers of Terengganu, Malaysia. Indian J. Sci. Technol. **9**, 1–6 (2016)
12. Taha, Z., Haque, M., Musa, R.M., Abdullah, M.R., Maliki, A.B.H.M., Alias, N., Kosni, N.A.: Intelligent prediction of suitable physical characteristics toward archery performance using multivariate techniques. J. Glob. Pharma Technol. **9**, 44–52 (2017)
13. Abdullah, M.R., Maliki, M., Husin, A.B., Musa, R.M., Kosni, N.A., Juahir, H., Haque, M.: Multi-hierarchical pattern recognition of athlete's relative performance as a criterion for predicting potential athletes. J. Young Pharm. **8**, 463–470 (2016)
14. Musa, R.M., Majeed, A.P.P.A., Kosni, N.A., Abdullah, M.R.: Key performance indicators in elite beach soccer. In: Machine Learning in Team Sports, pp. 13–19. Springer (2020)
15. Razali, M.R., Alias, N., Maliki, A.B.H.M., Musa, R.M., Kosni, L.A., Juahir, H.: Unsupervised pattern recognition of physical fitness related performance parameters among Terengganu youth female field hockey players. Int. J. Adv. Sci. Eng. Inf. Technol. **7**, 100–105 (2017). https://doi.org/10.18517/ijaseit.7.1.1145
16. Ollie—Definition of Ollie by Merriam-Webster. https://www.merriam-webster.com/dictionary/ollie. Accessed 12 Jan 2021

The Diagnosis of Diabetic Retinopathy: A Transfer Learning with Support Vector Machine Approach

Farhan Nabil Mohd Noor[1], Wan Hasbullah Mohd Isa[1(✉)], Ismail Mohd Khairuddin[1], Mohd Azraai Mohd Razman[1], Jessnor Arif Mat Jizat[1], Ahmad Fakhri Ab. Nasir[1,3], Rabiu Muazu Musa[2], and Anwar P. P. Abdul Majeed[1,3]

[1] Innovative Manufacturing, Mechatronics and Sports Laboratory, Faculty of Manufacturing and Mechatronic Engineering Technology, Universiti Malaysia Pahang (UMP), 26600 Pekan, Pahang Darul Makmur, Malaysia
wanhasbullah@ump.edu.my

[2] Centre for Fundamental and Continuing Education, Universiti Malaysia Terengganu, Kuala Nerus, Terengganu, Malaysia

[3] Centre for Software Development and Integrated Computing, Universiti Malaysia Pahang (UMP), 26000 Pekan, Pahang Darul Makmur, Malaysia

Abstract. Diabetic Retinopathy (DR) is one of the eye diseases that occur in patients due to diabetic mellitus complications. Without an earlier diagnosis and treatment, patients suffering from DR may lose their vision permanently. Hence, this preliminary work investigates the efficacy of the use of different transfer learning models, namely VGG16, VGG19 and InceptionV3 in extracting features from the DR images classify the disease via Support Vector Machine (SVM). It was demonstrated from the preliminary work that the InceptionV3-SVM pipeline could achieve a train and test classification accuracy of 89.8% and 86.1%, respectively. The present results suggest the promising employment for auto-diagnosis of DR upon further refinement.

Keywords: VGG16 · VGG19 · InceptionV3 · Transfer learning · SVM · Diabetic retinopathy

1 Introduction

Diabetic Retinopathy [DR] is one of the vision impairment triggered by diabetes mellitus complications. It happens due to the excessive degree of blood glucose that causes the eye's blood vessels to spill blood and fluids onto its surrounding tissues. The fluid leaking triggers hard exudates, microaneurysms, cotton wool spots, and hemorrhage [1]. Diabetic Retinopathy begins with moderate non-proliferative irregularities categorized by increased vascular permeability, severe non-proliferative diabetic retinopathy (NPDR), and vascular closure. It concludes with proliferative diabetic retinopathy (PDR), which is described by the development of additional blood vessels on the retina and the posterior surface of the vitreous [2]. Figure 1 shows the difference between normal retinal

J. A. Mat Jizat et al. (Eds.): iCITES 2020, AISC 1350, pp. 391–398, 2021.
https://doi.org/10.1007/978-3-030-70917-4_38

image and diabetic retinopathy image. It has been reported that the number of people suffering from DR would increase from 127 million in 2010 to 191 million in 2030, in the event that appropriate action is not taken [3]. Permanent vision loss can transpire if the DR is not treated for a long time. Therefore, its early detection and treatment of DR is non-trivial [4]. However, it is worth mentioning that more often than not, DR screening is typically based on human grading and is costly due to the limited numbers of trained human graders [5]. Therefore, plenty of attempts that have been made on the design and build of automated DR grading since 1982 [6].

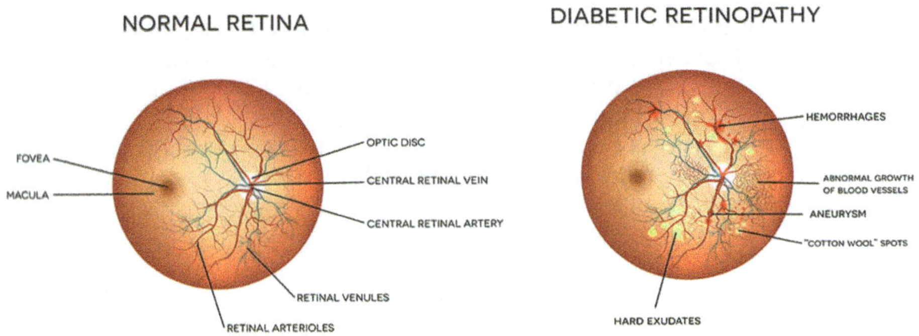

Fig. 1. The difference between normal retina and diabetic retinopathy retina

With the advent of technology, particularly the emergence of machine learning, researchers have investigated different approaches to facilitate the diagnosis of RD. For instance, Qomariah et al. proposed to classify the normal and DR retinal images by using pre-trained convolutional neural network (CNN) with support vector machine (SVM) [7]. The retinal images were obtained from the base 12 and 13 of the Messidor Database consists of two classes, which is normal and severe NPDR. Pre-trained CNN architecture such as AlexNet, VGG16, VGG19, Inception V2, Inception V3, Resnet 50, DenseNet201, and GoogleNet were used for the feature extraction purpose before it was fed into the SVM classifier. It was shown from the study that the Resnet 50 transfer learning model performed the highest classification accuracy (CA) on the dataset from the base 12 with 95.83%, while features from VGG19 and InceptionV3 both achieved 95.24% CA for the dataset from base 13 which are the highest amongst the other transfer learning models.

Lahmiri investigated and evaluated the efficiency of hybrid deep learning [8]. It consists of a CNN architecture for the features extraction process whilst pairing it with conventional machine learning models such as SVM, Naïve Bayes (NB), Linear Discriminant Analysis (LDA), and k-Nearest Neighbour (kNN) to classify the presence of hemorrhage in the retina. The dataset, which consists of 114 digital retinal images for both normal and abnormal that was taken from the STARE database. Ten best features were identified and chosen by using a student t-test as the feature selection technique. The dataset was also trained and tested using the 10-fold cross validation technique, and the SVM classifier model was optimized using the Bayesian Optimization technique. The experiment result showed that the CNN-SVM pipeline could achieve a CA of 99.11%.

In another study, Gayathri et al. proposed using a lightweight CNN for the DR classification. The authors conducted two different experiments. The first one is that the binary classification consists of DR and normal images, while for the second experiment, the multi-class classification consists of normal, mild, moderate, severe, and PDR images. All the images were acquired from different online databases such as IDRiD, Messidor, and Kaggle Databases. CNN-based features were fed to different conventional machine learning model such as SVM, Multi Linear Perceptron (MLP), Random Forest (RF) and J48 classifiers. It was shown from the study that the J48 classifier outperforms all the other classifiers for binary and multi-class classification [9].

The present study sought to evaluate the efficiency of using different transfer learning models such as VGG16, VGG19, and InceptionV3 in extracting features from images obtained from an online repository and subsequently classify the images from the features via an SVM classifier. The efficacy of the pipelines shall be evaluated by means of CA as well as observing the confusion matrix.

2 Methodology

2.1 Dataset

This study used the data set obtained from the online Kaggle database that was publicly used for the classification of DR from the digital images. A total of 1700 color fundus retina images, which consist of normal images (No DR) and DR images of any severity, were used for the evaluation and classification process. The color fundus retina images represent a varied cohort of patients with all grades of DR. These images were being resized into a new pixel dimensions to fit into the transfer learning models, in which VGG16 and VGG19 models requires a minimum image size of 224×224 pixels whilst the InceptionV3 requires 299×299 pixels. Figure 2 shows the color fundus retina images of No DR and DR that have already been resized into 224×224 pixels. The dataset has been divided into 70:30 ratio for the training and testing sets.

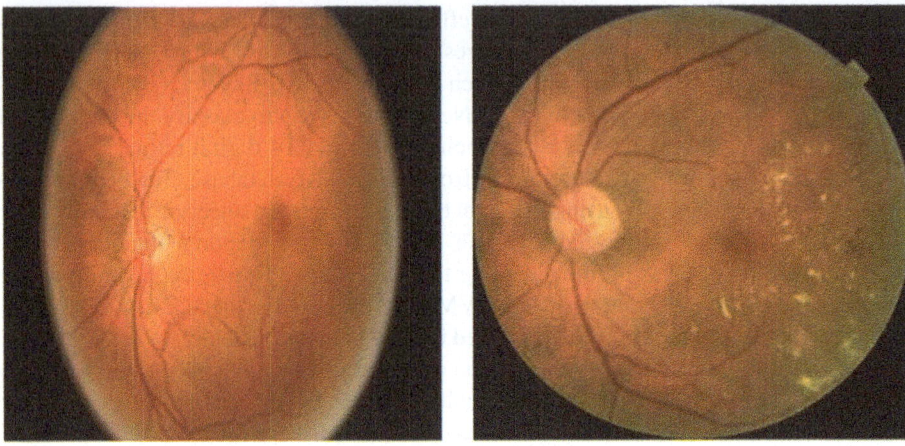

Fig. 2. Left is the normal retinal image, and right is the diabetic retinopathy image from the dataset

2.2 Feature Extraction: Transfer Learning Model

The general CNN architecture consists of an input layer, convolutional layer, fully connected layer as well as the output layer. In the present study, the convolutional layer is frozen as it is used for feature extraction purpose (transfer learning), whilst the fully connected layers are replaced with the SVM classifier. Contrary to conventional machine learning pipeline, the color fundus retina images shall be fed to the pre-trained CNN (a.k.a. transfer learning model) which will automatically learn complex representative features directly from the data itself [8]. The transfer learning models that are investigated in this study are VGG16, VGG19, and InceptionV3. Figure 3 shows the proposed framework utilized.

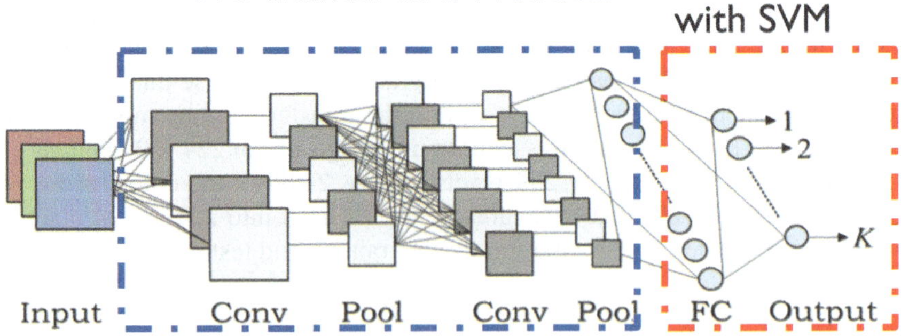

Fig. 3. The general example of CNN architecture

2.3 Classification: SVM

Support Vector Machine (SVM) is one of the most popular supervised Machine Learning algorithms that primarily draws a border between two classes to distinguish two different classes [10]. It exploits kernel parameters to transform input data to higher dimensional space by categorizing data by verifying a hyperplane with a maximum margin. It classifies all linear and nonlinearly separable data effectively by constructing a data hyperplane in the original space. The dot product of a vector hyperplane is constant in that space. Hyperplane can be described as

$$\vec{w}.\vec{x} - b = 0 \tag{1}$$

Where w is assumed to be a normalized hyperplane vector and the equation of distance from a point to a plane can calculate the x, which is a set of points and $\frac{b}{\|w\|}$. If the data points are divisible linearly, it is possible to have distance between the hyperplanes by taking two parallel hyperplanes. The small area is referred to as the margin. This means the data points belong to any of the data group [11]. In the present study, the default parameters of the SVM classifier from the sklearn library is used. The python-based Anaconda platform, particularly Orange v3 is used to evaluate the efficacy of the proposed pipelines. The classification accuracy as well as the confusion matrix are used as the performance measure for this study [12].

3 Results and Discussion

Figure 4 shows the classification accuracy obtained for all the proposed pipelines evaluated transfer learning models. It could be seen that the training CA for both VGG19 as well as InceptionV3 pipelines are comparable to each other. Nonetheless, with regards to the test CA, it is evident that the InceptionV3-SVM pipeline demonstrated a higher CA, with a CA of 86.1% in comparison to the other evaluated pipelines. Conversely, it from the bar chart depicted in Fig. 4, it is clear that the VGG16 pipeline is the worst performing pipeline evaluated. From the results, it appears that the Inception V3 model is able to extract the features better in comparison to other TL models evaluated. A close examination on the efficacy of the models is carried out through the observation on the confusion matrix of the test dataset for each pipeline as illustrated in Fig. 5, 6 and 7. It is evident that the lower misclassification of DR as No DR shown in Fig. 7, further suggest that the Inception V3 pipeline is much more conservative than the rest. It should be noted that in the present study, they hyperparameters of the SVM model is taken as default, hence a better CA is expected in the event that the hyperparameters are optimized.

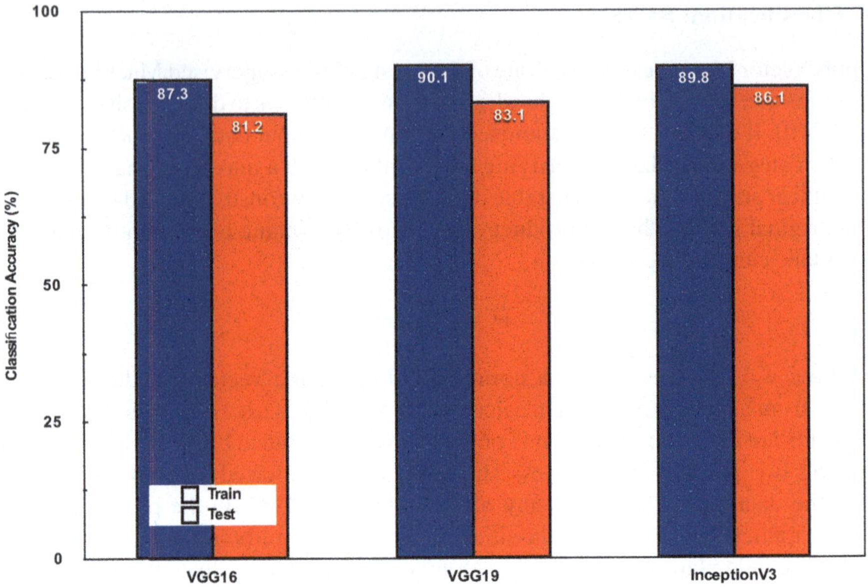

Fig. 4. Classification accuracy of the evaluated pipelines

Actual \ Predicted	DR	No DR	Σ
DR	202	38	240
No DR	58	212	270
Σ	260	250	510

Fig. 5. VGG16 pipeline confusion matrix of the test dataset.

Actual \ Predicted	DR	No DR	Σ
DR	211	29	240
No DR	57	213	270
Σ	268	242	510

Fig. 6. VGG19 pipeline confusion matrix of the test dataset.

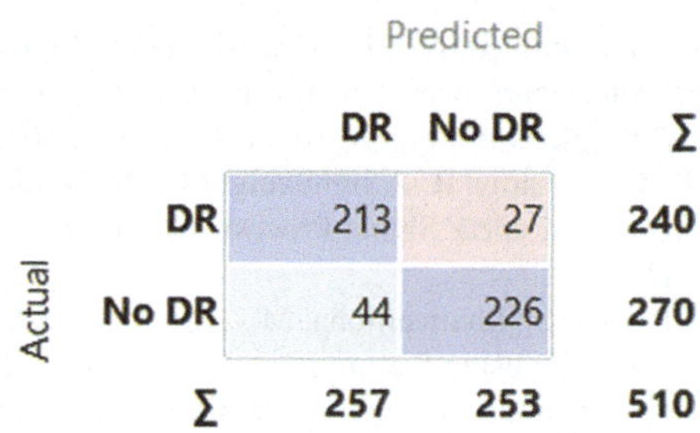

Fig. 7. InceptionV3 pipeline confusion matrix of the test dataset

4 Conclusion

In this study, the classification of Diabetic Retinopathy by means of different transfer learning pipelines were investigated. The InceptionV3-SVM pipeline was shown to be able to provide a favorable diagnosis of DR with a test CA of 86.1%. The findings are promising and could be further improved by tuning the hyperparameters of the SVM model used. Moreover, different transfer learning models as well as conventional classifiers should be examined to identify the best pipeline that could provide a reasonable judgement of DR for the ophthalmologist as a second opinion during the diagnosis process.

References

1. Carrera, E.V., Gonzalez, A., Carrera, R.: Automated detection of diabetic retinopathy using SVM. In: Proceedings of the 2017 IEEE 24th International Congress on Electronics Electrical Engineering and Computing INTERCON 2017, pp. 6–9 (2017). https://doi.org/10.1109/INTERCON.2017.8079692
2. Mahendran, G., Dhanasekaran, R.: Investigation of the severity level of diabetic retinopathy using supervised classifier algorithms. Comput. Electr. Eng. **45**, 312–323 (2015). https://doi.org/10.1016/j.compeleceng.2015.01.013
3. Washburn, P.S., Mahendran, Dhanasekharan, Periyasamy, Murugeswari: Investigation of severity level of diabetic retinopathy using adaboost classifier algorithm. Mater. Today Proc. 1–6 (2020). https://doi.org/10.1016/j.matpr.2020.03.199
4. Stewart, M.W.: Diabetic Retinopathy Current Pharmacologic Treatment and Emerging Strategies, 1st edn., no. 20. Adis, Florida (1974)
5. Heydon, P., Egan, C., Bolter, L., Chambers, R., Anderson, J., Aldington, S., Stratton, I.M., Scanlon, P.H., Webster, L., Mann, S., Du Chemin, A., Owen, C.G., Tufail, A., Rudnicka, A.R.: Prospective evaluation of an artificial intelligence-enabled algorithm for automated diabetic retinopathy screening of 30 000 patients. Br. J. Ophthalmol. 1–6 (2020). https://doi.org/10.1136/bjophthalmol-2020-316594
6. Hashim, M.F., Mohd Hashim, S.Z.: Diabetic retinopathy lesion detection using region-based approach. In: 2014 8th Malaysian Software Engineering Conference MySEC 2014, pp. 306–310 (2014). https://doi.org/10.1109/MySec.2014.6986034
7. Qomariah, D.U.N., Tjandrasa, H., Fatichah, C.: Classification of diabetic retinopathy and normal retinal images using CNN and SVM. In: Proceedings of the 2019 International Conference on Information and Communication Technology and System. ICTS 2019, pp. 152–157 (2019). https://doi.org/10.1109/ICTS.2019.8850940

8. Lahmiri, S.: Hybrid deep learning convolutional neural networks and optimal nonlinear support vector machine to detect presence of hemorrhage in retina. Biomed. Signal Process. Control **60**, 101978 (2020). https://doi.org/10.1016/j.bspc.2020.101978
9. Gayathri, S., Gopi, V.P., Palanisamy, P.: A lightweight CNN for diabetic retinopathy classification from fundus images. Biomed. Signal Process. Control **62** (2020). https://doi.org/10.1016/j.bspc.2020.102115
10. Huda, S.M.A., Ila, I.J., Sarder, S., Shamsujjoha, M., Ali, M.N.Y.: An improved approach for detection of diabetic retinopathy using feature importance and machine learning algorithms. In: 2019 7th International Conference on Smart Computing & Communications. ICSCC 2019, pp. 1–5 (2019). https://doi.org/10.1109/ICSCC.2019.8843676
11. Mohanty, K.K., Barik, P.K., Barik, R.C., Bhuyan, K.C.: An efficient prediction of diabetic from retinopathy using machine learning and signal processing approach. In: Proceedings of the 2019 International Conference on Information Technology. ICIT 2019, pp. 103–108 (2019). https://doi.org/10.1109/ICIT48102.2019.00025
12. Prem, S.S., Umesh, A.C.: Classification of exudates for diabetic retinopathy prediction using machine learning, pp. 357–362 (2020). https://doi.org/10.1109/iccca49541.2020.9250858
13. Lotlekar, K.S., Desai, S.D.: Multilevel classification model for diabetic retinopathy. In: Proceedings of the International Conference on Computational Techniques, Electronics and Mechanical Systems. CTEMS 2018, pp. 326–331 (2018). https://doi.org/10.1109/CTEMS.2018.8769207

Gearbox Fault Diagnostics: An Examination on the Efficacy of Different Feature Extraction Techniques

Md Jahid Hasan[1], Mamunur Rashid[1], Ahmad Fakhri Ab. Nasir[1,3], Muhammad Amirul Abdullah[1], Mohd Azraai Mohd Razman[1], Rabiu Muazu Musa[2], and Anwar P. P. Abdul Majeed[1,3(✉)]

[1] Innovative Manufacturing, Mechatronics and Sports Laboratory, Faculty of Manufacturing and Mechatronic Engineering Technology, Universiti Malaysia Pahang (UMP), 26600 Pekan, Pahang Darul Makmur, Malaysia
amajeed@ump.edu.my

[2] Centre for Fundamental and Continuing Education, Universiti Malaysia Terengganu, Kuala Nerus, Terengganu, Malaysia

[3] Centre for Software Development and Integrated Computing, Universiti Malaysia Pahang (UMP), 26000 Pekan, Pahang Darul Makmur, Malaysia

Abstract. Gearbox defects have been considered as one of the major causes of failure in rotating machinery. It is important to identify and diagnose the actual reasons behind the failure of gearbox for a reliable operation of equipment that relies such a system. In this paper, a fault diagnosis method based on entropy-based feature and support vector machine (SVM) has been proposed for detecting the faults in bearings and gear set in the gearbox. Initially, different features in terms of time domain as well as time-frequency domain have been extracted and classified via SVM. The proposed method has been validated by the publicly available online dataset which is consists of nine classes (eight types of faults and healthy) with load and unloaded conditions. The optimum validation accuracy (98.84% for 20-0 load condition and 99.87% for 30-2 load conditions) has been obtained by the entropy-based feature extracted from transformed continuous wavelet transform (CWT) signal. The outcome of this study is very encouraging since it emphasizes to avoid the computational complexity in feature extraction as well as classification.

Keywords: Support vector machine (SVM) · Continuous Wavelet Transform (CWT) · Machine learning · Gearbox fault detection

1 Introduction

In order to ensure a safe and reliable operation of complex mechanical systems, especially rotating machineries, the use of failure diagnostic techniques are of great importance. Amongst the elements that often require regular health monitoring are gears and rolling-element bearings as such elements are often susceptible to failure. Owing to the

J. A. Mat Jizat et al. (Eds.): iCITES 2020, AISC 1350, pp. 399–406, 2021.
https://doi.org/10.1007/978-3-030-70917-4_39

advancement of technology, particularly the advent of industry 4.0, signal processing as well as data-driven methods have received considerable attention [1].

Over the past few years, machine learning-based approaches in the area of rolling bearing fault analysis have gained due attention [2]. These methods include three main stages: extraction of functions, reduction in dimensionality, and identification of designs. First, it is important to remove insignificant or unwanted features from the vibration signal. During the growth of nonlinear technology, a growing number of complex nonlinear methods have been implemented for fault extraction based on statistical parameter estimates [3]. Approached entropy [3], sample entropy (SE) [4], entropy allowance [5], and fuzzy entropy [6] are amongst the various techniques that is most commonly been employed. Among them, sample entropy can efficiently calculate the regularity of signals and identify poor shifts as one of the most widely used entropy methods. The nonlinear dynamic complexity changes accordingly when failures occur in rolling covers. SE is therefore ideal for rolling bearings' functional extraction. Nonetheless, it is worth noting that SE only analyzes the time series signal on a single dimension.

Costa et al. proposed a multi-scale Sample Entropy (MSE) method to tackle this problem, which could quantify multi-scale system complexity [7]. The original time series is divided into non-overlapping segments initially of length s (called scale factor), according to the MSE algorithm. Next, by calculating an average value for each segment, the gross grained time series is obtained. Finally, the entropy of the ground grain series is calculated on each scale. In mechanical vibration signals, MSE has also been used to extract features. For instance, Zhang et al. have used MSE and neuro-fuzzy adaptive inference to detect and evaluate rolling bearing faults [8]. Conversely, Li and Jiang proposed a local mean decomposition and MSE feature extraction technique [9]. In a different study, Hsieh et al. employed the MSE curve to recognize several characteristic defects of the high-speed spindle [10].

This paper proposes a novel bearing and gear set in the gearbox fault diagnosis model, based on entropy and support vector machine (SVM). Furthermore, the experiment has been carried out under various load conditions to figure out the most efficient model. This paper has been organized in the following sections i.e. Sect. 2 and 3 describes the methodology employed in the present investigation as well as the results and discussion, respectively. Section 4 draws the conclusion of the investigation.

2 Methodology

Figure 1 illustrates the flow diagram of the methods taken to achieve the objective of the present study. The key sections are the data collection, pre-processing of the signal, selection of features, and classification of the signal. In the pre-processing stage, 1024 data points for one sample were chosen for the recorded vibration results. The characteristics of different statistical properties are extracted from the vibration data of the pre-processing. Finally, the features extracted were classified using the SVM classifier.

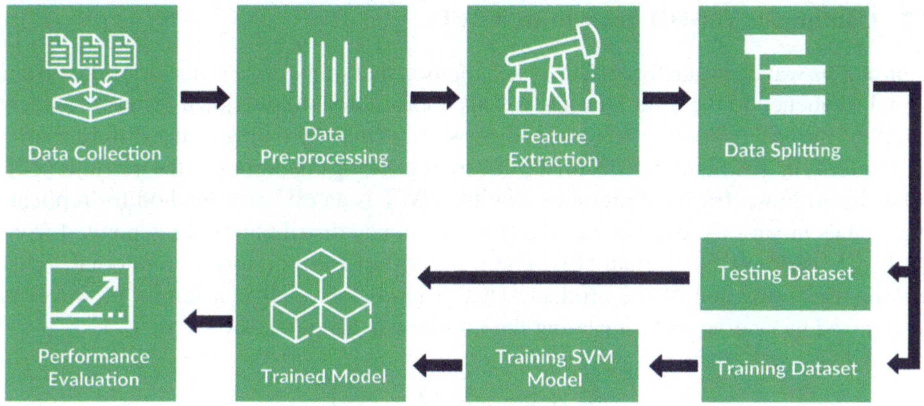

Fig. 1. Complete methodology of the proposed study.

2.1 Feature Extraction

The extraction of features is an essential part of machine learning as it provides certain meaning to the evaluated signal. Different types of features have been investigated in the literature, namely Mean, Root Mean Square (RMS), Standard Deviation (SD), Variance (Var), Kurtosis (Kur), Skewness (Skew), Entropy (E) as well as Power (P). These features are taken upon transforming the original sensor signal via Continuous Wavelet Transform (CWT). The statistical formulae are shown in Table 1 [11].

Table 1. Formulas for statistical features.

Parameter	Formula	Parameter	Formula
Mean	$\bar{y} = \frac{1}{N}\sum_{k=1}^{N}(y_k)$	Skewness (Skew)	$s_k = \frac{\Sigma_{k=1}^{N}(y_k-\bar{y})^3}{N\sigma^3}$
Root mean square (RMS)	$y_{rms} = \sqrt{\frac{1}{N}\left[\Sigma_{k=1}^{N}(y_k)^2\right]}$	Entropy (En)	$H(x) = -\sum_{k=1}^{N}P(y_k)logP(y_k)$
Standard deviation (SD)	$\sigma = \sqrt{\frac{1}{N}\sum_{k=1}^{N}(y_k-\bar{y})^2}$	Energy (E)	$E = \sqrt{\sum_{k=1}^{N}\lvert y_k\rvert^2}$
Variance (Var)	$\sigma^2 = \sqrt{\frac{1}{N}\sum_{k=1}^{N}(y_k-\bar{y})^2}$	Power (P)	$P = \frac{1}{N}\sum_{k=1}^{N}\lvert y_k\rvert^2$
Kurtosis (Kur)	$K = \frac{\sum_{k=1}^{N}(y_k-\bar{y})^2}{N\sigma^4}$		

2.2 Continuous Wavelet Transform (CWT)

Continuous wavelet transform (CWT) is a method that transforms time-series data from signal frequency to the time-frequency. This approach could provide a means of understanding further the nature of failure diagnostics, primarily owing to the different time rate patterns that are associated with the different working conditions of the machine [12]. It has been shown from the literature that the CWT is an effective method for replicating signals in various resolutions. The time-frequency distributions are generated from the CWT based on the original [13]. The wavelet transform performs an internal signal and wavelet operation of the product. This group of wavelets is a family of wavelets recognized by scaling and translating the mother wavelet ψ(t), shown as

$$\psi_{s,\tau}(t) = \frac{1}{\sqrt{S}}\psi\left(\frac{t-\tau}{S}\right)$$

Whereas is a frequency-related scale parameter, and τ is a Frequency translation. A CWT of the signal x(t) can be obtained by mathematically defined conversion of complex conjugates:

$$w_{(5,\tau)} = \langle x(t), \psi_{s,\tau}\rangle = \frac{1}{\sqrt{s}}\int x(t)\psi^*\left(\frac{t-\tau}{S}\right)dt$$

where ψ^* (·) denotes the complex conjugate of the above function ψ(·).

2.3 Support Vector Machine (SVM)

SVM is based on the notion of the decision-making planes and it has extensively used for both regression and classification problems. The functioning of SVM has been described in two different classes:

Consider N as a training data set, $D = \{(x_i, y_i)\}_{i=1}^{N}, x_i \in R^M$ is the ith training sample, sample label is $y_i \in \{1, -1\}$. The classification hyperplane $\omega^T x + b = 0$ is the optimization objective of the SVM model. Moreover, Lagrangian factor as well as the penalty factor is denoted as α and C, respectively. The κ (·, ·) is the kernel function to achieve linear separation of the samples. The objective function is given as

$$\min_{\alpha}\frac{1}{2}\sum_{i=1}^{N}\sum_{j=1}^{N} a_i a_j y_i y_j k(x_i, x_j) - \sum_{i=1}^{N} a_i$$

$$s \cdot t \sum_{i=1}^{N} a_i y_i = 0, 0 \le a_i \le C$$

A solution of the optimization goal α obtained the minimal optimization theory, the classification function:

$$f(x) = \sum_{i=1}^{N} a_i y_i k(x, x_i) + b$$

In the proposed system, initially, the performance has been evaluated by the classification accuracy which could be calculated via the following equation [14]:

$$Accuracy = \frac{TP + TN}{TP + TN + FP + FN} \times 100\%$$

3 Experimental Verification

3.1 Dataset

The data set from the drivetrain dynamic simulator (DDS) has been collected from [15]. The rotating load system is either 20 Hz–0 V or 30 Hz–2 V under two different working conditions. Table 2 demonstrates the numerous categories of failures for bearings and gearboxes. The dataset consists of five different working conditions for both the bearing and gearbox data. Each set has four distinct failure categories and one healthy state. To test the proposed method, nine classes are examined which includes four types of gear failure, four types of bearing failure, and one healthy state. A ratio of 50:50 was used for the train and test ratio, in which for each working state contains 500 samples.

Table 2. Fault type.

Location	Type	Description
Bearing	Ball	The crack occurs in the ball
	Inner	The crack occurs in the inner ring
	Outer	The crack occurs in the outer ring
	Combo	The crack occurs in the outer and inner ring
Gearbox	Chipped	The crack occurs in the gear feet
	Miss	Missing of feet in the gear
	Root	The crack occurs in the roof of gear feet
	Surface	Wear occurs on the surface of the gear

3.2 Results and Discussion

In this study, the significance of different features in its original raw form as well as transformed CWT form towards classification accuracy of the health state of both bearing and gearbox were examined. Table 3 and Table 4 presents the classification accuracy (CA) in terms of training as well as validation for two different working conditions dataset, i.e., 20 Hz–0 V and 30 Hz–2 V datasets, respectively. From the experimental results tabulated in Table 3, it is evident that the highest validation accuracy were derived from Root Mean Square (RMS), Standard Deviation (SD), Variance (Var) as well as Energy (E) with a CA of 98.56%, 98.62%, 98.62%, and 98.56%, respectively in the event that the raw data is considered. Whilst for the other working condition investigated as tabulated in Table 4, it could be seen that features such as Standard deviation (SD), and Variance (Var) could yield a validation CA of 98.78% and 98.69%, respectively through the use of raw data. However, it is worth noting that from both Table 3 and Table 4, the maximum validation accuracy is attained via entropy extracted features of the CWT transformed data with a CA of 98.84% and 99.87%, respectively.

Table 3. 20-0 data.

	Training accuracy		Validation accuracy	
	Raw	CWT	Raw	CWT
Mean	98.6	65.0	89.51	34.62
Root mean square (RMS)	99.7	96.9	98.56	86.49
Standard deviation (SD)	98.9	98.9	98.62	95.69
Variance (Var)	98.7	99.0	98.62	95.44
Kurtosis (Kur)	60.8	87.0	58.84	85.91
Skewness (Skew)	48.6	54.2	47.09	47.87
Entropy (En)	41.9	99.4	47.17	98.84
Energy (E)	99.7	97.6	98.56	94.67
Power (P)	99.7	96.8	97.78	92.49

Table 4. 30-2 data.

	Training accuracy		Validation accuracy	
	Raw	CWT	Raw	CWT
Mean	98.8	39.0	77.29	27.67
Root mean square (RMS)	99.5	99.7	93.73	99.71
Standard deviation (SD)	98.8	99.7	98.78	99.73
Variance (Var)	98.7	99.7	98.69	99.67
Kurtosis (Kur)	56.9	89.3	57.87	88.98
Skewness (Skew)	46.1	62.2	47.44	59.40
Entropy (En)	39.8	99.8	40.02	99.87
Energy (E)	99.5	99.5	93.91	99.58
Power (P)	97.4	99.6	93.91	98.49

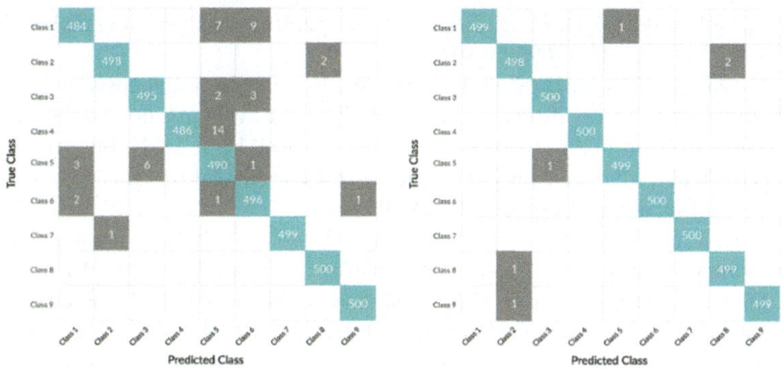

Fig. 2. Confusion matrix for the entropy-based feature extracted from CWT in two working conditions 20 Hz–0 V (left) and 30 Hz–2 V (right).

Figure 2 depicts the confusion matrix of the best feature (entropy) and signal transformation (CWT) on the two different working conditions evaluated. It is apparent from the presented CA and depicted confusion matrix, the proposed architecture can achieve accurate predictions of the faults detected on the gearbox dataset.

4 Conclusion

To conclude, we have established a method for mechanical fault diagnosis and classification based on entropy-based extracted CWT feature with a support vector machine (SVM) classifier. Our comparative analysis demonstrated that the approach proposed could achieve a reasonably-well classification across each of the evaluated datasets. The proposed approach could play an invaluable role in the identification and diagnosis of faults through various mechanical structures, as well as associated fields in the future.

References

1. Shao, S., McAleer, S., Yan, R., Baldi, P.: Highly accurate machine fault diagnosis using deep transfer learning. IEEE Trans. Ind. Inform. **15**, 2446–2455 (2019)
2. Worden, K., Staszewski, W., Hensman, J.: Natural computing for mechanical systems research: a tutorial overview. Mech. Syst. Signal Process. **25**(1), 4–111 (2011).
3. Han, Z., Xu, B., Zhu, X., Jiao, H.: Research on multi-fault diagnosis of rotor based on approximate entropy and EEMD. China Mech. Eng. **27**, 2186–2189 (2016)
4. Cui, J., Zheng, Q., Xin, Y., Zhou, C., Wang, Q., Zhou, N.: Feature extraction and classification method for switchgear faults based on sample entropy and cloud model. IET Gener. Transm. Distrib. **11**, 2938–2946 (2017)
5. Yan, X., Jia, M., Zhao, Z.: A novel intelligent detection method for rolling bearing based on IVMD and instantaneous energy distribution-permutation entropy. Meas. J. Int. Meas. Confed. **130**, 435–447 (2018)
6. Zheng, J., Pan, H., Cheng, J.: Rolling bearing fault detection and diagnosis based on composite multiscale fuzzy entropy and ensemble support vector machines. Mech. Syst. Signal Process. **85**, 746–759 (2017)

7. Costa, M., Goldberger, A.L., Peng, C.K.: Multiscale entropy analysis of complex physiologic time series. Phys. Rev. Lett. **89**, 068102 (2002)
8. Zhang, L., Xiong, G., Liu, H., Zou, H., Guo, W.: Bearing fault diagnosis using multi-scale entropy and adaptive neuro-fuzzy inference. Expert Syst. Appl. **37**, 6077–6085 (2010)
9. Li, Y., Jiang, J.: Fault diagnosis of bearing based on LMD and MSE. In: 2017 Prognostics and System Health Management Conference, PHM-Harbin 2017 - Proceedings. Institute of Electrical and Electronics Engineers Inc. (2017)
10. Hsieh, N.-K., Lin, W.-Y., Young, H.-T.: High-speed spindle fault diagnosis with the empirical mode decomposition and multiscale entropy method. Entropy **17**, 2170–2183 (2015)
11. Goyal, D., Choudhary, A., Pabla, B.S., Dhami, S.S.: Support vector machines based non-contact fault diagnosis system for bearings. J. Intell. Manuf. **31**, 1275–1289 (2019)
12. Boashash, B.: Time-Frequency Signal Analysis and Processing: A Comprehensive Reference. Elsevier Inc., Oxford (2015)
13. Yan, R., Gao, R.X., Chen, X.: Wavelets for fault diagnosis of rotary machines: a review with applications. Signal Process. **96**, 1–5 (2014)
14. Rashid, M., Sulaiman, N., Mustafa, M., Jadin, M.S., Najib, M.S., Bari, B.S., Khatun, S.: Analysis of EEG features for brain computer interface application. In: Lecture Notes in Electrical Engineering, pp. 529–540. Springer (2020)
15. Wang, J., Zhao, R., Wang, D., Yan, R., Mao, K., Shen, F.: Machine health monitoring using local feature-based gated recurrent unit networks. IEEE Trans. Ind. Electron. **65**, 1539–1548 (2017)

Minimizing Normal Vehicle Forces Effect During Cornering of a Two In-Wheel Vehicle Through the Identification of Optimum Speed via Particle Swarm Optimization (PSO)

Nurul Afiqah Zainal[1], Kamil Zakwan Mohd Azmi[1], Muhammad Aizzat Zakaria[1], and Anwar P. P. Abdul Majeed[1,2](✉)

[1] Innovative Manufacturing, Mechatronics and Sports (iMAMS) Laboratory, Faculty of Manufacturing and Mechatronics Engineering Technology, Universiti Malaysia Pahang, 26600 Pekan, Pahang, Malaysia
amajeed@ump.edu.my
[2] Centre for Software Development and Integrated Computing, Universiti Malaysia Pahang, Pekan, Pahang, Malaysia

Abstract. Driving comfort is a non-trivial issue as it could pose adverse health issues particularly to those who travel-long distances. In the present study, we identify the optimum speed during cornering in order to reduce the effect of normal vertical forces towards driving experience of a two-in wheel vehicle. A bio-inspired metaheuristic method, namely Particle Swarm Optimization (PSO) was employed to address the objective of the present investigation on a twelve degrees of freedom human biodynamic model that is fused to a two-in-wheel car model. It was established through the PSO algorithm, that the optimal speed during cornering for the given car configuration is approximately 19 km/h. The findings of the present study are essential in ensuring human comfort is not compromised that in turn minimizes possible untoward risk to the driver.

Keywords: PSO · Human comfort · Two-in-wheel vehicle · Optimization

1 Introduction

Ride comfort is a non-trivial issue towards evaluating a vehicle's performance. This topic has received due attention over the years primarily in facilitating the mitigation of adverse health effects that is associated with long-distance travelling and prolonged exposure to vibration. A handful of studies have reported that pointed out the aforesaid issue [1–3]. Nevertheless, the determination of ride comfort that is associated with vibration has been quite a contention between vehicle designers and no common or gold standard has been established thus far [4].

Hitherto, there are a number of studies that have been reported on the employment of optimization techniques to quantify human comfort based on the driving speed [5–8]. For

J. A. Mat Jizat et al. (Eds.): iCITES 2020, AISC 1350, pp. 407–412, 2021.
https://doi.org/10.1007/978-3-030-70917-4_40

instance, Papaioannou and Koulocheris have employed a Pareto-based Genetic Algorithm (GA) in ascertaining optimal vehicle suspension systems [9]. A Non-dominated Sort GeneticAlgorithm (NSGA-II) has also been utilized by Nagarkar et al. on a nonlinear quarter car model along with a 4 degrees of freedom (DoF) driver model to evaluate ride comfort as well as health criterion [10]. A Design of Experiment (DoE) method has also been employed in investigating ride comfort through the optimization of suspension system parameters on varying speeds [4].

The present study aims at identifying the optimal speed during cornering of a two-in-wheel vehicle that is coupled with a twelve degrees of freedom human biodynamic model [11] by means of Particle Swarm Optimization technique. It is worth noting that for the two in-wheel vehicle model, the effect of lateral and longitudinal forces are taken into consideration [12].

2 Methodology

The readers are referred to [11] for the detail modelling of the two-in-wheel vehicle (in-house developed) coupled with a twelve degrees of freedom human biodynamic model. Equation (1) is used to determine the vehicle mass shift force during cornering. The values of the parameters used in this equation is also referred to [11].

$$F_{zfr} = mg\frac{l_r}{l} - \frac{m_f a_y h_c}{l_w} - ma_x\frac{h_c}{l} \tag{1}$$

Figure 1 below depicts the wheel steering input and heading of the vehicle that is utilized as an input to the model. The road profile is used to mimic the driving behavior while cornering to the right.

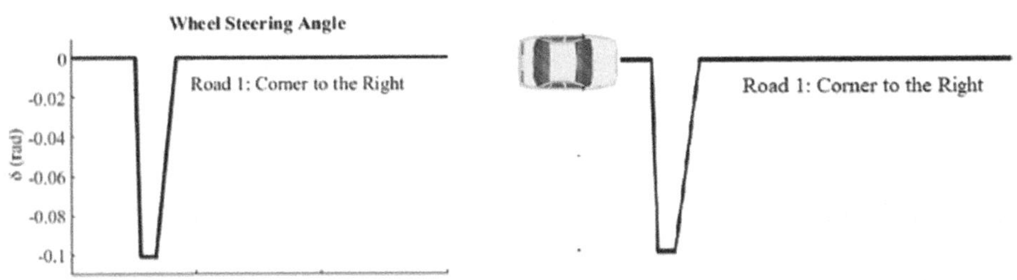

Fig. 1. Wheel steering input

PSO is employed to determine the optimal velocity value of vehicle in order to minimize the force exerted on driver's seat (F_{zfr}) and its flowchart is provided in Fig. 2. It is worth mentioning that the algorithm's inherent parameters as tabulated in Table 1 are initially initialized. The PSO parameter values include the number of particles, maximum inertia weight (w_{max}), minimum inertia weight (w_{min}), cognitive component (c_1) and social component (c_2). Moreover, the initial position of particles is randomly located in a search space herein between 10 km/h and 30 km/h. Upon the execution of the initialization stage, the objective function is defined, and it is essentially the minimization of the force exerted on driver's seat which is based on Eq. (1).

The personal best (*pbest*) is the best solution found by each particle while global best (*gbest*) is defined as the best *pbest*. It is worth mentioning that both the *pbest* and *gbest* are updated at each iteration. The rate of change of the displacement of the particles, i.e., the velocities of each particle are revised based on the summation of parameters as shown in Eq. (2).

$$v_i^t = \omega^t v_i^t + c_1 rand\left(p_{best} - x_i^t\right) + c_2 rand\left(g_{best} - x_i^t\right) \tag{2}$$

where x_i^t and v_i^t are the position and velocity of particle i at iteration t, respectively. The *rand* is a random number between 0 and 1, whilst c_1 and c_2 are denoted as the cognitive and social coefficients, respectively, while ω^t refers to the inertia weight at iteration t. In the present investigation, the linear dynamic inertia weight is employed and computed based on Eq. (3).

$$\omega^t = \omega_{max} - \left(\frac{\omega_{max} - \omega_{min}}{t_{max}}\right) \times t \tag{3}$$

where ω_{max} and ω_{min} denote the maximum and minimum values of inertia weight, respectively, whilst t_{max} is the maximum iteration. Then, the particle's new velocity, v_i^t is used to update the particle's position using Eq. (4)

$$x_i^t = x_i^t + v_i^t \tag{4}$$

This process of ascertaining the optimum solution continues until the stopping condition is obtained. The experimentation carried out in the present study is done through MATLAB.

3 Result and Discussion

Based on the parameter values configured in Table 1, the results obtained from the experiment are discussed in this section. The convergence curve is shown in Fig. 3. It could be seen that at the beginning of the iteration, the PSO takes into account the various speeds of the vehicle as inputs and examines the total force exerted on driver's seat (F_{zfr}) to make judgments. Then after several iterations, the PSO manages to attain the optimal speed value during cornering to reduce the total force. Through PSO, this experiment suggests that the optimal velocity for cornering is 19 km/h.

Figure 4 shows the force acting on front right of the vehicle based on the suggested speed (19 km/h). The normal forces are changed according to the steering input applied. Based on an earlier study reported by Zainal et al. [11], it could be established that the forces arising from different road profile conditions of a moving vehicle do somewhat affect parts of the human body to a certain degree. Moreover, with the optimal speed identified during cornering, the drivers could be cautious whilst speeding to avoid any unwanted skidding or roll-over owing to the physical capabilities of the vehicle and the road condition.

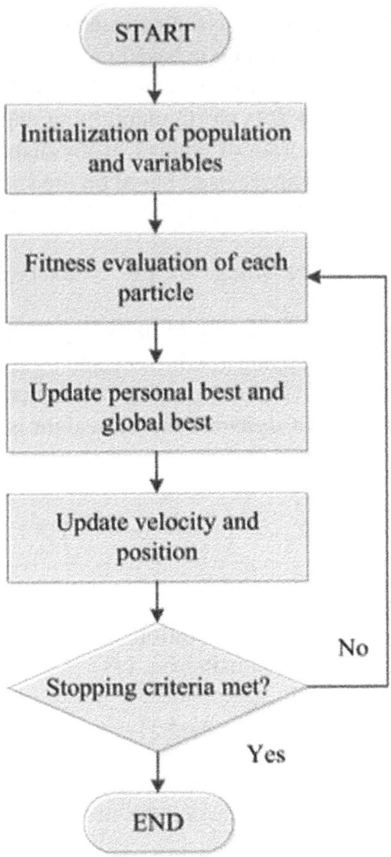

Fig. 2. The particle swarm optimization (PSO) algorithm

Table 1. Parameter values of PSO algorithm

Parameter	Value
Number of particles, n	5
Maximum inertia weight, $w_{\max}$	0.9
Minimum inertia weight, $w_{\min}$	0.4
Cognitive component, c_1	2
Social component, c_2	2
Maximum iterations, t_{max}	50

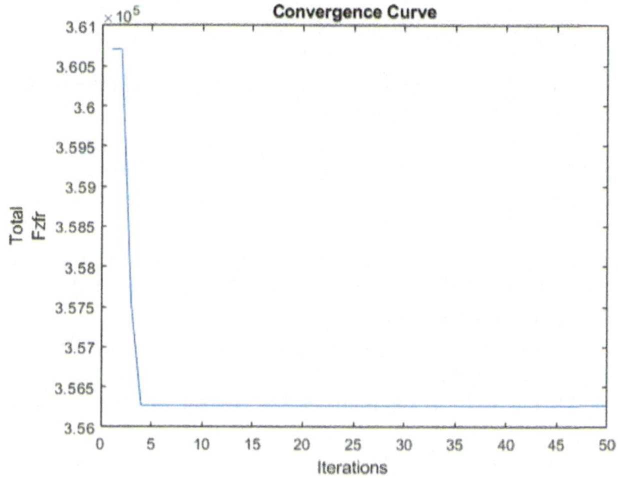

Fig. 3. Convergence curve

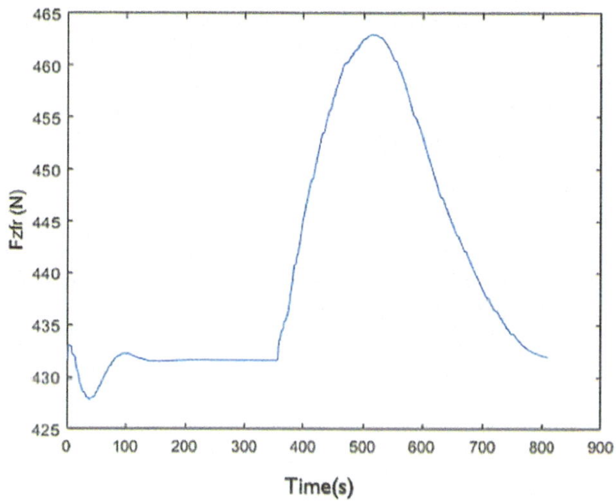

Fig. 4. Force acting on front right of vehicle (velocity = 19 km/h)

4 Conclusion

The present study employs the use of Particle Swarm Optimization technique in identifying optimal speed during cornering of a two-in-wheel vehicle. The findings are non-trivial towards the establishment of a gold standard for ride comfort and road safety. Future works shall incorporate other optimization techniques as well as other forms of road profile and vehicle input.

Acknowledgement. The authors would like to thank Universiti Malaysia Pahang for funding this study via RDU190328.

References

1. Katu, U.S., Desavale, R.G., Kanai, R.A.: Effect of vehicle vibration on human body – RIT experience. In: NaCoMM, pp. 1–9 (2003)
2. Azizan, M.A., Fard, M.: The influence of vibrations on vehicle occupant fatigue. In: INTERNOISE Conference, vol. 62, no. ISO 2631, pp. 1–15 (2014)
3. Azizan, A., et al.: The influence of vibration on seated human drowsiness. Ind. Health **54**(4), 296–307 (2016)
4. Marzbanrad, J., Mohammadi, M., Mostaani, S.: Optimization of a passive vehicle suspension system for ride comfort enhancement with different speeds based on design of experiment method (DOE) method. J. Mech. Eng. Res. **5**(3), 50–59 (2013)
5. Yan, Q., Zhang, B., Kezunovic, M.: Optimization of electric vehicle movement for efficient energy consumption, November 2014
6. Lu, D., Ouyang, M., Gu, J., Li, J.: Optimal velocity control for a battery electric vehicle driven by permanent magnet synchronous motors. Math. Probl. Eng. **2014**, 1–14 (2014). https://doi.org/10.1155/2014/193960
7. Vaz, W., Nandi, A., Landers, R., Koylu, U.: Electric vehicle range prediction for constant speed trip using multi-objective optimization. J. Power Sources **275**, 435–446 (2015). https://doi.org/10.1016/j.jpowsour.2014.11.043
8. Wang, J., Besselink, I., Nijmeijer, H.: Battery electric vehicle energy consumption prediction for a trip based on route information. Proc. Inst. Mech. Eng. Part D Automob. Eng. **232**(11), 1528–1542 (2018)
9. Papaioannou, G., Koulocheris, D.: An approach for minimizing the number of objective functions in the optimization of vehicle suspension systems. J. Sound Vib. **435**, 149–169 (2018)
10. Nagarkar, M.P., Vikhe Patil, G.J., Zaware Patil, R.N.: Optimization of nonlinear quarter car suspension–seat–driver model. J. Adv. Res. **7**(6), 991–1007 (2016)
11. Afiqah, N., Muhammad, Z., Zakaria, A., Anwar, K.B., Majeed, P.P.A.: The normal vehicle forces effects of a two in - wheel electric vehicle towards the human brain on different road profile maneuver. SN Appl. Sci. **2**, 223 (2020)
12. Baarath, K., Zakaria, M.A., Zainal, N.A.: An investigation on the effect of lateral motion on normal forces acting on each tires for nonholonomic vehicle (2018)

Author Index

J. A. Mat Jizat et al. (Eds.): iCITES 2020, AISC 1350, pp. 413–415, 2021.
https://doi.org/10.1007/978-3-030-70917-4

The manufacturer's authorised representative in the EU is Springer Nature Customer Service Centre GmbH, Europaplatz 3, 69115 Heidelberg, Germany. If you have any concerns regarding our products, please contact ProductSafety@springernature.com

Printed and bound by CPI Group (UK) Ltd, Croydon, CR0 4YY
30/09/2026
02225251-0013